Minimal Models and Extremal Rays (Kyoto, 2011)

ADVANCED STUDIES IN PURE MATHEMATICS 70

Chief Editor of the Series: Yoshihiro Tonegawa (Tokyo Institute of Technology)

Minimal Models and Extremal Rays (Kyoto, 2011)

Edited by

János Kollár (Princeton University)
Osamu Fujino (Osaka University)
Shigeru Mukai (RIMS, Kyoto University)
Noboru Nakayama (RIMS, Kyoto University)

Mathematical Society of Japan

This book was typeset by $\mathcal{A}_\mathcal{M}\mathcal{S}$-TEX and $\mathcal{A}_\mathcal{M}\mathcal{S}$-LATEX, the TEX macro systems of the American Mathematical Society, together with the style files aspm.sty *and* aspmfm.sty *for $\mathcal{A}_\mathcal{M}\mathcal{S}$-TEX written by Dr. Chiaki Tsukamoto and* aspmproc.sty *for $\mathcal{A}_\mathcal{M}\mathcal{S}$-LATEX written by Dr. Akihiro Munemasa and* aspm.cls *for $\mathcal{A}_\mathcal{M}\mathcal{S}$-LATEX provided by Livretech Co., Ltd.*

TEX is a trademark of the American Mathematical Society.

Edited by the Mathematical Society of Japan.

Published by the Mathematical Society of Japan.

Distributed by the Mathematical Society of Japan, the American Mathematical Society and the World Scientific Publishing Co., Ltd.
Distributed exclusively in North America by the American Mathematical Society.

Advanced Studies in Pure Mathematics 70
ISBN 978-4-86497-036-5

PRINTED IN JAPAN
by Livretech Co., Ltd.

2010 Mathematics Subject Classification.
Primary 14-06.
Secondary 14E30, 14B07, 14D05, 14E07, 14E15, 14F17, 14F18, 14J17, 14J28, 14J32, 14J45, 53D05.

Preface

Since the appearance of extremal rays and the minimal model program around 1980, we have seen the tremendous development of algebraic geometry. With this in mind, the conference "Minimal Models and Extremal Rays" was held at RIMS Kyoto University during June 20–24 as an activity of one of two RIMS International Project Researches in 2011. Our purpose was to review the past, examine the present and enjoy discussing the future. This volume is the outcome of this conference and consists of one survey article on the mathematical work of Professor Shigefumi Mori, who turned sixty in 2011, and thirteen research papers, which were rigorously refereed. We would like to thank all the referees for their contribution to this volume.

March 2015

Editorial Committee
János Kollár
Shigeru Mukai (Chief)
Osamu Fujino
Noboru Nakayama

All papers in this volume have been refereed and are in final form. No version of any of them will be submitted for publication elsewhere.

This volume is respectfully dedicated to

Prof. Shigefumi Mori

on his sixtieth birthday in 2011
from all of us who owe so much to him
through both his works and personal guidance.

RIMS International Project Research
— Minimal Models and Extremal Rays —

The conference "Minimal Models and Extremal Rays" was held at RIMS Kyoto University during June 20–24 as one of two RIMS International Project Researches in 2011, when Professor Shigefumi Mori turned sixty (*kanreki* in Japanese). The pre-conference "Summer School on Algebraic Geometry" was also held at the same place one week before, during June 13–17.

In the conference, there were 156 participants including 37 from overseas, and we have twenty one-hour lectures by invited speakers. In the pre-conference, there were 89 participants including 28 from overseas, and we have four series lectures and eight 45 minutes lectures by invited speakers. The present volume consists of fourteen articles, ten of which are contributed by invited speakers of the conference. We thank all the speakers and participants for their invaluable contributions, which made the (pre-)conference most successful.

The (pre-)conference was supported by Research Institute for Mathematical Sciences (RIMS), Kyoto University. We are grateful for their generous financial support.

March 2015

Organizing Committee
János Kollár
Takayuki Hayakawa
Masayuki Kawakita
Atsushi Moriwaki
Shigeru Mukai (Chief)
Osamu Fujino
Noboru Nakayama

A brief explanation of Prof. Mori's career

Professor Mori graduated from Kyoto University in 1973, and received his Master of Science and Doctor of Science in 1975 and 1978, respectively, from Kyoto University. He became a research associate of Kyoto University in 1975, and was a visiting professor at Harvard University during 1977–80, where he succeeded in solving Hartshorne's conjecture and initiated his theory of extremal rays, for which the readers are referred to Prof. Kollár's survey article in this volume. He became a lecturer at Nagoya University in 1980, and was later promoted to an associate professor and a professor there. In 1990 he moved to the Research Institute for Mathematical Sciences (RIMS), Kyoto University, where he has been professor since then. At the International Congress of Mathematics held in Kyoto in 1990, Professor Mori was awarded the Fields Medal for his work on the classification of 3-dimensional algebraic varieties culminating in proving the existence of minimal models. He visited many places overseas, including the Institute for Advanced Study, Max-Planck-Institut für Mathematik, Columbia University and the University of Utah, and has inspired many mathematicians and students internationally through his deep ideas and clear lectures. He was elected President of the International Mathematical Union in 2014 and has been taking the leadership as the first head of the organization from Asia since January 2015.

Professor Shigefumi Mori

MMP and extremal rays

June 20–24, 2011

RIMS, Kyoto University, Room 420

June 20 (Monday)

10:00–11:00 J. M[c]Kernan (Massachusetts Institute of Technology),
The importance of being big.

11:30–12:30 C. Birkar (Cambridge University),
Existence of log canonical flips and a special LMMP.

14:00–15:00 J. A. Chen (National Taiwan University),
Explicit three-dimensional minimal model program.

16:00–17:00 M. Mustaţă (University of Michigan),
Asymptotic invariants and valuations.

June 21 (Tuesday)

9:30–10:30 V. V. Shokurov (Johns Hopkins University/Steklov Institute of Mathematics),
Mobility and rigidity in birational algebraic geometry.

10:45–11:45 C. Xu (Massachusetts Institute of Technology),
Boundedness of algebraic varieties.

13:30–14:30 A. Corti (Imperial College London),
Extremal Laurent Polynomials and Fano Manifolds.

14:45–15:45 V. Lazić (Imperial College London),
MMP revisited.

16:15–17:15 M. Kawakita (RIMS),
Ideal-adic semi-continuity problem for minimal log discrepancies.

June 22 (Wednesday)

10:00–11:00 H. Clemems (Ohio State University),
The intersection number for pairs of (topological) normal functions and a height pairing for (disjoint, homologically trivial) submanifolds.

11:30–12:30 Y. Miyaoka (University of Tokyo),
Canonical degree of nodal curves on a surface of general type.

14:00–15:00 J. Kollár (Princeton University),
MMP and moduli of varieties.

16:00–17:00 S. Mori (RIMS),
Threefold $\mathbb{Q}$-conic bundles and a personal review of extremal rays.

June 23 (Thursday)

9:30–10:30 V. Alexeev (University of Georgia),
Extending Torelli maps.

10:45–11:45 S. Kovács (University of Washington),
Deformation of Du Bois singularities.

13:30–14:30 H. Takagi (University of Tokyo),
Mirror symmetry and projective geometry of Reye congruence.

14:45–15:45 J.-M. Hwang (Korea Institute for Advanced Study),
Varieties of minimal rational tangents of codimension one.

16:15–17:15 Y. Namikawa (Kyoto University),
Equivalence of symplectic singularities.

June 24 (Friday)

9:30–10:30 Y. Prokhorov (Moscow State Lomonosov University),
On the classification of singular Fano threefolds.

10:45–11:45 M. Reid (University of Warwick),
Rings and varieties.

Organizers: Osamu Fujino, János Kollár, Atsushi Moriwaki, Shigeru Mukai, Noboru Nakayama

Summer School on Algebraic Geometry

June 13–17, 2011
RIMS, Kyoto University, Room 420

June 13 (Monday)

9:30–10:30 P. Cascini (Imperial College London),
On the Minimal Model Program, I.

10:45–11:45 P. Hacking (University of Massachusetts Amherst),
Moduli spaces of surfaces of general type and their compactifications, I.

14:45–15:30 C.-Y. Chi (Harvard University),
Pseudonorms and Theorems of Torelli type.

15:45–16:30 D. Kim (Korea Institute for Advanced Study),
Some division problems implied by the openness conjecture for multiplier ideal sheaves.

June 14 (Tuesday)

10:45–11:45 J. Starr (Stony Brook University),
Spaces of rational curves on Fano manifolds, I.

13:30–14:30 P. Cascini,
On the Minimal Model Program, II.

14:45–15:30 Y. Gongyo (University of Tokyo),
Log pluricanonical representations and abundance conjecture.
15:45–16:30 C.-J. Lai (University of Utah),
Varieties fibered by good minimal models.

June 15 (Wednesday)
9:30–10:30 P. Hacking,
Moduli spaces of surfaces of general type and their compactifications, II.
10:45–11:45 J. Starr,
Spaces of rational curves on Fano manifolds, II.
12:00–13:00 M. Păun (Institut Élie Cartan, Université Henri Poincaré),
Extension theorems for pluricanonical forms, I.

June 16 (Thursday)
9:30–10:30 P. Cascini,
On the Minimal Model Program, III.
10:45–11:45 M. Păun,
Extension theorems for pluricanonical forms, II.
13:30–14:30 J. Starr,
Spaces of rational curves on Fano manifolds, III.
14:45–15:30 K. Watanabe (University of Tokyo),
Lengths of chains of minimal rational curves on Fano manifolds of Picard number 1.
15:45–16:30 A. Ito (University of Tokyo),
Seshadri constants and Okounkov bodies.

June 17 (Friday)
9:30–10:30 M. Păun,
Extension theorems for pluricanonical forms, III.
10:45–11:45 P. Hacking,
Moduli spaces of surfaces of general type and their compactifications, III.
13:30–14:15 S. Matsumura (University of Tokyo),
Asymptotic cohomology vanishing and a converse of the Andreotti–Grauert theorem on surfaces.
14:30–15:15 T. Okada (Kyoto University),
On the Mori fiber structures of $\mathbb{Q}$*-Fano 3-fold weighted complete intersections.*

Organizers: Osamu Fujino, Takayuki Hayakawa, Masayuki Kawakita, Noboru Nakayama.

Overseas Participants

Alexeev, Valery (University of Georgia)
Birkar, Caucher (University of Cambridge)
Cascini, Paolo (Imperial College London)
Chen, Hou-Yi (Institute of Mathematics, Academia Sinica)
Chen, Jheng-Jie (National Center for Theoretical Sciences, Taipei Office)
Chen, Jiun-Cheng (National Tsing Hua University)
Chen, Jungkai (National Taiwan University)
Chi, Chen-Yu (Harvard University)
Choi, Sung Rak (University of California, Riverside)
Clemens, Herbert (The Ohio State University)
Corti, Alessio (Imperial College London)
Feng, Tien-Shun (National Taiwan University)
Hacking, Paul (University of Massachusetts Amherst)
Hwang, DongSeon (Korea Institute for Advanced Study)
Hwang, Jun-Muk (Korea Institute for Advanced Study)
Kaloghiros, Anne-Sophie (University of Cambridge)
Kamenova, Ljudmila (Stony Brook University)
Kazanova, Anna (University of Massachusetts Amherst)
Keum, JongHae (Korea Institute for Advanced Study)
Kim, Dano (Korea Institute for Advanced Study)
Kim, Hosung (Korea Institute for Advanced Study)
Kollár, János (Princeton University)
Kovács, Sándor (University of Washington)
Lai, Ching-Jui (University of Utah)
Lazić, Vladimir (Imperial College London)
Lee, Tsung-Ju (National Taiwan University)
Lee, Yongnam (Sogang University)
Liu, Yu-Han (Princeton University)
McKernan, James (Massachusetts Institute of Technology)
Matsuki, Kenji (Purdue University)
Morrison, David (University of California, Santa Barbara)
Mustaţă, Mircea (University of Michigan)
Păun, Mihai (Institut Élie Cartan, Université Henri Poincaré)
Park, Jinhyung (Korea Advanced Institute of Science and Technology)
Prokhorov, Yuri (Moscow State Lomonosov University)
Reid, Miles (University of Warwick)
Sano, Taro (University of Warwick)
Shokurov, Vyacheslav (Johns Hopkins University / Steklov Institute

of Mathematics)
Spinaci, Marco (Institut Fourier)
Starr, Jason (Stony Brook University)
Su, Yu-Cheng (National Taiwan University)
Tziolas, Nicos (University of Cyprus)
Venkatram, Kartik (Massachusetts Institute of Technology)
Wang, Sz-Sheng (National Taiwan University)
Xu, Chenyang (Massachusetts Institute of Technology)
Zhang, De-Qi (National University of Singapore)
Zhu, Zhixian (University of Michigan)

渡辺 敬一　野間 淳

稲場 道明　渡邉 究　宮崎 誓　Zhang, De-Qi

岩見 智宏　池田 京司　臼井 三平

前田 博信　Choi, Sung Rak　Tziolas, Nicos　早川 貴之　Hwang, DongSeon

佐藤 拓　吉川 謙一　川ノ上 帆　Feng, Tien-Shun　Chen, Hou-Yi　Chen, Jheng-Jie

松村 慎一　鈴木 拓　永井 保成　藤澤 太郎　伊藤 敦　Su, Yu-Cheng　Lee, Tsung-Ju　高橋 慶祐　徳永 浩雄　Mustaţă, Mircea

Kim, Dano　楫 元　広瀬大輔　古川 勝久　中川 泰宏　高木 寛通　真瀬 真樹子

Kaloghiros, Anne-Sophie　大川 新之介　齋藤 夏雄　巴山竜来　安田 健彦

月岡 透　Prokhorov, Yuri　權業 善範　Wang, Sz-Sheng　桑田 孝泰　Lee, Yongnam　酒井 文雄

安武 和範　川口 周　佐野 太郎　廣門 正行　Liu, Yu-Han　Lai, Ching-Jui　松木 謙二　Kim, Hosung　Helmke, Stefan

田中 公　松下 大介　Kazanova, Anna

Park, Jinhyung　内田 幸寛　足利 正　Chi, Chen-Yu　土岡 俊介　中山 昇　那須 弘和　梅津 裕美子　青木 昌雄

Xu, Chenyang　辻 元　McKernan, James　Chen, Jiun-Cheng　Hacking, Paul

藤本 圭男　Nolla, Alvaro　小木曽 啓示　橋本 光靖　増田 佳代　Zhu, Zhixian　伊藤 哲史　高木 聡

伊藤 由佳理

並河 良典　Lazić, Vladimir　川北 真之　Starr, Jason

藤田 健人　尾高 悠志　岡田 拓三　Galkin, Sergey　齋藤 恭司　川北 素子　岡田 崇

鈴木 香織　Kamenova, Ljudmila

Kovács, Sándor

阿部 健　大橋 久範　藤野 修　Chen, Jungkai　Alexeev, Valery　Hwang, Jun-Muk　Birkar, Caucher　佐藤 栄一　高山 幸秀　石井 志保子　齋藤 政彦　森脇 淳　Morrison, David　Cascini, Paolo

Corti, Alessio　向井 茂　Kollár, János　柏原 正樹　Clemens, Herbert　森 重文　Reid, Miles　宮岡 洋一　桂 利行　宮西 正宜　小田 忠雄　塩田 徹治

Shokurov, Vyacheslav

Minimal models and extremal rays, June 2011, RIMS Kyoto Univ.

CONTENTS

I. Survey Article

II. Research Articles

Advanced Studies in Pure Mathematics 70, 2016
Minimal Models and Extremal Rays (Kyoto, 2011)
pp. 1–7

The mathematical works of Shigefumi Mori

János Kollár

Shigefumi Mori has been one of the most influential mathematicians of the past three decades. A proper survey of his achievements and influence—showing how his work fits in with that of his contemporaries and detailing the results proved by other mathematicians following the paths he laid out—would be a complete history of higher dimensional birational geometry, and thus far beyond the scope of this short summary.

The following overview groups his works according to themes rather than focusing on his main theorems, with the aim of laying out the fundamental ideas more clearly. However, this approach does result in some seemingly odd choices: one of the main topics (Mori's program) does not correspond to any of his publications and several important papers are omitted or mentioned only in passing.

I have had the privilege of knowing and working with Shigefumi Mori since 1983. This collaboration has been a fundamental influence on my mathematical point of view and development, thus I cannot pretend to offer an unbiased survey of Mori's contributions to algebraic geometry. Those who prefer to form their own opinion would be well-rewarded by reading his papers, especially [1] and [2]. These fundamentally important and enjoyable works illustrate the depth and originality of his ideas, as well as the clarity and precision of his writing.

§ Bend-and-break

It is a rare pleasure in mathematics to be surprised by a theorem whose proof is simple, yet its applications are profound. *Bend-and-break* is among them. The proof consists of three observations.

(i) Let C be a smooth, projective curve and $f_t : C \to X$ a 1-parameter family of non-constant morphisms. If the genus of C is ≥ 1

Received December 1, 2012.

and $f_t(c)$ stays fixed for some point $c \in C$, then a suitable limit of the 1-cycles $f_t(C)$ becomes reducible and acquires a rational component.

(ii) If $C \cong \mathbb{P}^1$, then one gets a reducible limit cycle if $f_t(c_i)$ stays fixed for at least 2 points $c_1, c_2 \in C$.

(iii) If $\deg f_0^* T_X > 0$, then one can construct such deformations in positive characteristic, reduce the degree of the resulting rational curves using (ii) and then lift the low-degree rational curves back to characteristic 0.

The bend-and-break method has several applications. The first, given in [1], characterizes the projective space $\mathbb{P}^n$ as the unique n-dimensional smooth, projective variety whose tangent bundle is ample. This solved conjectures of Frankel and Hartshorne.

Since we assume that the tangent bundle is ample, the condition $\deg f_0^* T_X > 0$ is automatic and the resulting rational curves turn out to be the lines in $\mathbb{P}^n$. Once we have the lines, the rest is relatively straightforward. These ideas have been used repeatedly to describe varieties whose tangent bundle is close to being ample; the conditions becoming more general every year.

A subsequent paper [9] shows that one can get low-degree rational curves by *one* application of a variant of (i) and (iii), making it possible to use the method for many singular varieties.

Another application, presented in [2] shows that the cone of curves of a Fano variety is polyhedral. This led to a classification of smooth, 3-dimensional Fano varieties [3, 4, 21] and is the cornerstone of most results on higher-dimensional Fano varieties.

§ The cone of curves and extremal rays

Let X be a projective variety, for simplicity over $\mathbb{C}$. Every algebraic curve $C \subset X$ has a homology class $[C] \in H_2(X, \mathbb{R})$. The *cone of curves* $\overline{NE}(X)$ is the closure of the cone generated by them. This cone was studied by Hironaka and Kleiman, but it was Mori who realized that if X is smooth, then the part of the cone lying in the open half-space where the intersection number with the canonical class is negative is locally finitely generated [2]. These generators are called the *extremal rays* of X or of $\overline{NE}(X)$.

Mori used bend-and-break to show that every extremal ray is spanned by (the homology class of) a rational curve. Then he showed, in dimension 3, that for every extremal ray $R \subset \overline{NE}(X)$ there is a unique morphism $g_R : X \to Z$, called the *contraction of* R such that an irreducible curve $C \subset X$ is mapped to a point by g_R iff $[C] \in R$. He also gave a complete description of all possible contractions.

This resulted in a completely different way of thinking about morphisms of varieties. Before [2], people were thinking in terms of linear systems; since then everyone imagines an extremal ray or face of a cone.

Later, cohomological methods were used to extend the existence results to all dimensions and to log terminal varieties and pairs, but the basic picture established by Mori remains unchanged.

An even more important consequence of [2] is that, as we discuss next, it led to a whole new approach to algebraic varieties.

§ Mori's program

Mori is, and probably will remain, best known for developing what is now known as *Mori's program, Mori theory* or *minimal model program.* Interestingly, he never wrote down the program and there is no published reference that one can point to.

It was obvious to experts that [2] gave the beginning steps of a program, but the continuation was not immediately clear. It seems that Mori had the whole plan by 1980 but he did not write it down; thus Reid and Tsunoda both discovered various parts independently. By 1982 the details were well-known to a rapidly growing group of researchers.

The program provides a recipe that allows one to write any birational contraction as a composite of two types of elementary maps: *divisorial contractions* and *flips.* As an end result, every smooth projective variety is birational either to a variety (with terminal singularities) whose canonical class is nef or to the total space of a family of Fano varieties. More generally, for many questions, the program allows us to find a birational model whose global and local geometry match optimally.

This viewpoint was new even for surfaces and it led to predictions about higher dimensions that were not even contemplated before. It determined the main direction of research in birational geometry for decades.

After a few years, two basic approaches started to diverge rapidly from each other. The schools of Kawamata and of Shokurov aimed at establishing methods that would work in all dimensions, culminating in the results of Hacon and McKernan. These results give existence theorems but rarely produce explicit descriptions of flips or of minimal models.

Mori's aim was to develop a deep and complete understanding of the 3-dimensional situation. As a consequence of his work, now we see that for 3-folds it is possible to understand and control every detail of the minimal model program; we discuss some of it next. By contrast, even the classification of 4-dimensional terminal singularities seems unlikely.

§ Extremal curve neighborhoods

As [2] showed, the contraction of an extremal ray of a smooth 3-fold sometimes results in a singular 3-fold. In order to continue with Mori's program, it was necessary to extend the theory of [2] to 3-folds with terminal singularities.

At the beginning the most important part was to understand flipping contractions. For later applications the key class turned out to be extremal contractions of 3-folds $f : X \to Y$ whose fiber dimension is ≤ 1. If X has terminal singularities and $C \subset X$ is a fiber, its neighborhood (formal or analytic) is called an *extremal curve neighborhood.*

Mori's greatest technical achievement was the development of a method to understand extremal curve neighborhoods. The full method—which should give a complete classification of extremal curve neighborhoods—was never written down; it exists only in his head and in thousands of pages of hand-written notes. The classification is exceedingly lengthy and it is not clear that every detail of it is needed in applications. Thus the published parts of the method give partial results, stopping when enough knowledge is assembled for the question at hand. This project, started in 1980, is still unfinished.

The paper [7] is preliminary; it gives a classification of terminal 3-fold singularities. The first, and most important, application of the method is the proof of the existence of 3-fold flips given in [10], completing the minimal model program in dimension 3.

Later uses of the method include the classification of 3-dimensional, irreducible, flipping curve neighborhoods [15, 20] and a (hopefully soon to be complete) classification of divisorial contractions with 1-dimensional fibers and $\mathbb{Q}$-conic bundles [22, 23, 24, 25].

§ Rationally connected varieties

Rational curves are the simplest among algebraic curves and rational surfaces form the simplest class among algebraic surfaces. Max Noether knew that rational 3-folds are too special and for over a century there was no agreement, nor even a sensible candidate, for what should constitute the correct class of "simplest varieties."

In characteristic 0, the definition of this class, called *rationally connected varieties,* was given in [14], together with numerous characterizations and good properties. Roughly speaking, a variety X is rationally connected if it has as many rational curves as rational varieties have, but we do not assume anything about the existence of rational surfaces on X.

Other papers of the same series [12, 13, 19] prove that smooth (and certain singular) Fano varieties are rationally connected. As a byproduct, they establish boundedness theorems for Fano varieties.

Since then the study of rationally connected varieties became a significant subfield of birational geometry. Most geometric questions have been solved in characteristic 0 but there are many open problems concerning the arithmetic of rationally connected varieties.

§ Miscellanea

The study of Fano 3-folds is intimately connected with the study of their hyperplane sections, which are K3 surfaces. Understanding curves on K3 surfaces turned out to be useful in proving the unirationality of the moduli of curves of genus 11 and in understanding space curves [5, 6].

One of the problems of dealing with minimal models is that many of the invariants that are positive integers for smooth varieties end up being positive rational numbers. In applications it is important to know that these are bounded away from 0 or, more generally, that they satisfy the descending chain condition. By now many existence results are known but [18] remains almost the only known effective estimate in this area.

The paper [16] shows that the geometric quotient of an algebraic space by a groupoid with finite stabilizer is again an algebraic space. This finished a problem that has been investigated since the sixties.

§ Books and surveys

For many of the younger generation, their introduction to Mori's works came through his books [11, 17] or his survey paper [8]. These still are the best introductions to higher dimensional birational geometry.

References

[1] *Projective manifolds with ample tangent bundles*, Ann. of Math. (2) **110** (1979), no. 3, 593–606. MR 554387 (81j:14010)

[2] *Threefolds whose canonical bundles are not numerically effective*, Ann. of Math. (2) **116** (1982), no. 1, 133–176. MR 662120 (84e:14032)

[3] *Classification of Fano 3-folds with $B_2 \geq 2$* (with Shigeru Mukai) Manuscripta Math. **36** (1981/82), no. 2, 147–162. MR 641971 (83f:14032)

[4] *On Fano 3-folds with $B_2 \geq 2$* (with Shigeru Mukai) Algebraic varieties and analytic varieties (Tokyo, 1981), Adv. Stud. Pure Math., vol. 1, North-Holland, Amsterdam, 1983, pp. 101–129. MR 715648 (85b:14055)

[5] *The uniruledness of the moduli space of curves of genus* 11 (with Shigeru Mukai) Algebraic geometry (Tokyo/Kyoto, 1982), Lecture Notes in Math., vol. 1016, Springer, Berlin, 1983, pp. 334–353. MR 726433 (85b:14033)

[6] *On degrees and genera of curves on smooth quartic surfaces in* $\mathbf{P}^3$, Nagoya Math. J. **96** (1984), 127–132. MR 771073 (86g:14023)

[7] *On 3-dimensional terminal singularities*, Nagoya Math. J. **98** (1985), 43–66. MR 792770 (86m:14003)

[8] *Classification of higher-dimensional varieties*, Algebraic geometry, Bowdoin, 1985 (Brunswick, Maine, 1985), Proc. Sympos. Pure Math., vol. 46, Amer. Math. Soc., Providence, RI, 1987, pp. 269–331. MR 927961 (89a:14040)

[9] *A numerical criterion for uniruledness* (with Yoichi Miyaoka) Ann. of Math. (2) **124** (1986), no. 1, 65–69. MR 847952 (87k:14046)

[10] *Flip theorem and the existence of minimal models for 3-folds*, J. Amer. Math. Soc. **1** (1988), no. 1, 117–253. MR 924704 (89a:14048)

[11] *Higher-dimensional complex geometry* (with C. Herbert Clemens and János Kollár) Astérisque (1988), no. 166, 144 pp. (1989). MR 1004926 (90j:14046)

[12] *Rational curves on Fano varieties* (with János Kollár and Yoichi Miyaoka) Classification of irregular varieties (Trento, 1990), Lecture Notes in Math., vol. 1515, Springer, Berlin, 1992, pp. 100–105. MR 1180339 (94f:14039)

[13] *Rational connectedness and boundedness of Fano manifolds* (with János Kollár and Yoichi Miyaoka) J. Differential Geom. **36** (1992), no. 3, 765–779. MR 1189503 (94g:14021)

[14] *Rationally connected varieties* (with János Kollár and Yoichi Miyaoka) J. Algebraic Geom. **1** (1992), no. 3, 429–448. MR 1158625 (93i:14014)

[15] *Classification of three-dimensional flips* (with János Kollár) J. Amer. Math. Soc. **5** (1992), no. 3, 533–703. MR 1149195 (93i:14015)

[16] *Quotients by groupoids* (with Seán Keel) Ann. of Math. (2) **145** (1997), no. 1, 193–213. MR 1432041 (97m:14014)

[17] *Birational geometry of algebraic varieties* (with János Kollár and with the collaboration of C. H. Clemens and A. Corti) Cambridge Tracts in Mathematics, vol. 134, Cambridge University Press, Cambridge, 1998, Translated from the 1998 Japanese original. MR 1658959 (2000b:14018)

[18] *Bounding singular surfaces of general type* (with Valery Alexeev) Algebra, arithmetic and geometry with applications (West Lafayette, IN, 2000), Springer, Berlin, 2004, pp. 143–174. MR 2037085 (2005f:14077)

[19] *Boundedness of canonical* $\mathbf{Q}$*-Fano 3-folds* (with János Kollár, Yoichi Miyaoka and Hiromichi Takagi) Proc. Japan Acad. Ser. A Math. Sci. **76** (2000), no. 5, 73–77. MR 1771144 (2001h:14053)

[20] *On semistable extremal neighborhoods*, Higher dimensional birational geometry (Kyoto, 1997), Adv. Stud. Pure Math., vol. 35, Math. Soc. Japan, Tokyo, 2002, pp. 157–184. MR 1929794 (2004e:14029)

[21] *Extremal rays and Fano 3-folds* (with Shigeru Mukai) The Fano Conference, Univ. Torino, Turin, 2004, pp. 37–50. MR 2112566 (2005k:14085)

[22] *On $\mathbb{Q}$-conic bundles* (with Yuri Prokhorov) Publ. Res. Inst. Math. Sci. **44** (2008), no. 2, 315–369. MR 2426350 (2009e:14062)
[23] *On $\mathbb{Q}$-conic bundles. II* (with Yuri Prokhorov) Publ. Res. Inst. Math. Sci. **44** (2008), no. 3, 955–971. MR 2451614 (2009h:14068)
[24] *On $\mathbb{Q}$-conic bundles. III* (with Yuri Prokhorov) Publ. Res. Inst. Math. Sci. **45** (2009), no. 3, 787–810. MR 2569567 (2011a:14080)
[25] *Threefold extremal contractions of type (IA)* (with Yuri Prokhorov) Kyoto J. Math. **51** (2011), no. 2, 393–438. MR 2793273 (2012e:14033)

Princeton University, Princeton NJ 08544-1000
E-mail address: kollar@math.princeton.edu

Advanced Studies in Pure Mathematics 70, 2016
Minimal Models and Extremal Rays (Kyoto, 2011)
pp. 9–28

Multiplier ideals via Mather discrepancy

Lawrence Ein, Shihoko Ishii and Mircea Mustaţă

Dedicated to Professor Shigefumi Mori on the occasion of his 60th birthday

Abstract.

We define a version of multiplier ideals, the *Mather multiplier ideals*, on a variety with arbitrary singularities, using the Mather discrepancy and the Jacobian ideal. In this context we prove a relative vanishing theorem, thus obtaining restriction, subadditivity and summation theorems. The Mather multiplier ideals also satisfy a Skoda type result. As an application, we obtain a Briançon-Skoda type formula for the integral closures of ideals on a variety with arbitrary singularities.

§1. Introduction

The main goal of this paper is to introduce and prove the basic properties of *Mather multiplier ideals*, a variant of the usual multiplier ideals which have found wide applications recently in higher-dimensional algebraic geometry. Typically, the definition of multiplier ideals is given in terms of a suitable resolution of singularities, and an ingredient that plays a fundamental role is the discrepancy (or relative canonical divisor). In order for the discrepancy to be well-defined, one needs to make some assumptions on the singularities, namely to assume that the ambient variety is normal and $\mathbf{Q}$-Gorenstein. The main advantage of Mather multiplier ideals is that they can be defined on any variety (by which we mean an integral scheme of finite type over an algebraically closed field of characteristic zero). This agrees with the usual multiplier ideals on normal, locally complete intersection varieties.

Given an arbitrary variety X and a nonzero ideal $\mathfrak{a}$ on X, we consider a log resolution $f\colon Y \longrightarrow X$ for the product of $\mathfrak{a}$ with the Jacobian ideal

---

Received July 6, 2011.

Ein was partially supported by NSF grant DMS-1001336, Ishii was partially supported by Grant-in-Aid (B) 22340004 and (S) 19104001, and Mustaţă was partially supported by NSF grant DMS-0758454, and by a Packard Fellowship.

$\mathcal{J}ac_X$ of X (see §1 for the relevant definitions). Such a resolution has the property that the image of the canonical map $f^*(\Omega_X^n) \longrightarrow \Omega_Y^n$ (where $n = \dim(X)$) can be written as $\mathcal{O}_Y(-\widehat{K}_{Y/X}) \cdot \Omega_Y^n$, for some effective divisor $\widehat{K}_{Y/X}$ on Y. This is the *Mather discrepancy divisor*, and if we write $\mathcal{J}ac_X \cdot \mathcal{O}_Y = \mathcal{O}_Y(-J_{Y/X})$, then the difference $\widehat{K}_{Y/X} - J_{Y/X}$ plays the role that the discrepancy plays in the usual definition of multiplier ideals. More precisely, if $\mathfrak{a} \cdot \mathcal{O}_Y = \mathcal{O}_Y(-Z_{Y/X})$ and if $t \in \mathbf{R}_{\geq 0}$, then the Mather multiplier ideal of $\mathfrak{a}$ with exponent t is given by

$$\widehat{\mathcal{J}}(X, \mathfrak{a}^t) := f_*\mathcal{O}_Y(\widehat{K}_{Y/X} - J_{Y/X} - \lfloor tZ_{Y/X} \rfloor).$$

A priori this is only a fractional ideal, but we show that it is, in fact, contained in $\mathcal{O}_X$. If X is normal and locally complete intersection, then the difference $\widehat{K}_{Y/X} - J_{Y/X}$ is equal to the usual discrepancy $K_{Y/X}$, hence the Mather multiplier ideals on X agree with the usual multiplier ideals.

The importance of multiplier ideals comes from the role they play in connection with vanishing theorems (see [11, Chapter 9]). The basic result that accounts for this connection is a relative vanishing theorem. We prove a similar result in the context of Mather multiplier ideals. With the above notation, this says that we have the vanishing of all higher direct images:

$$R^i f_*\mathcal{O}_Y(\widehat{K}_{Y/X} - J_{Y/X} - \lfloor tZ_{Y/X} \rfloor) = 0 \text{ for all } i > 0.$$

As in the "classical" case, this leads to some important properties of Mather multiplier ideals: restriction theorems (we prove three such results, in various settings), a Subadditivity Theorem and a Summation Theorem, as well as a Skoda-type Theorem (see [11] for the "classical" statements of these results). In proving these results, we make use of the approach developed by Eugene Eisenstein in [5] for proving a version of the Restriction Theorem, and a Subadditivity Theorem for multiplier ideals on singular varieties. We obtain as a corollary the following analogue of the Briançon-Skoda Theorem on arbitrary varieties: if $\mathfrak{a}$ is an ideal on X, and $\dim(X) = n$, then the integral closure of $\mathcal{J}ac_X \cdot \mathfrak{a}^m$ is contained in $\mathfrak{a}^{m-n+1}$ for every $m \geq n$.

The notion of Mather discrepancy is suggested by the approach to singularities via spaces of arcs (see [6]). This replacement for the usual discrepancy was first introduced in [3], where it was applied to relate divisorial valuations to contact loci in arc spaces for singular varieties. It was subsequently used in [8] to define new versions of some familiar invariants of singularities (such as the log canonical threshold and the minimal log discrepancies) on arbitrary varieties. Around the same time that our present work was done, related ideas have been developed in the paper [4] of Tommaso de Fernex and Roi Docampo, in which the authors also undertake a study of invariants of singularities defined from the point of view of Mather discrepancies, and relate these to rational and du Bois singularities.

Acknowledgement We are grateful to Takehiko Yasuda and Orlando Villamayor for useful conversations with the second author on Nash blow-ups and factorizing resolutions, respectively.

§2. Definition of Mather multiplier ideal

Recall that all our varieties are assumed to be irreducible and reduced, over a fixed algebraically closed field k of characteristic zero. We begin by recalling the definition of Nash blow-up and Mather discrepancy, following [3].

Let X be an n-dimensional variety. The sheaf $\Omega_X^n = \wedge^n \Omega_X$ is invertible over the smooth locus X_{reg} of X, hence the morphism

$$\pi \colon \mathbf{P}(\Omega_X^n) = \mathcal{P}roj(\operatorname{Sym}(\Omega_X^n)) \longrightarrow X$$

is an isomorphism over X_{reg}. The *Nash blow-up* $\widehat{X}$ is the closure of $\pi^{-1}(X_{\text{reg}})$ in $\mathbf{P}(\Omega_X^n)$ (with the reduced scheme structure). Note that by construction we have a projective birational morphism $\nu \colon \widehat{X} \longrightarrow X$ and a surjective morphism $\nu^*(\Omega_X^n) \longrightarrow \mathcal{O}_{\mathbf{P}(\Omega_X^n)}(1)|_{\widehat{X}}$. Furthermore, this is universal in the following sense: given a birational morphism of varieties $\phi \colon Y \longrightarrow X$, this factors through ν if and only if $\phi^*(\Omega_X^n)$ admits a morphism onto an invertible sheaf $\mathcal{L}$ on Y (which identifies $\mathcal{L}$ to the quotient of $\phi^*(\Omega_X^n)$ by its torsion).

Recall that a *resolution of singularities* of X is a projective birational morphism $f \colon Y \longrightarrow X$, with Y smooth. This is a *log resolution* of a nonzero ideal $\mathfrak{a}$ on X if $\mathfrak{a} \cdot \mathcal{O}_Y = \mathcal{O}_Y(-D)$ for an effective divisor D, the exceptional locus of f is an effective divisor E, and $D \cup E$ has simple normal crossings. Given several ideals $\mathfrak{a}_1, \ldots, \mathfrak{a}_r$, a log resolution of the product $\mathfrak{a}_1 \cdot \ldots \cdot \mathfrak{a}_r$ gives a simultaneous log resolution of all $\mathfrak{a}_i$, such that the divisors corresponding to each of the $\mathfrak{a}_i$ and the exceptional divisor have simple normal crossings.

Definition 2.1 ([3]). *Let X be a variety of dimension n and $f \colon Y \longrightarrow X$ a resolution of singularities that factors through the Nash blow-up of X. In this case the image of the canonical homomorphism*

$$f^*(\Omega_X^n) \longrightarrow \Omega_Y^n$$

is an invertible sheaf of the form $J \cdot \Omega_Y^n$*, where* J *is the invertible ideal sheaf on* Y *that defines an effective divisor supported on the exceptional locus of* f*. This divisor is called the* Mather discrepancy divisor *and we denote it by* $\widehat{K}_{Y/X}$*. We refer to [3] for further details about Nash blow-ups and Mather discrepancies.*

Definition 2.2. *Recall that the* Jacobian ideal $\mathcal{J}ac_X$ *of a variety* X *is the* n^{th} *Fitting ideal* $\mathrm{Fitt}_n(\Omega_X)$ *of* Ω_X*. This is an ideal whose support is the singular locus of* X*. If* $f\colon Y \longrightarrow X$ *is a log resolution of* $\mathcal{J}ac_X$*, we denote by* $J_{Y/X}$ *the effective divisor on* Y *such that* $\mathcal{J}ac_X \cdot \mathcal{O}_Y = \mathcal{O}_Y(-J_{Y/X})$*.*

Remark 2.1. Every log-resolution of $\mathcal{J}ac_X$ factors through the Nash blow-up. Indeed, it follows from [12, Lemma 1] that given a proper birational morphism $\varphi\colon Y \longrightarrow X$ of varieties of dimension n, the following are equivalent:

(1) The smallest Fitting ideal $\mathrm{Fitt}_n(\varphi^*\Omega_X)$ is invertible;
(2) The projective dimension $\mathrm{pd}(\varphi^*\Omega_X) \leq 1$ and $\varphi^*\Omega_X/(\varphi^*\Omega_X)^{tor}$ is locally free of rank n, where $(\varphi^*\Omega_X)^{tor}$ is the torsion part of $\varphi^*\Omega_X$.

If φ is a log resolution of $\mathcal{J}ac_X$, then $\mathrm{Fitt}_n(\varphi^*\Omega_X) = \mathcal{J}ac_X \cdot \mathcal{O}_Y$ is invertible. Therefore by Lipman's result $\varphi^*\Omega_X/(\varphi^*\Omega_X)^{tor}$ is locally free of rank n, and we deduce that φ factors through $\widehat{X}$ by the universal property of the Nash blow-up.

Definition 2.3. *Let* X *be a variety,* $\mathfrak{a} \subseteq \mathcal{O}_X$ *a nonzero ideal on* X*, and* $t \in \mathbf{R}_{\geq 0}$*. Given a log resolution* $f\colon Y \longrightarrow X$ *of* $\mathcal{J}ac_X \cdot \mathfrak{a}$*, we denote by* $Z_{Y/X}$ *the effective divisor on* Y *such that* $\mathfrak{a} \cdot \mathcal{O}_Y = \mathcal{O}_Y(-Z_{Y/X})$*. The Mather multiplier ideal of* $\mathfrak{a}$ *of exponent* t *is defined by*

$$\widehat{\mathcal{J}}(X, \mathfrak{a}^t) = f_*\left(\mathcal{O}_Y(\widehat{K}_{Y/X} - J_{Y/X} - \lfloor tZ_{Y/X} \rfloor)\right),$$

where for a real divisor D *we denote by* $\lfloor D \rfloor$ *the largest integral divisor* $\leq D$*.*

Remark 2.2. The Mather multiplier ideal does not depend on the choice of f. Indeed, arguing as in the proof of [11, Theorem 9.2.18], we see that it is enough to show that if $g\colon Y' \longrightarrow Y$ is such that $f' = f \circ g$ is another log resolution of $\mathcal{J}ac_X \cdot \mathfrak{a}$, then

(1)
$$\mathcal{O}_Y(\widehat{K}_{Y/X} - J_{Y/X} - \lfloor tZ_{Y/X} \rfloor) = g_*\left(\mathcal{O}_{Y'}(\widehat{K}_{Y'/X} - J_{Y'/X} - \lfloor tZ_{Y'/X} \rfloor)\right).$$

Since $\widehat{K}_{Y'/X} = g^*(\widehat{K}_{Y/X}) + K_{Y/X}$, where $K_{Y/X}$ is the (usual) discrepancy of h, and $J_{Y'/X} = g^*(J_{Y/X})$ and $Z_{Y'/X} = g^*(Z_{Y/X})$, then (1)

follows from [11, Lemma 9.2.19], which is the main ingredient in showing that the usual definition of multiplier ideals is independent of log resolution.

Remark 2.3. For nonzero ideals $\mathfrak{a}_1, \ldots, \mathfrak{a}_r$ on a variety X, one can similarly define a mixed Mather multiplier ideal $\widehat{\mathcal{J}}(X, \mathfrak{a}_1^{t_1} \cdots \mathfrak{a}_r^{t_r})$ for every $t_1, \ldots, t_r \in \mathbf{R}_{\geq 0}$. With the notation in Definition 2.3, if f is a log resolution of $\mathcal{J}ac_X \cdot \mathfrak{a}_1 \cdot \ldots \cdot \mathfrak{a}_r$, and if we put $\mathfrak{a}_i \cdot \mathcal{O}_Y = \mathcal{O}_Y(-Z_i)$, then

$$\widehat{\mathcal{J}}(X, \mathfrak{a}_1^{t_1} \cdots \mathfrak{a}_r^{t_r}) = f_* \left(\mathcal{O}_Y(\widehat{K}_{Y/X} - J_{Y/X} - \lfloor t_1 Z_1 + \ldots + t_r Z_r \rfloor) \right).$$

For simplicity, we will mostly consider Mather multiplier ideals for one ideal, but all statements have obvious generalizations to the mixed case.

Remark 2.4. If X is normal and locally a complete intersection, then $\widehat{\mathcal{J}}(X, \mathfrak{a}^t) = \mathcal{J}(X, \mathfrak{a}^t)$, where the right hand side is the usual multiplier ideal (see [11, Chapter 9] for definition). Indeed, in this case the image of the canonical map $\Omega_X^n \longrightarrow \omega_X$ is $\mathcal{J}ac_X \cdot \omega_X$, hence $\widehat{K}_{Y/X} - J_{Y/X} = K_{Y/X}$. In particular, we see that $\widehat{\mathcal{J}}(X, \mathfrak{a}^t) = \mathcal{J}(X, \mathfrak{a}^t)$ if X is smooth.

Remark 2.5. For now, the Mather multiplier ideal is only a fractional ideal of $\mathcal{O}_X$, with $\widehat{\mathcal{J}}(X, \mathfrak{a}^t) \subset i_*(\mathcal{O}_{X_{reg}})$, where $i \colon X_{reg} \hookrightarrow X$ is the inclusion of the smooth locus. This inclusion implies that $\widehat{\mathcal{J}}(X, \mathfrak{a}^t)$ is an ideal of $\mathcal{O}_X$ if X is normal. We will see in Corollary 2.10 that, in fact, the same is true on every variety X.

The assertions in the next proposition are an immediate consequence of the definition.

Proposition 2.6. *Let $\mathfrak{a}$ and $\mathfrak{b}$ be nonzero ideals on the variety X.*

(1) *If $\mathfrak{a} \subseteq \mathfrak{b}$, then $\widehat{\mathcal{J}}(X, \mathfrak{a}^t) \subseteq \widehat{\mathcal{J}}(X, \mathfrak{b}^t)$ for every $t \in \mathbf{R}_{\geq 0}$.*

(2) *If $s \leq t$ are in $\mathbf{R}_{\geq 0}$, then $\widehat{\mathcal{J}}(X, \mathfrak{a}^t) \subseteq \widehat{\mathcal{J}}(X, \mathfrak{a}^s)$.*

(3) *If $t \in \mathbf{R}_{\geq 0}$ and m is a positive integer, then $\widehat{\mathcal{J}}(X, (\mathfrak{a}^m)^t) = \widehat{\mathcal{J}}(X, \mathfrak{a}^{mt})$.*

Definition 2.4. *Let A be a variety and X a reduced closed subscheme of A. A morphism $\varphi_A \colon \overline{A} \longrightarrow A$ is a* factorizing resolution *of X inside A if the following hold:*

(1) *φ_A is an isomorphism at the generic point of every irreducible component of X. In particular, the strict transform $\overline{X}$ of X is defined.*

(2) *The morphisms φ_A and $\varphi_X := \varphi_A|_{\overline{X}}$ are resolutions of singularities of A and X, respectively, and the union of $\overline{X}$ with the exceptional locus $\mathrm{Exc}(\varphi_A)$ has simple normal crossings.*

(3) *If I_X and $I_{\overline{X}}$ are the defining ideals of X and $\overline{X}$ in A and $\overline{A}$, respectively, then there exists an effective divisor $R_{X/A}$ on $\overline{A}$ such that*

$$I_X \cdot \mathcal{O}_{\overline{A}} = I_{\overline{X}} \cdot \mathcal{O}_{\overline{A}}(-R_{X/A}).$$

Note that the divisor $R_{X/A}$ in (3) is supported on $\mathrm{Exc}(\phi_A)$, hence by assumption has simple normal crossings with $\overline{X}$. Bravo and Villamayor proved in [1] that if A is smooth and X is reduced and equidimensional, then factorizing resolutions of X in A exist. The following strengthening is due to Eisenstein [5, Lemma 3.1].

Lemma 2.7. *Let $X \subseteq A$ be a reduced closed subscheme of a variety A. Let $\varphi_1 \colon A' \longrightarrow A$ be a birational morphism from a smooth variety A' that is an isomorphism at the generic points of the irreducible components of X. Let X' be the strict transform of X in A', and let E be a divisor on A' with simple normal crossing support, such that no component of X' is contained in E. Then there exists a morphism $\varphi_2 \colon \overline{A} \longrightarrow A'$ such that $\varphi_A := \varphi_1 \circ \varphi_2$ is a factorizing resolution of X inside A, and $\overline{X} \cup \mathrm{Exc}(\varphi_A) \cup \mathrm{Supp}(\varphi_2^* E)$ has simple normal crossings, where $\overline{X}$ is the strict transform of X in $\overline{A}$.*

The following lemma is essentially [5, Lemma 4.4].

Lemma 2.8. *Let $X \subseteq A$ be a reduced equidimensional closed subscheme of a smooth variety A, and let $\varphi_A \colon \overline{A} \longrightarrow A$ be a factorizing resolution of X inside A. Let $c = \mathrm{codim}(X, A)$ and $I_X \cdot \mathcal{O}_{\overline{A}} = I_{\overline{X}} \cdot \mathcal{O}_{\overline{A}}(-R_{X/A})$. If $\varphi_A|_{\overline{X}}$ factors through the blow-up along the ideal $\mathcal{J}ac_X$, then*

$$(K_{\overline{A}/A} - cR_{X/A})|_{\overline{X}} = \widehat{K}_{\overline{X}/X} - J_{\overline{X}/X}.$$

Proposition 2.9. *Let X be a closed subvariety of the smooth variety A, with $\mathrm{codim}(X, A) = c$. If $\tilde{\mathfrak{a}} \subseteq \mathcal{O}_A$ is an ideal on A whose restriction $\mathfrak{a} = \tilde{\mathfrak{a}} \cdot \mathcal{O}_X$ to X is nonzero, then for every $t \in \mathbf{R}_{\geq 0}$*

$$\widehat{\mathcal{J}}(X, \mathfrak{a}^t) \subseteq \mathcal{J}(A, \tilde{\mathfrak{a}}^t) \cdot \mathcal{O}_X.$$

Proof. Let $\mathfrak{b} \subseteq \mathcal{O}_A$ be the pull-back of $\mathcal{J}ac_X$ by the canonical surjection $\mathcal{O}_A \longrightarrow \mathcal{O}_X$. We begin by choosing a log resolution $\varphi_1 \colon A' \longrightarrow A$ of $\mathfrak{b} \cdot \tilde{\mathfrak{a}}$. Let E be the divisor on A' such that $\mathcal{O}_{A'}(-E) = \tilde{\mathfrak{a}}\mathfrak{b} \cdot \mathcal{O}_{A'}$. Applying Lemma 2.7, we obtain a factorizing resolution $\varphi_A \colon \overline{A} \longrightarrow A$ of X inside A, that is also a log resolution of $\mathfrak{b} \cdot \tilde{\mathfrak{a}}$. Let $R_{X/A}$ be the divisor on $\overline{A}$ such that

$$I_X \cdot \mathcal{O}_{\overline{A}} = I_{\overline{X}} \cdot \mathcal{O}_{\overline{A}}(-R_{X/A}),$$

where $\overline{X}$ is the strict transform of X in $\overline{A}$. As φ_A factors through φ_1, the morphism $\varphi_X := \varphi_A|_{\overline{X}}$ is a log resolution of $\mathcal{J}ac_X$. Let $\tilde{\mathfrak{a}} \cdot \mathcal{O}_{\overline{A}} = \mathcal{O}_{\overline{A}}(-Z_{\overline{A}/A})$ and $D = K_{\overline{A}/A} - \lfloor tZ_{\overline{A}/A} \rfloor - cR_{X/A}$. The exact sequence

$$0 \longrightarrow I_{\overline{X}} \cdot \mathcal{O}_{\overline{A}}(D) \longrightarrow \mathcal{O}_{\overline{A}}(D) \longrightarrow \mathcal{O}_{\overline{X}}(D|_{\overline{X}}) \longrightarrow 0$$

yields the exact sequence

$$\text{(2)} \quad 0 \longrightarrow \varphi_{A*}(I_{\overline{X}} \cdot \mathcal{O}_{\overline{A}}(D)) \longrightarrow \varphi_{A*}(\mathcal{O}_{\overline{A}}(D)) \longrightarrow \varphi_{X*}(\mathcal{O}_{\overline{X}}(D|_{\overline{X}})) \longrightarrow R^1\varphi_{A*}(I_{\overline{X}} \cdot \mathcal{O}_{\overline{A}}(D)).$$

We claim that the last term in this sequence is zero. Indeed, let $\psi\colon \tilde{A} \longrightarrow \overline{A}$ be the blow-up along $\overline{X}$ with exceptional divisor T. If $\mathcal{F} = \mathcal{O}_{\tilde{A}}(-T + \psi^*D)$, then the projection formula gives $\psi_*\mathcal{F} = I_{\overline{X}} \cdot \mathcal{O}_{\overline{A}}(D)$. For $p > 0$, we have

$$R^p\psi_*(\mathcal{F}) = \mathcal{O}_{\overline{A}}(D) \otimes R^p\psi_*(\mathcal{O}_{\tilde{A}}(-T)) = 0, \quad R^p(\varphi_A \circ \psi)_*(\mathcal{F}) = 0.$$

The second vanishing follows from the Local Vanishing Theorem for multiplier ideals (see [11, Theorem 9.4.1]), using the fact that if $I_X \cdot \mathcal{O}_{\tilde{A}} = \mathcal{O}_{\tilde{A}}(-F)$, then we can write

$$-T + \psi^*(D) = K_{\tilde{A}/A} - \lfloor cF + t\psi^*(Z_{\overline{A}/A}) \rfloor.$$

Using the Leray spectral sequence

$$E_2^{p,q} = R^p\varphi_{A*}R^q\psi_*(\mathcal{F}) \Rightarrow R^{p+q}(\varphi_A \circ \psi)_*(\mathcal{F}),$$

we obtain $R^1\varphi_{A*}(\psi_*\mathcal{F}) = 0$, as claimed.

Since $R_{X/A}$ is effective, we have by definition $\varphi_{A*}(\mathcal{O}_{\overline{A}}(D)) \subseteq \mathcal{J}(A, \tilde{\mathfrak{a}}^t)$, hence the exact sequence (2) gives

$$\varphi_{X*}(\mathcal{O}_{\overline{X}}(D|_{\overline{X}})) \subseteq \mathcal{J}(A, \tilde{\mathfrak{a}}^t) \cdot \mathcal{O}_X.$$

By Lemma 2.8 we have

$$\varphi_{X*}(\mathcal{O}_{\overline{X}}(D|_{\overline{X}})) = \varphi_{X*}\left(\mathcal{O}_{\overline{X}}(\widehat{K}_{\overline{X}/X} - J_{\overline{X}/X} - \lfloor tZ_{\overline{X}/X} \rfloor\right) = \widehat{\mathcal{J}}(X, \mathfrak{a}^t),$$

which completes the proof of the proposition. Q.E.D.

Corollary 2.10. *Given a variety X, a nonzero ideal $\mathfrak{a}$ on X, and $t \in \mathbf{R}_{\geq 0}$, the following hold:*

(1) *The Mather multiplier ideal $\widehat{\mathcal{J}}(X, \mathfrak{a}^t)$ is an ideal of $\mathcal{O}_X$.*

(2) *The ideal $\widehat{\mathcal{J}}(X, \mathfrak{a}^t)$ is integrally closed.*

(3) *If $\nu\colon X^\nu \longrightarrow X$ is the normalization of X, then the integral closure $\overline{\mathcal{J}ac_X}$ of $\mathcal{J}ac_X$ satisfies*

$$\overline{\mathcal{J}ac_X} \subseteq (\mathcal{O}_X : \nu_*\mathcal{O}_{X^\nu}).$$

Proof. (1) follows from Proposition 2.9, since X can be locally embedded in a smooth variety, and the result is known (and easy to prove) for usual multiplier ideals on smooth varieties. For (2), it is enough to note that if $f\colon Y \longrightarrow X$ is a birational morphism, with Y normal, and E is a Weil divisor on Y such that $\mathcal{I} := f_*\mathcal{O}_Y(E)$ is an ideal of $\mathcal{O}_X$, then $\mathcal{I}$ is integrally closed.

In order to show (3), let $\varphi\colon Y \longrightarrow X$ be a log resolution of $\mathcal{J}ac_X$. We have the inclusions

$$\overline{\mathcal{J}ac_X} \subseteq \varphi_*\mathcal{O}_Y(-J_{Y/X}) \subseteq \varphi_*\mathcal{O}_Y(\widehat{K}_{Y/X} - J_{Y/X}) = \widehat{\mathcal{J}}(X, \mathcal{O}_X) \subseteq \mathcal{O}_X.$$

Since φ factors through ν, it follows that $\varphi_*\mathcal{O}_Y(-J_{Y/X})$ is a $\nu_*\mathcal{O}_{X^\nu}$-module, and we deduce the inclusion in (3). Q.E.D.

Example 2.11. If $X = \operatorname{Spec} k[x^2, x^3]$, then

$$\widehat{\mathcal{J}}(X, \mathcal{O}_X) = (x^2) \subset k[x^2, x^3].$$

Note that X is a curve with a cusp at the origin and has the normalization $\nu\colon \overline{X} = \operatorname{Spec} k[x] \longrightarrow X$. The morphism ν is a resolution of singularities and factors through the Nash blow-up (since the normalization of a curve factors through every blow-up). As $\nu^*\Omega_X$ is generated by dx^2, dx^3, the image of the canonical map $\nu^*\Omega_X \longrightarrow \Omega_{\overline{X}}$ is generated by xdx and x^2dx, hence $\mathcal{O}_{\overline{X}}(-\widehat{K}_{\overline{X}/X}) = (x)$. On the other hand, $\mathcal{J}ac_X = (x^3, x^4) \subset k[x^2, x^3]$ and $\mathcal{J}ac_X\mathcal{O}_{\overline{X}} = (x^3) \subset k[x]$. Therefore $\mathcal{O}_{\overline{X}}(\widehat{K}_{\overline{X}/X} - J_{\overline{X}/X}) = (x^2) \subset k[x]$, so that $\widehat{\mathcal{J}}(X, \mathcal{O}_X) = (x^2)$, as an ideal of $k[x^2, x^3]$.

§3. The local vanishing theorem and applications

We first recall that for every variety X of dimension n we have a canonical morphism $\eta_X\colon \Omega_X^n \longrightarrow \omega_X$, where ω_X is the canonical sheaf of X. This is clear when X is normal, since in that case $\omega_X \simeq i_*(\Omega^n_{X_{\rm reg}})$, where $i\colon X_{\rm reg} \hookrightarrow X$ is the inclusion of the smooth locus. In the general case, let $\nu\colon X' \longrightarrow X$ be the normalization, so we have $\eta_{X'}\colon \Omega^n_{X'} \longrightarrow \omega_{X'}$. Since ν is finite and surjective, we have an isomorphism $\nu_*(\omega_{X'}) \simeq \mathcal{H}om_{\mathcal{O}_X}(\nu_*(\mathcal{O}_{X'}), \omega_X)$, and the inclusion $\mathcal{O}_X \hookrightarrow \nu_*(\mathcal{O}_{X'})$ induces a canonical morphism $\nu_*(\omega_{X'}) \longrightarrow \omega_X$. We then obtain η_X as the composition

$$\Omega_X^n \longrightarrow \nu_*(\Omega_{X'}^n) \stackrel{\nu_*(\eta_{X'})}{\longrightarrow} \nu_*(\omega_{X'}) \longrightarrow \omega_X.$$

The following is a consequence of [6, Proposition 9.1].

Lemma 3.1. *Let $X \subseteq \mathbf{A}^N$ be a closed subvariety of dimension n, defined by the ideal $I_X = (f_1, \ldots, f_d)$, and let $c = N - n$. If the subscheme M defined by c functions $f_{i_1}, \ldots, f_{i_c}$ has pure codimension c, and $M = X$ at the generic point of X, then there exists an isomorphism of ω_X with an ideal of $\mathcal{O}_X$*

$$\omega_X \simeq ((I_M : I_X) + I_X)/I_X.$$

Under this isomorphism, the image of the canonical map $\eta_X \colon \Omega_X^n \longrightarrow \omega_X$ corresponds to $\mathcal{J}ac_M|_X$.

Remark 3.2. Suppose that $X \subseteq \mathbf{A}^N$ is as in Lemma 3.1. If $F_i := \sum_{j=1}^d a_{ij} f_j$, with a_{ij} general elements of k, then $I_X = (F_1, \ldots, F_d)$ and every set of distinct c elements $F_{i_1}, \ldots, F_{i_c}$ defines a subscheme M that satisfies the conditions in Lemma 3.1.

Lemma 3.3. *For every n-dimensional affine variety X, there exist a non-zero regular function h on X and ideals $I, G \subseteq \mathcal{O}_X$ such that there is a surjective morphism $\Omega_X^n \longrightarrow I$ and we have*

$$h \cdot \mathcal{J}ac_X = I \cdot G. \tag{3}$$

Proof. We apply Lemma 3.1 using the notation in Remark 3.2. For every subset $J \subseteq \{1, \ldots, d\}$ with c elements, we obtain a subscheme M_J of $\mathbf{A}^N$ defined by $(F_i \mid i \in J)$, that satisfies the conditions in Lemma 3.1. In particular, we obtain an isomorphism $\alpha_J \colon \omega_X \longrightarrow \mathfrak{q}_J \subseteq \mathcal{O}_X$ that maps the image of $\Omega_X^n \longrightarrow \omega_X$ onto $\mathcal{J}ac_{M_J}|_X$.

Let us fix a subset J_0 as above, so α_{J_0} induces a surjective map from Ω_X^n onto $I = \mathcal{J}ac_{M_{J_0}}|_X$. Note that the isomorphism $\alpha_J \circ \alpha_{J_0}^{-1} \colon \mathfrak{q}_{J_0} \longrightarrow \mathfrak{q}_J$ is given by $\alpha_J \circ \alpha_{J_0}(u) = \frac{b_J}{a_J} u$ for some nonzero $a_J, b_J \in \mathcal{O}(X)$ (if $\mathfrak{q}$ is an ideal in a domain R, and $\varphi \colon \mathfrak{q} \longrightarrow R$ is an R-linear map, then for every nonzero $u, v \in \mathfrak{q}$ we have $\frac{\varphi(u)}{u} = \frac{\varphi(u)v}{uv} = \frac{\varphi(uv)}{uv} = \frac{u\phi(v)}{uv} = \frac{\varphi(v)}{v}$). Furthermore, after multiplying each a_J by a suitable factor, we may assume that all a_J are equal, and we denote the common value by h.

In particular, by considering the image of Ω_X^n, we deduce $h \cdot \mathcal{J}ac_{M_J}|_X = b_J \cdot I$ for every J. Since $\mathcal{J}ac_X = \sum_J \mathcal{J}ac_{M_J}|_X$, we obtain (3) by taking $G = \sum_J (b_J)$. Q.E.D.

Remark 3.4. This lemma gives another proof for Remark 2.1.

The following theorem gives an analogue for the Local Vanishing Theorem for multiplier ideals (see [11, Theorem 9.4.1]) in the context of Mather multiplier ideals.

Theorem 3.5. *Let $\mathfrak{a}$ be a nonzero ideal on the variety X, and let $t \in \mathbf{R}_{\geq 0}$. If $f\colon Y \longrightarrow X$ is a log-resolution of $\mathfrak{a} \cdot \mathcal{J}ac_X$, and we put $\mathfrak{a} \cdot \mathcal{O}_Y = \mathcal{O}_Y(-Z_{Y/X})$, then*

$$R^i f_* \left(\mathcal{O}_Y(\widehat{K}_{Y/X} - J_{Y/X} - \lfloor t Z_{Y/X} \rfloor) \right) = 0 \quad \textit{for all } i > 0.$$

Proof. Since the assertion is local on X, we may assume that X is affine. It follows from Remark 2.1 that we can factor f through the Nash blow-up as $Y \xrightarrow{\psi} \widehat{X} \xrightarrow{\varphi} X$. Let $\mathcal{O}_{\widehat{X}}(1) = \mathcal{O}_{\mathbf{P}(\Omega^n_X)}(1)|_{\widehat{X}}$. With I as in Lemma 3.3, we have a surjection of graded algebras $\mathrm{Sym}(\Omega^n_X) \longrightarrow \oplus_{i=0}^{\infty} I^i$. This induces a closed immersion $\mathrm{Bl}_I(X) \hookrightarrow \mathbf{P}(\Omega^n_X)$, hence an isomorphism $\mathrm{Bl}_I(X) \simeq \widehat{X}$ over X, which yields $\mathcal{O}_{\widehat{X}}(1) \simeq I \cdot \mathcal{O}_{\widehat{X}}$.

It follows from [3, Proposition 1.7] that

$$\mathcal{O}_Y(\widehat{K}_{Y/X}) \simeq \mathcal{O}_Y(K_Y) \otimes \psi^*(\mathcal{O}_{\widehat{X}}(1))^{-1} \simeq \mathcal{O}_Y(K_Y) \otimes (I \cdot \mathcal{O}_Y)^{-1}.$$

Since f is a log resolution of $\mathcal{J}ac_X$, we deduce from Lemma 3.3 that we may write $G \cdot \mathcal{O}_Y = \mathcal{O}_Y(-B)$ for a divisor B on Y, and we have

$$\begin{aligned} \mathcal{O}_Y(\widehat{K}_{Y/X} - J_{Y/X} - \lfloor t Z_{Y/X} \rfloor) &\simeq \mathcal{O}_Y(K_Y) \otimes (I \cdot \mathcal{O}_Y)^{-1} \otimes IG \cdot \mathcal{O}_Y \otimes \mathcal{O}_Y(-\lfloor t Z_{Y/X} \rfloor) \\ &\simeq \mathcal{O}_Y(K_Y - B - \lfloor t Z_{Y/X} \rfloor). \end{aligned}$$

Since both $-B$ and $-Z_{Y/X}$ are relatively nef over X, the Kawamata-Viehweg Vanishing Theorem (see for example [10, Theorem 1-2-3]) implies the vanishing in the theorem. Q.E.D.

Remark 3.6. The vanishing in Theorem 3.5 holds, with the same proof, for mixed Mather multiplier ideals. More precisely, if $\mathfrak{a}_1, \ldots, \mathfrak{a}_r$ are nonzero ideals on X, and $f\colon Y \longrightarrow X$ is a log resolution of $\mathcal{J}ac_X \cdot \mathfrak{a}_1 \cdot \ldots \cdot \mathfrak{a}_r$ such that $\mathfrak{a}_i \cdot \mathcal{O}_Y = \mathcal{O}_Y(-Z_i)$, then

$$R^i f_* \left(\mathcal{O}_Y(\widehat{K}_{Y/X} - J_{Y/X} - \lfloor t_1 Z_1 + \ldots + t_r Z_r \rfloor) \right) = 0$$

for every $t_1, \ldots, t_r \in \mathbf{R}_{\geq 0}$ and every $i > 0$.

According to our principle to replace the usual discrepancy $K_{Y/X}$ by $\widehat{K}_{Y/X} - J_{Y/X}$ we can define "canonical singularities" for a not necessarily $\mathbf{Q}$-Gorenstein variety. The paper [4] introduces the notion of J-canonical singularities. For an arbitrary variety X, we say that X has J-canonical singularities if $\widehat{K}_{Y/X} - J_{Y/X} \geq 0$ for a log resolution $f : Y \longrightarrow X$ of $\mathcal{J}ac_X$ (and automatically for all log resolutions of $\mathcal{J}ac_X$). They prove the following statement ([4, Theorem 7.5]), which we also obtain as a corollary of our results.

Corollary 3.7. *If a variety X has J-canonical singularities, then X is normal and has rational singularities.*

For the proof of the corollary, we use the following generalization of Fujita's vanishing theorem (the original Fujita's vanishing theorem is in [7]).

Proposition 3.8 ([10], Theorem 1-3-1)**.** *Let $f : Y \longrightarrow X$ be a proper birational morphism from a non-singular variety Y onto a variety X with divisors $L, \tilde{L}$ on Y. Assume that there exist $\mathbf{Q}$-divisors $D, \tilde{D}$ and an effective divisor E on Y such that the following conditions are satisfied:*

(1) $\operatorname{supp} D$ *and* $\operatorname{supp} \tilde{D}$ *are divisors with only simple normal crossings and* $\lfloor D \rfloor = \lfloor \tilde{D} \rfloor = 0$*;*
(2) *both* $-L - D$ *and* $-\tilde{L} - \tilde{D}$ *are* f*-nef;*
(3) $K_Y \sim L + \tilde{L} + E$*; and*
(4) E *is exceptional for* f*.*

Then, $R^i f_* \mathcal{O}_Y(L) = 0$ *for* $i > 0$*.*

Proof of Corollary 3.7. Let $\nu : X^\nu \longrightarrow X$ be the normalization and $f : Y \longrightarrow X$ be a log resolution of $\mathcal{J}ac_X$. By the assumption $\widehat{K}_{Y/X} - J_{Y/X} \geq 0$, we obtain

$$\widehat{\mathcal{J}}(X, \mathcal{O}_X) = f_*\mathcal{O}_Y(\widehat{K}_{Y/X} - J_{Y/X}) \supseteq f_*\mathcal{O}_Y = \nu_*\mathcal{O}_{X^\nu}.$$

Since $\widehat{\mathcal{J}}(X, \mathcal{O}_X)$ is an ideal of $\mathcal{O}_X$, we obtain $X^\nu = X$. Now, we will prove $R^i f_* \mathcal{O}_Y = 0$ for $i > 0$. Note that $\widehat{K}_{Y/X} - J_{Y/X} = K_Y - B$ with $-B$ being f-nef, which we obtained in the proof of Theorem 3.5. Let $D = \tilde{D} = L = 0$, $E = \widehat{K}_{Y/X} - J_{Y/X}$ and $\tilde{L} = B$, then $-L - D = 0$ and $-\tilde{L} - \tilde{D} = -B$ are both f-nef. The condition $K_Y \sim L + \tilde{L} + E$ is also satisfied. Then, we can apply Fujita's vanishing theorem and obtain $R^i f_* \mathcal{O}_Y = 0$ for $i > 0$. □

We apply the above Local Vanishing Theorem (Theorem 3.5) to prove two more restriction results. We say that a proper subscheme X of a variety B is a hyperplane section if B can be embedded as a closed subvariety of a smooth variety A, and if there is a smooth hypersurface H in A such that $X = B \cap H$.

Theorem 3.9 (Restriction theorem for hyperplane sections)**.** *Let X be a closed subvariety of a variety B which is a hyperplane section, such that $X \cap B_{\rm reg} \neq \emptyset$. If $\tilde{\mathfrak{a}} \subseteq \mathcal{O}_B$ is an ideal on B such that its restriction $\mathfrak{a} = \tilde{\mathfrak{a}} \cdot \mathcal{O}_X$ is non-zero, then for every $t \in \mathbf{R}_{\geq 0}$ we have*

$$\widehat{\mathcal{J}}(X, \mathfrak{a}^t) \subseteq \widehat{\mathcal{J}}(B, \tilde{\mathfrak{a}}^t) \cdot \mathcal{O}_X.$$

Proof. Let $B \hookrightarrow A$ be a closed embedding in a smooth variety, and H a smooth hypersurface in A such that $X = B \cap H$. We denote by $\tilde{\tilde{\mathfrak{a}}}$ and $\mathfrak{b}_B$ the pull back of $\tilde{\mathfrak{a}}$ and $\mathcal{J}ac_B$, respectively, by the canonical projection $\mathcal{O}_A \longrightarrow \mathcal{O}_B$, and by $\mathfrak{b}_X$ the pull back of $\mathcal{J}ac_X$ by the canonical projection $\mathcal{O}_A \longrightarrow \mathcal{O}_X$.

We claim that there is a projective, birational morphism $\varphi_A \colon \overline{A} \longrightarrow A$ that is an isomorphism at the generic points of B, H and X, such that if $\overline{B}$, $\overline{H}$ and $\overline{X}$ denote the proper transforms in $\overline{A}$ of B, H and X, respectively, then the following hold:

(i) φ_A is a log resolution of $\tilde{\tilde{\mathfrak{a}}} \cdot \mathfrak{b}_B \cdot \mathfrak{b}_X$,

(ii) φ_A is a factorizing resolution of B, and also of H, inside A,

(iii) $\varphi_H := \varphi_A|_{\overline{H}} \colon \overline{H} \longrightarrow H$ is a factorizing resolution of X inside H and $\varphi_B := \varphi_A|_{\overline{B}} \colon \overline{B} \longrightarrow B$ is a factorizing resolution of X inside B,

(iv) φ_B and $\varphi_X := \varphi_A|_{\overline{X}} \colon \overline{X} \longrightarrow X$ are log resolutions of $\tilde{\mathfrak{a}} \cdot \mathcal{J}ac_B$ and $\mathfrak{a} \cdot \mathcal{J}ac_X$, respectively.

We construct φ_A as follows. We first take a log resolution $\varphi' \colon A' \longrightarrow A$ for the product of $\tilde{\tilde{\mathfrak{a}}}$, $\mathfrak{b}_B$, $\mathfrak{b}_X$, and $I_{H/A}$, where $I_{H/A}$ is the defining ideal of H in A. Since $X \cap B_{\rm reg} \neq \emptyset$, we may and will assume that φ' is an isomorphism at the generic points of B, H and X. Let E be the simple normal crossings divisor on A' such that $\mathcal{O}_{A'}(-E) = \tilde{\tilde{\mathfrak{a}}} \cdot \mathfrak{b}_B \cdot \mathfrak{b}_X \cdot I_{H/A} \cdot \mathcal{O}_{A'}$. We now apply Lemma 2.7 to $(A,\ A',\ E,\ B)$ to obtain a factorizing resolution $\varphi_A \colon \overline{A} \longrightarrow A' \longrightarrow A$ of B inside A. Note that φ_A is also a factorizing resolution of H inside A, since it is an embedded resolution of (A, H) and the factorizing condition (3) in Definition 2.4 is automatic, as H is a hypersurface. Therefore φ_A satisfies condition (ii) above, and conditions (i) and (iv) also follow by construction. As φ_A is a factorizing resolution of B inside A, we have

$$I_{B/A} \cdot \mathcal{O}_{\overline{A}} = I_{\overline{B}/\overline{A}} \cdot \mathcal{O}_{\overline{A}}(-R_{\overline{B}/\overline{A}})$$

for an effective divisor $R_{\overline{B}/\overline{A}}$ on $\overline{A}$, where $I_{B/A}$ and $I_{\overline{B}/\overline{A}}$ are the defining ideals of B in A and $\overline{B}$ in $\overline{A}$, respectively. It is clear that $\overline{X}$ is a connected component of $\overline{B} \cap \overline{H}$. After blowing-up the other components, we may assume that $\overline{X} = \overline{B} \cap \overline{H}$. Since $I_{X/H} = I_{B/A} \cdot \mathcal{O}_H$ and $I_{\overline{X}/\overline{H}} = I_{\overline{B}/\overline{A}} \cdot \mathcal{O}_{\overline{H}}$, we obtain

$$I_{X/H} \cdot \mathcal{O}_{\overline{H}} = I_{\overline{X}/\overline{H}} \cdot \mathcal{O}_{\overline{H}}(-R_{B/A}|_{\overline{H}}),$$

which shows the first part of (iii), and that we may take $R_{X/H} = R_{B/A}|_{\overline{H}}$. Using similarly that $I_{X/B} = I_{H/A} \cdot \mathcal{O}_B$ and $I_{\overline{X}/\overline{B}} = I_{\overline{H}/\overline{A}} \cdot \mathcal{O}_{\overline{B}}$, we obtain

$$I_{X/B} \cdot \mathcal{O}_{\overline{B}} = I_{\overline{X}/\overline{B}} \cdot \mathcal{O}_{\overline{B}}(-R_{H/A}|_{\overline{B}}).$$

This gives the second part of (iii), and that we may take $R_{X/B} = R_{H/A}|_{\overline{B}}$. Therefore φ_A satisfies all our requirements.

Since φ_A is a factorizing resolution of B, Lemma 2.8 gives

$$(K_{\overline{A}/A} - cR_{B/A})|_{\overline{B}} = \widehat{K}_{\overline{B}/B} - J_{\overline{B}/B}, \tag{4}$$

where $c = \operatorname{codim}(B, A)$. As φ_A is a factorizing resolution of the smooth hypersurface H, Lemma 2.8 implies

$$(K_{\overline{A}/A} - R_{H/A})|_{\overline{H}} = K_{\overline{H}/H}. \tag{5}$$

Finally, since φ_H is a factorizing resolution of X inside H, and we have $\operatorname{codim}(X, H) = c$, Lemma 2.8 gives

$$(K_{\overline{H}/H} - cR_{X/H})|_{\overline{X}} = \widehat{K}_{\overline{X}/X} - J_{\overline{X}/X}. \tag{6}$$

Using (4), (5), (6), and the formulas for $R_{X/H}$ and $R_{X/B}$, we obtain

$$(\widehat{K}_{\overline{B}/B} - J_{\overline{B}/B} - R_{X/B})|_{\overline{X}} = \widehat{K}_{\overline{X}/X} - J_{\overline{X}/X}. \tag{7}$$

Let $\tilde{\mathfrak{a}} \cdot \mathcal{O}_{\overline{B}} = \mathcal{O}_{\overline{B}}(-Z_{\overline{B}/B})$ and $D = \widehat{K}_{\overline{B}/B} - J_{\overline{B}/B} - \lfloor tZ_{\overline{B}/B} \rfloor - R_{X/B}$. We have an exact sequence:

$$\begin{aligned} (8)\quad 0 \longrightarrow {\varphi_B}_*(I_{\overline{X}/\overline{B}} \cdot \mathcal{O}_{\overline{B}}(D)) \longrightarrow {\varphi_B}_*(\mathcal{O}_{\overline{B}}(D)) \longrightarrow {\varphi_X}_*(\mathcal{O}_{\overline{X}}(D|_{\overline{X}})) \\ \longrightarrow R^1 {\varphi_B}_*(I_{\overline{X}/\overline{B}} \cdot \mathcal{O}_{\overline{B}}(D)). \end{aligned}$$

Arguing as in the proof of Proposition 2.9, we see that the last term in this sequence is zero: instead of using the Local Vanishing Theorem for multiplier ideals, we use Theorem 3.5 (in the version mentioned in Remark 3.6). Since $R_{X/B}$ is effective, we have an inclusion ${\varphi_B}_*(\mathcal{O}_{\overline{B}}(D)) \subseteq \widehat{\mathcal{J}}(B, \tilde{\mathfrak{a}}^t)$, hence the exact sequence (8) gives

$${\varphi_X}_*(\mathcal{O}_{\overline{X}}(D|_{\overline{X}})) \subseteq \widehat{\mathcal{J}}(B, \tilde{\mathfrak{a}}^t) \cdot \mathcal{O}_X.$$

On the other hand, it follows from (7) that

$$\varphi_{X*}(\mathcal{O}_{\overline{X}}(D|_{\overline{X}})) = \varphi_{X*}\left(\mathcal{O}_{\overline{X}}(\widehat{K}_{\overline{X}/X} - J_{\overline{X}/X} - \lfloor tZ_{\overline{X}/X}\rfloor)\right) = \widehat{\mathcal{J}}(X, \mathfrak{a}^t),$$

and this completes the proof of the theorem. Q.E.D.

Corollary 3.10 (Restriction theorem for Cohen-Macaulay varieties with proper intersection). *Let B and H be Cohen-Macaulay subvarieties of a smooth ambient variety A, such that $X = B \cap H$ is irreducible and reduced, and the intersection is proper, that is,* $\dim(X) = \dim(B) + \dim(H) - \dim(A)$. *If $X \cap B_{\mathrm{reg}} \neq \emptyset$, and $\tilde{\mathfrak{a}} \subset \mathcal{O}_B$ is an ideal on B whose restriction $\mathfrak{a} = \tilde{\mathfrak{a}} \cdot \mathcal{O}_X$ to X is nonzero, then for every $t \in \mathbf{R}_{\geq 0}$ we have*

$$\widehat{\mathcal{J}}(X, \mathfrak{a}^t) \subseteq \widehat{\mathcal{J}}(B, \tilde{\mathfrak{a}}^t) \cdot \mathcal{O}_X. \tag{9}$$

Proof. Consider the diagonal $\Delta \subset A \times A$. Since H and B intersect properly, it follows that $X \simeq \Delta \cap (H \times B)$ is of codimension $N = \dim(A)$ in $H \times B$. Both $A \times A$ and Δ are smooth, hence Δ is locally defined in $A \times A$ by N equations. Since the inclusion (9) is a local statement, we may replace A by a suitable open cover, and therefore we may assume that A is affine, and the ideal of Δ in $A \times A$ is generated by N equations $f_1, \ldots, f_N \in \mathcal{O}(A \times A)$.

We claim that there is a sequence of varieties

$$X = X_0 \subset X_1 \subset \cdots \subset X_N = H \times B$$

such that X_i is a hyperplane section of X_{i+1} for each i with $0 \leq i < N$. Let $H_1, \ldots, H_N$ be general elements of the linear system $\Lambda = \{\sum_i \lambda_i f_i \mid \lambda_i \in k\}$, and let $X_i = (H \times B) \cap H_1 \cap \ldots \cap H_{N-i}$. Note that $X_0 = X$, and we make the convention $X_N = H \times B$. By Bertini's Theorem, each $H_1 \cap \ldots \cap H_i$ is smooth, hence X_i is a hyperplane section in X_{i+1} for $i < N$. Furthermore, $H \times B$ is Cohen-Macaulay by assumption, hence each X_i is Cohen-Macaulay. In particular, no X_i has embedded components. We now prove by descending induction that X_i is irreducible and reduced for every i with $1 \leq i \leq N$. The case $i = N$ is clear, and let us assume that X_{i+1} is irreducible and reduced for some $i \geq 1$. The base locus of $\Lambda|_{H \times B}$ is equal to X, hence another application of Bertini's theorem implies that $X_i \smallsetminus X$ is irreducible and reduced. Since X is not an irreducible component of X_i by dimension considerations, it follows that X_i is irreducible and generically reduced, hence reduced (recall that X_i has no embedded components). This completes the proof of our claim.

Let $p_2 \colon H \times B \longrightarrow B$ be the canonical projection. Applying successively Theorem 3.9 for the above sequence of hyperplane sections, we obtain

$$\widehat{\mathcal{J}}(X, \mathfrak{a}^t) \subseteq \widehat{\mathcal{J}}(H \times B, p_2^*(\tilde{\mathfrak{a}})^t) \cdot \mathcal{O}_X. \tag{10}$$

On the other hand, if we consider an appropriate resolution of $H \times B$ of the form $\overline{H} \times \overline{B} \longrightarrow H \times B$, and if we denote by $q_1 \colon \overline{H} \times \overline{B} \longrightarrow \overline{H}$ and $q_2 \colon \overline{H} \times \overline{B} \longrightarrow \overline{B}$ the two projections, then we have

$$\widehat{K}_{(\overline{H}\times\overline{B})/(H\times B)} = q_1^*\widehat{K}_{\overline{H}/H} + q_2^*\widehat{K}_{\overline{B}/B}, \; J_{(\overline{H}\times\overline{B})/(H\times B)} = q_1^* J_{\overline{H}/H} + q_2^* J_{\overline{B}/B}.$$

Using also the Künneth formula, these imply

$$\widehat{\mathcal{J}}(H \times B, q_2^*(\tilde{\mathfrak{a}})^t) = q_1^*\widehat{\mathcal{J}}(H, \mathcal{O}_H) \otimes q_2^*\widehat{\mathcal{J}}(B, \tilde{\mathfrak{a}}^t),$$

and we thus obtain

$$\widehat{\mathcal{J}}(H\times B, p_2^*(\tilde{\mathfrak{a}})^t)\cdot\mathcal{O}_X \subseteq (\widehat{\mathcal{J}}(H,\mathcal{O}_H)\cdot\mathcal{O}_X)\cdot(q_2^*\widehat{\mathcal{J}}(B,\tilde{\mathfrak{a}}^t)\cdot\mathcal{O}_X) \subseteq \widehat{\mathcal{J}}(B,\tilde{\mathfrak{a}}^t)\cdot\mathcal{O}_X.$$

Combining this with (10), we obtain (9). Q.E.D.

Question 3.11. *Does the assertion in Corollary 3.10 hold if we drop the assumption that H and B are Cohen-Macaulay?*

On the other hand, the following example shows that one does not have a restriction theorem when X is an arbitrary subvariety of a singular[1] variety B.

Example 3.12. Let B be the hypersurface defined by $x_1^d + \cdots + x_n^d = 0$ in $\mathbf{A}^n$, with $d \geq n \geq 3$. Note that $(B, 0)$ is a non-canonical normal isolated singularity. B is a complete intersection, hence

$$\widehat{\mathcal{J}}(B, \mathcal{O}_B) = \mathcal{J}(B, \mathcal{O}_B) = f_*(K_{Y/X}),$$

where $f \colon Y \longrightarrow X$ is a log resolution of $\mathcal{J}ac_B$. Since B is non-canonical, we see that $\widehat{\mathcal{J}}(B, \mathcal{O}_B)$ is contained in the maximal ideal defining 0. Suppose now that $X \subset B$ is a line through 0. Since X is smooth, we have $\widehat{\mathcal{J}}(X, \mathcal{O}_X) = \mathcal{J}(X, \mathcal{O}_X) = \mathcal{O}_X$. Therefore the inclusion $\widehat{\mathcal{J}}(X, \mathcal{O}_X) \subseteq \widehat{\mathcal{J}}(B, \mathcal{O}_B)_X$ does not hold.

We can now apply our vanishing result in Theorem 3.5 to deduce a version of the Subadditivity Theorem for Mather multiplier ideals. We refer to [11, §9.5.B] and [5, Theorem 6.5] for the corresponding result in the setting of usual multiplier ideals. In particular, our approach follows closely the one in [5], which treats the case of singular varieties. We start with the following adjunction-type formula, which is [5, Lemma 6.3].

[1]When the ambient variety B is smooth, the restriction theorem is provided by Proposition 2.9.

Lemma 3.13. *Let X be an n-dimensional variety and $Y \longrightarrow X$ a log resolution of $\mathcal{J}ac_X$. Consider $A = X \times X$ and $A' = Y \times Y$, and let $\varphi\colon \overline{A} \longrightarrow A' \longrightarrow A$ be a factorizing resolution of the diagonal Δ inside A, provided by Lemma 2.7. If $\overline{\Delta}$ is the strict transform of Δ in $\overline{A}$, then we have the following inequality of divisors on $\overline{A}$:*

$$\left(\widehat{K}_{\overline{A}/A} - J_{\overline{A}/A} - nR_{\Delta/A}\right)|_{\overline{\Delta}} \geq \widehat{K}_{\overline{\Delta}/\Delta} - 2J_{\overline{\Delta}/\Delta}.$$

Theorem 3.14. *If $\mathfrak{a}$ and $\mathfrak{b}$ are nonzero ideals on a variety X, then for every $s, t \in \mathbf{R}_{\geq 0}$ we have*

$$\widehat{\mathcal{J}}(X, \mathcal{J}ac_X \mathfrak{a}^s \mathfrak{b}^t) \subseteq \widehat{\mathcal{J}}(X, \mathfrak{a}^s) \cdot \widehat{\mathcal{J}}(X, \mathfrak{b}^t).$$

Proof. Let $g\colon Y \longrightarrow X$ be a log resolution of $\mathcal{J}ac_X \cdot \mathfrak{a} \cdot \mathfrak{b}$. We write $\mathfrak{a} \cdot \mathcal{O}_Y = \mathcal{O}_Y(-Z_{Y/X})$ and $\mathfrak{b} \cdot \mathcal{O}_Y = \mathcal{O}_Y(-W_{Y/X})$. Let $g \times g\colon A' = Y \times Y \longrightarrow A := X \times X$ be the product map. We denote by $\tilde{p}$ and $\tilde{q}$ the first and the second projections from $Y \times Y$ to Y, and by p and q the first and the second projections from $X \times X$ to X. Let E be the divisor on A' such that $\mathcal{O}_{A'}(-E) = p^*(\mathfrak{a} \cdot \mathcal{J}ac_X) \cdot q^*(\mathfrak{b} \cdot \mathcal{J}ac_X)$. We consider a factorizing resolution $\varphi\colon \overline{A} \longrightarrow A' \longrightarrow A$ of the diagonal Δ in A, obtained by applying Lemma 2.7 with respect to E. Let $\overline{\Delta}$ be the strict transform of Δ in $\overline{A}$, and let $\varphi_\Delta = \varphi|_{\overline{\Delta}}$. If we write $\mathfrak{a} \cdot \mathcal{O}_{\overline{\Delta}} = \mathcal{O}_{\overline{\Delta}}(-Z_{\overline{\Delta}/\Delta})$ and $\mathfrak{b} \cdot \mathcal{O}_{\overline{\Delta}} = \mathcal{O}_{\overline{\Delta}}(-W_{\overline{\Delta}/\Delta})$, then after identifying X and Δ, we obtain

$$\widehat{\mathcal{J}}(X, \mathcal{J}ac_X \mathfrak{a}^s \mathfrak{b}^t) = \varphi_{\Delta *}\left(\widehat{K}_{\overline{\Delta}/\Delta} - 2J_{\overline{\Delta}/\Delta} - \lfloor sZ_{\overline{\Delta}/\Delta} \rfloor - \lfloor tW_{\overline{\Delta}/\Delta} \rfloor\right),$$

and by Lemma 3.13, the right hand side is contained in

$$\varphi_{\Delta *}\left(\widehat{K}_{\overline{A}/A} - J_{\overline{A}/A} - \lfloor s\tilde{p}^* Z_{\overline{Y}/X} \rfloor - \lfloor t\tilde{q}^* W_{\overline{Y}/X} \rfloor - nR_{\Delta/A}|_{\overline{\Delta}}\right), \tag{11}$$

where $n = \dim(X)$. Arguing as in the proof of Theorem 3.9, we see using Theorem 3.5 that the ideal in (11) is contained in

$$\widehat{\mathcal{J}}(X \times X, (p^*\mathfrak{a})^s \cdot (q^*\mathfrak{b})^t) \cdot \mathcal{O}_\Delta.$$

Considering the log resolution $g \times g\colon A' \longrightarrow A$ of $\mathcal{J}ac_A \cdot p^*(\mathfrak{a}) \cdot q^*(\mathfrak{b})$, and noting that $\widehat{K}_{A'/A} = \tilde{p}^* \widehat{K}_{Y/X} + \tilde{q}^* \widehat{K}_{Y/X}$ and $J_{A'/A} = \tilde{p}^* J_{Y/X} + \tilde{q}^* J_{Y/X}$, we obtain

$$\widehat{\mathcal{J}}(X \times X, (p^*\mathfrak{a})^s \cdot (q^*\mathfrak{b})^t) \cdot \mathcal{O}_\Delta = \widehat{\mathcal{J}}(X, \mathfrak{a}^s) \cdot \widehat{\mathcal{J}}(X, \mathfrak{b}^t),$$

and we obtain the inclusion in the statement of the theorem. Q.E.D.

We now give a Skoda-type formula for Mather multiplier ideas (see [11, §11.1.A] for the statement in the context of usual multiplier ideals). The proof follows the one in *loc. cit.*, using our vanishing result in Theorem 3.5.

Theorem 3.15. *If X is a variety of dimension n, and $\mathfrak{a} \subseteq \mathcal{O}_X$ is a nonzero ideal on X, then for every $t \geq n$ we have*

$$\widehat{\mathcal{J}}(X, \mathfrak{a}^t) = \mathfrak{a} \cdot \widehat{\mathcal{J}}(X, \mathfrak{a}^{t-1}).$$

In particular, if t is an integer, then

$$\widehat{\mathcal{J}}(X, \mathfrak{a}^t) = \mathfrak{a}^{t-n+1} \cdot \widehat{\mathcal{J}}(X, \mathfrak{a}^{n-1})$$

Proof. The inclusion $\mathfrak{a} \cdot \widehat{\mathcal{J}}(X, \mathfrak{a}^{t-1}) \subseteq \widehat{\mathcal{J}}(X, \mathfrak{a}^t)$ holds for every $t \geq 1$, and it is an immediate consequence of the definition of the Mather multiplier ideal. Therefore from now on we focus on the reverse inclusion.

If X is covered by open subsets U_i, then it is clearly enough to prove the theorem for each U_i. Since X is n-dimensional, after taking a suitable open cover, we may assume that X is affine, and there is a reduction $\mathfrak{b}$ of $\mathfrak{a}$ generated by n elements $s_1, \ldots, s_n$ (see [2, Proposition 4.6.8]). This means that $\mathfrak{b} \subseteq \mathfrak{a}$, and there is $m \geq 1$ such that $\mathfrak{b}\mathfrak{a}^m = \mathfrak{a}^{m+1}$.

Let $\varphi\colon Y \longrightarrow X$ be a log-resolution of $\mathfrak{a} \cdot \mathfrak{b} \cdot \mathcal{J}ac_X$. We write $\mathfrak{a} \cdot \mathcal{O}_Y = \mathfrak{b} \cdot \mathcal{O}_Y = \mathcal{O}_Y(-Z_{Y/X})$ and consider $\mathcal{L}_s = \mathcal{O}_Y(\widehat{K}_{Y/X} - J_{Y/X} - \lfloor sZ_{Y/X} \rfloor)$ for every $s \in \mathbf{R}_{\geq 0}$. It follows from Theorem 3.5 that $R^p\varphi_*\mathcal{L}_s = 0$ for $p > 0$.

If V is the k-vector space generated by $s_1, \ldots, s_n$, then we have on Y an exact Koszul complex

$$0 \longrightarrow \wedge^n V \otimes_k \mathcal{O}_Y(nZ_{Y/X}) \longrightarrow \cdots \longrightarrow V \otimes_k \mathcal{O}_Y(Z_{Y/X}) \longrightarrow \mathcal{O}_Y \longrightarrow 0.$$

By tensoring with $\mathcal{L}_t$ we get the exact complex

$$0 \longrightarrow \wedge^n V \otimes_k \mathcal{L}_{t-n} \longrightarrow \cdots \longrightarrow V \otimes_k \mathcal{L}_{t-1} \longrightarrow \mathcal{L}_t \longrightarrow 0.$$

Since $R^p\varphi_*\mathcal{L}_{t-i} = 0$ for $0 \leq i \leq n$ and for all $p > 0$, we deduce that we have an exact complex on X

$$0 \longrightarrow \wedge^n V \otimes_k \varphi_*\mathcal{L}_{t-n} \longrightarrow \cdots \longrightarrow V \otimes_k \varphi_*\mathcal{L}_{t-1} \longrightarrow \varphi_*\mathcal{L}_t \longrightarrow 0.$$

In particular, the map $V \otimes_k \varphi_*\mathcal{L}_{t-1} \longrightarrow \varphi_*\mathcal{L}_t$ is surjective, hence

$$\widehat{\mathcal{J}}(X, \mathfrak{a}^t) = \mathfrak{b} \cdot \widehat{\mathcal{J}}(X, \mathfrak{a}^{t-1}) \subseteq \mathfrak{a} \cdot \widehat{\mathcal{J}}(X, \mathfrak{a}^{t-1}).$$

This completes the proof of the theorem. Q.E.D.

Corollary 3.16 (Briançon-Skoda type formula). *If $\mathfrak{a}$ is an ideal on an n-dimensional variety X, then for every integer $m \geq n$ we have*

$$\overline{\mathcal{J}ac_X \cdot \mathfrak{a}^m} \subseteq \mathfrak{a}^{m-n+1}.$$

Proof. We may assume that $\mathfrak{a}$ is nonzero, since otherwise the assertion is trivial. We keep the notation in the proof of Theorem 3.15. Since the divisor $\widehat{K}_{Y/X}$ is effective, we have an inclusion

$$\mathcal{O}_Y(-J_{Y/X} - mZ_{Y/X}) \subseteq \mathcal{O}_Y(\widehat{K}_{Y/X} - J_{Y/X} - mZ_{Y/X}).$$

We deduce

$$\overline{\mathcal{J}ac_X \cdot \mathfrak{a}^m} \subseteq \varphi_*(\mathcal{O}_Y(-J_{Y/X} - mZ_{Y/X}))$$
$$\subseteq \varphi_*(\mathcal{O}_Y(\widehat{K}_{Y/X} - J_{Y/X} - mZ_{Y/X})) = \widehat{\mathcal{J}}(X, \mathfrak{a}^m).$$

Using Theorem 3.15, we get

$$\overline{\mathcal{J}ac_X \cdot \mathfrak{a}^m} \subseteq \widehat{\mathcal{J}}(X, \mathfrak{a}^m) = \mathfrak{a}^{m-n+1} \cdot \widehat{\mathcal{J}}(X, \mathfrak{a}^{n-1}) \subseteq \mathfrak{a}^{m-n+1}.$$

Q.E.D.

We also have a version of the Summation Theorem (see [9, Corollary 2]) for Mather multiplier ideals.

Theorem 3.17. *If $\mathfrak{a}_1, \ldots, \mathfrak{a}_r, \mathfrak{b}$ are nonzero ideals on the variety X, then for every $s, t \in \mathbf{R}_{\geq 0}$ we have*

$$\widehat{\mathcal{J}}\left(X, (\mathfrak{a}_1 + \ldots + \mathfrak{a}_r)^s \mathfrak{b}^t\right) = \sum_{s_1 + \ldots + s_r = s} \widehat{\mathcal{J}}(X, \mathfrak{a}_1^{s_1} \cdots \mathfrak{a}_r^{s_r} \mathfrak{b}^t). \tag{12}$$

Proof. Let $\varphi \colon Y \longrightarrow X$ be a log resolution of the product of $\mathfrak{b}$, $\mathcal{J}ac_X$, and of all $\mathfrak{a}_i + \mathfrak{a}_j$, for $1 \leq i, j \leq r$. It follows from [9, Lemma 3.1] that φ is also a log resolution of $\mathfrak{b} \cdot (\mathfrak{a}_1 + \ldots + \mathfrak{a}_r) \cdot \mathcal{J}ac_X$. Furthermore, if we write $\mathfrak{a}_i \cdot \mathcal{O}_Y = \mathcal{O}_Y(-A_i)$, $\mathfrak{b} \cdot \mathcal{O}_Y = \mathcal{O}_Y(-B)$, and $(\mathfrak{a}_1 + \ldots + \mathfrak{a}_r) \cdot \mathcal{O}_Y = \mathcal{O}_Y(-A)$, then A is the largest effective divisor on Y with $A \leq A_i$ for all i. For every $s_1, \ldots, s_r$ with $\sum_i s_i = s$, we have

$$\lfloor s_1 A_1 + \ldots + s_r A_r \rfloor \geq \sum_{i=1}^{r} \lfloor sA \rfloor,$$

and the inclusion "$\supseteq$" in (12) then follows from definition of Mather multiplier ideals.

We now consider the reverse inclusion. It is shown in [9, §3] that there is an exact complex on Y

$$\mathcal{E}_\bullet : \ 0 \longrightarrow \mathcal{E}_{r-1} \longrightarrow \ldots \longrightarrow \mathcal{E}_1 \longrightarrow \mathcal{E}_0 \longrightarrow 0$$

with the following properties:

(1) $\mathcal{E}_0 = \mathcal{O}_Y(-A)$.

(2) $\mathcal{E}_1$ is a finite direct sum of sheaves of the form $\mathcal{O}_Y(-\lfloor s_1 A_1 + \ldots + s_r A_r + tB \rfloor)$, with $\sum_i s_i = s$.

(3) Every $\mathcal{E}_i$ with $2 \leq i \leq r-1$ is a finite direct sum of sheaves of the form $\mathcal{O}_Y(-\lfloor a_1 A_1 + \ldots + a_r A_r + bB \rfloor)$, for some $a_1, \ldots, a_r, b \in \mathbf{R}_{\geq 0}$.

It follows from the above properties and Theorem 3.5 that

$$R^i \varphi_*(\mathcal{E}_j \otimes \mathcal{O}_Y(\widehat{K}_{Y/X} - J_{Y/X})) = 0 \text{ for all } i > 0 \text{ and } j \geq 0.$$

This implies that the push-forward $\varphi_*(\mathcal{E}_\bullet \otimes \mathcal{O}_Y(\widehat{K}_{Y/X} - J_{Y/X}))$ is exact. The surjectivity of the last map in this complex, and properties (1) and (2) give the inclusion "$\subseteq$" in (12). Q.E.D.

We end with some considerations concerning an asymptotic version of Mather multiplier ideals. For basic facts about graded sequences of ideals and their multiplier ideals, we refer to [11, Chapter 11]. Let X be a variety, and let $\mathfrak{a}_\bullet = (\mathfrak{a}_m)_{m \geq 1}$ be a graded sequence of ideals on X, that is, $\mathfrak{a}_p \cdot \mathfrak{a}_q \subseteq \mathfrak{a}_{p+q}$ for all $p, q \geq 1$. We assume that some $\mathfrak{a}_p$ is nonzero. Given $t \in \mathbf{R}_{\geq 0}$ and $p, m \geq 1$ such that $\mathfrak{a}_p$ is nonzero, we obtain using Proposition 2.6

$$\widehat{\mathcal{J}}(X, \mathfrak{a}_p^{t/p}) = \widehat{\mathcal{J}}(X, (\mathfrak{a}_p^m)^{t/mp}) \subseteq \widehat{\mathcal{J}}(X, \mathfrak{a}_{pm}^{t/mp}).$$

It follows from the Noetherian property that there is an ideal $\widehat{\mathcal{J}}(X, \mathfrak{a}_\bullet^t)$ such that $\widehat{\mathcal{J}}(X, \mathfrak{a}_p^{t/p}) = \widehat{\mathcal{J}}(X, \mathfrak{a}_\bullet^t)$ whenever p is divisible enough. This is the *asymptotic Mather multiplier ideal* of $\mathfrak{a}_\bullet$. It is straightforward to extend Theorems 3.9, 3.10, 3.14, and 3.17 to the setting of graded sequences of ideals. We leave for the reader the task of formulating these extensions. We only note that if $\mathfrak{a}_\bullet^{(1)}, \ldots, \mathfrak{a}_\bullet^{(r)}$ are graded sequences of ideals, then their product and sum have as their p^{th} terms $\prod_{i=1}^r \mathfrak{a}_p^{(i)}$, and respectively, $\sum_{i_1 + \ldots + i_r = p} \mathfrak{a}_{i_1}^{(1)} \cdot \ldots \cdot \mathfrak{a}_{i_r}^{(r)}$.

References

[1] A. Bravo and O. Villamayor, A strengthening of resolution of singularities in characteristic zero, Proc. London Math. Soc. (3) **86** (2003), 327–357.

[2] W. Bruns and J. Herzog, *Cohen-Macaulay rings*, Cambridge Studies in Advanced Mathematics **39**, Cambridge University Press, Cambridge, revised edition 1998.

[3] T. de Fernex, L. Ein and S. Ishii, Divisorial valuations via arcs, Publ. RIMS, **44** (2008), 425–448.

[4] T. de Fernex and R. Docampo, Jacobian discrepancies and rational singularities, J. Eur. Math. Soc. (JEMS) **16** (2014), 165–199.
[5] E. Eisenstein, Generalizations of the restriction theorem for multiplier ideals, preprint, arXiv:1001.2841v1.
[6] L. Ein and M. Mustaţă, Jet schemes and singularities, In: Algebraic Geometry–Seattle 2005, Part 2, 505–546, Proc. Sympos. Pure Math. **80**, Part 2, Amer. Math. Soc., Providence, RI, 2009.
[7] T. Fujita, A relative version of Kawamata-Viehweg's vanishing theorem, unpublished Proceedings of Symposium of Algebraic Geometry, Hiroshima Kyouiku-Kaikan (1985) 107–119.
[8] S. Ishii, Mather discrepancy and the arcs spaces, Ann. Inst. Fourier (Grenoble) **63** (2013), 89–111.
[9] S.-Y. Jow and E. Miller, Multiplier ideals of sums via cellular resolutions, Math. Res. Lett. **15** (2008), 359–373.
[10] Y. Kawamata, K. Matsuda, and K. Matsuki, Introduction to the miniml model problem, In: Algebraic geometry, Sendai, 1985, 283–360, Adv. Stud. Pure Math. **10**, North-Holland, Amsterdam, 1987.
[11] R. Lazarsfeld, *Positivity in Algebraic Geometry*, II, Ergebnisse der Mathematik und ihrer Grenzgebiete, 3. Folge, Vol. **49**, Springer-Verlag, Berlin, 2004.
[12] J. Lipman, On the Jacobian ideal of the module of differentials, Proc. Amer. Math. Soc. **21** (1969), 422–426.

Department of Mathematics, University of Illinois at Chicago, Chicago, IL 60607-7045, USA
E-mail address: ein@math.uic.edu

Department of Mathematical Science, University of Tokyo, Meguro, Tokyo, Japan
E-mail address: shihoko@ms.u-tokyo.ac.jp

Department of Mathematics, University of Michigan, Ann Arbor, MI 48109-1043, USA
E-mail address: mmustata@umich.edu

Advanced Studies in Pure Mathematics 70, 2016
Minimal Models and Extremal Rays (Kyoto, 2011)
pp. 29–48

Sources of log canonical centers

János Kollár

Abstract.

Given a log canonical pair (X, Δ) and a log canonical center $Z \subset X$, we define a Calabi–Yau fiber space $(S, \Delta_S) \to Z$, called the *source* of Z. We believe that the source carries – and makes accessible – all the relevant information about the log canonical center Z. There is a natural Poincaré residue map from X to S which is used to solve several problems in higher-codimension adjunction. The main application is to the construction of semi-log-canonical pairs.

§1. Introduction

Let X be a smooth variety and $S \subset X$ a smooth hypersurface. The *Poincaré residue map* is an isomorphism

$$\mathcal{R} : \omega_X(S)|_S \cong \omega_S.$$

In additive form it gives the *adjunction formula* $(K_X + S)|_S \sim K_S$, but this variant does not show that $\mathcal{R}$ is a *canonical* isomorphism.

Its generalization to log canonical pairs $(X, S + \Delta)$ has been an important tool in birational geometry; see, for instance, [Kol92, KM98]. One defines a twisted version of the restriction of Δ to S, called the *different* and, for $m > 0$ sufficiently divisible, one gets a Poincaré residue map

$$\mathcal{R}^m : \big(\omega_X^{[m]}(mS + m\Delta)\big)|_S \cong \omega_S^{[m]}\big(m \operatorname{Diff}_S \Delta\big),$$

where the exponent $[m]$ denotes the double dual of the mth tensor power. As before, it is frequently written as a $\mathbb{Q}$-linear equivalence of divisors

$$\big(K_X + S + \Delta\big)|_S \sim_{\mathbb{Q}} K_S + \operatorname{Diff}_S \Delta.$$

There have been several attempts to extend these formulas to the case when S is replaced by a higher codimension log canonical center of a pair (X, Δ) [Kaw97, Kaw98, Kol07]. None of these have been completely

Received September 15, 2011.

successful; the main difficulty is understanding what kind of object the different should be.

Let $Z \subset X$ be a log canonical center of a pair (X, Δ). We can choose a resolution $f : X' \to X$ such that if we write $f^*\big(K_X + \Delta\big) \sim_{\mathbb{Q}} K_{X'} + \Delta'$ then there is a divisor $S \subset X'$ that dominates Z and appears in Δ' with coefficient 1. The usual adjunction formula now gives

$$\big(K_{X'} + \Delta'\big)|_S \sim_{\mathbb{Q}} K_S + \operatorname{Diff}_S(\Delta' - S) =: K_S + \Delta_S.$$

Note further that $K_{X'} + \Delta'$ is trivial on the fibers of f, hence so is $K_S + \Delta_S$. Thus

$$f|_S : (S, \Delta_S) \to Z$$

is a fiber space whose (possibly disconnected) fibers have (numerically) trivial (log) canonical class. The aim of previous attempts was to generalize Kodaira's canonical bundle formula for elliptic surfaces (cf. [BPV84, Sec.V.12]) to this setting. The difficulty is to make sure that we do not lose information in the summand that corresponds to the j-invariant of the fibers in the classical case. (For families of elliptic curves this could be achieved by keeping the corresponding variation of Hodge structures as part of our data.)

This suggests that it could be better to view the pair (S, Δ_S) as the answer to the problem. However, in general there are many divisors $S_j \subset X'$ that satisfy our requirements and they do not seem to be related to each other in any nice way.

Our aim is to remedy this problem, essentially by looking at the smallest possible intersections of the various divisors S_j on a dlt model of (X, Δ). There can be many of these models and intersections, but they turn out to be birational to each other and have several unexpectedly nice properties. These are summarized in the next theorem. For the rest of this note we work over a field of characteristic 0.

Dlt models, the different and crepant birational equivalence are recalled in Definitions 4–6.

Theorem 1. *Let (X, Δ) be an lc pair, $Z \subset X$ an lc center and $n : Z^n \to Z$ its normalization. Let $f : \big(X^m, \Delta^m\big) \to (X, \Delta)$ be a dlt model and $S \subset X^m$ a minimal (with respect to inclusion) lc center of $\big(X^m, \Delta^m\big)$ that dominates Z. Set $\Delta_S := \operatorname{Diff}^*_S \Delta^m$ and $f_S := f|_S$. Let $f^n_S : S \to \tilde{Z}_S \to Z^n$ denote the Stein factorization.*

(1) *(Uniqueness of sources) The crepant birational equivalence class of (S, Δ_S) does not depend on the choice of X^m and S. It is called the* source *of Z and denoted by* $\operatorname{Src}(Z, X, \Delta)$.

(2) *(Uniqueness of springs) The isomorphism class of $\tilde{Z}_S$ does not depend on the choice of X^m and S. It is called the* spring *of Z and denoted by* $\mathrm{Spr}(Z, X, \Delta)$.

(3) *(Crepant log structure) (S, Δ_S) is dlt, $K_S + \Delta_S \sim_{\mathbb{Q}} f_S^*(K_X + \Delta)$ and (S, Δ_S) is klt on the generic fiber of f_S.*

(4) *(Poincaré residue map) For $m > 0$ sufficiently divisible, there are well defined isomorphisms*

$$\begin{aligned} f^*\big(\omega_X^{[m]}(m\Delta)\big)|_S &\cong \omega_S^{[m]}(m\Delta_S) \quad \textit{and} \\ n^*\big(\omega_X^{[m]}(m\Delta)|_Z\big) &\cong \Big(\big(f_S^n\big)_*\omega_S^{[m]}(m\Delta_S)\Big)^{\mathrm{inv}} \end{aligned}$$

where the exponent inv *denotes the invariants under the action of the group of crepant birational self-maps $\mathrm{Bir}_Z^c(S, \Delta_S)$.*

(5) *(Galois property) The extension $\tilde{Z}_S \to Z$ is Galois and the natural map $\mathrm{Bir}_Z(S, \Delta_S) \twoheadrightarrow \mathrm{Gal}\big(\tilde{Z}_S/Z\big)$ is surjective.*

(6) *(Adjunction) Assume $\Delta = D + \Delta_1$. Let $n_D : D^n \to D$ be the normalization and $Z_D \subset D^n$ an lc center of $\big(D^n, \mathrm{Diff}_{D^n}\Delta_1\big)$ such that $n_D(Z_D) = Z$. Then there is a commutative diagram*

$$\begin{array}{ccc} \mathrm{Src}\big(Z_D, D^n, \mathrm{Diff}_{D^n}\Delta_1\big) & \overset{cbir}{\sim} & \mathrm{Src}\big(Z, X, D + \Delta_1\big) \\ \downarrow & & \downarrow \\ Z_D & \overset{n_D}{\to} & Z. \end{array}$$

Crepant log structures are defined in Section 2. Theorem 10 shows that minimal lc centers are birational to each other; this proves (1.1) and it also establishes (1.6). Its consequences for the Poincaré residue map are derived in Section 3. Sources and springs are formally defined in Section 4 and (1.5) is proved in Proposition 19.

Section 5 contains the main application, Theorems 23–24. We show that normalization gives a one-to-one correspondence:

$$\left\{ \begin{array}{c} \text{slc pairs } (X, \Delta) \\ \text{such that} \\ K_X + \Delta \text{ is ample} \end{array} \right\} \cong \left\{ \begin{array}{l} \text{lc pairs } \big(\bar{X}, \bar{D} + \bar{\Delta}\big) \text{ such that} \\ K_{\bar{X}} + \bar{D} + \bar{\Delta} \text{ is ample plus an} \\ \text{involution } \tau \text{ of } \big(\bar{D}^n, \mathrm{Diff}_{\bar{D}^n}\bar{\Delta}\big) \end{array} \right\}.$$

The papers [Oda13, OX12] contain further applications to K-stability and to slc models of deminormal schemes.

Shokurov informed me that his forthcoming paper [Sho13] contains another approach to Theorem 1.

§2. Crepant log structures

Definition 2. Let Z be a normal variety. A *crepant log structure* on Z is a proper, surjective morphism $f : (X, \Delta) \to Z$ such that

(1) f has connected fibers,
(2) (X, Δ) is lc and
(3) $K_X + \Delta \sim_{f,\mathbb{Q}} 0$.

A proper morphism $f : (X, \Delta) \to Z$ is called a *weak crepant log structure* on Z if it satisfies (1) and (3) but Δ is allowed to be a non-effective sub-boundary.

Any lc pair (Z, Δ_Z) has a trivial crepant log structure where $(X, \Delta) = (Z, \Delta_Z)$. Conversely, if f is birational then $(Z, \Delta_Z := f_*\Delta)$ is lc.

An irreducible subvariety $W \subset Z$ is a *log canonical center* or *lc center* of a weak crepant log structure $f : (X, \Delta) \to Z$ iff it is the image of an lc center $W_X \subset X$ of (X, Δ). A weak crepant log structure has only finitely many lc centers.

Let (Z, Δ_Z) be an lc pair and $f : X \to Z$ a proper, birational morphism. Write $K_X + \Delta_X \sim_{\mathbb{Q}} f^*(K_Z + \Delta_Z)$. Then $f : (X, \Delta_X) \to Z$ is a weak crepant log structure. The lc centers of $f : (X, \Delta_X) \to Z$ are the same as the lc centers of (Z, Δ_Z).

By Proposition 5 we can choose f such that $f : (X, \Delta_X) \to (Z, \Delta_Z)$ is a crepant log structure, X is $\mathbb{Q}$-factorial and (X, Δ_X) is dlt.

Let $f : (X, \Delta_X) \to Z$ be a dlt crepant log structure and $Y \subset X$ an lc center. Consider the Stein factorization

$$f|_Y : Y \xrightarrow{f_Y} Z_Y \xrightarrow{\pi} Z$$

and set $\Delta_Y := \operatorname{Diff}_Y^* \Delta_X$. Then (Y, Δ_Y) is dlt and $f_Y : (Y, \Delta_Y) \to Z_Y$ is a crepant log structure.

Definition 3 (Divisorial log terminal)**.** A pair $(X, \sum a_i D_i)$ is called *simple normal crossing* (abbreviated as *snc*) if X is smooth and for every $p \in X$ one can choose an open neighborhood $p \in U$ and local coordinates x_i such that for every i there is an index $a(i)$ such that $D_i \cap U = (x_{a(i)} = 0)$.

As key examples, I emphasize that the pair $(\mathbb{A}^2_k, (x^2 = y^2 + y^3))$ is not snc and $(\mathbb{A}^2_k, (x^2 + y^2 = 0))$ is snc iff $\sqrt{-1} \in k$. Thus being snc is a Zariski local (but not étale local) property.

Given any pair (X, Δ), there is a largest open subset $X^{snc} \subset X$ such that $(X^{snc}, \Delta|_{X^{snc}})$ is snc.

A pair (X, Δ) is called *divisorial log terminal* (abbreviated as *dlt*) if the discrepancy $a(E, X, \Delta)$ is > -1 for every divisor whose center is contained in $X \setminus X^{snc}$.

Definition 4 (Different)**.** Let (X, Δ) be a dlt pair and $Y \subset X$ an lc center. Generalizing the usual notion of the different [Kol92, Sec.16], there is a naturally defined $\mathbb{Q}$-divisor $\operatorname{Diff}_Y^* \Delta$, called the *different* of Δ

on Y such that

$$\big(K_X + \Delta\big)|_Y \sim_{\mathbb{Q}} K_Y + \operatorname{Diff}^*_Y \Delta.$$

The traditional different [Kol92, Sec.16] is defined such that if $Y = D$ is a divisor then

$$\big(K_X + D + \Delta\big)|_D \sim_{\mathbb{Q}} K_D + \operatorname{Diff}_D \Delta.$$

Thus, in this case, $\operatorname{Diff}^*_D(D+\Delta) = \operatorname{Diff}_D \Delta$. This inductively defines $\operatorname{Diff}^*_Y \Delta$ whenever Y is an irreducible component of a complete intersection of divisors in $\lfloor \Delta \rfloor$. In the dlt case, this takes care of every lc center by [Fuj07, Sec.3.9]; see also [Kol13, Sec.4.1] for details.

The following result was proved by Hacon (and published in [KK10]). A simplified proof is in [Fuj11].

Proposition 5. *Let (Z, Δ_Z) be an lc pair. Then it has a $\mathbb{Q}$-factorial, crepant, dlt model $p : (X, \Delta_X) \to (Z, \Delta_Z)$. That is, X is $\mathbb{Q}$-factorial, (X, Δ_X) is dlt, $K_X + \Delta_X$ is p-nef and $\Delta_X = E + p_*^{-1}\Delta_Z$ where E contains all p-exceptional divisors with multiplicity 1.* *Q.E.D.*

6 (Birational weak crepant log structures)**.**
Let $f : (X, \Delta) \to Z$ be a weak crepant log structure. If f factors as $X \stackrel{g}{\to} X' \stackrel{f'}{\to} Z$ where g is birational, then $f' : (X', \Delta' := g_*\Delta) \to Z$ also a weak crepant log structure. We say that $f : (X, \Delta) \to Z$ *birationally dominates* $f' : (X', \Delta') \to Z$.

Conversely, assume that $f' : (X', \Delta') \to Z$ is a weak crepant log structure and $g : X \to X'$ is a proper birational morphism. Write $K_X + \Delta \sim_{\mathbb{Q}} g^*\big(K_{X'} + \Delta'\big)$. Then $f := f' \circ g : (X, \Delta) \to Z$ is also a weak crepant log structure.

By Proposition 5 every (weak) crepant log structure $f : (X, \Delta) \to Z$ is dominated by another (weak) crepant log structure $f^* : (X^*, \Delta^*) \to Z$ such that (X^*, Δ^*) is dlt and $\mathbb{Q}$-factorial. If Δ is effective then we can choose Δ^* to be effective.

Two weak crepant log structures $f_i : (X_i, \Delta_i) \to Z$ are called *crepant birational* if there is a third weak crepant log structure $h : (Y, \Delta_Y) \to Z$ which birationally dominates both of them. Crepant birational equivalence is denoted by $\stackrel{cbir}{\sim}$.

The group of crepant birational self-maps of a weak crepant log structure $f : (X, \Delta) \to Z$ is denoted by $\operatorname{Bir}^c_Z(X, \Delta)$. By also allowing k-automorphisms, we get the larger group $\operatorname{Bir}^c_k(X, \Delta)$.

Let $f : (X, \Delta) \to Z$ be a weak crepant log structure and $f' : X' \to Z$ a proper morphism. Assume that there is a birational map $\phi : X \dashrightarrow X'$ such that $f' \circ \phi = f$. By the above, there is a unique $\mathbb{Q}$-divisor Δ' such that $f' : (X', \Delta') \to Z$ is a weak crepant log structure that is birational to $f : (X, \Delta) \to Z$. If ϕ^{-1} has no exceptional divisors, then $\Delta' = \phi_* \Delta$ and hence Δ' is effective if Δ is.

Let $f_i : (X_i, \Delta_i) \to S$ be weak crepant log structures and $\phi : X_1 \dashrightarrow X_2$ a birational map. Let $Z_1 \subset X_1$ an lc center such that, at the generic point of Z_1, the pair (X_1, Δ_1) is dlt and ϕ is a local isomorphism. Then $Z_2 := \phi_* Z_1$ is also an lc center and

$$\phi|_{Z_1} : \big(Z_1, \operatorname{Diff}^*_{Z_1} \Delta_1\big) \dashrightarrow \big(Z_2, \operatorname{Diff}^*_{Z_2} \Delta_1\big) \quad \text{is crepant birational.}$$

Theorem 7. [NU73, Uen75, Gon13, FG14] *Let $f : (X, \Delta_X) \to Z$ be a crepant log structure. Then:*

(1) *The $\operatorname{Bir}^c_Z(X, \Delta_X)$ action on $\omega_X^{[m]}(m\Delta_X)$ is finite for every $m \geq 0$.*

(2) *If Z is projective and $K_X + \Delta_X \sim_{\mathbb{Q}} f^*(\text{ample } \mathbb{Q}\text{-divisor})$ then the $\operatorname{Bir}^c_k(X, \Delta_X)$ action on Z is finite.* *Q.E.D.*

8 (Minimal dominating lc centers)**.** Let $f : (X, \Delta) \to S$ be a dlt, weak crepant log structure. Let $W \subset S$ be an lc center and $\{W_i : i \in I(W)\}$ the minimal (with respect to inclusion) lc centers of (X, Δ) that dominate W. We claim that the set of their crepant birational isomorphism classes

$$\big\{\big(W_i, \operatorname{Diff}^*_{W_i} \Delta\big) : i \in I(W)\big\} \tag{8.1}$$

is a birational invariant of $f : (X, \Delta) \to S$.

To see this note that by [Sza94] we can assume that (X, Δ) is snc. Then it is enough to check birational invariance for one smooth blow up. If we blow up $V \subset X$ that is not an lc center, then the set of lc centers is unchanged.

If V is an lc center that is the complete intersection of say $D_1, \dots, D_r \subset \lfloor \Delta \rfloor$, then we get an exceptional divisor E_V that is a $\mathbb{P}^{r-1}$-bundle over V. Locally on V, we get a direct product

$$\big(E_V, \operatorname{Diff}^*_{E_V} \Delta_{B_V X}\big) \cong \big(V, \operatorname{Diff}^*_V \Delta\big) \times \big(\mathbb{P}^{r-1}, (x_1 \cdots x_r = 0)\big),$$

thus every minimal lc center of $\big(V, \operatorname{Diff}^*_V \Delta\big)$ corresponds to r isomorphic copies of itself among the minimal lc centers of $\big(E_V, \operatorname{Diff}^*_{E_V} \Delta_{B_V X}\big)$, hence among the minimal lc centers of $\big(B_V X, \Delta_{B_V X}\big)$. Q.E.D.

Our next aim is to prove that for crepant log structures, the invariant defined in (8.1) consist of a single birational equivalence class.

$\mathbb{P}^1$-linking of minimal lc centers

Definition 9 ($\mathbb{P}^1$-linking). A *standard* $\mathbb{P}^1$*-link* is a dlt, $\mathbb{Q}$-factorial, pair $\bigl(X, D_1 + D_2 + \Delta\bigr)$ whose sole lc centers are D_1, D_2 (hence D_1 and D_2 are disjoint) plus a proper morphism $\pi : X \to S$ such that $K_X + D_1 + D_2 + \Delta \sim_{\mathbb{Q},\pi} 0$, $\pi : D_i \to S$ are both isomorphisms and every reduced fiber $\operatorname{red} X_s$ is isomorphic to $\mathbb{P}^1$.

Let F denote a general smooth fiber. Then $\bigl((K_X+D_1+D_2)\cdot F\bigr) = 0$, hence $(\Delta \cdot F) = 0$. That is, Δ is a vertical divisor, the projection gives an isomorphism $\bigl(D_1, \operatorname{Diff}_{D_1} \Delta\bigr) \cong \bigl(D_2, \operatorname{Diff}_{D_2} \Delta\bigr)$ and these pairs are klt.

The simplest example of a standard $\mathbb{P}^1$-link is a product

$$\bigl(S \times \mathbb{P}^1, S \times \{0\} + S \times \{\infty\} + \Delta_S \times \mathbb{P}^1\bigr)$$

for some $\mathbb{Q}$-divisor Δ_S.

It turns out that every standard $\mathbb{P}^1$-link is locally the quotient of a product. To see this note that $\bigl((D_1 - D_2) \cdot F\bigr) = 0$, thus every point $s \in S$ has an open neighborhood U such that $D_1 - D_2 \sim_{\mathbb{Q}} 0$ on $\pi^{-1}(U)$. Taking the corresponding cyclic cover we get another standard $\mathbb{P}^1$-link

$$\tilde{\pi} : \bigl(\tilde{X}_U, \tilde{D}_1 + \tilde{D}_2 + \tilde{\Delta}\bigr) \to \tilde{U}$$

where the $\tilde{D}_i$ are now Cartier divisors and $\tilde{\Delta} = \tilde{\pi}^* \tilde{\Delta}_U$ for some $\mathbb{Q}$-divisor $\tilde{\Delta}_U$. Here $\tilde{D}_1 \sim \tilde{D}_2$, hence the linear system $|\tilde{D}_1, \tilde{D}_2|$ maps $\tilde{X}_U$ to $\mathbb{P}^1$. Together with $\tilde{\pi}$ this gives an isomorphism

$$\bigl(\tilde{U} \times \mathbb{P}^1, \tilde{U} \times \{0\} + \tilde{U} \times \{\infty\} + \tilde{\Delta}_U \times \mathbb{P}^1\bigr) \cong \bigl(\tilde{X}_U, \tilde{D}_1 + \tilde{D}_2 + \tilde{\Delta}\bigr).$$

Let $g : (X, \Delta) \to S$ be a crepant, dlt log structure and $Z_1, Z_2 \subset X$ two lc centers. We say that Z_1, Z_2 are *directly* $\mathbb{P}^1$*-linked* if there is an lc center $W \subset X$ containing the Z_i such that $g(W) = g(Z_1) = g(Z_2)$ and $\bigl(W, \operatorname{Diff}^*_W \Delta\bigr)$ is crepant birational to a standard $\mathbb{P}^1$-link with Z_i mapping to D_i.

We say that $Z_1, Z_2 \subset X$ are $\mathbb{P}^1$*-linked* if there is a sequence of lc centers $Z'_1, \dots, Z'_m$ such that $Z'_1 = Z_1, Z'_m = Z_2$ and Z'_i is directly $\mathbb{P}^1$-linked to Z'_{i+1} for $i = 1, \dots, m-1$ (or $Z_1 = Z_2$).

The following strengthening of [KK10, 1.7] was the reason to introduce the notion of $\mathbb{P}^1$-linking.

Theorem 10. *Let k be a field and S essentially of finite type over k. Let $f : (X, \Delta) \to S$ be a proper morphism such that $K_X + \Delta \sim_{\mathbb{Q},f} 0$ and (X, Δ) is dlt. Let $s \in S$ be a point such that $f^{-1}(s)$ is connected (as a $k(s)$-scheme). Let $Z \subset X$ be minimal (with respect to inclusion)* *among the lc centers of* (X,Δ) *such that* $s \in f(Z)$. *Let* $W \subset X$ *be an lc center of* (X,Δ) *such that* $s \in f(W)$.

Then there is an lc center $Z_W \subset W$ *such that* Z *and* Z_W *are* $\mathbb{P}^1$-*linked.*

In particular, all the minimal (with respect to inclusion) lc centers $Z_i \subset X$ *such that* $s \in f(Z_i)$ *are* $\mathbb{P}^1$-*linked to each other.*

Remarks. For the applications it is crucial to understand the case when $k(s)$ is not algebraically closed.

Each $\mathbb{P}^1$-linking defines a birational map $Z \dashrightarrow Z_W$, but different $\mathbb{P}^1$-linkings can give different birational maps.

Proof. We use induction on $\dim X$ and on $\dim Z$.

Write $\lfloor \Delta \rfloor = \sum D_i$. By passing to a suitable étale neighborhood of $s \in S$ we may assume that each $D_i \to Y$ has connected fiber over s and every lc center of (X,Δ) intersects $f^{-1}(s)$. (We need to do this without changing the residue field so that $f^{-1}(s)$ stays connected, cf. [Mil80, I.4.2].)

Assume first that $f^{-1}(s) \cap \sum D_i$ is connected. By suitable indexing, we may assume that $Z \subset D_1$, $W \subset D_r$ and $f^{-1}(s) \cap D_i \cap D_{i+1} \neq \emptyset$ for $i = 1, \dots, r-1$.

By induction, we can apply Theorem 10 to $D_1 \to S$ with Z as Z and $D_1 \cap D_2$ as W. We get that there is an lc center $Z_2 \subset W$ such that Z and Z_2 are $\mathbb{P}^1$-linked. As we noted in Definition 9, Z_2 is also minimal (with respect to inclusion) among the lc centers of (X,Δ) such that $s \in f(Z_2)$. Note that Z_2 is an lc center of $\bigl(D_1, \mathrm{Diff}^*_{D_1}(\Delta)\bigr)$. By adjunction, it is an lc center of (X,Δ) and also an lc center of $\bigl(D_2, \mathrm{Diff}^*_{D_2}(\Delta)\bigr)$.

Next we apply Theorem 10 to $D_2 \to S$ with Z_2 as Z and $D_2 \cap D_3$ as W, and so on. At the end we work on $D_r \to S$ with Z_r as Z and W as W to get an lc center $Z_W \subset W$ such that Z and Z_W are $\mathbb{P}^1$-linked. This proves the first claim if $f^{-1}(s) \cap \sum D_i$ is connected.

If $f^{-1}(s) \cap \sum D_i$ is disconnected, then write $\Delta = \sum_{i=1}^m D_i + \Delta_1$. We claim that in this case $m = 2$ and $D_1, D_2 \subset X$ are directly $\mathbb{P}^1$-linked (by $W = X$). We may assume that X is $\mathbb{Q}$-factorial.

First we show that $\sum D_i$ dominates S. Indeed, consider the exact sequence

$$0 \to \mathcal{O}_X(-\textstyle\sum D_i) \to \mathcal{O}_X \to \mathcal{O}_{\sum D_i} \to 0$$

and its push-forward

$$\mathcal{O}_S \cong f_*\mathcal{O}_X \to f_*\mathcal{O}_{\sum D_i} \to R^1 f_*\mathcal{O}_X(-\textstyle\sum D_i).$$

Since $-\sum D_i \sim_{\mathbb{Q},f} K_X + \Delta_1$, the sheaf $R^1 f_*\mathcal{O}_X(-\sum D_i)$ is torsion free by [Kol86] (see [KK10] for the extension to the klt case that we use).

Thus $\mathcal{O}_S \twoheadrightarrow f_*\mathcal{O}_{\sum D_i}$ is surjective hence $\sum D_i \to S$ has connected fibers, a contradiction.

This implies that $K_X + \Delta_1$ is not f-pseudo-effective and so by [BCHM10, 1.3.2] one can run the (X, Δ_1)-MMP over S.

Every step is numerically $K_X + \sum D_i + \Delta_1$-trivial, hence $\sum D_i$ is ample on every extremal ray. Therefore a connected component of $\sum D_i$ can never be contracted by a birational contraction. By the Connectedness Theorem [Kol92, 17.4], the connected components of $\sum D_i$ are unchanged for birational contractions and flips. Thus, at some point, we must encounter a Fano contraction $p : (X^*, \Delta_1^*) \to V$ where $\sum D_i^*$ is p-ample. So there is an irreducible component, say D_1^* that has positive intersection with the contracted ray. Therefore D_1^* is p-ample. By assumption, there is another irreducible component, say D_2^* that is disjoint from D_1^*. Let $F_v \subset X^*$ be any fiber that intersects D_2^*. Since D_2^* is disjoint from D_1^*, we see that D_2^* does not contain F_v. Thus D_2^* also has positive intersection with the contracted ray, hence D_2^* is also p-ample.

Thus D_1^* and D_2^* are both relatively ample (possibly multi-) sections of p and they are disjoint. This is only possible if p has fiber dimension 1, the generic fiber is a smooth rational curve and D_1^* and D_2^* are sections of p.

Since p is an extremal contraction, $R^1p_*\mathcal{O}_{X^*} = 0$, which implies that every fiber of p is a tree of smooth rational curves. Both D_1^* and D_2^* intersects every fiber in a single point and they both intersect every contracted curve. Thus every fiber is irreducible and so $p : (X^*, \Delta^*) \to V$ is a standard $\mathbb{P}^1$-link with D_1^*, D_2^* as sections. As we noted in Definition 9, the rest of Δ^* consists of vertical divisors. Thus any other D_i^* would make $f^{-1}(s) \cap \sum D_i$ connected. Therefore D_1^*, D_2^* are the only lc centers of $(X^*, D_1^* + D_2^* + \Delta_1^*)$ and so D_1, D_2 are the only lc centers of (X, Δ). As noted at the end of Definition 6, $D_1, D_2 \subset X$ are directly $\mathbb{P}^1$-linked (by $W = X$). Q.E.D.

Corollary 11. *Let $f : (X, \Delta_X) \to S$ be a dlt, crepant log structure. Let $Y \subset X$ be an lc center. Consider the Stein factorization $f|_Y : Y \xrightarrow{f_Y} S_Y \xrightarrow{\pi} S$ and set $\Delta_Y := \operatorname{Diff}_Y^* \Delta_X$. Then*

(1) *$f_Y : (Y, \Delta_Y) \to S_Y$ is a dlt, crepant log structure.*

(2) *Let $W_Y \subset S_Y$ be an lc center of $f_Y : (Y, \Delta_Y) \to S_Y$. Then $\pi(W_Y) \subset S$ is an lc center of $f : (X, \Delta_X) \to S$ and every minimal lc center of (Y, Δ_Y) dominating W_Y is also a minimal lc center of (X, Δ_X) dominating $\pi(W_Y)$.*

(3) *Let* $W \subset S$ *be an lc center of* $f : (X, \Delta_X) \to S$. *Then every irreducible component of* $\pi^{-1}(W)$ *is an lc center of* $f_Y : (Y, \Delta_Y) \to S_Y$.

Proof. (1) is clear. To see (2), note that W_Y is dominated by an lc center V_Y of $(Y, \operatorname{Diff}_Y^* \Delta)$. Thus, by adjunction, V_Y is also an lc center of (X, Δ), hence $\pi(W_Y) = f(V_Y)$ is an lc center of S. By Theorem 10, a minimal lc center of Y that dominates W_Y is also a minimal lc center of X that dominates $\pi(W_Y)$. Thus $\operatorname{Src}(W_Y, Y, \Delta_Y) \sim \operatorname{Src}(\pi(W_Y), X, \Delta_X)$.

Finally let $W \subset S$ be an lc center of $f : (X, \Delta_X) \to S$ and $w \in W$ the generic point. Let $V_X \subset X$ be a minimal lc center that dominates W. By Theorem 10, there is an lc center $V_Y \subset Y$ that is $\mathbb{P}^1$-linked to V_X. By adjunction, V_Y is also an lc center of $(Y, \operatorname{Diff}_Y^* \Delta)$. Thus $f_Y(V_Y) \subset S_Y$ is an lc center of $f_Y : (Y, \Delta_Y) \to S_Y$ and it is also one of the irreducible components of $\pi^{-1}(W)$.

In order to get (3), after replacing S by an étale neighborhood of w, we may assume that $Y = \cup Y_j$ such that each $f^{-1}(w) \cap Y_j$ is connected. By the previous argument, each Y_j yields an lc center $f_{Y_j}(V_{Y_j}) \subset S_{Y_j}$ and together these show that every irreducible component of $\pi^{-1}(W)$ is an lc center of $f_Y : (Y, \Delta_Y) \to S_Y$. Q.E.D.

Example 12. Fix $m \geq 3$ and ϵ a primitive mth root of unity. On $\mathbb{P}^{m-1}$ consider the μ_m-action generated by

$$\tau_1 : (x_0 : x_1 : \cdots : x_{m-1}) \mapsto (x_0 : \epsilon x_1 : \cdots : \epsilon^{m-1} x_{m-1}).$$

The action moves the divisor $D_0 := (x_0 + x_1 + \cdots + x_{m-1} = 0)$ into m different divisors $D_0, \ldots, D_{m-1}$. One easily checks that $(\mathbb{P}^{m-1}, D_0 + \cdots + D_{m-1})$ is snc (if ϵ is in our base field) and has trivial log canonical class.

Let A be an abelian variety with a μ_m-action τ_2. On

$$\left(\mathbb{P}^{m-1} \times A, \Delta := D_0 \times A + \cdots + D_{m-1} \times A\right)$$

we have a μ_m-action generated by $\tau := (\tau_1, \tau_2)$.

Let $X_1 := (\mathbb{P}^{m-1} \times A)/\langle \tau \rangle$. The quotient of the boundary Δ has only 1 component but it has complicated self-intersections, hence it is not dlt. Let (X, Δ_X) be a dlt model.

We see that the minimal lc centers are isomorphic to $(A, 0)$ and the different $\mathbb{P}^1$-linkings between them differ from each other by a power of τ_2.

§3. Poincaré residue map

Definition 13. Let (X, Δ) be a dlt pair and $Z \subset X$ an lc center. As in Definition 4, if $\omega_X^{[m]}(m\Delta)$ is locally free, then, by iterating the usual Poincaré residue maps for divisors, we get a *Poincaré residue map*

$$\mathcal{R}^m_{X\to Z} : \omega_X^{[m]}(m\Delta)|_Z \xrightarrow{\cong} \omega_Z^{[m]}(m \cdot \operatorname{Diff}^*_Z \Delta). \tag{13.1}$$

(This is well defined if m is even, defined only up to sign if m is odd.)

Let $f : (X, \Delta) \to Y$ be a dlt, weak crepant log structure. Choose $m > 0$ even such that $\omega_X^{[m]}(m\Delta) \sim f^*L$ for some line bundle L on Y. Let $Z \subset X$ be an lc center of (X, Δ). We can view the Poincaré residue map as

$$\mathcal{R}^m_{X\to Z} : f^*L|_Z \cong \omega_X^{[m]}(m\Delta)|_Z \xrightarrow{\cong} \omega_Z^{[m]}(m \cdot \operatorname{Diff}^*_Z \Delta). \tag{13.2}$$

The following result shows, that, for minimal lc centers, (13.2) is essentially independent of the choice of Z.

Proposition 14. *Let $f : (X, \Delta) \to Y$ be a dlt crepant log structure. Choose $m > 0$ even such that $\omega_X^{[m]}(m\Delta) \cong f^*L$ for some line bundle L on Y. Let Z_1, Z_2 be minimal lc centers of (X, Δ) such that $f(Z_1) = f(Z_2)$. Then there is a birational map $\phi : Z_2 \dashrightarrow Z_1$ such that the following diagram commutes*

$$\begin{array}{ccc} \omega_X^{[m]}(m\Delta) & \cong f^*L \cong & \omega_X^{[m]}(m\Delta) \\ \mathcal{R}^m_{X\to Z_1} \downarrow & & \downarrow \mathcal{R}^m_{X\to Z_2} \\ \omega_{Z_1}^{[m]}\big(m \operatorname{Diff}^*_{Z_1} \Delta\big) & \xrightarrow{\phi^*} & \omega_{Z_2}^{[m]}\big(m \operatorname{Diff}^*_{Z_2} \Delta\big) \end{array} \tag{14.1}$$

Proof. By Theorem 10 it is sufficient to prove this in case there is an lc center W that is birational to a $\mathbb{P}^1$-bundle $\mathbb{P}^1 \times U$ with Z_1, Z_2 as sections. Thus projection to U provides a birational isomorphism $\phi : Z_2 \dashrightarrow Z_1$.

Since $\mathcal{R}^m_{X\to Z_i} = \mathcal{R}^m_{W\to Z_i} \circ \mathcal{R}^m_{X\to W}$, we may assume that $X = W$. The sheaves in (14.1) are torsion free, hence it is enough to check commutativity after localizing at the generic point of U. This reduces us to the case when $W = \mathbb{P}^1_L$ with coordinates $(x{:}y)$, $Z_1 = (0{:}1)$ and $Z_2 = (1{:}0)$. A generator of $H^0\big(\mathbb{P}^1, \omega_{\mathbb{P}^1}(Z_1 + Z_2)\big)$ is dx/x which has residue 1 at Z_1 and -1 at Z_2. Thus (14.1) commutes for m even and anti-commutes for m odd. Q.E.D.

Remark 15. By Proposition 14 we get a Poincaré residue map as stated in (1.4) but it is not yet completely canonical. We think of (Z, Δ_Z)

as an element of a crepant birational equivalence class, thus so far $\mathcal{R}^m$ is defined only up to the action of $\operatorname{Bir}^c_Y(Z,\Delta_Z)$. However, by Theorem 7, the image of this action is a finite group of rth roots of unity for some r. Thus the $\operatorname{Bir}^c_Y(Z,\Delta_Z)$ action is trivial on $\omega_Z^{[mr]}(mr\Delta_Z)$ hence

$$\mathcal{R}^{mr}:\omega_X^{[mr]}(mr\Delta)|_Z \cong \omega_Z^{[mr]}(mr\Delta_Z) \tag{15.1}$$

is completely canonical. Assume next that $\omega_X^{[mr]}(mr\Delta)\sim f^*L$. Let us factor $f|_Z: Z\to f(Z)$ using $g: Z\to W$ and the normalization $n: W\to f(Z)$. Then we can push forward (15.1) to get an isomorphism

$$n^*L\cong\bigl(g_*\omega_Z^{[m]}(m\Delta_Z)\bigr)^{\text{inv}} \tag{15.2}$$

where the exponent inv denotes the invariants under the action of the group of birational self-maps $\operatorname{Bir}_Y(Z,\Delta_Z)$. This shows the second isomorphism in (1.4).

Notation 16. Let (Y,Δ_Y) be lc and $(X,\Delta_X)\to(Y,\Delta_Y)$ a crepant, dlt model. Let $W\subset Y$ be an lc center of (Y,Δ_Y) and $Z\subset X$ minimal (with respect to inclusion) among the lc centers of (X,Δ_X) that dominate W. By Definition 13, we obtain a Poincaré residue map $\mathcal{R}_{X\to Z}$.

Let $D\subset\lfloor\Delta_Y\rfloor$ be a divisor with normalization $\pi: D^n\to D$. Let $D_X\subset X$ be its birational transform on X and set $\Delta_{D_X}:=\operatorname{Diff}^*_{D_X}\Delta_X$. Let $W_D\subset D^n$ be an lc center of $\bigl(D^n,\operatorname{Diff}^*_{D^n}\Delta_Y\bigr)$. Then $W_X:=\pi(W_D)$ is an lc center of $\bigl(Y,\Delta_Y\bigr)$. Choose minimal lc centers $Z_X\subset X$ (resp. $Z_D\subset D_X$) dominating W_X (resp. W_D).

Theorem 17. *Notation and assumptions as above. Then there is a birational map $\phi: Z_D \dashrightarrow Z_X$ such that for m sufficiently divisible, the following diagram commutes*

$$\begin{array}{ccc}
\omega_X^{[m]}(m\Delta_X) & \xrightarrow{\mathcal{R}^m_{X\to D_X}} & \omega_{D_X}^{[m]}(m\Delta_{D_X})\\
\mathcal{R}^m_{X\to Z_X}\downarrow & & \downarrow\mathcal{R}^m_{D_X\to Z_D}\\
\omega_{Z_X}^{[m]}(m\operatorname{Diff}^*_{Z_X}\Delta_X) & \xrightarrow{\phi^*} & \omega_{Z_D}^{[m]}(m\operatorname{Diff}^*_{Z_D}\Delta_{D_X})
\end{array}$$

Proof. If we choose Z_X as the image of Z_D, this holds by the definition of the higher codimension residue maps. This and Proposition 14 proves the claim for every other choice of Z_X. Q.E.D.

§4. Sources and Springs

Definition 18. Let $f:(X,\Delta)\to S$ be a crepant, dlt log structure and $Z\subset S$ an lc center. An lc center Z' of (X,Δ) is called a *source of*

Z if $f(Z') = Z$ and Z' is minimal (with respect to inclusion) among the lc centers that dominate Z. By restriction we have

$$f|_{Z'} : \big(Z', \operatorname{Diff}^*_{Z'} \Delta\big) \to Z \quad \text{and} \quad K_{Z'} + \operatorname{Diff}^*_{Z'} \Delta \sim_{f,\mathbb{Q}} 0.$$

By adjunction, there is a one-to-one correspondence between lc centers of $\big(Z', \operatorname{Diff}^*_{Z'} \Delta\big)$ and lc centers of (X, Δ) that are contained in Z'. Thus Z' is a source of Z iff the general fiber of $\big(Z', \operatorname{Diff}^*_{Z'} \Delta\big) \to Z$ is klt.

By Theorem 10 all sources of Z are birational to each other (as weak crepant log structures over Z). Any representative of their birational equivalence class will be denoted by $\operatorname{Src}(Z, X, \Delta)$. One can choose a representative $(S^t, \Delta^t) \to Z$ whose generic fiber is terminal. Such models are still not unique, but their generic fibers are isomorphic outside codimension 2 sets. However, if there is an irreducible component of Δ^t whose coefficient is 1 (these can not dominate Z) then it does not seem possible to choose a sensible subclass of models that are isomorphic to each other outside codimension 2 sets.

Note further that by Remark 8, if two crepant log structures $f_i : (X_i, \Delta_i) \to Y$ are crepant birational over Y, then $\operatorname{Src}(Z, X_1, \Delta_1)$ is crepant birational to $\operatorname{Src}(Z, X_2, \Delta_2)$.

One can uniquely factor $f|_{Z'}$ as

$$f|_{Z'} : \big(Z', \operatorname{Diff}^*_{Z'} \Delta'\big) = \operatorname{Src}(Z, X, \Delta) \xrightarrow{c_Z} \tilde{Z}' \xrightarrow{p_Z} Z \qquad (18.1)$$

where $\tilde{Z}'$ is normal, p_Z is finite and c_Z has connected fibers.

Thus in (18.1), $\tilde{Z}'$ is uniquely defined up to isomorphism over Z. Its isomorphism class will be denoted by $\operatorname{Spr}(Z, X, \Delta)$ and called the *spring* of Z.

Define the group of *source-automorphisms* of $\operatorname{Spr}(Z, X, \Delta)$ as

$$\operatorname{Aut}^s \operatorname{Spr}(Z, X, \Delta) := \operatorname{im}\big[\operatorname{Bir}^c_k \operatorname{Src}(Z, X, \Delta) \to \operatorname{Aut}_k \operatorname{Spr}(Z, X, \Delta)\big].$$

By Theorem 7, if $K_X + \Delta$ is ample then $\operatorname{Aut}^s \operatorname{Spr}(Z, X, \Delta)$ is finite for every lc center $Z \subset X$.

Let (Y, Δ) be lc and $f : (X, \Delta_X) \to (Y, \Delta)$ a dlt model. Let $Z \subset Y$ be an lc center of (Y, Δ). As noted above, the source $\operatorname{Src}(Z, X, \Delta_X)$ of Z depends only on (Y, Δ) but not on the choice of (X, Δ_X). Thus we also use $\operatorname{Src}(Z, Y, \Delta)$ (resp. $\operatorname{Spr}(Z, Y, \Delta)$) to denote the source (resp. spring) of Z.

Next we prove (1.5).

Proposition 19. *Let $f : (X, \Delta) \to Y$ be a crepant log structure and $Z \subset Y$ an lc center. Then the field extension $k\big(\operatorname{Spr}(Z, X, \Delta)\big)/k(Z)$ is*

Galois and

$$\operatorname{Gal}\bigl(\operatorname{Spr}(Z, X, \Delta)/Z\bigr) \subset \operatorname{Aut}^s \operatorname{Spr}(Z, X, \Delta).$$

Proof. We may localize at the generic point of Z. Thus we may assume that Z is a point and then prove the following more precise result.

Lemma 20. *Let $g : (X, \Delta) \to Y$ be a weak crepant log structure over a field k. Assume that (X, Δ) is dlt and X is $\mathbb{Q}$-factorial. Let $z \in Y$ be an lc center such that $g^{-1}(z)$ is connected (as a $k(z)$-scheme). Then there is a unique smallest finite field extension $K(z) \supset k(z)$ such that the following hold.*

(1) *Every lc center of $(X_{\bar{k}}, \Delta_{\bar{k}})$ that intersects $g^{-1}(z)$ is defined over $K(z)$.*
(2) *Let $W_{\bar{z}} \subset Y_{\bar{k}}$ be a minimal lc center contained in $g^{-1}(z)$. Then $K(z) = k_{ch}(W_{\bar{z}})$, the field of definition of $W_{\bar{z}}$.*
(3) *$K(z) \supset k(z)$ is a Galois extension.*
(4) *Let W_z be a minimal lc center contained in $g^{-1}(z)$. Then*

$$\operatorname{Bir}^c_{k(z)}\bigl(W_z, \operatorname{Diff}^*_{W_z} \Delta\bigr) \to \operatorname{Gal}\bigl(K(z))/k(z)\bigr) \quad \textit{is surjective.}$$

Proof. There are only finitely many lc centers and a conjugate of an lc center is also an lc center. Thus the field of definition of any lc center is a finite extension of k. Since $K(z)$ is the composite of some of them, it is finite over $k(z)$.

Let $W_{\bar{z}} \subset X_{\bar{k}}$ be a minimal lc center contained in $g^{-1}(z)$ and $k_{ch}(W_{\bar{z}})$ its field of definition. Let $D_i \subset \lfloor \Delta \rfloor$ be the irreducible components that contain $W_{\bar{z}}$. Each D_i is smooth at the generic point of $W_{\bar{z}}$, hence the $\bar{k}$-irreducible component of D_i that contains $W_{\bar{z}}$ is also defined over $k_{ch}(W_{\bar{z}})$. Thus every lc center of $(X_{\bar{k}}, \Delta_{\bar{k}})$ containing $W_{\bar{z}}$ is also defined over $k_{ch}(W_{\bar{z}})$. Therefore, any lc center that is $\mathbb{P}^1$-linked to $W_{\bar{z}}$ is defined over $k_{ch}(W_{\bar{z}})$. By Theorem 10 this implies that every lc center of $(X_{\bar{k}}, \Delta_{\bar{k}})$ that intersects $g^{-1}(z)$ is defined over $k_{ch}(W_{\bar{z}})$, hence $k_{ch}(W_{\bar{z}}) \supset K(z)$. By construction, $k_{ch}(W_{\bar{z}}) \subset K(z)$, thus $k_{ch}(W_{\bar{z}}) = K(z)$.

A conjugate of $W_{\bar{z}}$ over $k(z)$ is defined over the corresponding conjugate field of $k_{ch}(W_{\bar{z}})$. By the above, every conjugate of the field of $k_{ch}(W_{\bar{z}})$ over $k(z)$ is itself, hence $k_{ch}(W_{\bar{z}}) = K(z)$ is Galois over $k(z)$.

Finally, in order to see (4), fix $\sigma \in \operatorname{Gal}\bigl(K(z)/k(z)\bigr)$ and let $W^{\sigma}_{\bar{z}}$ be the corresponding conjugate of $W_{\bar{z}}$. By Theorem 10, $W^{\sigma}_{\bar{z}}$ and $W_{\bar{z}}$ are $\mathbb{P}^1$-linked over $K(z)$; fix one such $\mathbb{P}^1$-link. The union of the conjugates of this $\mathbb{P}^1$-link over $k(z)$ define an element of $\operatorname{Bir}^c_{k(z)}\bigl(W_z, \operatorname{Diff}^*_{W_z} \Delta\bigr)$ which induces σ on $K(z)/k(z)$. (The $\mathbb{P}^1$-link is not unique, hence the lift is not unique. Thus in (4) we only claim surjectivity, not a splitting.) Q.E.D.

We also note the following direct consequence of Corollary 11.

Corollary 21 (Adjunction for sources). *Let $(X, D+\Delta)$ be lc and $n : D^n \to D$ the normalization. Let $Z_D \subset D^n$ be an lc center of $\bigl(D^n, \operatorname{Diff}_{D^n}\Delta\bigr)$ and $Z_X := n(Z_D)$ its image in X. Then*

(1) $\operatorname{Src}\bigl(Z_D, D^n, \operatorname{Diff}_{D^n}\Delta\bigr) \overset{cbir}{\sim} \operatorname{Src}\bigl(Z_X, X, D+\Delta\bigr)$ *and*
(2) $\operatorname{Spr}\bigl(Z_D, D^n, \operatorname{Diff}_{D^n}\Delta\bigr) \cong \operatorname{Spr}\bigl(Z_X, X, D+\Delta\bigr)$. *Q.E.D.*

§5. Applications to slc pairs

22 (Normalization of slc pairs). Let (X, Δ) be a semi log canonical pair. Let $\pi : \bar{X} \to X$ denote the normalization of X, $\bar{\Delta}$ the divisorial part of $\pi^{-1}(\Delta)$ and $\bar{D} \subset \bar{X}$ the conductor of π. Since X is seminormal, $\bar{D}$ is reduced. X has an ordinary node at a codimension 1 singular point, thus interchanging the two preimages of the node gives an involution τ of the normalization $n : \bar{D}^n \to \bar{D}$. This gives an injection

$$\bigl\{\ \text{slc pairs } (X,\Delta)\ \bigr\} \quad \hookrightarrow \quad \left\{ \begin{array}{c} \text{lc pairs } \bigl(\bar{X}, \bar{D}+\bar{\Delta}\bigr) \\ \text{plus an involution } \tau \text{ of } \bar{D}^n \end{array} \right\}. \tag{22.1}$$

For many purposes, it is important to understand the image of this map. That is, we would like to know which quadruples $\bigl(\bar{X}, \bar{D}+\bar{\Delta}, \tau\bigr)$ correspond to an slc pair (X, Δ). An easy condition to derive is that τ is an involution not just of the variety $\bar{D}^n$ but of the lc pair $\bigl(\bar{D}^n, \operatorname{Diff}_{\bar{D}^n}\bar{\Delta}\bigr)$. Thus we obtain a refined version of the map

$$\bigl\{\ \text{slc pairs } (X,\Delta)\ \bigr\} \quad \hookrightarrow \quad \left\{ \begin{array}{c} \text{lc pairs } \bigl(\bar{X}, \bar{D}+\bar{\Delta}\bigr) \\ \text{plus an involution } \tau \\ \text{of } \bigl(\bar{D}^n, \operatorname{Diff}_{\bar{D}^n}\bar{\Delta}\bigr) \end{array} \right\}. \tag{22.2}$$

For surfaces, the above constructions are discussed in [Kol92, Sec.12]. The higher dimensional generalizations are straightforward; see [Kol13, Chap.5].

There are three major issues involved in trying to prove that the map (22.2) is surjective.

22.3.1. *Does τ generate a finite equivalence relation?*

The normalization $n : \bar{D}^n \to \bar{D} \to \bar{X}$ and τ generate an equivalence relation $R(\tau)$, called the *gluing relation*, on the points of $\bar{X}$ by declaring $n(p) \sim n(\tau(p))$ for every $p \in \bar{D}^n$. It is easy to see (cf. [Kol12]) that $R(\tau)$

is a set-theoretic, pro-finite, algebraic equivalence relation. That is, one can give $R(\tau)$ by countably many subschemes

$$\{R_i \subset \bar{X} \times \bar{X} : i \in I\}$$

such that $\cup_i R_i(K) \subset \bar{X}(K) \times \bar{X}(K)$ is an equivalence relation on $\bar{X}(K)$ for every algebraically closed field K and the coordinate projections induce finite morphisms

$$\pi_1 : R_i \to \bar{X} \quad \text{and} \quad \pi_2 : R_i \to \bar{X}.$$

(One can make the R_i unique if we choose them irreducible, reduced and assume that none of them contains another.)

It is clear that if X exists then every equivalence class of $R(\tau)$ is contained in a fiber of $\pi : \bar{X} \to X$. In particular, if X exists then the $R(\tau)$-equivalence classes are finite. Equivalently, I is a finite set.

In general the $R(\tau)$-equivalence classes need not be finite. Moreover, non-finiteness can appear in high codimension. This is the question that we will study here using the sources of lc centers, especially their Galois property (1.5).

A closely related example is given by [BT09]: there is a smooth curve D of genus ≥ 2 and a finite relation $R_0 \subset D \times D$ such that both projections $R_0 \rightrightarrows D$ are étale yet R_0 generates a non-finite equivalence relation.

22.3.2. *Constructing* (X, Δ) *from* $\big(\bar{X}, \bar{D} + \bar{\Delta}, \tau\big)$.

Following the method of [Kol12], it is proved in [Kol13, Chap.5], that if the $R(\tau)$-equivalence classes are finite, then (X, Δ) exists.

22.3.3. *Is* $K_X + \Delta$ *a* $\mathbb{Q}$*-Cartier divisor?*

The answer turns out to be yes, see [Kol13, Chap.5], but my proof, using Poincaré residue maps and Theorem 7, is somewhat indirect.

As a consequence we obtain that (22.2) is one-to-one for pairs with ample log canonical class.

Theorem 23. *Taking the normalization gives a one-to-one correspondence between the following two sets, where* $X, \bar{X}$ *are projective schemes over a field.*

$$\left\{ \begin{array}{c} \textit{slc pairs } (X, \Delta) \\ \textit{such that} \\ K_X + \Delta \textit{ is ample} \end{array} \right\} \cong \left\{ \begin{array}{c} \textit{lc pairs } \big(\bar{X}, \bar{D} + \bar{\Delta}\big) \textit{ such that} \\ K_{\bar{X}} + \bar{D} + \bar{\Delta} \textit{ is ample plus an} \\ \textit{involution } \tau \textit{ of } \big(\bar{D}^n, \operatorname{Diff}_{\bar{D}^n} \bar{\Delta}\big) \end{array} \right\}.$$

This can be extended to the relative case as follows.

Theorem 24. *Let S be a scheme which is essentially of finite type over a field. Taking the normalization gives a one-to-one correspondence between the following two sets.*

(1) *Slc pairs (X, Δ) such that X/S is proper and $K_X + \Delta$ is ample on the generic fiber of $W \to S$ for every lc center $W \subset X$.*

(2) *Lc pairs $(\bar{X}, \bar{D} + \bar{\Delta})$ such that $\bar{X}/S$ is proper and $K_{\bar{X}} + \bar{D} + \bar{\Delta}$ is ample on the generic fiber of $\bar{W} \to S$ for every lc center $\bar{W} \subset \bar{X}$, plus an involution τ of $(\bar{D}^n, \operatorname{Diff}_{\bar{D}^n} \bar{\Delta})$.*

Furthermore, the cases when $K_X + \Delta$ is ample on X/S correspond to the cases when $K_{\bar{X}} + \bar{D} + \bar{\Delta}$ is ample on $\bar{X}/S$.

As we noted in (22.3), the following result implies Theorem 23.

Proposition 25. *Let $(\bar{X}, \bar{D} + \bar{\Delta})$ be an lc pair and τ an involution of $(\bar{D}^n, \operatorname{Diff}_{\bar{D}^n} \bar{\Delta})$.*

Assume that X is proper over a base scheme S that is essentially of finite type over a field. Assume furthermore that $K_{\bar{X}} + \bar{D} + \bar{\Delta}$ is ample on the generic fiber of $\bar{W} \to S$ for every lc center $\bar{W} \subset \bar{X}$.

Then the gluing relation $R(\tau)$, defined in (22.3.1), is finite.

This in turn will be derived from Theorem 28 on the gluing relation $R(\tau)$ which applies whether $K_{\bar{X}} + \bar{D} + \bar{\Delta}$ is ample or not.

Definition 26. Let Y be a normal scheme and $R = \cup_{i \in I} R_i \subset Y \times Y$ a set-theoretic, pro-finite, algebraic equivalence relation where the R_i are irreducible.

R is called a *groupoid* if every R_i is the graph of an isomorphism between two irreducible components of Y.

Let $Y^j \subset Y$ be an irreducible component. The restriction of R to Y^j is $R^j := R \cap (Y^j \times Y^j)$. If R is a groupoid then one can identify R^j with a subgroup of $\operatorname{Aut}(Y^j)$ called the *stabilizer* of Y^j in R.

We are now ready to formulate and prove a structure theorem for gluing relations. Roughly speaking, we prove that for every lc center $\bar{W} \subset \bar{X}$ there is a "canonically" defined finite cover $p : \tilde{W} \to \bar{W}$ such that $(p \times p)^{-1}(R(\tau) \cap (\bar{W} \times \bar{W}))$ is a groupoid and the stabilizer action W is compatible with $p^*(K_{\bar{X}} + \bar{D} + \bar{\Delta})$. The compatibility condition is somewhat delicate to state. Thus I give the actual construction of $\tilde{W}$ and then specify the compatibility condition for that particular case.

Notation 27. Let (X, Δ) be lc. Let $S_i^*(X, \Delta)$ be the union of all $\leq i$-dimensional lc centers of (X, Δ) and set $S_i(X, \Delta) := S_i^*(X, \Delta) \setminus S_{i-1}^*(X, \Delta)$. Let $Z_{ij}^0 \subset S_i(X, \Delta)$ be the irreducible components. The closure Z_{ij} of Z_{ij}^0 is an lc center of (X, Δ), hence it has a spring p_{ij} :

$\mathrm{Spr}(Z_{ij}, X, \Delta) \to Z_{ij}$. Set $\mathrm{Spr}(Z^0_{ij}, X, \Delta) := p_{ij}^{-1} Z^0_{ij}$ and

$$\mathrm{Spr}_i\bigl(X, \Delta\bigr) := \amalg_j \mathrm{Spr}(Z^0_{ij}, X, \Delta).$$

Let $p_i : \mathrm{Spr}_i\bigl(X, \Delta\bigr) \to S_i(X, \Delta)$ be the induced morphism. Then p_i is finite, surjective and universally open since $S_i(X, \Delta)$ is normal. Furthermore, p_i is Galois over every Z_{ij} by Proposition 19.

Theorem 28. *Let $\bigl(X, D+\Delta\bigr)$ be an lc pair, τ an involution of $\bigl(D^n, \mathrm{Diff}_{D^n} \Delta\bigr)$ and $R(\tau) \subset X \times X$ the corresponding equivalence relation as in (22.3.1). Let $p_i : \mathrm{Spr}_i\bigl(X, D+\Delta\bigr) \to S_i$ be as above. Then*

1. *$(p_i \times p_i)^{-1}\bigl(R(\tau) \cap (S_i(X, \Delta) \times S_i(X, \Delta))\bigr)$ is a groupoid on $\mathrm{Spr}_i\bigl(X, D+\Delta\bigr)$.*
2. *For every irreducible component $Z^0_{ij} \subset S_i(X, \Delta)$, the stabilizer of its spring $\mathrm{Spr}(Z^0_{ij}, X, D+\Delta) \subset \mathrm{Spr}_i\bigl(X, D+\Delta\bigr)$ is a subgroup of the source-automorphism group $\mathrm{Aut}^s \mathrm{Spr}(Z_{ij}, X, D+\Delta)$.*

Proof. We need to describe how the generators of $R(\tau)$ pull back to the spring $\mathrm{Spr}_i\bigl(X, D+\Delta\bigr)$.

First, the preimage of the diagonal of $Z^0_{ij} \times Z^0_{ij}$ is a group $\Gamma(G_{ij})$ and $G_{ij} = \mathrm{Gal}\bigl(\mathrm{Spr}(Z_{ij}, X, D+\Delta)/Z_{ij}\bigr)$ is a subgroup of $\mathrm{Aut}^s \mathrm{Spr}(Z_{ij}, X, D+\Delta)$ by Proposition 19.

Second, let $Z_{ijk} \subset D^n$ be an irreducible component of the preimage of Z_{ij}. Then Z_{ijk} is an lc center of $\bigl(D^n, \mathrm{Diff}_{D^n} \Delta\bigr)$ and

$$\mathrm{Src}\bigl(Z_{ijk}, D^n, \mathrm{Diff}_{D^n} \Delta\bigr) \overset{cbir}{\sim} \mathrm{Src}\bigl(Z_{ij}, X, D+\Delta\bigr)$$

by Corollary 21. Thus, for each ijk, the isomorphism $\tau : D^n \cong D^n$ lifts to isomorphisms

$$\tau_{ijkl} : \mathrm{Spr}(Z^0_{ij}, X, D+\Delta) \cong \mathrm{Spr}(Z^0_{il}, X, D+\Delta).$$

Given ijk, the value of l is determined by $Z_{il} := n\bigl(\tau(Z_{ijk})\bigr)$, but the lifting is defined only up to left and right multiplication by elements of G_{ij} and G_{il}.

Thus $(p_i \times p_i)^{-1}\bigl(R(\tau) \cap (S_i(X, \Delta) \times S_i(X, \Delta))\bigr)$ is the groupoid generated by the G_{ij} and the τ_{ijkl}, hence the stabilizer of $\mathrm{Spr}(Z^0_{ij}, X, D+\Delta)$ is generated by the groups $\tau_{ijkl}^{-1} G_{il} \tau_{ijkl}$. The latter are all subgroups of $\mathrm{Aut}^s \mathrm{Spr}(Z_{ij}, X, D+\Delta)$. Q.E.D.

29 (Proof of Proposition 25). Since $\mathrm{Spr}_i(X, D+\Delta)$ has finitely many irreducible components, the groupoid is finite iff the stabilizer of each $\mathrm{Spr}(Z^0_{ij}, X, D+\Delta)$ is finite. By Theorem 28 this holds if the groups $\mathrm{Aut}^s \mathrm{Spr}(Z_{ij}, X, D+\Delta)$ are finite.

The automorphism group of a variety $\tilde{Z}$ over a base scheme S injects into the automorphism group of the generic fiber $\tilde{Z}_{gen}$.

By assumption, $K_{\bar{X}} + \bar{D} + \bar{\Delta}$ is ample on the generic fiber of $Z_{ij} \to S$, thus Theorem 7 implies that each $\operatorname{Aut}^s \operatorname{Spr}(Z_{ij}, X, D + \Delta)$ is finite. Q.E.D.

Acknowledgments. This paper was written while I visited RIMS, Kyoto University. I thank S. Mori and S. Mukai for the invitation and their hospitality. I am grateful to O. Fujino, C. Hacon, S. Kovács, S. Mori, Y. Odaka, V. Shokurov and C. Xu for many comments and corrections. Partial financial support was provided by the NSF under grant number DMS-0758275.

References

[BCHM10] Caucher Birkar, Paolo Cascini, Christopher D. Hacon, and James McKernan, *Existence of minimal models for varieties of log general type*, J. Amer. Math. Soc. **23** (2010), no. 2, 405–468.

[BPV84] W. Barth, C. Peters, and A. Van de Ven, *Compact complex surfaces*, Ergebnisse der Mathematik und ihrer Grenzgebiete (3), vol. 4, Springer-Verlag, Berlin, 1984.

[BT09] Fedor Bogomolov and Yuri Tschinkel, *Co-fibered products of algebraic curves*, Cent. Eur. J. Math. **7** (2009), no. 3, 382–386. MR 2534459 (2010i:14023)

[FG14] Osamu Fujino and Yoshinori Gongyo, *Log pluricanonical representations and the abundance conjecture*, Compos. Math. **150** (2014), no. 4, 593–620. MR 3200670

[Fuj07] Osamu Fujino, *What is log terminal?*, Flips for 3-folds and 4-folds, Oxford Lecture Ser. Math. Appl., vol. 35, Oxford Univ. Press, Oxford, 2007, pp. 49–62. MR 2359341

[Fuj11] ———, *Semi-stable minimal model program for varieties with trivial canonical divisor*, Proc. Japan Acad. Ser. A Math. Sci. **87** (2011), no. 3, 25–30. MR 2802603 (2012j:14023)

[Gon13] Yoshinori Gongyo, *Abundance theorem for numerically trivial log canonical divisors of semi-log canonical pairs*, J. Algebraic Geom. **22** (2013), no. 3, 549–564. MR 3048544

[Kaw97] Yujiro Kawamata, *Subadjunction of log canonical divisors for a subvariety of codimension* 2, Birational algebraic geometry (Baltimore, MD, 1996), Contemp. Math., vol. 207, Amer. Math. Soc., Providence, RI, 1997, pp. 79–88. MR 1462926 (99a:14024)

[Kaw98] ———, *Subadjunction of log canonical divisors. II*, Amer. J. Math. **120** (1998), no. 5, 893–899. MR 1646046 (2000d:14020)

[KK10] János Kollár and Sándor J. Kovács, *Log canonical singularities are Du Bois*, J. Amer. Math. Soc. **23** (2010), no. 3, 791–813. MR 2629988

[KM98] János Kollár and Shigefumi Mori, *Birational geometry of algebraic varieties*, Cambridge Tracts in Mathematics, vol. 134, Cambridge University Press, Cambridge, 1998, With the collaboration of C. H. Clemens and A. Corti, Translated from the 1998 Japanese original.

[Kol86] János Kollár, *Higher direct images of dualizing sheaves. I*, Ann. of Math. (2) **123** (1986), no. 1, 11–42. MR 825838 (87c:14038)

[Kol92] János Kollár (ed.), *Flips and abundance for algebraic threefolds*, Société Mathématique de France, 1992, Papers from the Second Summer Seminar on Algebraic Geometry held at the University of Utah, Salt Lake City, Utah, August 1991, Astérisque No. 211 (1992).

[Kol07] ———, *Kodaira's canonical bundle formula and adjunction*, Flips for 3-folds and 4-folds, Oxford Lecture Ser. Math. Appl., vol. 35, Oxford Univ. Press, Oxford, 2007, pp. 134–162.

[Kol12] ———, *Quotients by finite equivalence relations*, Current developments in algebraic geometry, Math. Sci. Res. Inst. Publ., vol. 59, Cambridge Univ. Press, Cambridge, 2012, With an appendix by Claudiu Raicu, pp. 227–256. MR 2931872

[Kol13] ———, *Singularities of the minimal model program*, Cambridge Tracts in Mathematics, vol. 200, Cambridge University Press, Cambridge, 2013, With the collaboration of Sándor Kovács.

[Mil80] James S. Milne, *Étale cohomology*, Princeton Mathematical Series, vol. 33, Princeton University Press, Princeton, N.J., 1980. MR 559531 (81j:14002)

[NU73] Iku Nakamura and Kenji Ueno, *An addition formula for Kodaira dimensions of analytic fibre bundles whose fibre are Moišezon manifolds*, J. Math. Soc. Japan **25** (1973), 363–371. MR 0322213 (48 #575)

[Oda13] Yuji Odaka, *The GIT stability of polarized varieties via discrepancy*, Ann. of Math. (2) **177** (2013), no. 2, 645–661. MR 3010808

[OX12] Yuji Odaka and Chenyang Xu, *Log-canonical models of singular pairs and its applications*, Math. Res. Lett. **19** (2012), no. 2, 325–334.

[Sho13] V. V. Shokurov, *Log adjunction: effectiveness and positivity* (arXiv:1308.5160), 2013.

[Sza94] Endre Szabó, *Divisorial log terminal singularities*, J. Math. Sci. Univ. Tokyo **1** (1994), no. 3, 631–639. MR 1322695 (96f:14019)

[Uen75] Kenji Ueno, *Classification theory of algebraic varieties and compact complex spaces*, Lecture Notes in Mathematics, Vol. 439, Springer-Verlag, Berlin, 1975, Notes written in collaboration with P. Cherenack.

Princeton University, Princeton NJ 08544-1000
E-mail address: kollar@math.princeton.edu

Advanced Studies in Pure Mathematics 70, 2016
Minimal Models and Extremal Rays (Kyoto, 2011)
pp. 49–65

Du Bois singularities deform

Sándor J. Kovács and Karl Schwede

Abstract.

Let X be a variety and H a Cartier divisor on X. We prove that if H has Du Bois (or DB) singularities, then X has Du Bois singularities near H. As a consequence, if $X \longrightarrow S$ is a proper flat family over a smooth curve S whose special fiber has Du Bois singularities, then the nearby fibers also have Du Bois singularities. We prove this by obtaining an injectivity theorem for certain maps of canonical modules. As a consequence, we also obtain a restriction theorem for certain non-lc ideals.

§1. Introduction

Du Bois singularities, or henceforth simply DB singularities, were introduced by Steenbrink in [Ste81]. They may be considered a generalization of the notion of rational singularities. The definition and its simple consequences makes DB singularities the natural class to consider in many situations including vanishing theorems and moduli theory. More precisely it is important and useful that the singularities considered in these situation are Du Bois. For instance, Steenbrink showed that families over smooth curves whose fibers have DB singularities possess particularly nice properties; this maxim and its consequences have been further explored in [KK10, Section 7]. These applications imply that the question of whether DB singularities are invariant under small

Received November 3, 2011.
2010 *Mathematics Subject Classification.* 14B07, 14B05, 14F17, 14F18.
Key words and phrases. Du Bois singularities, DB singularities, deformation, log canonical singularities, non-lc ideal.
The first named author was supported in part by NSF Grants DMS #0856185 and #1301888, by a grant from the Simons Foundation #304043, and the Craig McKibben and Sarah Merner Endowed Professorship in Mathematics at the University of Washington.
The second named author was partially supported by the NSF grant DMS #1064485 and an NSF Postdoctoral Fellowship.

deformations, that is whether the property of having DB singularities is open in flat families, is very important. In this paper we settle this question in the affirmative.

As both rational singularities [Kov99] and log canonical singularities [KK10] are DB, it is interesting to note that rational singularities are invariant under small deformations by [Elk78], while log canonical singularities are not unless the total space has a $\mathbb{Q}$-Cartier canonical divisor compatible with the canonical divisors of the family members. In this latter case the statement follows from inversion of adjunction [Kaw07].

Our main result is the following:

Main Theorem [Theorem 4.1]. *Let X be a scheme of finite type over $\mathbb{C}$ and H a reduced Cartier divisor on X. If H has DB singularities, then X has DB singularities near H.*

The openness of the Du Bois locus in proper flat families follows immediately, see Corollary 4.2.

In [Ish86], Ishii proved this result for isolated Gorenstein singularities, and it follows for normal Gorenstein singularities from a combination of [Kov99] and [Kaw07]. The first named author claimed a proof of the same statement in general in [Kov00]. That proof unfortunately is incomplete and only works under an additional condition. The problem lies in the first paragraph of the proof, namely that one may not always reduce to the case when the non-Du Bois locus of X is contained in $H = X_s$. For additional discussion of this issue see [KS11, Section 12].

In this paper we correct that proof by showing a more general injectivity theorem, Theorem 3.3, which should be viewed as playing the same role for Du Bois singularities that Grauert-Riemenschneider vanishing plays for rational singularities, see Corollary 3.5. Using this injectivity, we can follow the strategy of [Kov00] and mimic Elkik's proof [Elk78] that rational singularities deform in families to obtain the main result.

As another corollary of this injectivity theorem, we also prove a restriction theorem for the so-called *maximal non-lc ideals* defined in [FST11], at least in the case of a Gorenstein ambient variety, see Theorem 7.1.

Acknowledgements:

The authors would like to thank the referee, Kazuma Shimomoto and Burt Totaro for pointing out typos in previous drafts of this paper. We would also like to thank Florin Ambro for pointing out some references to us.

§2. Preliminaries on DB singularities

Throughout this paper, all schemes are assumed to be separated and of finite type over $\mathbb{C}$, and all morphisms are defined over $\mathbb{C}$. A *variety* here means a reduced connected scheme.

We use $D^b_{\mathrm{coh}}(X)$ to denote the bounded derived category of $\mathscr{O}_X$-modules with coherent cohomology. Given an object $C^{\bullet} \in D^b_{\mathrm{coh}}(X)$, its ith cohomology is denoted by $h^i(C^{\bullet})$. For any scheme X of finite type over $\mathbb{C}$, we use $\omega_X^{\bullet}$ to denote the dualizing complex of X which is defined as $\epsilon^!\mathbb{C}$ where $\epsilon : X \longrightarrow \mathbb{C}$ is the structure map of X. We will repeatedly use Grothendieck duality in the following form: For any proper map of schemes $f : Y \longrightarrow X$, and any $C^{\bullet} \in D^b_{\mathrm{coh}}(Y)$, there exists a functorial quasi-isomorphism:

$$\mathcal{R}f_*\mathcal{R}\mathcal{H}om^{\bullet}_Y(C^{\bullet}, \omega_Y^{\bullet}) \simeq \mathcal{R}\mathcal{H}om^{\bullet}_X(\mathcal{R}f_*C^{\bullet}, \omega_X^{\bullet}).$$

For an introduction to derived categories and Grothendieck duality in the context used in this paper, see [Har66].

Recall that given a variety X, a *resolution of singularities* $\pi : \widetilde{X} \longrightarrow X$ is a proper birational[1] map from a smooth variety $\widetilde{X}$. Given a closed subscheme $Z \subseteq X$ with associated ideal sheaf $\mathscr{I}_Z$, we say that $\pi : \widetilde{X} \longrightarrow X$ is a *log resolution of* $Z \subseteq X$ if π is a resolution of singularities and if in addition $\pi^*\mathscr{I}_Z \simeq \mathscr{O}_{\widetilde{X}}(G)$ where G is a divisor, the exceptional set of π, $\mathrm{Exc}(\pi) \subseteq \widetilde{X}$, is also a divisor, and the divisor $\mathrm{Exc}(\pi) \cup \mathrm{supp}(G)$ has simple normal crossings. Note that resolutions of singularities, and log resolutions, exist by [Hir64].

We briefly recall some common objects used in the study of DB singularities. For a more extensive discussion of DB singularities, please see [KS11], [HK10, Section 3.I], or [PS08].

Lemma 2.1. *Given a variety X, one may associate to X an object $\underline{\Omega}^0_X \in D^b_{\mathrm{coh}}(X)$ defined as follows: let $\pi_{\bullet} : X_{\bullet} \longrightarrow X$ be a (cubic or simplicial) hyperresolution of X, see* [GNPP88, Car85, Del74], *then*

$$\underline{\Omega}^0_X := \mathcal{R}\pi_{\bullet\, *}\mathscr{O}_{X_{\bullet}}.$$

This object has the following properties:

(i) *$\underline{\Omega}^0_X$ is functorial with respect to morphisms of varieties, i.e., given a morphism of varieties $f : Y \longrightarrow X$, there is an induced morphism $\underline{\Omega}^0_X \longrightarrow \mathcal{R}f_*\underline{\Omega}^0_Y$.*

[1]birational = there exists a bijection of irreducible components with an induced isomorphism of fields of fractions.

(ii) *There is a natural morphism* $\mathcal{O}_X \longrightarrow \underline{\Omega}^0_X$ *compatible with (i) in the obvious way.*

(iii) *If in addition* X *is proper, then the composition*

$$H^i(X^{\mathrm{an}}, \mathbb{C}) \longrightarrow H^i(X, \mathcal{O}_X) \longrightarrow \mathbb{H}^i(X, \underline{\Omega}^0_X)$$

is surjective.

Proof. See [DB81] and [Ste81] for the original definitions and proofs and [KS11] for a survey on DB singularities. Property (iii) follows directly from the E_1-degeneration of the Deligne-Du Bois variant of the Hodge-to-De Rham spectral sequence. Q.E.D.

Definition 2.2. We say that X has *DB singularities* if the morphism $\mathcal{O}_X \longrightarrow \underline{\Omega}^0_X$ from (ii) above is a quasi-isomorphism.

We also recall the following fact about DB singularities.

Lemma 2.3 (*cf.* [Kol95, Proof of Theorem 12.8]). *If* X *has DB singularities and* H *is a general member of a base-point-free linear system* δ *on* X*, then* H *also has DB singularities.*

In this paper, we will repeatedly use the Grothendieck dual of $\underline{\Omega}^0_X$. To make that easier we introduce the notation

$$\underline{\omega}^{\bullet}_X := \mathcal{R}\mathcal{H}om^{\bullet}_X(\underline{\Omega}^0_X, \omega^{\bullet}_X).$$

We will also use the fact that there exists a morphism $\Phi : \underline{\omega}^{\bullet}_X \longrightarrow \omega^{\bullet}_X$, which is dual to the natural morphism $\mathcal{O}_X \longrightarrow \underline{\Omega}^0_X$.

Remark 2.4. Note that X has DB singularities if and only if Φ is a quasi-isomorphism since applying the Grothendieck duality functor again yields a morphism $\mathcal{O}_X \longrightarrow \underline{\Omega}^0_X$ which can be identified with the morphism from Lemma 2.1(ii) up to quasi-isomorphism.

§3. The key injectivity

In Theorem 3.3 below, we prove the following injectivity. For every integer $j \in \mathbb{Z}$,

$$\Phi^j : h^j(\underline{\omega}^{\bullet}_X) \hookrightarrow h^j(\omega^{\bullet}_X)$$

is injective. In the case that $x \in X$ is a closed point such that $X \setminus \{x\}$ is DB, the injectivity of this morphism played a key role in proving that rational, log canonical and F-injective singularities are DB, see [Kov99, KK10, Sch09].

Because of its potential usefulness it has been asked several times whether this injectivity holds. In particular, it was asked in [Sch09, Question 8.3] and [KS11, Question 5.2].

First we prove a lemma that is interesting on its own.

Lemma 3.1. *Let X be a variety and $\mathscr{L}$ a semi-ample line bundle. Choose $s \in \mathscr{L}^n$ a general global section for some $n \gg 0$ and take the nth-root of this section as in* [KM98, 2.50]*:*

$$\eta : Y = \mathbf{Spec} \bigoplus_{i=0}^{n-1} \mathscr{L}^{-i} \longrightarrow X.$$

Then $\eta_* = \mathcal{R}\eta_*$,

$$\eta_* \underline{\Omega}^0_Y \simeq \underline{\Omega}^0_X \otimes \eta_* \mathscr{O}_Y \simeq \bigoplus_{i=0}^{n-1} (\underline{\Omega}^0_X \otimes \mathscr{L}^{-i}),$$

and this direct sum decomposition is compatible with the decomposition $\eta_* \mathscr{O}_Y = \bigoplus_{i=0}^{n-1} \mathscr{L}^{-i}$.

Proof. We fix $\pi_{\bullet} : X_{\bullet} \longrightarrow X$ a finite cubic (or simplicial) hyperresolution of X as in [GNPP88]. On each component X_i of $X_{\bullet}$, $\mathscr{L}$ pulls back to a semi-ample line bundle and further s is still a general member of the base-point free linear subsystem of $\pi_i^* \mathscr{L}^n$. Thus we obtain a cyclic cover $\eta_i : Y_i \longrightarrow X_i$ for each i as well. Furthermore, each Y_i is smooth since it is ramified over a general element of a base-point free linear system. Obviously, these Y_i's glue to give a diagram of smooth $\mathbb{C}$-schemes $Y_{\bullet}$ with an augmentation morphism $\rho_{\bullet} : Y_{\bullet} \longrightarrow Y$. From the construction of a cubic hyperresolution, it is easy to see that $Y_{\bullet}$ is also a cubic hyperresolution.

We briefly sketch the idea of this last claim: if $X' \longrightarrow X$ is a resolution of singularities, then the induced $Y' \longrightarrow Y$ is also a resolution of singularities. Furthermore, if $X' \longrightarrow X$ is an isomorphism outside of $\Sigma \subseteq X$, then $Y' \longrightarrow Y$ is also an isomorphism outside of $\eta^{-1}(\Sigma)$, which is itself the induced cyclic cover of Σ.

Therefore,

$$\begin{aligned}
\mathcal{R}\eta_* \underline{\Omega}^0_Y &\simeq \mathcal{R}\eta_* \mathcal{R}\rho_{\bullet\,*} \mathscr{O}_{Y_{\bullet}} \simeq \mathcal{R}\pi_{\bullet\,*} \mathcal{R}\eta_{\bullet\,*} \mathscr{O}_{Y_{\bullet}} \\
&\simeq \mathcal{R}\pi_{\bullet\,*} \left(\oplus_{i=0}^{n-1} (\mathscr{O}_{X_{\bullet}} \otimes \pi_{\bullet}^* \mathscr{L}^{-i}) \right) \\
&\simeq \oplus_{i=0}^{n-1} \left((\mathcal{R}\pi_{\bullet\,*} \mathscr{O}_{X_{\bullet}}) \otimes \mathscr{L}^{-i} \right) \\
&\simeq \oplus_{i=0}^{n-1} (\underline{\Omega}^0_X \otimes \mathscr{L}^{-i}) \\
&\simeq \underline{\Omega}^0_X \otimes (\oplus_{i=0}^{n-1} \mathscr{L}^{-i}) \\
&\simeq \underline{\Omega}^0_X \otimes \eta_* \mathscr{O}_Y .
\end{aligned}$$

and the result follows, the compatibility statement following by construction.

Alternatively, if one wishes to avoid hyperresolutions one may proceed as follows. By restricting to an open set, we may assume that X embeds as a closed subscheme in a smooth scheme U such that $\mathscr{L}$ is the restriction of a globally generated line-bundle $\mathscr{M}$ on U. Further set $\pi : U' \longrightarrow U$ to be a log resolution of $X \subseteq U$ where we use $\overline{X}$ to denote the reduced divisor $\pi^{-1}(X)_{\mathrm{red}}$. Then $\mathcal{R}\pi_* \mathscr{O}_{\overline{X}} \simeq \underline{\Omega}^0_X$. Choosing a general section s of the globally generated line bundle $\mathscr{M}^n$, we obtain a diagram of cyclic covers:

$$\begin{array}{ccc} \overline{Y} & \hookrightarrow & W' \\ \downarrow & & \downarrow \\ Y & \hookrightarrow & W \end{array}$$

where Y, W, W' and $\overline{Y}$ are the induced cyclic covers of X, U, U' and $\overline{X}$ respectively. It is clear that W and W' are smooth and that $\overline{Y}$ is the reduced-preimage of Y and has simple normal crossings. Thus the result follows again since $\mathcal{R}\pi_* \mathscr{O}_{\overline{Y}} \simeq \underline{\Omega}^0_Y$ by [Sch07], also see [Esn90]. Q.E.D.

Before proving our main injectivity, we need one more result.

Proposition 3.2. *Let X be a proper variety over $\mathbb{C}$ and $\mathscr{L}$ a semi-ample line bundle on X. Then the natural map*

$$H^j(X, \mathscr{L}^{-i}) \longrightarrow \mathbb{H}^j(X, \underline{\Omega}^0_X \otimes \mathscr{L}^{-i})$$

is surjective for all $j, i \geq 0$.

Proof. Choose $n > i$ such that $\mathscr{L}^n$ is base-point-free and choose a general section $s \in \Gamma(X, \mathscr{L}^n)$. Consider the induced cyclic cover $\eta : Y \longrightarrow X$ and note that Y is also proper. Now, we have the following factorization

$$H^i(Y^{\mathrm{an}}, \mathbb{C}) \longrightarrow H^i(Y, \mathscr{O}_Y) \longrightarrow \mathbb{H}^i(Y, \underline{\Omega}^0_Y).$$

This composition is surjective by Lemma 2.1(iii). Thus $H^i(Y, \mathscr{O}_Y) \longrightarrow \mathbb{H}^i(Y, \underline{\Omega}^0_Y)$ is also surjective. Then the statement follows by Lemma 3.1. Q.E.D.

Now we are ready to prove the main result of the section.

Theorem 3.3. *Let X be a variety over $\mathbb{C}$. Then the natural map*

$$\Phi^j : h^j(\underline{\omega}^{\bullet}_X) \hookrightarrow h^j(\omega^{\bullet}_X)$$

is injective for every $j \in \mathbb{Z}$.

Proof. The statement is local and compatible with restriction to an open subset. Therefore we may assume that X is projective. Let $j \in \mathbb{Z}$ and $\mathscr{L}$ an ample line bundle on X. It follows from Proposition 3.2 that $H^{-j}(X, \mathscr{L}^{-i}) \longrightarrow \mathbb{H}^{-j}(X, \underline{\Omega}^0_X \otimes \mathscr{L}^{-i})$ is surjective. Next, apply $\mathrm{Hom}_{\mathbb{C}}(\underline{\quad}, \mathbb{C})$ and observe that then

$$\mathbb{H}^{-j}(X, \underline{\Omega}^0_X \otimes \mathscr{L}^{-i})^{\vee} \hookrightarrow H^{-j}(X, \mathscr{L}^{-i})^{\vee}$$

is injective. However,

$$\begin{aligned} H^{-j}(X, \mathscr{L}^{-i})^{\vee} &\simeq h^j(\mathcal{R}\Gamma(X, \mathcal{R}\mathcal{H}om_{\mathscr{O}_X}(\mathscr{L}^{-i}, \omega^{\bullet}_X))) \\ &\simeq \mathbb{H}^j(X, \omega^{\bullet}_X \otimes \mathscr{L}^i) \end{aligned}$$

by Grothendieck duality applied to the structure map $\epsilon : X \longrightarrow \mathbb{C}$. Likewise,

$$\mathbb{H}^{-j}(X, \underline{\Omega}^0_X \otimes \mathscr{L}^{-i})^{\vee} \simeq \mathbb{H}^j(X, \underline{\omega}^{\bullet}_X \otimes \mathscr{L}^i).$$

Thus we get that

$$\mathbb{H}^j(X, \underline{\omega}^{\bullet}_X \otimes \mathscr{L}^i) \hookrightarrow \mathbb{H}^j(X, \omega^{\bullet}_X \otimes \mathscr{L}^i)$$

is injective. Notice that

$$\mathbb{H}^j(X, \omega^{\bullet}_X \otimes \mathscr{L}^i) \simeq H^0(X, h^j(\omega^{\bullet}_X) \otimes \mathscr{L}^i)$$

for $i \gg 0$ by Serre-vanishing and the associated Grothendieck spectral sequence. Likewise,

$$\mathbb{H}^j(X, \omega^{\bullet}_X \otimes \mathscr{L}^i) \simeq H^0(X, h^j(\underline{\omega}^{\bullet}_X) \otimes \mathscr{L}^i)$$

for $i \gg 0$. Therefore,

$$H^0(X, h^j(\underline{\omega}^{\bullet}_X) \otimes \mathscr{L}^i) \hookrightarrow H^0(X, h^j(\omega^{\bullet}_X) \otimes \mathscr{L}^i) \tag{3.3.1}$$

is injective for $i \gg 0$. Observe that since $\mathscr{L}$ is ample, both $h^j(\underline{\omega}^{\bullet}_X) \otimes \mathscr{L}^i$ and $h^j(\omega^{\bullet}_X) \otimes \mathscr{L}^i$ are generated by global sections for $i \gg 0$. Therefore the injectivity of equation (3.3.1) implies, that

$$\Phi^j : h^j(\underline{\omega}^{\bullet}_X) \longrightarrow h^j(\omega^{\bullet}_X)$$

is also injective for every j. This completes the proof. Q.E.D.

We also have the following local-dual version of Theorem 3.3.

Corollary 3.4 (*cf.* [Kov99, Lemma 2.2]). *Let X be a variety and $P \in X$ is a point (not necessarily closed). Then the natural map*

$$H^i_P(X, \mathscr{O}_{X,P}) \twoheadrightarrow \mathbb{H}^i_P(X, \underline{\Omega}^0_X \otimes \mathscr{O}_{X,P})$$

is surjective for all $i \geq 0$.

Proof. We have the injection $h^i(\underline{\omega}^{\bullet}_X)_P \longrightarrow h^i(\omega^{\bullet}_X)_P$ for all i. After shifting (in case P is not a closed point), we have that $h^i(\underline{\omega}^{\bullet}_{\mathscr{O}_{X,P}}) \longrightarrow h^i(\omega^{\bullet}_{\mathscr{O}_{X,P}})$ also injects for all i. Let E be the injective hull of the residue field $\mathscr{O}_{X,P}/\mathfrak{m}_{X,P}$ and apply the (faithful and exact) functor $\mathrm{Hom}_{\mathscr{O}_{X,P}}(\_\_, E)$. Local duality in the form of [Har66, IV, Theorem 6.2] then yields the corollary. Q.E.D.

With respect for deciding whether X has DB singularities, the complex $\underline{\Omega}^0_X$ plays the same role as the complex $\mathscr{R}\pi_*\mathscr{O}_{\widetilde{X}}$ does for detecting rational singularities, here $\pi : \widetilde{X} \longrightarrow X$ is a resolution of singularities.

However, in many applications what makes $\mathscr{R}\pi_*\mathscr{O}_{\widetilde{X}}$ a useful object is the Grauert-Riemenschneider vanishing theorem [GR70] applied to its Grothendieck dual, $\mathscr{R}\pi_*\omega^{\bullet}_{\widetilde{X}} \simeq \mathscr{R}\mathscr{H}om^{\bullet}_{\mathscr{O}_X}(\mathscr{R}\pi_*\mathscr{O}_{\widetilde{X}}, \omega^{\bullet}_X)$ implying that it is a complex with non-zero cohomology in only one spot:

$$\mathscr{R}\pi_*\omega^{\bullet}_{\widetilde{X}} \simeq \pi_*\omega_{\widetilde{X}}[\dim X].$$

For X Cohen-Macaulay, Theorem 3.3 yields an analogous vanishing for DB singularities.

Corollary 3.5. *Let X be a Cohen-Macaulay variety of dimension d. Then*

$$\underline{\omega}^{\bullet}_X \simeq h^{-d}(\underline{\omega}^{\bullet}_X)[d].$$

If additionally X is normal and $\pi : \widetilde{X} \longrightarrow X$ is a log resolution of singularities with reduced exceptional divisor E, then

$$\underline{\omega}^{\bullet}_X \simeq \pi_*\omega_{\widetilde{X}}(E)[d]$$

Proof. Since X is Cohen-Macaulay and connected, it is equidimensional. The first statement is immediate since a submodule of the zero-module is zero and because $h^i(\omega^{\bullet}_X) = 0$ for $i \neq -d$. For the second statement, use the fact that $h^{-d}(\underline{\omega}^{\bullet}_X) \simeq \pi_*\omega_{\widetilde{X}}(E)$ by [KSS10, Theorem 3.8]. Q.E.D.

Remark 3.6. Notice that if X is DB, then $\underline{\omega}^{\bullet}_X \simeq \omega^{\bullet}_X$ and hence the statement is equivalent to X being Cohen-Macaulay.

A slight reinterpretation of the previous result also gives us the following corollary.

Corollary 3.7. *Let Y be a smooth n-dimensional variety and $X \subseteq Y$ a Cohen-Macaulay subvariety of pure dimension d. Let $\pi : \widetilde{Y} \longrightarrow Y$ be a log resolution of $X \subseteq Y$. Set $E \subseteq Y$ to be the reduced pre-image of X in Y (which is a divisor since π is a log resolution). Then*

$$R^i\pi_*\omega_{\widetilde{Y}}(E) = 0$$

for all $i \neq 0, n-d-1$.

Proof. Consider the long exact sequence

$$R^i\pi_*\omega_{\widetilde{Y}} \longrightarrow R^i\pi_*\omega_{\widetilde{Y}}(E) \longrightarrow R^i\pi_*\omega_E \longrightarrow R^{i+1}\pi_*\omega_{\widetilde{Y}}$$

and notice first that $R^i\pi_*\omega_{\widetilde{Y}} = 0$ for all $i \neq 0$ by [GR70]. Since $\omega_E[n-1] \simeq \omega_E^{\bullet}$ we have $R^{j+n-1}\pi_*\omega_E \simeq h^j(\mathcal{R}\pi_*\omega_E^{\bullet})$. However, $\mathcal{R}\pi_*\omega_E^{\bullet} \simeq \underline{\omega}^{\bullet}_X$ by [Sch07]. Therefore, since $h^j(\underline{\omega}^{\bullet}_X) = 0$ for $j \neq -d$ by Corollary 3.5, we see that $R^{j+n-1}\pi_*\omega_E = 0$ for $j \neq -d$. Thus $R^i\pi_*\omega_E = 0$ for $i \neq n-d-1$ and the result follows. Q.E.D.

Remark 3.8. The previous two corollaries do not hold if X is not Cohen-Macaulay. In fact they automatically fail for any non-Cohen-Macaulay variety with Du Bois singularities. For example, they fail for the affine cone over an Abelian variety of dimension > 1.

Theorem 3.3 also provides slightly simpler proofs of existing results.

Corollary 3.9 ([Kov99], *cf.* [Kol95, Section 12]). *If the morphism $\mathcal{O}_X \longrightarrow \underline{\Omega}^0_X$ has a left-inverse in $D^b_{\mathrm{coh}}(X)$, then X has DB singularities.*

Proof. The hypothesis implies that $\Phi^i : h^i(\underline{\omega}^{\bullet}_X) \longrightarrow h^i(\omega^{\bullet}_X)$ is surjective for every i. Thus Φ^i is an isomorphism by Theorem 3.3 and hence $\Phi : \underline{\omega}^{\bullet}_X \longrightarrow \omega^{\bullet}_X$ is a quasi-isomorphism and so X has DB singularities by Remark 2.4. Q.E.D.

§4. Deformation of DB singularities

We now prove the main result of the paper. In fact, simply using Corollary 3.4 fills in the gap in the first author's proof of this statement in [Kov00, Theorem 3.2]. For completeness, we provide a proof below. This proof (as well as the proof of [Kov00, Theorem 3.2]) was inspired by Elkik's proof of the fact that rational singularities deform [Elk78].

Theorem 4.1. *Let X be a scheme of finite type over $\mathbb{C}$ and H a reduced effective Cartier divisor (if X is not normal, by a Cartier divisor we mean a subscheme locally defined by a single non-zero-divisor at each stalk). If H has DB singularities, then X has DB singularities near H.*

Proof. Choose hyperresolutions $\pi_\bullet : X_\bullet \longrightarrow X$ and $\mu_\bullet : H_\bullet \longrightarrow H$ with a map $H_\bullet \longrightarrow X_\bullet$ factoring through the diagram of schemes $Z_\bullet := X_\bullet \times_X H$ as pictured below, *cf.* [GNPP88].

$$\begin{array}{ccccc} H_\bullet & \longrightarrow & Z_\bullet & \longrightarrow & X_\bullet \\ & {\scriptstyle \mu_\bullet}\searrow & \downarrow{\scriptstyle \varepsilon_\bullet} & & \downarrow{\scriptstyle \pi_\bullet} \\ & & H & \hookrightarrow & X \end{array}$$

Note that the components of $Z_\bullet$ need not be smooth or even reduced.

Choose a closed point $\mathfrak{q}$ of X contained within H. It is sufficient to prove that X is DB at $\mathfrak{q}$. Let R denote the stalk $\mathscr{O}_{X,\mathfrak{q}}$ and choose $f \in R$ to denote a defining equation of H in R. We also define $\underline{\Omega}^0_R := \underline{\Omega}^0_X \otimes R$ and $\underline{\omega}^\bullet_R := \mathscr{R}\mathrm{Hom}^\bullet_R(\underline{\Omega}^0_R, \omega^\bullet_R)$. Consider the following diagram whose rows are exact triangles in $D^b_{\mathrm{coh}}(X)$:

$$\begin{array}{ccccccc} R & \xrightarrow{\times f} & R & \longrightarrow & R/(f) & \xrightarrow{+1} & \\ \downarrow & & \downarrow & & {\scriptstyle\rho}\downarrow & & \\ \underline{\Omega}^0_R & \xrightarrow[\times f]{} & \underline{\Omega}^0_R & \longrightarrow & (\mathscr{R}\varepsilon_{\bullet *}\mathscr{O}_{Z_\bullet}) \otimes R & \xrightarrow{+1} & \\ & & & & {\scriptstyle\tau}\downarrow & & \\ & & & & \underline{\Omega}^0_H \otimes R & & \end{array}$$

where $\tau \circ \rho$ is a quasi-isomorphism by hypothesis. Next we apply the functor $\mathscr{R}\mathrm{Hom}^\bullet_R(\underline{\quad}, \omega^\bullet_R)$. Using the notation $\widetilde{\omega}^\bullet_{Z_\bullet} = \mathscr{R}\mathrm{Hom}^\bullet_R((\mathscr{R}\varepsilon_{\bullet *}\mathscr{O}_{Z_\bullet}) \otimes R, \omega^\bullet_R)$ and taking cohomology we obtain the following diagram of long exact sequences:

$$\begin{array}{ccccccccc} h^i(\omega^\bullet_R) & \xleftarrow{\times f} & h^i(\omega^\bullet_R) & \xleftarrow{\delta_i} & h^i(\omega^\bullet_{R/f}) & \xleftarrow{\alpha_i} & h^{i-1}(\omega^\bullet_R) & \xleftarrow{\times f} & h^{i-1}(\omega^\bullet_R) \\ \uparrow{\scriptstyle \Phi^i} & & \uparrow{\scriptstyle \Phi^i} & & \uparrow{\scriptstyle \gamma_i} & & \uparrow{\scriptstyle \Phi^{i-1}} & & \uparrow{\scriptstyle \Phi^{i-1}} \\ h^i(\underline{\omega}^\bullet_R) & \xleftarrow[\times f]{} & h^i(\underline{\omega}^\bullet_R) & \longleftarrow & h^i(\widetilde{\omega}^\bullet_{Z_\bullet}) & \xleftarrow[\beta_i]{} & h^{i-1}(\underline{\omega}^\bullet_R) & \xleftarrow[\times f]{} & h^{i-1}(\underline{\omega}^\bullet_R) \end{array}$$

where the vertical Φ maps are injective because of Theorem 3.3 and the morphism γ_i is surjective because $\tau \circ \rho$ is an isomorphism.

Fix $z \in h^{i-1}(\omega^\bullet_R)$. Pick $w \in h^i(\widetilde{\omega}^\bullet_{Z_\bullet})$ such that $\alpha_i(z) = \gamma_i(w)$. Since $\delta_i(\alpha_i(z)) = 0$ and Φ^i is injective, it follows that there exists a $u \in h^{i-1}(\underline{\omega}^\bullet_R)$ such that $\beta_i(u) = w$. Therefore, $\alpha_i(\Phi^{i-1}(u)) = \alpha_i(z)$ and so

$$z - \Phi^{i-1}(u) \in f \cdot h^{i-1}(\omega^\bullet_R). \tag{4.1.1}$$

Now, fix C_{i-1} to be the cokernel of Φ^{i-1} and set $\overline{z} \in C_{i-1}$ to be the image of z. Equation (4.1.1) then guarantees that $\overline{z} \in f \cdot C_{i-1}$.

But z was arbitrary and so the multiplication map $C_{i-1} \xrightarrow{\times f} C_{i-1}$ is surjective. But this contradicts Nakayama's lemma unless $C_{i-1} = 0$. Therefore $C_{i-1} = 0$ and Φ^{i-1} is also surjective. This holds for all i and so the natural morphism $\underline{\omega}^{\bullet}_X \longrightarrow \omega^{\bullet}_X$ is a quasi-isomorphism. Thus X has DB singularities by Remark 2.4. Q.E.D.

Corollary 4.2. *Let $f : X \longrightarrow S$ be a proper flat family of varieties over a smooth curve S and $s \in S$ a closed point. If the fiber X_s has DB singularities, then so do the other fibers near s.*

Proof. By Theorem 4.1, X has DB singularities near X_s. Let Σ denote the non-Du Bois locus of X. Since f is proper, $f(\Sigma)$ is a closed subset of S not containing $s \in S$. Thus by restricting S to an open set, we may assume that X has DB singularities. By Lemma 2.3, all fibers over nearby points of $s \in S$ have DB singularities. Q.E.D.

§5. DB pairs

In [Kov11], the first author defined a notion of Du Bois (or simply DB) pairs. Indeed, given a (possibly non-reduced) subscheme $Z \subseteq X$ one has an induced map in $D^b_{\mathrm{coh}}(X)$,

$$\underline{\Omega}^0_X \longrightarrow \underline{\Omega}^0_Z,$$

noting that by definition $\underline{\Omega}^0_Z = \underline{\Omega}^0_{Z_{\mathrm{red}}}$. Then $\underline{\Omega}^0_{X,Z}$ to be the object in the derived category making the following an exact triangle:

$$\underline{\Omega}^0_{X,Z} \longrightarrow \underline{\Omega}^0_X \longrightarrow \underline{\Omega}^0_Z \xrightarrow{+1} .$$

If $\mathscr{I}_Z$ is the ideal sheaf of Z, then it is easy to see that there is a natural map $\mathscr{I}_Z \longrightarrow \underline{\Omega}^0_{X,Z}$, [Kov11, Section 3.D].

Definition 5.1. [Kov11, Definition 3.13] The *Du Bois defect of* (X, Z), denoted $\underline{\Omega}^{\times}_{X,Z}$, is the mapping cone of the morphism $\mathscr{I}_Z \longrightarrow \underline{\Omega}^0_{X,Z}$, so that there is an exact triangle

$$\mathscr{I}_Z \longrightarrow \underline{\Omega}^0_{X,Z} \longrightarrow \underline{\Omega}^{\times}_{X,Z} \xrightarrow{+1}$$

We say that (X, Z) has *Du Bois singularities* if $\underline{\Omega}^{\times}_{X,Z}$ is quasi-isomorphic to zero. In other words, if $\mathscr{I}_Z \longrightarrow \underline{\Omega}^0_{X,Z}$ is a quasi-isomorphism.

We now mimic our approach before:

Lemma 5.2 (*cf.* Lemma 3.1). *Let X be a variety and $\mathscr{L}$ a semi-ample line bundle. Choose $s \in \mathscr{L}^n$ a general global section for some $n \gg 0$*

and take the nth-root of this section as in [KM98, 2.50]*:*

$$\eta : Y = \mathbf{Spec}\bigoplus_{i=0}^{n-1}\mathscr{L}^{-i} \longrightarrow X.$$

Set $W = \eta^{-1}(Z)$ *(with the induced scheme structure). Note that we have* $\eta|_W : W = \mathbf{Spec}\bigoplus_{i=0}^{n-1}\mathscr{L}^{-i}|_Z \longrightarrow Z$*. Then as before* $\eta_* = \mathcal{R}\eta_*$*,*

$$\eta_*\underline{\Omega}^0_{Y,W} \simeq \underline{\Omega}^0_{X,Z}\otimes\eta_*\mathscr{O}_Y \simeq \bigoplus_{i=0}^{n-1}(\underline{\Omega}^0_{X,Z}\otimes\mathscr{L}^{-i}),$$

and this direct sum decomposition is compatible with the decomposition $\eta_*\mathscr{O}_Y = \bigoplus_{i=0}^{n-1}\mathscr{L}^{-i}$.

Proof. This can be proven just as in Lemma 3.1 or alternately follows formally from Lemma 3.1 via the functoriality of the construction. Q.E.D.

Just as in Proposition 3.2, we also obtain that

$$H^j(X, \mathscr{I}_Z\otimes\mathscr{L}^{-i}) \longrightarrow \mathbb{H}^j(X, \underline{\Omega}^0_{X,Z}\otimes\mathscr{L}^{-i})$$

simply by using [Kov11, Theorem 4.1] in place of Lemma 2.1(iii).

If we set $\underline{\omega}^{\bullet}_{X,Z} = \mathcal{R}\mathcal{H}om^{\bullet}_{\mathscr{O}_X}(X, \omega^{\bullet}_X)$, then we easily obtain.

Theorem 5.3. *Let* X *be a variety over* $\mathbb{C}$*. Then the natural map*

$$\Phi^j : h^j(\underline{\omega}^{\bullet}_{X,Z}) \hookrightarrow h^j(\mathcal{R}\mathcal{H}om_{\mathscr{O}_X}(\mathscr{I}_Z, \omega^{\bullet}_X))$$

is injective for every $j \in \mathbb{Z}$*.*

Proof. The proof is the same as in Theorem 3.3 Q.E.D.

§6. Tranversality

Lemma 6.1. *Let* X *be a reduced scheme and* $\Sigma \subseteq X$ *a reduced subscheme with ideal sheaf* $\mathscr{I}_\Sigma$*. Further let* $H \subseteq X$ *be a Cartier divisor with ideal sheaf* $\mathscr{I}_H$ *such that* H *does not contain any irreducible components of either* X *or* Σ*. Then*

$$\mathscr{I}_H \cap \mathscr{I}_\Sigma = \mathscr{I}_H \cdot \mathscr{I}_\Sigma.$$

Proof. The statement is local, so we may assume that $X = \operatorname{Spec} A$. Let $I \subseteq A$ be the ideal of Σ, i.e., $\mathscr{I}_\Sigma = \widetilde{I}$. Since Σ is reduced, $I = \sqrt{I}$ and hence $I = \cap_{i=1}^r \mathfrak{p}_i$ with prime ideals $\mathfrak{p}_i \subset A$. Assume that this is an economic decomposition, i.e., none of the $\mathfrak{p}_i$ are redundant. Further let $f \in A$ a local equation for H, i.e., $\widetilde{\mathscr{I}_H = (f)}$. The assumption that H does not contain any irreducible components of either X or Σ imply that

(6.1.1) f is not contained in any minimal primes of A, and
(6.1.2) f is not contained in any of the $\mathfrak{p}_i$.

CLAIM 6.2. For any prime ideal $\mathfrak{p} \subseteq A$ such that $f \notin \mathfrak{p}$,

$$(f) \cap \mathfrak{p} = f\mathfrak{p}.$$

Proof. $(f) \cap \mathfrak{p} \supseteq f\mathfrak{p}$ trivially, so we only need to prove the opposite containment. Let $fg \in (f) \cap \mathfrak{p}$. Since $f \notin \mathfrak{p}$, it follows that $g \in \mathfrak{p}$, so $fg \in f\mathfrak{p}$ as desired. Q.E.D.

Applying this to the $\mathfrak{p}_i$ we obtain that

$$(f) \cap I = (f) \bigcap (\cap_{i=1}^{r} \mathfrak{p}_i) = \bigcap_{i=1}^{r} ((f) \cap \mathfrak{p}_i) = \bigcap_{i=1}^{r} f\mathfrak{p}_i \tag{6.2.1}$$

CLAIM 6.3. Assume that f is not contained in any minimal primes of A. Then for any set of prime ideals $\{\mathfrak{p}_i \subseteq A\}$,

$$\bigcap_{i=1}^{r} f\mathfrak{p}_i = f\left(\cap_{i=1}^{r} \mathfrak{p}_i\right). \tag{6.3.1}$$

Proof. Let $x \in \bigcap_{i=1}^{r} f\mathfrak{p}_i$ and let $g_i \in \mathfrak{p}_i$ such that $x = fg_i$ for all i. We claim that $g_i = g_j$ for any i, j. Indeed, $fg_i = x = fg_j$ so

$$f(g_i - g_j) = 0 \in \bigcap_{\substack{\mathfrak{p} \subseteq A \\ \text{is a minimal prime}}} \mathfrak{p}$$

By assumption $f \notin \mathfrak{p}$ for any of the $\mathfrak{p}$, so we must have $g_i - g_j \in \mathfrak{p}$ for all $\mathfrak{p}$. However, since X is reduced,

$$\bigcap_{\substack{\mathfrak{p} \subseteq A \\ \text{is a minimal prime}}} \mathfrak{p} = 0,$$

so it follows that $g_i = g_j =: g$. Finally this implies that $x = fg \in f\left(\cap_{i=1}^{r} \mathfrak{p}_i\right)$. Q.E.D.

Combining (6.2.1) and (6.3.1) implies that $(f) \cap I = f \cdot I$. Q.E.D.

§7. Application to restriction theorems for maximal non-LC ideals

In this section we assume the reader is familiar with log canonical singularities; see [KM98] for an introduction. Let X be a normal variety, Δ an effective $\mathbb{Q}$-divisor on X such that $K_X+\Delta$ is $\mathbb{Q}$-Cartier and $\pi : \widetilde{X} \longrightarrow X$ is a log resolution for (X,Δ). Write $K_{\widetilde{X}} - \pi^*(K_X+\Delta) = \sum a_i E_i$ and set $E^{=-1} = \sum_{a_i=-1} E_i$. The following ideal

$$\mathcal{J}_{\mathrm{NLC}}(X,\Delta) := \pi_*\mathscr{O}_{\widetilde{X}}(\lceil K_{\widetilde{X}} - \pi^*(K_X+\Delta) + E^{=-1}\rceil)$$

is defined to be the *non-log canonical ideal of* X.

This ideal was first defined by F. Ambro in [Amb03, Definition 4.1] where it was denoted by $\mathcal{I}_{X_{-\infty}}$. The study of this ideal as an object similar to the multiplier ideal, was recently initiated by O. Fujino in [Fuj10]. One of the main facts about this ideal is that the zero set of $\mathcal{J}_{\mathrm{NLC}}$ is exactly the locus where (X,Δ) does not have log canonical singularities. Fujino proved the following restriction theorem for $\mathcal{J}_{\mathrm{NLC}}(X,\Delta)$ (in fact, he proved a more general result):

Theorem. [Fuj10, Theorem 1.2] *If H is a normal Cartier divisor on a $\mathbb{Q}$-Gorenstein variety X, then $\mathcal{J}_{\mathrm{NLC}}(X,H)\otimes\mathscr{O}_H \simeq \mathcal{J}_{\mathrm{NLC}}(H,0)$.*

However, there are other natural ideals that define the non-lc locus. With notation as above, set $E=\sum E_i$ and set $E^{\mathbb{Z}} = \sum_{a_i\in\mathbb{Z}} E_i$. Then consider the ideal

$$\begin{aligned}\mathcal{J}'(X,\Delta) &:= \pi_*\mathscr{O}_{\widetilde{X}}(\lceil K_{\widetilde{X}} - \pi^*(K_X+\Delta) + E^{\mathbb{Z}}\rceil)\\ &= \pi_*\mathscr{O}_{\widetilde{X}}(\lceil K_{\widetilde{X}} - \pi^*(K_X+\Delta) + \varepsilon E\rceil)\end{aligned}$$

where we choose $1 \gg \varepsilon > 0$. This is the largest ideal which canonically defines the non-log canonical locus of (X,Δ) and as such is called the *maximal non-lc ideal.* In [FST11], the authors explored this ideal (and other non-lc-ideals). In particular, they obtained restriction theorems in special cases [FST11, Theorem 12.7, Theorem 13.13]. As an application of Theorem 3.3, we obtain the following restriction theorem for $\mathcal{J}'(X,H)$ in the case that X is Gorenstein.

Theorem 7.1. *If X is a normal d-dimensional Gorenstein variety and H is a normal Cartier divisor on X, then $\mathcal{J}'(X,H)|_H \simeq \mathcal{J}'(H,0)$.*

The proof strategy is the same as in [FST11, Section 13]

Proof. By working sufficiently locally, we may assume that $K_X \sim 0$ and $H = V(f) \sim 0$ for some $f \in \Gamma(X,\mathscr{O}_X)$. Shrinking X again if necessary, we embed $X \subseteq Y$ as a closed subscheme in a smooth scheme

Y. Let $\pi : \widetilde{Y} \longrightarrow Y$ be a log resolution of $H \subseteq Y$ which is simultaneously an embedded resolution of $X \subseteq Y$. Let $\overline{X} = \pi^{-1}(X)_{\mathrm{red}}$, $\widetilde{X}$ the strict transform of X, and $\overline{H} = \pi^{-1}(H)_{\mathrm{red}}$. We may assume that π is an isomorphism outside of $\operatorname{Sing} X \cup H$ and write $\overline{X} = \widetilde{X} \cup E \cup \overline{H}$ where $E = \pi^{-1}(\operatorname{Sing} X)_{\mathrm{red}}$. Finally, we may also assume that $E \cup \overline{H}$ is a reduced simple normal crossings divisor which intersects $\widetilde{X}$ with normal crossings so that $(E \cup \overline{H}) \cap \widetilde{X}$ is a reduced simple normal crossings divisor on $\widetilde{X}$. We have the following short exact sequence:

$$0 \longrightarrow \mathscr{O}_{\widetilde{X}}(-E \cup \overline{H}) \longrightarrow \mathscr{O}_{\overline{X}} \longrightarrow \mathscr{O}_{E \cup \overline{H}} \longrightarrow 0.$$

By pushing forward and using [Sch07], we obtain the exact triangle,

$$\mathcal{R}\pi_* \mathscr{O}_{\widetilde{X}}(-E \cup \overline{H}) \longrightarrow \underline{\Omega}^0_X \longrightarrow \underline{\Omega}^0_{H \cup \operatorname{Sing} X} \xrightarrow{+1} .$$

Applying $\mathcal{R}\mathcal{H}om^{\bullet}_{\mathscr{O}_X}(\_\_, \omega^{\bullet}_X)$ gives

$$\underline{\omega}^{\bullet}_{H \cup \operatorname{Sing} X} \longrightarrow \underline{\omega}^{\bullet}_X \longrightarrow \mathcal{R}\pi_* \mathscr{O}_{\widetilde{X}}(K_{\widetilde{X}} + E \cup \overline{H})[d] \xrightarrow{+1} ,$$

and by taking cohomology, we arrive at the exact sequence

$$\begin{aligned} (7.1.1) \quad 0 \longrightarrow h^{-d}(\underline{\omega}^{\bullet}_X) \longrightarrow \pi_* \mathscr{O}_{\widetilde{X}}(K_{\widetilde{X}} + E \cup \overline{H}) \longrightarrow \\ \longrightarrow h^{-d+1}(\underline{\omega}^{\bullet}_{H \cup \operatorname{Sing} X}) \longrightarrow h^{-d+1}(\underline{\omega}^{\bullet}_X) = 0. \end{aligned}$$

The vanishing on the right follows by Corollary 3.5 since X is Gorenstein and thus Cohen-Macaulay.

By [FST11, Lemma 13.11],

$$h^{-d+1}(\underline{\omega}^{\bullet}_{H \cup \operatorname{Sing} X}) \simeq h^{-d+1}(\underline{\omega}^{\bullet}_H).$$

Furthermore, by [KSS10, Theorem 3.8] we have that

$$\mathcal{J}'(X, 0) \cong h^{-d}(\underline{\omega}^{\bullet}_X) \otimes \mathscr{O}_X(-K_X)$$

and

$$\mathcal{J}'(H, 0) \cong h^{-d+1}(\underline{\omega}^{\bullet}_H) \otimes \mathscr{O}_X(-K_X - H).$$

Hence twisting (7.1.1) by $\mathscr{O}_X(-K_X - H)$ we obtain the following short exact sequence: *cf.* [FST11, Lemma 13.8] [KSS10, Lemma 4.14],

$$\begin{aligned} 0 \longrightarrow \mathcal{J}'(X, 0) \otimes \mathscr{O}_X(-H) \longrightarrow \\ \longrightarrow \pi_* \mathscr{O}_{\widetilde{X}}(K_{\widetilde{X}} - \pi^*(K_X + H) + E \cup \overline{H}) \longrightarrow \\ \longrightarrow \mathcal{J}'(H, 0) \longrightarrow 0. \end{aligned}$$

This completes the proof. Q.E.D.

References

[Amb03] F. Ambro: *Quasi-log varieties*, Tr. Mat. Inst. Steklova **240** (2003), no. Biratsion. Geom. Linein. Sist. Konechno Porozhdennye Algebry, 220–239. MR1993751 (2004f:14027)

[Car85] J. A. Carlson: *Polyhedral resolutions of algebraic varieties*, Trans. Amer. Math. Soc. **292** (1985), no. 2, 595–612. MR808740 (87i:14008)

[Del74] P. Deligne: *Théorie de Hodge. III*, Inst. Hautes Études Sci. Publ. Math. (1974), no. 44, 5–77. MR0498552 (58 #16653b)

[DB81] P. Du Bois: *Complexe de de Rham filtré d'une variété singulière*, Bull. Soc. Math. France **109** (1981), no. 1, 41–81. MR613848 (82j:14006)

[Elk78] R. Elkik: *Singularités rationnelles et déformations*, Invent. Math. **47** (1978), no. 2, 139–147. MR501926 (80c:14004)

[Esn90] H. Esnault: *Hodge type of subvarieties of* $\mathbf{P}^n$ *of small degrees*, Math. Ann. **288** (1990), no. 3, 549–551. 1079878 (91m:14075)

[Fuj10] O. Fujino: *Theory of non-lc ideal sheaves: basic properties*, Kyoto J. Math. **50** (2010), no. 2, 225–245. 2666656 (2011f:14031)

[FST11] O. Fujino, K. Schwede, and S. Takagi: *Supplements to non-lc ideal sheaves*, Higher Dimensional Algebraic Geometry, RIMS Kôkyûroku Bessatsu, B24, Res. Inst. Math. Sci. (RIMS), Kyoto, 2011, pp. 1–47.

[GR70] H. Grauert and O. Riemenschneider: *Verschwindungssätze für analytische Kohomologiegruppen auf komplexen Räumen*, Invent. Math. **11** (1970), 263–292. MR0302938 (46 #2081)

[GNPP88] Guillén, F., Navarro Aznar, V., Pascual Gainza, P., and Puerta, F.: *Hyperrésolutions cubiques et descente cohomologique*, Lecture Notes in Mathematics, vol. 1335, Springer-Verlag, Berlin, 1988, Papers from the Seminar on Hodge-Deligne Theory held in Barcelona, 1982. MR972983 (90a:14024)

[HK10] C. D. Hacon and S. J. Kovács: *Classification of higher dimensional algebraic varieties*, Oberwolfach Seminars, vol. 41, Birkhäuser Verlag, Basel, 2010. 2675555 (2011f:14025)

[Har66] R. Hartshorne: *Residues and duality*, Lecture notes of a seminar on the work of A. Grothendieck, given at Harvard 1963/64. With an appendix by P. Deligne. Lecture Notes in Mathematics, No. 20, Springer-Verlag, Berlin, 1966. MR0222093 (36 #5145)

[Hir64] H. Hironaka: *Resolution of singularities of an algebraic variety over a field of characteristic zero. I, II*, Ann. of Math. (2) 79 (1964), 109–203; ibid. (2) **79** (1964), 205–326. MR0199184 (33 #7333)

[Ish86] S. Ishii: *Small deformations of normal singularities*, Math. Ann. **275** (1986), no. 1, 139–148. MR849059 (87i:14003)

[Kaw07] M. Kawakita: *Inversion of adjunction on log canonicity*, Invent. Math. **167** (2007), no. 1, 129–133. MR2264806 (2008a:14025)

[Kol95] J. Kollár: *Shafarevich maps and automorphic forms*, M. B. Porter Lectures, Princeton University Press, Princeton, NJ, 1995. MR1341589 (96i:14016)

[KK10] J. Kollár and S. J. Kovács: *Log canonical singularities are Du Bois*, J. Amer. Math. Soc. **23** (2010), no. 3, 791–813. MR2629988

[KM98] J. Kollár and S. Mori: *Birational geometry of algebraic varieties*, Cambridge Tracts in Mathematics, vol. 134, Cambridge University Press, Cambridge, 1998, With the collaboration of C. H. Clemens and A. Corti, Translated from the 1998 Japanese original. MR1658959 (2000b:14018)

[Kov99] S. J. Kovács: *Rational, log canonical, Du Bois singularities: on the conjectures of Kollár and Steenbrink*, Compositio Math. **118** (1999), no. 2, 123–133. MR1713307 (2001g:14022)

[Kov00] S. J. Kovács: *Rational, log canonical, Du Bois singularities. II. Kodaira vanishing and small deformations*, Compositio Math. **121** (2000), no. 3, 297–304. MR1761628 (2001m:14028)

[Kov11] S. J. Kovács: *Du Bois pairs and vanishing theorems*, Kyoto J. Math. **51** (2011), no. 1, 47–69. 2784747

[KS11] S. J. Kovács and K. Schwede: *Hodge theory meets the minimal model program: a survey of log canonical and Du Bois singularities*, Topology of Stratified Spaces (G. Friedman, E. Hunsicker, A. Libgober, and L. Maxim, eds.), Math. Sci. Res. Inst. Publ., vol. 58, Cambridge Univ. Press, Cambridge, 2011, pp. 51–94.

[KSS10] S. J. Kovács, K. Schwede, and K. E. Smith: *The canonical sheaf of Du Bois singularities*, Adv. Math. **224** (2010), no. 4, 1618–1640. 2646306

[PS08] C. A. M. Peters and J. H. M. Steenbrink: *Mixed Hodge structures*, Ergebnisse der Mathematik und ihrer Grenzgebiete. 3. Folge. A Series of Modern Surveys in Mathematics [Results in Mathematics and Related Areas. 3rd Series. A Series of Modern Surveys in Mathematics], vol. 52, Springer-Verlag, Berlin, 2008. MR2393625

[Sch07] K. Schwede: *A simple characterization of Du Bois singularities*, Compos. Math. **143** (2007), no. 4, 813–828. MR2339829

[Sch09] K. Schwede: *F-injective singularities are Du Bois*, Amer. J. Math. **131** (2009), no. 2, 445–473. MR2503989

[Ste81] J. H. M. Steenbrink: *Cohomologically insignificant degenerations*, Compositio Math. **42** (1980/81), no. 3, 315–320. MR607373 (84g:14011)

SJK:
Department of Mathematics, University of Washington
Seattle, WA, 98195, USA
E-mail address: skovacs@uw.edu

KS:
Department of Mathematics, University of Utah
155 South 1400 East, Salt Lake City, UT 48112-0090, USA
E-mail address: schwede@math.utah.edu

Advanced Studies in Pure Mathematics 70, 2016
Minimal Models and Extremal Rays (Kyoto, 2011)
pp. 67–113

Singular spaces with trivial canonical class

Daniel Greb, Stefan Kebekus and Thomas Peternell

Dedicated to Professor Shigefumi Mori on the occasion of his 60th birthday

Abstract.

The classical Beauville-Bogomolov Decomposition Theorem asserts that any compact Kähler manifold with numerically trivial canonical bundle admits an étale cover that decomposes into a product of a torus, and irreducible, simply-connected Calabi-Yau– and holomorphic-symplectic manifolds. The decomposition of the simply-connected part corresponds to a decomposition of the tangent bundle into a direct sum whose summands are integrable and stable with respect to any polarisation.

Building on recent extension theorems for differential forms on singular spaces, we prove an analogous decomposition theorem for the tangent sheaf of projective varieties with canonical singularities and numerically trivial canonical class.

In view of recent progress in minimal model theory, this result can be seen as a first step towards a structure theory of manifolds with Kodaira dimension zero. Based on our main result, we argue that the natural building blocks for any structure theory are two classes of canonical varieties, which generalise the notions of irreducible Calabi-Yau– and irreducible holomorphic-symplectic manifolds, respectively.

Received December 12, 2011.
Revised July 29, 2012.
2010 *Mathematics Subject Classification.* 14J32, 14E30, 32J27.
Key words and phrases. Beauville-Bogomolov Decomposition Theorem, Varieties of Kodaira Dimension Zero, Calabi-Yau, holomorphic-symplectic Varieties, Minimal Model Theory.

All three authors were supported in part by the DFG-Forschergruppe 790 "Classification of Algebraic Surfaces and Compact Complex Manifolds". During the preparation of parts of this paper, the first named author enjoyed the hospitality of the Mathematics Department at Princeton University and the Department of Pure Mathematics & Mathematical Statistics at University of Cambridge. He gratefully acknowledges the support of the Baden–Württemberg–Stiftung through the "Eliteprogramm für Postdoktorandinnen und Postdoktoranden".

Contents

§1. Introduction

1.A. Introduction and main result

The minimal model program aims to reduce the birational study of projective manifolds with Kodaira dimension zero to the study of associated *minimal models*, that is, normal varieties X with terminal singularities whose canonical divisor is numerically trivial. The ideal case, where the minimal variety X is smooth, is described in the fundamental Decomposition Theorem of Beauville and Bogomolov.

Theorem 1.1 (Beauville-Bogomolov Decomposition, [Bea83] and references there). *Let X be a compact Kähler manifold whose canonical divisor is numerically trivial. Then there exists a finite étale cover, $X' \to X$ such that X' decomposes as a product*

$$X' = T \times \prod_{\nu} X_{\nu} \tag{1.1.1}$$

where T is a compact complex torus, and where the X_ν are irreducible and simply-connected Calabi-Yau– or holomorphic-symplectic manifolds.

Remark 1.2. The decomposition (1.1.1) induces a decomposition of the tangent bundle $T_{X'}$ into a direct sum whose summands have vanishing Chern class, and are integrable in the sense of Frobenius' theorem. Those summands that correspond to the X_ν are slope-stable with respect to any ample polarisation.

In view of recent progress in minimal model theory, an analogue of Theorem 1.1 for minimal models would clearly be a substantial step towards a complete structure theory for varieties of Kodaira dimension zero. However, since the only known proof of Theorem 1.1 heavily uses Kähler-Einstein metrics and the solution of the Calabi conjecture, a full generalisation of Beauville-Bogomolov Decomposition Theorem 1.1 to the singular setting is difficult.

Main result The main result of our paper is the following Decomposition Theorem for the tangent sheaf of minimal varieties with vanishing first Chern class. Presenting the tangent sheaf as a direct sum of integrable subsheaves which are stable with respect to any polarisation, Theorem 1.3 can be seen as an infinitesimal analogue of the Beauville-Bogomolov Decomposition 1.1 in the singular setting.

Theorem 1.3 (Decomposition of the tangent sheaf). *Let X be a normal projective variety with at worst canonical singularities, defined over the complex numbers. Assume that the canonical divisor of X is numerically trivial: $K_X \equiv 0$. Then there exists an Abelian variety A as well as a projective variety $\widetilde{X}$ with at worst canonical singularities, a finite cover $f : A \times \widetilde{X} \to X$, étale in codimension one, and a decomposition*

$$\mathscr{T}_{\widetilde{X}} \cong \bigoplus \mathscr{E}_i$$

such that the following holds.

(1.3.1) *The $\mathscr{E}_i$ are integrable saturated subsheaves of $\mathscr{T}_{\widetilde{X}}$, with trivial determinants.*

Further, if $g : \widehat{X} \to \widetilde{X}$ is any finite cover, étale in codimension one, then the following properties hold in addition.

(1.3.2) *The sheaves $(g^*\mathscr{E}_i)^{**}$ are slope-stable with respect to any ample polarisation on $\widehat{X}$.*

(1.3.3) *The irregularity of $\widehat{X}$ is zero, $h^1(\widehat{X}, \mathscr{O}_{\widehat{X}}) = 0$.*

The decomposition found in Theorem 1.3 satisfies an additional uniqueness property. For a precise statement, see Remark 7.5 on page 93. Taking g to be the identity, we see that the irregularity of $\widetilde{X}$ is zero, and that the summands $\mathscr{E}_i$ are stable with respect to any polarisation.

Other results In the course of the proof, we show the following two additional results, pertaining to stability of the tangent bundle, and to wedge products of differentials forms that are defined on the smooth locus of a minimal model. We feel that these results might be of independent interest.

Proposition 1.4 (Stability of $\mathscr{T}_X$ does not depend on polarisation, Proposition 5.7). *Let X be a normal projective variety having at worst canonical singularities. Assume that K_X is numerically equivalent to zero. If the tangent sheaf $\mathscr{T}_X$ is slope-stable with respect to one polarisation, then it is also stable with respect to any other polarisation.* Q.E.D.

Proposition 1.5 (Non-degeneracy of the wedge product, Proposition 6.1). *Let X be a normal n-dimensional projective variety X having at worst canonical singularities. Denote the smooth locus of X by X_{reg}. Suppose that the canonical divisor is trivial. If $0 \leq p \leq n$ is any number, then the natural pairing given by the wedge product on X_{reg},*

$$\bigwedge : H^0\big(X_{\mathrm{reg}},\, \Omega^p_{X_{\mathrm{reg}}}\big) \times H^0\big(X_{\mathrm{reg}},\, \Omega^{n-p}_{X_{\mathrm{reg}}}\big) \longrightarrow \underbrace{H^0\big(X_{\mathrm{reg}},\, \omega_{X_{\mathrm{reg}}}\big)}_{\cong H^0\left(X,\, \omega_X\right) \cong \mathbb{C}},$$

is non-degenerate. *Q.E.D.*

Singular analogues of Calabi-Yau and irreducible symplectic manifolds Based on the Decomposition Theorem 1.3, we will argue in Section 8 that the natural building blocks for any structure theory of projective manifolds with Kodaira dimension zero are canonical varieties with *strongly stable* tangent sheaf. Strong stability, introduced in Definition 7.2 on page 92, is a formalisation of condition (1.3.2) that appears in the Decomposition Theorem 1.3.

In dimension no more than five, we show that canonical varieties with strongly stable tangent sheaf fall into two classes, which naturally generalise the notions of irreducible Calabi-Yau– and irreducible holomorphic-symplectic manifolds, respectively. There is ample evidence to suggest that this dichotomy should hold in arbitrary dimension.

Outline of the paper The proof of Theorem 1.3 relies on recent extension results for differential forms on singular spaces, which we recall in Section 2 below. There are three additional preparatory sections, Sections 3–5, where we recall structure results for varieties with trivial canonical bundle, and discuss stability properties of the tangent sheaf on varieties with numerically trivial canonical divisor. Some of the material in these sections is new.

Using the extension result together with Hodge-theoretic arguments, we will show in Section 6 that the wedge-product induces perfect pairings of reflexive differential forms. This will later on be used to split the inclusion of certain subsheaves of the tangent sheaf. The results obtained there generalise ideas of Bogomolov [Bog74], but are new in the singular setting, to the best of our knowledge. With these preparations in place, Theorem 1.3 is then shown in Section 7.

Based on our main results, the concluding Section 8 discusses possible approaches towards a more complete structure theory of singular varieties with trivial canonical bundle, and proves first results in this direction.

Global Convention Throughout the paper we work over the complex number field. In the discussion of sheaves, the word "stable" will always mean "slope-stable with respect to a given polarisation". Ditto for semistability.

1.B. Acknowledgements

The authors would like to thank Tommaso de Fernex, Alex Küronya and Keiji Oguiso for a number of discussions, and for answering our questions. The authors thank the referee for reading the manuscript with great care. The first named author wants to thank János Kollár and Burt Totaro for their invitations and for useful comments concerning the topics discussed in this paper.

§2. Reflexive differentials on normal spaces

2.A. Differentials, reflexive tensor operations

Given a normal variety X, we denote the sheaf of Kähler differentials by Ω^1_X. The tangent sheaf will be denoted by $\mathscr{T}_X = (\Omega^1_X)^*$. Reflexive differentials, as defined below, will play an important role in the discussion.

Definition 2.1 (Reflexive differential forms, cf. [GKKP11, Sect. 2.E]). Let X be a normal variety, let X_{reg} be the smooth locus of X and $\imath : X_{\mathrm{reg}} \hookrightarrow X$ its open embedding into X. If $0 \le p \le \dim X$ is any number, we denote the reflexive hull of the pth exterior power of Ω^1_X by

$$\Omega^{[p]}_X := \left(\wedge^p \Omega^1_X\right)^{**} = \imath_* \Omega^p_{X_{\mathrm{reg}}}.$$

We refer to sections in $\Omega^{[p]}_X$ as *reflexive p-forms on X*.

Remark 2.2 (Reflexive differentials and dualising sheaf). In the setting of Definition 2.1, recall that the Grothendieck dualising sheaf ω_X is always reflexive. We obtain that

$$\Omega^{[\dim X]}_X = \omega_X = \mathscr{O}_X\big(K_X\big).$$

Notation 2.3 (Reflexive tensor operations). Let X be a normal variety and $\mathscr{A}$ a coherent sheaf of $\mathscr{O}_X$-modules, of rank r. For $n \in \mathbb{N}$, set $\mathscr{A}^{[n]} := (\mathscr{A}^{\otimes n})^{**}$. Further, set $\det \mathscr{A} = \left(\wedge^r \mathscr{A}\right)^{**}$. If $\pi : X' \to X$ is a morphism of normal varieties, set $\pi^{[*]}(\mathscr{A}) := \left(\pi^* \mathscr{A}\right)^{**}$.

2.B. Extension results for differential forms on singular spaces

One of the main ingredients for the proof of the Decomposition Theorem 1.3 is the following extension result for differential forms on singular spaces, recently shown in joint work of the authors and Sándor Kovács. In its simplest form, the extension theorem asserts the following.

Theorem 2.4 (Extension Theorem, [GKKP11, Thm. 1.5]). *Let X be a quasi-projective complex algebraic variety and D an effective $\mathbb{Q}$-divisor on X such that the pair (X, D) is Kawamata log terminal (klt). If $\pi : \widetilde{X} \to X$ is any resolution of singularities and $0 \leq p \leq \dim X$ any number, then the push-forward sheaf $\pi_* \Omega^p_{\widetilde{X}}$ is reflexive.* *Q.E.D.*

Using Definition 2.1, the Extension Theorem 2.4 can be reformulated, saying that $\pi_* \Omega^p_{\tilde{X}} = \Omega^{[p]}_X$ for all $p \leq \dim X$. Equivalently, if $E \subset \widetilde{X}$ denotes the π-exceptional set, then the Extension Theorem asserts that for any open set $U \subset X$, any p-form on $\pi^{-1}(U) \setminus E$ extends across E, to give a p-form on $\pi^{-1}(U)$. In other words, it asserts that the natural restriction map

$$\Omega^p_{\widetilde{X}}\big(\pi^{-1}(U)\big) \to \Omega^p_{\widetilde{X}}\big(\pi^{-1}(U) \setminus E\big)$$

is surjective. We refer to the original papers [GKKP11, Sect. 1] and [GKK10] or to the survey [Keb13] for an in-depth discussion.

§3. Irregularity and Albanese map of canonical varieties

The Albanese map is one important tool in the study of varieties with trivial canonical divisor. The following invariant is relevant in its investigation.

Definition 3.1 (Augmented irregularity). Let X be a normal projective variety. We denote the irregularity of X by $q(X) := h^1\big(X, \mathcal{O}_X\big)$ and define the *augmented irregularity* $\widetilde{q}(X) \in \mathbb{N} \cup \{\infty\}$ as

$$\widetilde{q}(X) := \max\big\{q(\widetilde{X}) \mid \widetilde{X} \to X \text{ a finite cover, étale in codimension one}\big\}.$$

Remark 3.2 (Irregularity and the Albanese map). If X is a projective variety with canonical singularities, recall from [Kaw85, Sect. 8] that the Albanese map $\alpha_X : X \to \mathrm{Alb}(X)$ is well defined, and that $\dim \mathrm{Alb}(X) = q(X)$. Better still, if $\pi : \widetilde{X} \to X$ is any resolution of singularities, then the Albanese map $\alpha_{\widetilde{X}}$ of $\widetilde{X}$ agrees with α_X. In other words, there exists a commutative diagram as follows,

$$\begin{array}{ccc} \widetilde{X} & \xrightarrow[\text{Albanese map}]{\alpha_{\widetilde{X}}} & \mathrm{Alb}(\widetilde{X}) \\ {\scriptstyle \pi,\ \text{desing.}}\downarrow & & \| \\ X & \xrightarrow[\text{Albanese map}]{\alpha_X} & \mathrm{Alb}(X). \end{array}$$

The following result of Kawamata describes the Albanese map of varieties whose canonical divisor is numerically trivial. As we will see in Corollary 3.6, this often reduces the study of varieties with trivial canonical class to those with $\widetilde{q}(X) = 0$.

Proposition 3.3 (Fibre space structure of the Albanese map, [Kaw85, Prop. 8.3]). *Let X be a normal n-dimensional projective variety X with at worst canonical singularities. Assume that K_X is numerically trivial. Then K_X is torsion, the Albanese map $\alpha : X \to \mathrm{Alb}(X)$ is surjective and has the structure of an étale-trivial fiber bundle.*

In other words, there exists a positive number $m \in \mathbb{N}^+$ such that $\mathscr{O}_X(m \cdot K_X) \cong \mathscr{O}_X$. Furthermore, there exists a finite étale cover $B \to \mathrm{Alb}(X)$ from an Abelian variety B to $\mathrm{Alb}(X)$ such that the fiber product over $\mathrm{Alb}(X)$ decomposes as a product

$$X \times_{\mathrm{Alb}(X)} B \cong F \times B,$$

where F is a normal projective variety. *Q.E.D.*

Remark 3.4. If X is a projective variety with canonical singularities and numerically trivial canonical class, Proposition 3.3 implies that $q(X) = \dim \mathrm{Alb}(X) \leq \dim X$. If $\widetilde{X} \to X$ is any finite cover, étale in codimension one, then $\widetilde{X}$ will likewise have canonical singularities [KM98, Prop. 5.20] and numerically trivial canonical class. In summary, we see that $\widetilde{q}(X) \leq \dim X$. The augmented irregularity of canonical varieties with numerically trivial canonical class is therefore finite.

Remark 3.5. In the setting of Proposition 3.3, the canonical map

$$F \times B \cong X \times_{\mathrm{Alb}(X)} B \to X$$

is étale. The variety $F \times B$ is thus canonical by [KM98, Prop. 5.20]. Since B is smooth, this automatically implies that F is canonical. If the canonical divisor of X is trivial, then F will likewise have a trivial canonical divisor.

A variant of the following corollary has appeared as [Pet94, Thm. 4.2].

Corollary 3.6 (Structure of varieties with numerically trivial canonical class, cf. [Kaw85, Cor. 8.4]). *Let X be a normal n-dimensional projective variety with at worst canonical singularities. Assume that K_X is numerically trivial. Then there exist projective varieties A, Z and a morphism $\nu : A \times Z \to X$ such that the following holds.*

(3.6.1) *The variety A is Abelian.*

(3.6.2) *The variety Z is normal and has at worst canonical singularities.*

(3.6.3) *The canonical class of Z is trivial, $\omega_Z \cong \mathscr{O}_Z$.*

(3.6.4) *The augmented irregularity of Z is zero, $\widetilde{q}(Z) = 0$.*

(3.6.5) *The morphism ν is finite, surjective and étale in codimension one.*

Proof. We construct a sequence of finite surjective morphisms,

$$F \times B \xrightarrow[\text{étale}]{\gamma} X^{(2)} \xrightarrow[\text{étale in codim. one}]{\beta} X^{(1)} \xrightarrow[\text{index-one cover}]{\alpha} X,$$

as follows. Recall from Proposition 3.3 that the canonical divisor of X is torsion, and let $\alpha : X^{(1)} \to X$ be the associated index-one cover, cf. [KM98, Sect. 5.2] or [Rei87, Sect. 3.5]. We have seen in Remark 3.4 that the variety $X^{(1)}$ has a trivial canonical divisor and at worst canonical singularities. Remark 3.4 also shows that $\widetilde{q}(X^{(1)})$ is finite. This implies that there exists a finite morphism $\beta : X^{(2)} \to X^{(1)}$, étale in codimension one, such that

$$\widetilde{q}\left(X^{(1)}\right) = q\left(X^{(2)}\right) = \widetilde{q}\left(X^{(2)}\right).$$

Again, $X^{(2)}$ has trivial canonical divisor and canonical singularities. Next, let $\gamma : F \times B \to X^{(2)}$ be the étale morphism obtained by applying Proposition 3.3 and Remark 3.5 to the variety $X^{(2)}$. The variety B then satisfies the following,

$$\dim B = \dim \operatorname{Alb} X^{(2)} = q\left(X^{(2)}\right) = \widetilde{q}\left(X^{(2)}\right). \tag{3.6.6}$$

Remark 3.5 also asserts that F has a trivial canonical divisor and canonical singularities.

To finish the proof, it suffices to show that $\widetilde{q}(F) = 0$. If not, we could apply Proposition 3.3 and Remark 3.5 to F, obtaining an étale map $\delta : (F' \times B') \times B \to F \times B$, where B' is Abelian, and of positive dimension. The composed morphism $\gamma \circ \delta : F' \times (B' \times B) \to X^{(2)}$ is again étale, showing that

$$\widetilde{q}\left(X^{(2)}\right) \geq q\big(F' \times (B' \times B)\big) \geq \dim B + \dim B' > \dim B,$$

thus contradicting Equation (3.6.6) above. This shows that $\widetilde{q}(F) = 0$ and finishes the proof of Corollary 3.6. Q.E.D.

§4. A criterion for numerical triviality

The proof of the Decomposition Theorem 1.3 of rests on an analysis of the tangent sheaf of Kawamata log terminal spaces and of its destabilising subsheaves. We will show that the determinant of any destabilising subsheaf is trivial, at least after passing to a suitable cover. This part of the proof is based on a criterion for numerical triviality, formulated in Proposition 4.2. The criterion generalises the following result which goes back to Kleiman.

Lemma 4.1 (Kleiman's criterion for numerical triviality). *Let Z be an irreducible, normal projective variety of dimension $n \geq 2$ and D a $\mathbb{Q}$-Cartier divisor on Z. If $D \cdot H_1 \cdots H_{n-1} = 0$ for all $(n-1)$-tuples of ample divisors $H_1, \ldots, H_{n-1}$ on Z, then D is numerically trivial.*

Proof. Passing to a sufficiently high multiple of D, we can assume without loss of generality that D is Cartier, and thus linearly equivalent to the difference of two ample Cartier divisors, $D \sim H_{1,1} - H_{1,2}$. Let $H_2, \ldots, H_{n-1}$ be arbitrary ample divisors. Recall from [Kle66, Prop. 3 on page 305] that to prove numerical triviality of D, it suffices to show that the following two equalities

$$D \cdot H_1 \cdot H_2 \cdots H_{n-1} = 0 \qquad \text{and} \tag{4.1.1}$$

$$D \cdot D \cdot H_2 \cdots H_{n-1} = 0 \tag{4.1.2}$$

hold. Equation (4.1.1) holds by assumption. For Equation (4.1.2), observe that

$$D \cdot D \cdot H_2 \cdots H_{n-1} = D \cdot H_{1,1} \cdot H_2 \cdots H_{n-1} - D \cdot H_{1,2} \cdot H_2 \cdots H_{n-1},$$

where both summands are zero, again by assumption. Q.E.D.

Proposition 4.2, the main result of this section, is a variant of this criterion, adapted to the discussion of $\mathbb{Q}$-factorialisations, where the ample divisors H_i are replaced by big and nef divisors which are obtained as the pull-back of ample divisors via the $\mathbb{Q}$-factorialisation map.

Proposition 4.2 (Criterion for numerical triviality on $\mathbb{Q}$-factorialisations). *Let $\phi : Z' \to Z$ be a small birational morphism of irreducible, normal projective varieties of dimension $n \geq 2$.*

Let D' be a pseudoeffective $\mathbb{Q}$-Cartier $\mathbb{Q}$-divisor on Z' and assume that there are ample Cartier divisors $H_1, \ldots, H_{n-1}$ on Z such that

$$D' \cdot \phi^*(H_1) \cdots \phi^*(H_{n-1}) = 0. \tag{4.2.1}$$

If Z' is $\mathbb{Q}$-factorial, then D' is numerically trivial.

Remark 4.3 (Small morphisms). In Proposition 4.2 and elsewhere in this paper, we call a birational morphism ψ of normal, irreducible projective varieties *small* if its exceptional set has codimension at least two.

Proof of Proposition 4.2. Let $B' \subsetneq Z'$ be the ϕ-exceptional set, and $B := \phi(B') \subsetneq Z$ its image. It suffices to prove Proposition 4.2 for a multiple of D'. We will therefore assume without loss of generality that D' is an integral Cartier divisor. To show that D' is numerically trivial, we aim to apply Kleiman's criterion for numerical triviality, Lemma 4.1. To this end, let $A_1, \ldots, A_{n-1}$ be arbitrary ample Cartier divisors on Z'. Choose numbers $a_1, \ldots, a_{n-1} \in \mathbb{N}^+$ such that the $a_i A_i$ are very ample, choose general elements $\Theta_i \in |a_i A_i|$ and consider the complete intersection curve $\Gamma' := \Theta_1 \cap \cdots \cap \Theta_{n-1} \subsetneq Z'$. By general choice, the curve Γ' is smooth and will not intersect with the small set B'. Lemma 4.1 asserts that in order to establish numerical triviality of D' it suffices to show that

$$\Gamma' \cdot D' = a_1 \cdots a_{n-1} \cdot A_1 \cdots A_{n-1} \cdot D' = 0. \tag{4.3.1}$$

To establish (4.3.1), consider the image $\Gamma := \phi(\Gamma')$, which is a smooth curve contained in $Z \setminus B$. The curve Γ is not necessarily a complete intersection curve for $H_1, \ldots, H_{n-1}$, but can be completed to become a complete intersection curve, as follows. Choosing sufficiently large numbers $m_1, \ldots, m_{n-1} \in \mathbb{N}^+$, we can assume that the linear sub-systems

$$V_i := \{\Delta \in |m_i H_i| \,:\, \Gamma \subset \operatorname{supp} \Delta\}$$

are positive-dimensional, basepoint-free outside of Γ, and separate points outside of Γ. We can also assume that the sheaves $\mathscr{I}_\Gamma \otimes \mathscr{O}_Z(m_i H_i)$ are spanned. Choose general elements $\Delta_i \in V_i$ and consider the complete intersection curve

$$\Gamma_{\text{complete}} := \Delta_1 \cap \cdots \cap \Delta_{n-1} \subset Z.$$

The curve Γ_{complete} is reduced, avoids B and clearly contains Γ. We can thus write

$$\Gamma_{\text{complete}} = \Gamma \cup \Gamma_{\text{rest}} \quad \text{and} \quad \phi^{-1}(\Gamma_{\text{complete}}) = \Gamma' \cup \phi^{-1}(\Gamma_{\text{rest}}),$$

where Γ_{rest} is an irreducible movable curve on Z. To end the argument, observe that

$$\begin{aligned} 0 &= m_1 \cdots m_{n-1} \cdot D' \cdot \phi^*(H_1) \cdots \phi^*(H_{n-1}) && \text{by Assumption (4.2.1)} \\ &= D' \cdot \phi^*(\Gamma_{\text{complete}}) = \underbrace{D' \cdot \Gamma'}_{\geq 0} + \underbrace{D' \cdot \phi^*(\Gamma_{\text{rest}})}_{\geq 0}. \end{aligned}$$

Since D' is pseudoeffective, and since both Γ' and $\phi^{-1}(\Gamma_{\text{rest}})$ are movable, it follows that both summands are non-negative, hence zero. This shows Equation (4.3.1) and finishes the proof of Proposition 4.2. Q.E.D.

§5. The tangent sheaf of varieties with trivial canonical class

5.A. Semistability of the tangent sheaf

To prepare for the proof of the Decomposition Theorem 1.3, we study stability notions of the tangent sheaf. For canonical varieties with numerically trivial canonical divisor, we show that the tangent sheaf $\mathscr{T}_X$ is semistable with respect to any polarisation, and that $\mathscr{T}_X$ is stable with respect to one polarisation if and only if it is stable with respect to any other.

The following result of Miyaoka is crucial. We will refer to Theorem 5.1 at several places throughout the present paper.

Theorem 5.1 (Generic semipositivity, [Miy87a, Miy87b]). *Let X be a normal projective variety of dimension $n > 1$. Assume that X is* not *uniruled. Let $H_1, \ldots, H_{n-1}$ be ample line bundles on X. Then there exists a number $M \in \mathbb{N}^+$ such that for all $m_1, \ldots, m_{n-1} > M$ the following holds.*

(5.1.1) *The linear systems $|m_j H_j|$ are basepoint-free.*

(5.1.2) *If $C = D_1 \cap \cdots \cap D_{n-1} \subsetneq X$ is a curve cut out by general elements $D_j \in |m_j H_j|$, then C is smooth, X is smooth along C, and $\Omega_X^{[1]}|_C$ is a nef vector bundle on C.* Q.E.D.

Notation 5.2. The conclusion of Theorem 5.1 is often rephrased by saying that $\Omega_X^{[1]}$ is *generically nef with respect to* $H_1, \ldots, H_{n-1}$.

Theorem 5.3 (Mehta-Ramanathan theorem, cf. [MR82, Rem. 6.2] and [Fle84]). *In the setup of Theorem 5.1, if one chooses the number M large enough, then $\mathscr{T}_X$ is semistable with respect to the polarisation $(H_1, \ldots, H_{n-1})$ if and only if its restriction $\mathscr{T}_X|_C$ is semistable as a vector bundle on the curve C.* Q.E.D.

The well-known semistability of the tangent bundle is an immediate consequence of Miyaoka's generic semipositivity result.

Proposition 5.4 (Semistability of the tangent sheaf). *Let X be a normal projective variety having at worst canonical singularities. If K_X is numerically trivial, then $\mathscr{T}_X$ is semistable with respect to any polarisation.*

Proof. Let $(H_1, \ldots, H_{n-1})$ be arbitrary ample Cartier divisors on X. Choose numbers $M, m_1, \ldots, m_{n-1}$ and construct a general complete intersection curve $C \subsetneq X$ as in Theorems 5.1 and 5.3.

It follows from numerical triviality of K_X and from the assumption on the singularities that X is not uniruled. Theorem 5.1 therefore asserts that the restriction $\Omega_X^{[1]}|_C$ is nef. Since additionally $K_X \cdot C = 0$ by assumption, it follows that the bundle $\mathscr{T}_X|_C$ is semistable. The Mehta-Ramanathan Theorem 5.3 then shows that $\mathscr{T}_X$ is semistable with respect to $(H_1, \ldots, H_{n-1})$. Q.E.D.

Remark 5.5. Although semistable, the tangent sheaf of a variety with trivial canonical bundle might not be stable. To give an easy example, the tangent sheaf of the product of two such varieties is not stable.

5.B. Pseudoeffectivity of quotients of $\Omega_X^{[p]}$

The proof of our main result uses the following criterion for the pseudoeffectivity of Weil divisors on $\mathbb{Q}$-factorial spaces. In case where X is smooth, this has been shown by Campana-Peternell.

Proposition 5.6 (Pseudoeffectivity of quotients of $\Omega_X^{[p]}$, cf. [CP11, Thm. 0.1]). *Let X be a normal, $\mathbb{Q}$-factorial projective variety and let D be a Weil divisor on X. Assume that there exists a number $0 \leq p \leq \dim X$ and a non-trivial sheaf morphism $\psi : \Omega_X^{[p]} \to \mathscr{O}_X(D)$. If X is not uniruled, then D is pseudoeffective.*

Proof. Assume that X is not uniruled, and that a non-trivial sheaf morphism $\psi : \Omega_X^{[p]} \to \mathscr{O}_X(D)$ is given. The existence of a resolution of singularities combined with a classical result of Rossi [Ros68, Thm. 3.5] shows that there exists a strong log resolution of singularities $\pi : \widetilde{X} \to X$ with the additional property that $\pi^{[*]}\mathscr{O}_X(D)$ is locally free. The π-exceptional set $E \subsetneq \widetilde{X}$ is then of pure codimension one, and has simple normal crossing support. Finally, write $\pi^{[*]}\mathscr{O}_X(D) = \mathscr{O}_{\widetilde{X}}(\widetilde{D})$, where $\widetilde{D}$ is a divisor on $\widetilde{X}$ that agrees with the strict transform $\pi_*^{-1}(D)$ outside of the π-exceptional set E.

Since the two sheaves $\pi^{[*]}\Omega_X^{[p]}$ and $\Omega_{\widetilde{X}}^p$ are isomorphic outside of E, there exists a number $m \in \mathbb{N}^+$ and a sheaf morphism

$$\widetilde{\psi} : \Omega_{\widetilde{X}}^p \to \mathscr{O}_{\widetilde{X}}(\widetilde{D} + mE)$$

which agrees on $\widetilde{X} \setminus E$ with the pull-back of ψ. If $\mathscr{L} := (\mathrm{Image}\,\widetilde{\psi})^{**}$ denotes the reflexive hull of the image sheaf, then $\mathscr{L}$ is invertible by [OSS80, Lem. 1.1.15]. Recalling that $\Omega_{\widetilde{X}}^p$ is a quotient of $(\Omega_{\widetilde{X}}^1)^{\otimes p}$, it follows from [CP11, Thm. 0.1] that the line bundle $\mathscr{L}$ is in fact pseudoeffective. It follows that $\widetilde{D} + mE$ is pseudoeffective as well, since it contains the pseudoeffective subsheaf $\mathscr{L}$.

Since X is assumed to be $\mathbb{Q}$-factorial, it is clear that the cycle-theoretic push-forward of any pseudoeffective divisor on $\widetilde{X}$ is $\mathbb{Q}$-Cartier and again pseudoeffective. Using that $\pi_*\widetilde{D} = D$, this shows our claim. Q.E.D.

5.C. Stability of the tangent sheaf

While Theorem 5.1 and Proposition 5.4 are fairly standard today, the following result, which shows that stability of the tangent bundle is independent of the chosen polarisation, is new to the best of our knowledge. We feel that it might be of independent interest.

Proposition 5.7 (Stability of $\mathscr{T}_X$ does not depend on polarisation)**.** *Let X be a normal projective variety having at worst canonical singularities. Assume that K_X is numerically equivalent to zero. Let $h_1 := (H_1^{(1)}, \ldots, H_{n-1}^{(1)})$ and $h_2 := (H_1^{(2)}, \ldots, H_{n-1}^{(2)})$ be two sets of ample polarisations. If $\mathscr{T}_X$ is h_1-stable, then it is also h_2-stable.*

Proof. Suppose that $\mathscr{T}_X$ is *not* h_2-stable. Even though not stable, recall from Proposition 5.4 that $\mathscr{T}_X$ is semistable with respect to h_2. Choose a Jordan-Hölder filtration of $\mathscr{T}_X$ with respect to h_2 and let $0 \subsetneq \mathscr{S} \subsetneq \mathscr{T}_X$ be its first term, cf. [HL97, Sect. 1.5]. The sheaf $\mathscr{S}$ is then h_2-stable, and saturated as a subsheaf of $\mathscr{T}_X$. Semistability of $\mathscr{T}_X$ implies that $\mathscr{S}$ has h_2-slope zero. In other words, $\mu_{h_2}(\mathscr{S}) = 0$.

Next, let $\phi : X' \to X$ be a $\mathbb{Q}$-factorialisation, that is, a small birational morphism where X' is $\mathbb{Q}$-factorial and has at worst canonical singularities. The existence of ϕ is established in [BCHM10, Lem. 10.2]. Consider the reflexive pull-back sheaf $\mathscr{S}' := \phi^{[*]}\mathscr{S}$. Since $\mathscr{S}'$ injects into $\mathscr{T}_{X'}$ outside of the small π-exceptional set, and since both sheaves are reflexive, we obtain an injection $\mathscr{S}' \hookrightarrow \mathscr{T}_{X'} = \phi^{[*]}\mathscr{T}_X$.

We claim that $\det \mathscr{S}'$ is numerically trivial on X'. To this end, recall from Proposition 5.6 that $\left(\det \mathscr{S}'\right)^*$ is pseudoeffective. On the other hand, since $\mathscr{S}'$ and $\phi^*(\mathscr{S})$ agree outside of the π-exceptional set, we obtain

$$\left(\det \mathscr{S}'\right)^* \cdot \phi^*\left(H_1^{(2)}\right) \cdots \phi^*\left(H_{n-1}^{(2)}\right) = -\mu_{h_2}(\mathscr{S}) = 0.$$

Together with Proposition 4.2, these two observations show that $(\det \mathscr{S}')^*$ and $\det \mathscr{S}'$ are numerically trivial, as claimed.

Using numerical triviality of $\det \mathscr{S}'$, the same line of reasoning now gives

$$\mu_{h_1}(\mathscr{S}) = \left(\det \mathscr{S}'\right) \cdot \phi^*\left(H_1^{(1)}\right) \cdots \phi^*\left(H_{n-1}^{(1)}\right) = 0,$$

showing that $\mathscr{T}_X$ is *not* h_1-stable. This contradiction concludes the proof. Q.E.D.

§6. Differential forms on varieties with trivial canonical class

6.A. Non-degeneracy of the wedge product

Differential forms on smooth varieties with trivial canonical class were studied by Bogomolov [Bog74]. In this section we apply the Extension Theorem 2.4 to study reflexive differential forms on singular varieties with trivial canonical classes, following an approach discussed in [Pet94]. The results obtained in this section will play an important role in the proof of the Decomposition Theorem 1.3, which is given in the subsequent Section 7.

Proposition 6.1 (Non-degeneracy of the wedge product)**.** *Let X be a normal n-dimensional projective variety X having at worst canonical singularities. Suppose that $\omega_X \cong \mathscr{O}_X$. If $0 \le p \le n$ is any number, then the natural pairing given by the wedge product,*

$$\bigwedge : H^0\left(X,\, \Omega_X^{[p]}\right) \times H^0\left(X,\, \Omega_X^{[n-p]}\right) \longrightarrow H^0\left(X,\, \omega_X\right) \cong \mathbb{C},$$

is non-degenerate.

Remark 6.2. If $\imath : X_{\mathrm{reg}} \hookrightarrow X$ is the inclusion of the smooth locus into X, recall from Definition 2.1 that $\Omega_X^{[p]} = \imath_* \Omega^p_{X_{\mathrm{reg}}}$. Given a non-zero form $\eta \in H^0\left(X_{\mathrm{reg}},\, \Omega^p_{X_{\mathrm{reg}}}\right)$, Proposition 6.1 simply says that there exists a "complementary" form $\phi \in H^0\left(X_{\mathrm{reg}},\, \Omega^{n-p}_{X_{\mathrm{reg}}}\right)$, such that $\eta \wedge \phi$ extends to a non-zero, hence nowhere vanishing section of $\mathscr{O}_X \cong \omega_X \cong \Omega_X^{[n]}$.

Remark 6.3. Proposition 6.1 has been shown in relevant cases in [Pet94, Prop. 5.8]. Our proof of Proposition 6.1 follows [Pet94] closely.

6.A.1. *Proof of Proposition 6.1* We end the present Section 6.A with a proof of Proposition 6.1. To improve readability, the proof is subdivided into six, mostly independent steps.

Step 1 in the proof of Proposition 6.1: Setup of notation We choose a desingularisation $\pi : \widetilde{X} \to X$ of X. Denote the π-exceptional set by $E \subset \widetilde{X}$ and fix a non-zero section $\sigma \in H^0\big(X, \omega_X\big)$. Since ω_X is invertible, and since X has canonical singularities, it follows immediately from the definition that the pull-back of σ is a holomorphic n-form on $\widetilde{X}$, possibly with zeroes along the exceptional set, say

$$\tau := \pi^*(\sigma) \in H^0\big(\widetilde{X}, \omega_{\widetilde{X}}\big).$$

Because $H^0\big(\widetilde{X}, \omega_{\widetilde{X}}\big) \cong H^0\big(X, \omega_X\big) = \mathbb{C} \cdot \sigma$, the form τ clearly spans the vector space $H^0\big(\widetilde{X}, \omega_{\widetilde{X}}\big)$. By the Extension Theorem 2.4 we have $\pi_* \Omega^p_{\widetilde{X}} = \Omega^{[p]}_X$. To prove Proposition 6.1, it is therefore sufficient to prove the following claim.

Claim 6.4. Given any holomorphic p-form $\alpha \in H^0\big(\widetilde{X}, \Omega^p_{\widetilde{X}}\big)$ there exists a "complementary" form $\beta \in H^0\big(\widetilde{X}, \Omega^{n-p}_{\widetilde{X}}\big)$ such that $\alpha \wedge \beta = \tau$.

Step 2 in the proof of Proposition 6.1: Dolbeault cohomology on $\widetilde{X}$ Following standard notation, let $\mathcal{A}^{a,b}$ denote the sheaf of $\mathbb{C}$-valued differentiable forms of type (a, b) on $\widetilde{X}$. Taking products and wedge products with σ and τ, respectively, we obtain sheaf morphisms,

$$\begin{array}{rccc} \psi_X : & \mathscr{O}_X & \to & \omega_X \\ & f & \mapsto & f \cdot \sigma \end{array} \qquad \begin{array}{rccc} \psi_{\widetilde{X}} : & \mathscr{O}_{\widetilde{X}} & \to & \omega_{\widetilde{X}} \\ & f & \mapsto & f \cdot \tau \end{array}$$

$$\begin{array}{rccc} \psi_0 : & \mathcal{A}^{0,0} & \to & \mathcal{A}^{n,0} \\ & f & \mapsto & f \cdot \tau \end{array} \qquad \begin{array}{rccc} \psi_q : & \mathcal{A}^{0,q} & \to & \mathcal{A}^{n,q} \\ & \alpha & \mapsto & \alpha \wedge \tau. \end{array}$$

where $0 < q \leq n$. Observe that ψ_X is isomorphic by assumption.

Since τ is holomorphic, its exterior derivative vanishes $\overline{\partial}\tau = 0$. This immediately implies relations

$$\overline{\partial} \circ \psi_q = \psi_{q+1} \circ \overline{\partial} \quad \text{for all } 0 \leq q \leq n. \tag{6.5.1}$$

In particular, the sheaf morphisms ψ_q induce well-defined morphisms between Dolbeault cohomology groups,

$$\phi_q : H^{0,q}\big(\widetilde{X}\big) \to H^{n,q}\big(\widetilde{X}\big) \quad \text{for all } 0 \leq q \leq n.$$

We will later see in Step 4 of this proof that the morphisms ϕ_q are in fact isomorphic.

Step 3 in the proof of Proposition 6.1: Dolbeault and sheaf cohomology on $\widetilde{X}$ The sheaf morphisms ψ_X and $\psi_{\widetilde{X}}$ induce additional morphisms between sheaf cohomology groups,

$$\begin{aligned} &H^q(\psi_X) : H^q(X, \mathscr{O}_X) \to H^q(X, \omega_X) && \text{and} \\ &H^q(\psi_{\widetilde{X}}) : H^q(\widetilde{X}, \mathscr{O}_{\widetilde{X}}) \to H^q(\widetilde{X}, \omega_{\widetilde{X}}), \end{aligned}$$

for all $0 \leq q \leq n$. Again, observe that the morphisms $H^q(\psi_X)$ are isomorphic by assumption. We will see in Step 4 of this proof that the morphisms $H^q(\psi_{\widetilde{X}})$ are isomorphic as well.

The morphisms ϕ_q and $H^q(\psi_{\widetilde{X}})$ are closely related. In fact, it follows from (6.5.1) that the sheaf morphisms $\psi_\bullet$ align to give a morphism between the Dolbeault resolutions of $\mathscr{O}_{\widetilde{X}}$ and $\omega_{\widetilde{X}}$, respectively. In other words, there exists a commutative diagram as follows,

$$\tag{6.5.2} \begin{array}{ccccccccccc} \mathscr{O}_{\widetilde{X}} & \hookrightarrow & \mathscr{A}^{0,0} & \xrightarrow{\bar\partial} & \mathscr{A}^{0,1} & \xrightarrow{\bar\partial} & \mathscr{A}^{0,2} & \xrightarrow{\bar\partial} & \mathscr{A}^{0,3} & \xrightarrow{\bar\partial} & \cdots \\ \downarrow{\scriptstyle \psi_{\widetilde{X}}} & & \downarrow{\scriptstyle \psi_0} & & \downarrow{\scriptstyle \psi_1} & & \downarrow{\scriptstyle \psi_2} & & \downarrow{\scriptstyle \psi_3} & & \\ \omega_{\widetilde{X}} & \hookrightarrow & \mathscr{A}^{n,0} & \xrightarrow[\bar\partial]{} & \mathscr{A}^{n,1} & \xrightarrow[\bar\partial]{} & \mathscr{A}^{n,2} & \xrightarrow[\bar\partial]{} & \mathscr{A}^{n,3} & \xrightarrow[\bar\partial]{} & \cdots \end{array}$$

The following is then a standard consequence of homological algebra, see for instance [Dem09, Ch. IV, §6, eq. (6.3)].

Conclusion 6.6. There exist commutative diagrams

$$\begin{array}{ccc} H^{0,q}(\widetilde{X}) & \xrightarrow{\phi_q} & H^{n,q}(\widetilde{X}) \\ {\scriptstyle \text{Dolbeault isom.}}\downarrow{\scriptstyle \cong} & & {\scriptstyle \cong}\downarrow{\scriptstyle \text{Dolbeault isom.}} \\ H^q(\widetilde{X}, \mathscr{O}_{\widetilde{X}}) & \xrightarrow{H^q(\psi_{\widetilde{X}})} & H^q(\widetilde{X}, \omega_{\widetilde{X}}) \end{array}$$

for all indices $0 \leq q \leq n$. Q.E.D.

Step 4 in the proof of Proposition 6.1: cohomology on $\widetilde{X}$ and on X Next, we aim to compare cohomology groups on $\widetilde{X}$ with those on X. More precisely, we claim the following.

Claim 6.7. Given any index $0 \leq q \leq n$, there exist morphisms $\rho_{\mathscr{O}}$, ρ_ω forming a commutative diagram as follows,

$$
\begin{array}{ccc}
H^q\big(\widetilde{X},\, \mathscr{O}_{\widetilde{X}}\big) & \xrightarrow{\ H^q(\psi_{\widetilde{X}})\ } & H^q\big(\widetilde{X},\, \omega_{\widetilde{X}}\big) \\
{\scriptstyle \rho_{\mathscr{O}}}\big\uparrow{\scriptstyle \cong} & & {\scriptstyle \cong}\big\uparrow{\scriptstyle \rho_\omega} \\
H^q\big(X,\, \mathscr{O}_X\big) & \xrightarrow[\cong]{\ H^q(\psi_X)\ } & H^q\big(X,\, \omega_X\big).
\end{array}
\tag{6.7.1}
$$

In particular, the morphisms $H^q(\psi_{\widetilde{X}})$ are isomorphic for all $0 \leq q \leq n$.

Proof. Since π has connected fibers, and since X has only canonical singularities, we have canonical identifications

$$\pi_* \mathscr{O}_{\widetilde{X}} \cong \mathscr{O}_X \qquad \text{and} \qquad \pi_* \omega_{\widetilde{X}} \cong \omega_X .$$

Observe that the section σ, seen as a section in $\pi_* \omega_{\widetilde{X}}$, will be identified with the differential form τ. Using these identifications, we need to show that there exist two morphisms

$$\rho_{\mathscr{O}} : H^q\big(X,\, \mathscr{O}_X\big) \to H^q\big(\widetilde{X},\, \mathscr{O}_{\widetilde{X}}\big) \quad \text{and} \quad \rho_\omega : H^q\big(X,\, \omega_X\big) \to H^q\big(\widetilde{X},\, \omega_{\widetilde{X}}\big),$$

which make Diagram (6.7.1) commutative and are isomorphic. While this can be concluded from universal properties and spectral sequences, we found it more instructive to give an elementary construction using Čech cohomology.

To this end, choose an open affine cover $(U_i)_{i \in I}$ of X, which will be acyclic for any coherent sheaf, and let $\rho_{\mathscr{O}}$, ρ_ω be the compositions of the vertical arrows in the following natural diagram.

$$
\begin{array}{ccc}
\check{H}^q\big(\widetilde{X},\, \mathscr{O}_{\widetilde{X}}\big) & \xrightarrow{\check{H}^q(\psi_{\widetilde{X}})} & \check{H}^q\big(\widetilde{X},\, \omega_{\widetilde{X}}\big) \\
\uparrow {\scriptstyle r_{\widetilde{X},\mathscr{O}}} & & \uparrow {\scriptstyle r_{\widetilde{X},\omega}} \\
\check{H}^q\big((\pi^{-1}U_i)_{i\in I},\, \mathscr{O}_{\widetilde{X}}\big) & \xrightarrow{\check{H}^q((\pi^{-1}U_i)_{i\in I},\, \psi_{\widetilde{X}})} & \check{H}^q\big((\pi^{-1}U_i)_{i\in I},\, \omega_{\widetilde{X}}\big) \\
\updownarrow \cong & & \updownarrow \cong \\
\check{H}^q\big((U_i)_{i\in I},\, \pi_*\mathscr{O}_{\widetilde{X}}\big) & \xrightarrow{\check{H}^q((U_i)_{i\in I},\, \pi_*\psi_{\widetilde{X}})} & \check{H}^q\big((U_i)_{i\in I},\, \pi_*\omega_{\widetilde{X}}\big) \\
\updownarrow \cong & & \updownarrow \cong \\
\check{H}^q\big((U_i)_{i\in I},\, \mathscr{O}_X\big) & \xrightarrow{\check{H}^q((U_i)_{i\in I},\, \psi_X)} & \check{H}^q\big((U_i)_{i\in I},\, \omega_X\big) \\
{\scriptstyle r_{X,\mathscr{O}}} \downarrow \cong & & \cong \downarrow {\scriptstyle r_{X,\omega}} \\
\check{H}^q\big(X,\, \mathscr{O}_X\big) & \xrightarrow[\cong]{\check{H}^q(\psi_X)} & \check{H}^q\big(X,\, \omega_X\big)
\end{array}
\qquad \check{\rho}_{\mathscr{O}} : \check{H}^q(X,\mathscr{O}_X) \to \check{H}^q(\widetilde{X},\mathscr{O}_{\widetilde{X}}), \quad \check{\rho}_\omega : \check{H}^q(X,\omega_X) \to \check{H}^q(\widetilde{X},\omega_{\widetilde{X}})
$$

Here, the morphisms $r_{\bullet,\bullet}$ are the standard refinement morphisms that map Čech cohomology groups defined with respect to a specific open covering into Čech cohomology. Identifying Čech and sheaf cohomology, commutativity of Diagram (6.7.1) is then immediate.

To prove that $\rho_{\mathscr{O}}$ and ρ_ω are isomorphic, it suffices to show that the refinement morphisms, $r_{\widetilde{X},\mathscr{O}}$ and $r_{\widetilde{X},\omega}$ are isomorphic. We do that by showing that the covering $(\pi^{-1}U_i)_{i\in I}$ is acyclic for both $\mathscr{O}_{\widetilde{X}}$ and $\omega_{\widetilde{X}}$. That, however, follows immediately from the following two well-known vanishing results which hold for all indices $q > 0$,

$R^q\pi_*\mathscr{O}_{\widetilde{X}} = 0$ because X has rational singularities, [KM98, Thm. 5.22]

$R^q\pi_*\omega_{\widetilde{X}} = 0$ Grauert-Riemenschneider vanishing, [KM98, Cor. 2.68].

The finishes the proof of Claim 6.7. Q.E.D.

Combining Conclusion 6.6 and Claim 6.7, we arrive at the following statement, which summarises the results obtained so far.

Conclusion 6.8. The morphisms $\phi_q : H^{0,q}\big(\widetilde{X}\big) \to H^{n,q}\big(\widetilde{X}\big)$ are isomorphic for all $0 \le q \le n$. Q.E.D.

Step 5 in the proof of Proposition 6.1: Serre duality Given any index $0 \le p \le n$, consider the complex bilinear form ρ obtained as the composition of the following morphisms,

$$
\begin{aligned}
H^0\big(\widetilde{X},\,\Omega^p_{\widetilde{X}}\big)\times H^0\big(\widetilde{X},\,\Omega^{n-p}_{\widetilde{X}}\big) &\xrightarrow{\text{Dolbeault isom.}} H^{p,0}(\widetilde{X})\times H^{n-p,0}(\widetilde{X})\\
&\xrightarrow{\text{Id}\times\text{conjugation}} H^{p,0}(\widetilde{X})\times H^{0,n-p}(\widetilde{X})\\
&\xrightarrow{\text{Id}\times\phi_{n-p}} H^{p,0}(\widetilde{X})\times H^{n,n-p}(\widetilde{X})\\
&\xrightarrow{\text{Id}\times\text{conjugation}} H^{p,0}(\widetilde{X})\times H^{n-p,n}(\widetilde{X})\\
&\xrightarrow{\;s\;} \mathbb{C},
\end{aligned}
$$

where s is the perfect pairing given by Serre duality, cf. [Dem09, Ch. VI, Thm. 7.3].

Recall from Conclusion 6.8 that with the exception of s all maps used in the definition of ρ are isomorphisms. It follows that ρ is a perfect pairing. Unwinding the definition, ρ is given in elementary terms as follows,

$$\rho : H^0\big(\widetilde{X},\,\Omega^p_{\widetilde{X}}\big)\times H^0\big(\widetilde{X},\,\Omega^{n-p}_{\widetilde{X}}\big)\to\mathbb{C},\qquad (\alpha,\beta)\mapsto\int_X \alpha\wedge\beta\wedge\overline{\tau}.$$

Step 6 in the proof of Proposition 6.1: End of proof We are now ready to prove Claim 6.4. Assume we are given a non-zero form $\alpha\in H^0\big(\widetilde{X},\,\Omega^p_{\widetilde{X}}\big)$. Using that ρ is a perfect pairing, we can therefore find a form $\beta\in H^0\big(\widetilde{X},\,\Omega^{n-p}_{\widetilde{X}}\big)$ such that

$$\rho(\alpha,\beta)=\int_X \alpha\wedge\beta\wedge\overline{\tau}=1 \tag{6.8.1}$$

Equation (6.8.1) implies that $\alpha\wedge\beta$ is a non-vanishing element of $H^0\big(\widetilde{X},\,\omega_{\widetilde{X}}\big)$. Since $H^0\big(\widetilde{X},\,\omega_{\widetilde{X}}\big)$ is one-dimensional, there exists a scalar $\lambda\in\mathbb{C}^*$ such that

$$\tau=\lambda\cdot(\alpha\wedge\beta)=\alpha\wedge(\lambda\cdot\beta).$$

This finishes the proof of Claim 6.4 and hence of Proposition 6.1. Q.E.D.

6.B. Hodge duality for klt spaces

If X is a projective manifold, Hodge theory gives a complex-linear isomorphism between the spaces $H^0\big(X,\,\Omega^p_X\big)$ and $\overline{H^p\big(X,\,\mathscr{O}_X\big)}$. We show that the same statement holds for reflexive differentials if X has canonical singularities, or more generally if X is the base space of a klt pair.

Proposition 6.9 (Hodge duality for klt spaces). *Let X be a normal n-dimensional projective variety X. Suppose that there exists an effective $\mathbb{Q}$-divisor D on X such that (X, D) is klt. Given any number $0 \leq p \leq n$, there are complex-linear isomorphisms*

$$H^0\big(X, \Omega_X^{[p]}\big) \cong H^0\left(X_{\mathrm{reg}}, \Omega^p_{X_{\mathrm{reg}}}\right) \cong \overline{H^p\big(X, \mathscr{O}_X\big)}.$$

Proof. Fix a resolution of singularities $\pi : \widetilde{X} \to X$ of X. Then we have the following chain of complex-linear isomorphisms

$$\begin{aligned} H^0\big(X, \Omega_X^{[p]}\big) &\cong H^0\big(\widetilde{X}, \Omega^p_{\widetilde{X}}\big) && \text{see below} \\ &\cong H^{p,0}\big(\widetilde{X}\big) && \text{Dolbeault isomorphism} \\ &\cong \overline{H^{0,p}\big(\widetilde{X}\big)} && \text{Conjugation} \\ &\cong \overline{H^p\big(\widetilde{X}, \mathscr{O}_{\widetilde{X}}\big)} && \text{Dolbeault isomorphism} \\ &\cong \overline{H^p\big(X, \mathscr{O}_X\big)} && \text{see below.} \end{aligned}$$

The first isomorphism exists because $\pi_* \Omega^p_{\widetilde{X}} = \Omega_X^{[p]}$ by the Extension Theorem 2.4. The last isomorphism exists because X has rational singularities, cf. [KM98, Thm. 5.22]. Q.E.D.

We list a few immediate consequences of the results obtained so far.

Corollary 6.10. *Let X be a normal n-dimensional projective variety X having at worst canonical singularities. Suppose that the canonical sheaf of X is trivial, $\omega_X \cong \mathscr{O}_X$. Then the following holds.*

(6.10.1) *Non-zero forms $\eta \in H^0\big(X_{\mathrm{reg}}, \Omega^q_{X_{\mathrm{reg}}}\big)$ do not have any zeroes.*

(6.10.2) *For all $0 \leq p \leq n$, we have complex-linear isomorphisms $H^0\big(X, \Omega_X^{[p]}\big) \cong H^0\big(X, \Omega_X^{[n-p]}\big)^*$, canonically given up to multiplication with a constant.*

(6.10.3) *If the dimension of X is odd, then $\chi\big(X, \mathscr{O}_X\big) = 0$.* *Q.E.D.*

Corollary 6.11 (Existence of forms on canonical varieties with $K_X \equiv 0$). *Let X be a normal n-dimensional projective variety X having at worst canonical singularities. Assume that $\widetilde{q}(X) = 0$ and that the canonical divisor K_X is numerically trivial. Then*

$$h^0\big(X, \Omega_X^{[1]}\big) = h^0\big(X, \Omega_X^{[n-1]}\big) = 0.$$

Proof. We show that $H^0\big(X, \Omega_X^{[n-1]}\big) = 0$. Assume to the contrary, and let σ be a non-zero reflexive $(n-1)$-form on X. Recalling from

Kawamata's analysis of the Albanese map, Proposition 3.3, that K_X is torsion, let $f : \widetilde{X} \to X$ be the associated index-one cover. The morphism f is finite and étale in codimension one, the space $\widetilde{X}$ has canonical singularities, and trivial canonical sheaf $\omega_{\widetilde{X}} \cong \mathscr{O}_{\widetilde{X}}$, cf. [KM98, 5.19 and 5.20]. In particular, the reflexive form σ pulls back to a give non-vanishing reflexive $(n-1)$-form $\widetilde{\sigma}$ on $\widetilde{X}$. Furthermore, observe that the covering space $\widetilde{X}$ satisfies all requirements made in Proposition 6.9 and Corollary 6.10. This shows

$$\begin{aligned} 0 = \widetilde{q}(X) \geq q(\widetilde{X}) &= h^1\big(\widetilde{X},\, \mathscr{O}_{\widetilde{X}}\big) \\ &= h^0\big(\widetilde{X},\, \Omega^{[1]}_{\widetilde{X}}\big) && \text{by Proposition 6.9} \\ &= h^0\big(\widetilde{X},\, \Omega^{[n-1]}_{\widetilde{X}}\big) && \text{by Corollary (6.10.2)}, \end{aligned}$$

contradicting the existence of $\widetilde{\sigma}$. The same argument also shows $H^0\big(X,\, \Omega^{[1]}_X\big) = 0$, finishing the proof of Corollary 6.11. Q.E.D.

6.C. Existence of complementary sheaves

We conclude the present Section 6 with a final corollary which generalises [Pet94, Lem. 5.11]; see also [Bog74, p. 581]. It shows that saturated subsheaves of $\mathscr{T}_X$ with trivial determinant often have a complementary subsheaf which presents $\mathscr{T}_X$ as a direct product. Corollary 6.12 is thus an important ingredient in the proof of our main result, the Decomposition Theorem 1.3.

Corollary 6.12 (Existence of complementary subsheaves in $\mathscr{T}_X$). *Let X be a normal projective variety with trivial canonical sheaf $\omega_X \cong \mathscr{O}_X$, having at worst canonical singularities. Let $\mathscr{E} \subsetneq \mathscr{T}_X$ be a saturated subsheaf with trivial determinant, $\det \mathscr{E} \cong \mathscr{O}_X$. Then there exists a subsheaf $\mathscr{F} \subsetneq \mathscr{T}_X$ with trivial determinant such that*

$$\mathscr{T}_X \cong \mathscr{E} \oplus \mathscr{F}.$$

We will prove Corollary 6.12 in the remainder of the present Section 6.C. For convenience, the proof is subdivided into four steps.

Step 1 in the proof of Corollary 6.12: Setup We consider the obvious quotient sequence

$$0 \longrightarrow \mathscr{E} \xrightarrow{\ \alpha\ } \mathscr{T}_X \xrightarrow{\ \beta\ } \underbrace{\mathscr{T}_X/\mathscr{E}}_{=:\mathscr{Q}} \longrightarrow 0. \tag{6.12.1}$$

Since $\mathscr{E}$ is saturated in the reflexive sheaf $\mathscr{T}_X$, it is itself reflexive. Further, the associated quotient $\mathscr{Q}$ is a torsion free sheaf, say of rank $r > 0$.

We aim to split sequence (6.12.1) in codimension one. To be more precise, let $Z \subsetneq X$ be the smallest set such that $X^\circ := X \setminus Z$ is smooth and $\mathscr{Q}|_{X^\circ}$ is locally free. Since X is normal, and since torsion-free sheaves on manifolds are locally free in codimension one, [OSS80, p. 148], it follows that Z is small, that is, $\operatorname{codim}_X Z \geq 2$. If we can find a splitting of Sequence (6.12.1) on X° and write $\mathscr{T}_{X^\circ} \cong \mathscr{E}|_{X^\circ} \oplus \mathscr{Q}|_{X^\circ}$, it will follow from reflexivity that $\mathscr{T}_X \cong \mathscr{E} \oplus \mathscr{Q}^{**}$, and the proof of Corollary 6.12 will be finished.

Step 2 in the proof of Corollary 6.12: Construction of the splitting In order to construct the splitting, recall the assumptions that $\det \mathscr{E} \cong \mathscr{O}_X$ and $\omega_X \cong \mathscr{O}_X$. As a consequence, we have triviality of determinants, $\det \mathscr{Q} \cong \det \mathscr{Q}^* \cong \mathscr{O}_X$, see [Kob87, Ch. V, Prop. 6.9] for details. Let $\eta_{\mathscr{Q}} \in H^0\big(X, \det \mathscr{Q}^*\big)$ be any non-vanishing section.

Taking duals on X°, Sequence (6.12.1) gives injections

$$\begin{array}{rccc} \beta^* : & \mathscr{Q}^*|_{X^\circ} & \to & \Omega^1_{X^\circ} \\ \wedge^r \beta^* : & \wedge^r \mathscr{Q}^*|_{X^\circ} & \to & \Omega^r_{X^\circ} \\ \det \beta^* : & \det \mathscr{Q}^* & \to & \Omega^{[r]}_X. \end{array}$$

We obtain a non-trivial reflexive form

$$\eta := \big(\det \beta^*\big)(\eta_{\mathscr{Q}}) \in H^0\big(X, \Omega^{[r]}_X\big) \setminus \{0\}.$$

Denoting the dimension of X by n, Proposition 6.1 asserts the existence of a complementary reflexive form $\mu \in H^0\big(X, \Omega^{[n-r]}_X\big)$ such that $\eta \wedge \mu$ gives a nowhere-vanishing section of ω_X. The triviality of $\det \mathscr{Q}|_{X^\circ}$ and of $\omega_{X^\circ} = \det \mathscr{T}^*_{X^\circ}$ thus gives isomorphisms of sheaves,

$$\text{(6.12.2)} \qquad \begin{array}{rccc} \delta_{\mathscr{Q}} : & \mathscr{Q}|_{X^\circ} & \to & \wedge^{r-1} \mathscr{Q}^*|_{X^\circ} \\ & q & \mapsto & \eta_{\mathscr{Q}}(q, \cdot) \\ & & & \\ \delta_{\mathscr{T}_X} : & \mathscr{T}_{X^\circ} & \to & \Omega^{n-1}_{X^\circ} \\ & \vec{v} & \mapsto & \big(\eta \wedge \mu\big)(\vec{v}, \cdot). \end{array}$$

Remark 6.13. If $r = 1$, then $\wedge^{r-1} \mathscr{Q}^*|_{X^\circ} = \mathscr{O}_{X^\circ}$ is simply the sheaf of functions.

Using the isomorphisms (6.12.2) and the complementary form μ, we can now define a sheaf morphism $\phi : \mathscr{Q}|_{X^\circ} \to \mathscr{T}_{X^\circ}$ as the composite of the following natural maps

$$\text{(6.13.1)} \quad \mathscr{Q}|_{X^\circ} \xrightarrow{\delta_{\mathscr{Q}}} \wedge^{r-1} \mathscr{Q}^*|_{X^\circ} \xrightarrow{\wedge^{r-1}\beta^*} \Omega^{r-1}_{X^\circ} \xrightarrow{\wedge \mu} \Omega^{n-1}_{X^\circ} \xrightarrow{\delta^{-1}_{\mathscr{T}_X}} \mathscr{T}_{X^\circ}.$$

Remark 6.14. In case where $r = 1$, the sheaves $\wedge^{r-1}\mathscr{Q}^*|_{X^\circ}$ and $\Omega^{r-1}_{X^\circ}$ both equal the trivial sheaf $\mathscr{O}_{X^\circ}$. The morphism $\wedge^{r-1}\beta^*$ is then the identity map.

To end the proof of Corollary 6.12, it will now suffice to prove the following claim.

Claim 6.15. The morphism $\phi : \mathscr{Q}|_{X^\circ} \to \mathscr{T}_{X^\circ}$ defines a splitting of Sequence (6.12.1) over the open set X°.

Step 3 in the proof of Corollary 6.12: preparation for proof of Claim 6.15 It suffices to show Claim 6.15 locally, over sufficiently small open sets $U \subseteq X^\circ$. We will prove Claim 6.15 by explicit computation, choosing frames for the bundles $\mathscr{E}$, $\mathscr{T}_X$ and $\mathscr{Q}$ to write down the morphism ϕ and all relevant differential forms. Indeed, choosing U small enough, we can find frames

$$
\begin{aligned}
&e_1, \ldots, e_{n-r}, && \ldots \text{frame of } \mathscr{E}|_U, \\
&\vec{q}_1, \ldots, \vec{q}_r, \alpha(e_1), \ldots, \alpha(e_{n-r}) && \ldots \text{frame of } \mathscr{T}_X|_U, \text{ and} \\
&\beta(\vec{q}_1), \ldots, \beta(\vec{q}_{n-r}) && \ldots \text{frame of } \mathscr{Q}|_U.
\end{aligned}
$$

To simplify notation, set

$$\vec{e}_i := \alpha(e_i) \in \mathscr{T}_X(U) \quad \text{and} \quad q_j := \beta(\vec{q}_j) \in \mathscr{Q}(U).$$

We denote the dual frames by

$$
\begin{aligned}
&e^*_1, \ldots, e^*_{n-r} && \ldots \text{frame of } \mathscr{E}^*|_U, \\
&\vec{q}^{\,*}_1 \ldots, \vec{q}^{\,*}_r, \vec{e}^{\,*}_1, \ldots, \vec{e}^{\,*}_{n-r} && \ldots \text{frame of } \Omega^1_X|_U, \text{ and} \\
&q^*_1, \ldots, q^*_r && \ldots \text{frame of } \mathscr{Q}^*|_U.
\end{aligned}
$$

Observe that $\alpha^*(\vec{e}^{\,*}_i) = e^*_i$ and $\beta^*(q^*_j) = \vec{q}^{\,*}_j$, for all indices i and j. Scaling the frame $\vec{q}_1, \ldots, \vec{q}_r$ appropriately, we may assume that

$$\eta_{\mathscr{Q}}|_U = q^*_1 \wedge \cdots \wedge q^*_r \quad \text{and} \quad \eta|_U = \vec{q}^{\,*}_1 \wedge \cdots \wedge \vec{q}^{\,*}_r. \tag{6.15.1}$$

Scaling the frame $e_1, \ldots, e_{n-r}$, we can then find forms $\sigma_1, \ldots \sigma_r \in \Omega^{n-r-1}_X(U)$ such that the complementary form μ can be written as

$$\mu|_U = \vec{e}^{\,*}_1 \wedge \cdots \wedge \vec{e}^{\,*}_{n-r} + \sum_{i=1}^{r} \vec{q}^{\,*}_i \wedge \sigma_i. \tag{6.15.2}$$

Remark 6.16. We do not claim that Equation (6.15.2) defines the forms σ_i uniquely. In fact, there will almost always be several ways to write $\mu|_C$ in this way. If $n - r - 1 = 0$, then the σ_i are just functions.

Remark 6.17. On the open set U, Equations (6.15.1) and (6.15.2) together imply that the globally defined form $\eta \wedge \mu$ is given as

$$(\eta \wedge \mu)|_U = \vec{q}_1^{\,*} \wedge \cdots \wedge \vec{q}_r^{\,*} \wedge \vec{e}_1^{\,*} \wedge \cdots \wedge \vec{e}_{n-r}^{\,*}.$$

If $n - r - 1 > 0$, then the forms σ_i of Equation (6.15.2) can be decomposed further, writing them as sums of pure tensors that only involve $\vec{e}_1^{\,*}, \ldots, \vec{e}_{n-r}^{\,*}$, and tensors that involve $\vec{q}_1^{\,*}, \ldots, \vec{q}_r^{\,*}$,

$$\sigma_i = \sum_{j=1}^{n-r} a_{ij} \cdot \vec{e}_1^{\,*} \wedge \cdots \cancel{\vec{e}_j^{\,*}} \cdots \wedge \vec{e}_{n-r}^{\,*} + \sum_{k=1}^{r} \vec{q}_k^{\,*} \wedge \tau_{ik}, \tag{6.17.1}$$

for suitable functions $a_{ij} \in \mathscr{O}_{X^\circ}(U)$ and forms $\tau_{ik} \in \Omega_X^{n-r-2}(U)$.

Remark 6.18. Again, we do not claim that Equation (6.17.1) defines the forms τ_{ik} uniquely. In contrast, note that the functions a_{ij} are uniquely determined by (6.15.2) and (6.17.1).

Step 4 in the proof of Corollary 6.12: proof of Claim 6.15 and end of proof We will prove Claim 6.15 only in case where $n - r - 1 > 0$. The case where $n - r = 1$ follows exactly the same pattern, but is easier. To be precise, we will prove that

$$\begin{array}{rccc} \phi|_U : & \mathscr{Q}|_U & \to & \mathscr{T}_X|_U \\ & q_\ell & \mapsto & \vec{q}_\ell + \sum_{j=1}^{n-r} \pm a_{\ell j} \cdot \vec{e}_j \end{array} \tag{6.18.1}$$

where the $a_{\ell j} \in \mathscr{O}_{X^\circ}(U)$ are the functions introduced in Equation (6.17.1) above. Therefore, $\beta|_U \circ \phi|_U = \mathrm{id}_{\mathscr{Q}|_U}$, establishing Claim 6.15. By definition of ϕ, (6.18.1) is equivalent to showing that (6.18.2)

$$\delta_{\mathscr{T}_X}^{-1}\left(\left(\left(\wedge^{r-1}\beta^*\right)\left(\delta_{\mathscr{Q}}(q_\ell)\right)\right) \wedge \mu\right) = \vec{q}_\ell + \sum_{j=1}^{n-r} \pm a_{\ell j} \cdot \vec{e}_j \quad \text{for all indices } \ell.$$

The computation proving (6.18.2) uses the following elementary observation.

Observation 6.19. It follows immediately from (6.15.1) and from Remark 6.17 that the sheaf morphisms $\delta_{\mathscr{Q}}$ and $\delta_{\mathscr{T}_X}$ introduced in (6.12.2) have the following explicit description on U,

$$\begin{array}{llll} \delta_{\mathscr{Q}}|_U : & \mathscr{Q}|_U & \to & \wedge^{r-1}\mathscr{Q}^*|_U \\ & q_\ell & \mapsto & (-1)^{\ell+1}\cdot q_1^* \wedge \cdots \cancel{q_\ell^*} \cdots \wedge q_r^* \end{array}$$

$$\begin{array}{lllll} \delta_{\mathscr{T}_X}|_U : & \mathscr{T}_X|_U & \to & & \Omega_X^{n-1}|_U \\ & \vec{q}_\ell & \mapsto & (-1)^{\ell+1} & \cdot \vec{q}_1^{\,*} \wedge \cdots \cancel{\vec{q}_\ell^{\,*}} \cdots \wedge \vec{q}_r^{\,*} \wedge \vec{e}_1^{\,*} \wedge \cdots \wedge \vec{e}_{n-r}^{\,*} \\ & \vec{e}_\ell & \mapsto & (-1)^{r+\ell+1} & \cdot \vec{q}_1^{\,*} \wedge \cdots \wedge \vec{q}_r^{\,*} \wedge \vec{e}_1^{\,*} \wedge \cdots \cancel{\vec{e}_\ell^{\,*}} \cdots \wedge \vec{e}_{n-r}^{\,*}. \end{array}$$

With Observation 6.19 in place, Equation (6.18.2) is now shown easily by direct computation as follows.

$$\begin{aligned} \mathsf{A} &:= \delta_{\mathscr{Q}}(q_\ell) \\ &\overset{\text{Obs. 6.19}}{=} (-1)^{\ell+1}\cdot q_1^* \wedge \cdots \cancel{q_\ell^*} \cdots \wedge q_r^* \end{aligned}$$

$$\begin{aligned} \mathsf{B} &:= \big(\wedge^{r-1}\beta^*\big)(\mathsf{A}) \\ &\overset{\text{Defn. of } \beta^*}{=} (-1)^{\ell+1}\cdot \vec{q}_1^{\,*} \wedge \cdots \cancel{\vec{q}_\ell^{\,*}} \cdots \wedge \vec{q}_r^{\,*} \end{aligned}$$

$$\begin{aligned} \mathsf{C} &:= \mathsf{B} \wedge \mu \\ &\overset{\text{by (6.15.2)}}{=} (-1)^{\ell+1}\cdot \vec{q}_1^{\,*} \wedge \cdots \cancel{\vec{q}_\ell^{\,*}} \cdots \wedge \vec{q}_r^{\,*} \wedge \left(\vec{e}_1^{\,*} \wedge \cdots \wedge \vec{e}_{n-r}^{\,*} + \sum_{i=1}^{r} \vec{q}_i^{\,*} \wedge \sigma_i \right) \\ &= (-1)^{\ell+1}\cdot \vec{q}_1^{\,*} \wedge \cdots \cancel{\vec{q}_\ell^{\,*}} \cdots \wedge \vec{q}_r^{\,*} \wedge \big(\vec{e}_1^{\,*} \wedge \cdots \wedge \vec{e}_{n-r}^{\,*} + \vec{q}_\ell^{\,*} \wedge \sigma_i \big) \\ &\overset{\text{by (6.17.1)}}{=} (-1)^{\ell+1}\cdot \vec{q}_1^{\,*} \wedge \cdots \cancel{\vec{q}_\ell^{\,*}} \cdots \wedge \vec{q}_r^{\,*} \wedge \Bigg(\vec{e}_1^{\,*} \wedge \cdots \wedge \vec{e}_{n-r}^{\,*} \\ &\qquad + \vec{q}_\ell^{\,*} \wedge \left(\sum_{j=1}^{n-r} a_{\ell j}\cdot \vec{e}_1^{\,*} \wedge \cdots \cancel{\vec{e}_j^{\,*}} \cdots \wedge \vec{e}_{n-r}^{\,*} + \sum_{k=1}^{r} \vec{q}_k^{\,*} \wedge \tau_{ik} \right) \Bigg) \\ &= (-1)^{\ell+1}\cdot \vec{q}_1^{\,*} \wedge \cdots \cancel{\vec{q}_\ell^{\,*}} \cdots \wedge \vec{q}_r^{\,*} \wedge \vec{e}_1^{\,*} \wedge \cdots \wedge \vec{e}_{n-r}^{\,*} \\ &\qquad + \sum_{j=1}^{n-r} \pm a_{\ell j}\cdot \vec{q}_1^{\,*} \wedge \cdots \wedge \vec{q}_r^{\,*} \wedge \vec{e}_1^{\,*} \wedge \cdots \cancel{\vec{e}_j^{\,*}} \cdots \wedge \vec{e}_{n-r}^{\,*} \end{aligned}$$

and finally

$$\mathsf{D} := \delta_{\mathscr{T}_X}^{-1}(\mathsf{C}) \overset{\text{Obs. 6.19}}{=} \vec{q}_\ell + \sum_{j=1}^{n-r} \pm a_{\ell j}\cdot \vec{e}_j.$$

This finishes the proof of Equations (6.18.2), (6.18.1), Claim 6.15, and hence of Corollary 6.12. Q.E.D.

§7. Proof of Theorem 1.3

We have divided the proof of Theorem 1.3 into a sequence of steps, each formulated as a separate result. Some of these statements might be of independent interest. The proof of Theorem 1.3 follows quickly from these preliminary steps and is given in Section 7.C.

Theorem 7.1 (Splitting the tangent sheaf of varieties with trivial canonical bundle). *Let X be a normal n-dimensional projective variety with at worst canonical singularities. Assume that $\omega_X \cong \mathscr{O}_X$ and that $\widetilde{q}(X) = 0$. Let $h = (H_1, \ldots, H_{n-1})$ be ample divisors on X and assume that $\mathscr{T}_X$ is not h-stable. Let $0 \subsetneq \mathscr{E} \subsetneq \mathscr{T}_X$ be a saturated destabilising subsheaf, that is, a proper subsheaf with non-negative slope $\mu_h(\mathscr{E}) \geq 0$ and torsion free quotient $\mathscr{T}_X/\mathscr{E}$.*

Then there exists a number $M \in \mathbb{N}^+$ such that $(\det \mathscr{E})^{[M]} \cong \mathscr{O}_X$. Further, there exists a finite cover $f : \widetilde{X} \to X$, étale in codimension one, and a proper subsheaf $\mathscr{F} \subsetneq \mathscr{T}_{\widetilde{X}}$ such that the following holds.

(7.1.1) *The tangent sheaf of $\widetilde{X}$ decomposes as a direct sum, $\mathscr{T}_{\widetilde{X}} \cong (f^{[*]}\mathscr{E}) \oplus \mathscr{F}$.*

(7.1.2) *Both summands in (7.1.1) have trivial determinant. In other words, $\det f^{[*]}\mathscr{E} \cong \mathscr{O}_{\widetilde{X}}$ and $\det \mathscr{F} \cong \mathscr{O}_{\widetilde{X}}$.*

Before proving Theorem 7.1 in Section 7.A below, we note an important corollary, obtained from Theorem 7.1 by iterated application. The following notation, which summarises the conditions spelled out in Condition (1.3.2) of the Decomposition Theorem 1.3, is used in its formulation.

Definition 7.2 (Strong stability). Let X be a normal projective variety of dimension n, and $\mathscr{F}$ a reflexive coherent sheaf of $\mathscr{O}_X$-modules. We call $\mathscr{F}$ *strongly stable*, if for any finite morphism $f : \widetilde{X} \to X$ that is étale in codimension one, and for any choice of ample divisors $\widetilde{H}_1, \ldots, \widetilde{H}_{n-1}$ on $\widetilde{X}$, the reflexive pull-back $f^{[*]}\mathscr{F}$ is stable with respect to $(\widetilde{H}_1, \ldots, \widetilde{H}_{n-1})$.

Corollary 7.3 (Existence of a decomposition). *Let X be a normal projective variety having at worst canonical singularities. Assume that $\omega_X \cong \mathscr{O}_X$. Then there exists a finite cover $f : \widetilde{X} \to X$, étale in codimension one, and a decomposition*

$$\mathscr{T}_{\widetilde{X}} \cong \bigoplus \mathscr{E}_i,$$

where the $\mathscr{E}_i$ are strongly stable subsheaves of $\mathscr{T}_{\widetilde{X}}$ with trivial determinants, $\det \mathscr{E}_i \cong \mathscr{O}_{\widetilde{X}}$.

Remark 7.4. We note that the summands $\mathscr{E}_i$ in the decomposition established in Corollary 7.3 are automatically saturated. Indeed, as a subsheaf of the torsion-free sheaf $\mathscr{T}_{\widetilde{X}}$ each $\mathscr{E}_i$ is torsion-free. The quotient of $\mathscr{T}_{\widetilde{X}}$ by any of the summands is a direct sum of the remaining summands, hence torsion-free.

Proof of Corollary 7.3. We need to find a cover $f : \widetilde{X} \to X$ and a decomposition $\mathscr{T}_{\widetilde{X}} \cong \bigoplus \mathscr{E}_i$, such that all factors $\mathscr{E}_i$ are strongly stable. Since the rank of $\mathscr{T}_X$ is finite, there exists a finite cover $f : \widetilde{X} \to X$, étale in codimension one, with a proper decomposition

$$\mathscr{T}_{\widetilde{X}} \cong \bigoplus_{i \geq 1} \mathscr{E}_i, \tag{7.4.1}$$

in which the number of direct summands is maximal. We claim that each summand is then automatically strongly stable.

We argue by contradiction and assume that there exists a further finite cover, $g : \widehat{X} \to \widetilde{X}$, étale in codimension one, and a list of ample divisors $\widehat{h} = (\widehat{H}_1, \dots, \widehat{H}_{n-1})$ on $\widehat{X}$ such that the reflexive pull-back of one of the summands, say $\widehat{\mathscr{E}}_1 := g^{[*]}\mathscr{E}_1$, is not stable with respect to $\widehat{h}$. Let $0 \subsetneq \widehat{\mathscr{S}} \subsetneq \widehat{\mathscr{E}}_1$ be a $\widehat{h}$-destabilising subsheaf. By [Kob87, Prop. 7.6(b)], whose proof carries over without change from the smooth to the singular setup, we may assume that $\widehat{\mathscr{S}}$ is saturated in $\widehat{\mathscr{E}}_1$ and therefore also in $\mathscr{T}_{\widehat{X}}$. Since $\widehat{\mathscr{E}}_1$ and $\mathscr{T}_{\widehat{X}}$ both have vanishing $\widehat{h}$-slope, the sheaf $\widehat{\mathscr{S}}$ is also a destabilising subsheaf for $\mathscr{T}_{\widehat{X}}$. Replacing $\widehat{X}$ by a further cover, if necessary, Theorem 7.1 therefore allows to assume without loss of generality that the tangent sheaf splits, say $\mathscr{T}_{\widehat{X}} = \widehat{\mathscr{S}} \oplus \widehat{\mathscr{Q}}$. Since the sheaves $\mathscr{T}_{\widehat{X}}$, $\widehat{\mathscr{E}}_1$, $\widehat{\mathscr{S}}$ and $\widehat{\mathscr{Q}}$ are all locally free on the smooth locus of $\widehat{X}$, elementary linear algebra gives a decomposition

$$\widehat{\mathscr{E}}_1|_{\widehat{X}_{\mathrm{reg}}} = \widehat{\mathscr{S}}|_{\widehat{X}_{\mathrm{reg}}} \oplus (\widehat{\mathscr{E}}_1 \cap \widehat{\mathscr{Q}})|_{\widehat{X}_{\mathrm{reg}}}.$$

Taking double duals, we obtain a decomposition $\widehat{\mathscr{E}}_1 = \widehat{\mathscr{S}}^{**} \oplus (\widehat{\mathscr{E}}_1 \cap \widehat{\mathscr{Q}})^{**}$, which contradicts maximality of the decomposition (7.4.1) and therefore finishes the proof of Corollary 7.3. Q.E.D.

Remark 7.5 (Uniqueness of the decomposition). Given a variety X as in Corollary 7.3, let $f_1 : \widetilde{X}_1 \to X$ and $f_2 : \widetilde{X}_1 \to X$ be two finite morphisms, étale in codimension one, such that the tangent bundles split into strongly stable summands,

$$\mathscr{T}_{\widetilde{X}_1} \cong \bigoplus_{i=1}^{N} \mathscr{E}_i^1 \quad \text{and} \quad \mathscr{T}_{\widetilde{X}_2} \cong \bigoplus_{j=1}^{M} \mathscr{E}_j^2.$$

Let $\widehat{X}$ be an irreducible component of the normalisation of the fibered product $\widetilde{X}_1 \times_X \widetilde{X}_2$. We obtain a diagram

$$\begin{array}{ccc} \widehat{X} & \xrightarrow{g_1} & \widetilde{X}_1 \\ {\scriptstyle g_2}\downarrow & & \downarrow{\scriptstyle f_1} \\ \widetilde{X}_2 & \xrightarrow[f_2]{} & X, \end{array}$$

where g_1, g_2 are again finite and étale in codimension one. Since $\mathscr{T}_{\widehat{X}} \cong g_1^{[*]} \mathscr{T}_{\widetilde{X}_1} \cong g_2^{[*]} \mathscr{T}_{\widetilde{X}_2}$, we obtain decompositions

$$\mathscr{T}_{\widehat{X}} \cong \bigoplus_{i=1}^{N} g_1^{[*]} \mathscr{E}_i^1 \cong \bigoplus_{j=1}^{M} g_2^{[*]} \mathscr{E}_j^2. \tag{7.5.1}$$

Choosing any ample polarisation on $\widehat{X}$, stability of the summands implies that any morphism $\widehat{\mathscr{E}}_i^1 \to \widehat{\mathscr{E}}_j^2$ must either be trivial, or an isomorphism. It follows that the decompositions (7.5.1) satisfy the following extra conditions

(7.5.2) the number of summands in the decompositions agrees, $N = M$, and

(7.5.3) up to permutation of the summands we have isomorphisms $g_1^{[*]} \mathscr{E}_i^1 \cong g_2^{[*]} \mathscr{E}_i^2$ for all $i \in \{1, \ldots, N\}$.

In that sense, the decomposition found in Corollary 7.3 is unique.

7.A. Proof of Theorem 7.1

Since the proof of Theorem 7.1 is somewhat long, we have subdivided it into four relatively independent steps, given in Sections 7.A.1–7.A.4 below. Figure 7.1 on the facing page gives an overview of the spaces and morphisms constructed in the course of the proof.

7.A.1. *Step 1: The subsheaf* $\mathscr{E} \subsetneq \mathscr{T}_X$ To start the proof of Theorem 7.1, we discuss the structure of the saturated destabilising sheaf $\mathscr{E} \subsetneq \mathscr{T}_X$. First note that due to torsion-freeness of $\mathscr{E}$ and of $\mathscr{T}_X/\mathscr{E}$, the sheaf $\mathscr{E}$ is a sub-vectorbundle of $\mathscr{T}_X$ outside of a set of codimension two. Next, we compute its slope.

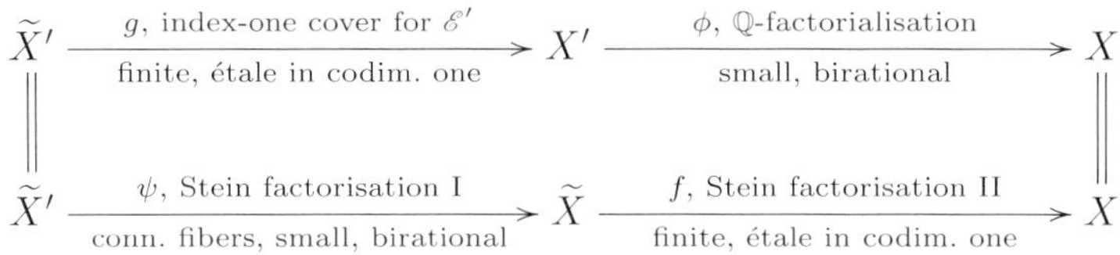

Fig. 7.1. Spaces and morphisms constructed in the proof of Theorem 7.1

Lemma 7.6 (Slopes of destabilising subsheaves). *In the setup of Theorem 7.1, any destabilising subsheaf of $\mathscr{T}_X$ has slope zero. In particular, $\mu_h(\mathscr{E}) = 0$.*

Proof. Let $\mathscr{G}$ be any destabilising subsheaf of $\mathscr{T}_X$. Since K_X is assumed to be trivial, it follows that $\mathscr{G}$ has non-negative slope, $\mu_h(\mathscr{G}) \geq 0$. On the other hand, we know from Proposition 5.4 that $\mathscr{T}_X$ is h-semistable. Consequently, we have $\mu_h(\mathscr{G}) = 0$, as claimed. Q.E.D.

7.A.2. *Step 2: The $\mathbb{Q}$-factorialisation of X* Let $\phi : X' \to X$ be a $\mathbb{Q}$-factorialisation of X, that is, a small birational morphism where X' is $\mathbb{Q}$-factorial and has only canonical singularities. The existence of ϕ is established in [BCHM10, Lem. 10.2]. Since ϕ is small, it is clear that $\omega_{X'} \cong \phi^*\omega_X \cong \mathscr{O}_{X'}$, and that $\widetilde{q}(X') = 0$. Set $\mathscr{E}' := \phi^{[*]}(\mathscr{E})$. Since $\mathscr{E}'$ injects into the tangent sheaf $\mathscr{T}_{X'}$ away from a set of codimension two, it follows from reflexivity that $\mathscr{E}'$ injects into $\mathscr{T}_{X'}$ everywhere. We can therefore view it as a proper subsheaf $\mathscr{E}' \subsetneq \mathscr{T}_{X'}$. Notice that $\mathscr{E}'$ is saturated in $\mathscr{T}_{X'}$, since ϕ is small.

Claim 7.7. To prove statements (7.1.1) and (7.1.2) of Theorem 7.1, it suffices to find a finite cover $g : \widetilde{X}' \to X'$, étale in codimension one, and a decomposition

$$\mathscr{T}_{\widetilde{X}'} \cong \mathscr{F}' \oplus \left(g^{[*]}\mathscr{E}'\right) \tag{7.7.1}$$

where $\det \mathscr{F}' \cong \det\left(g^{[*]}\mathscr{E}'\right) \cong \mathscr{O}_{\widetilde{X}'}$.

Proof. As indicated in Figure 7.1, consider the Stein factorisation of the composed morphism $\phi \circ g$,

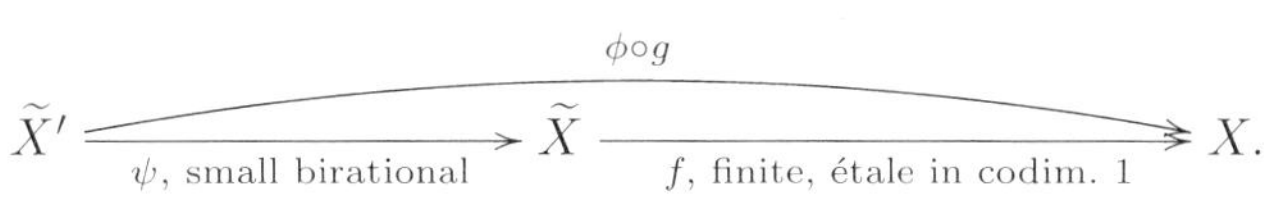

The reflexive push-forward of (7.7.1),

$$\mathscr{T}_{\widetilde{X}} \cong \left(\psi_* \mathscr{T}_{\widetilde{X}'}\right)^{**} \cong \left(\psi_*\left(\mathscr{F}' \oplus g^{[*]}\mathscr{E}'\right)\right)^{**} \cong \underbrace{\left(\psi_* \mathscr{F}'\right)^{**}}_{=:\mathscr{F}} \oplus \underbrace{\left(\psi_*(g^{[*]}\mathscr{E}')\right)^{**}}_{\cong f^{[*]}\mathscr{E}},$$

then yields a decomposition on $\widetilde{X}$ that satisfies both (7.1.1) and (7.1.2), thus finishing the proof of Claim 7.7. Q.E.D.

7.A.3. *Step 3: High reflexive powers of* $\det \mathscr{E}$ *and* $\det \mathscr{E}'$ *are trivial* We need to show that high reflexive powers of $\det \mathscr{E}$ and $\det \mathscr{E}'$ are trivial. To this end, recall that the reflexive sheaf $\det \mathscr{E}'$ is a Weil divisorial sheaf on X'. In other words, there exists a Weil divisor D' on X' so that $\det \mathscr{E}' \cong \mathscr{O}_{X'}(D')$. Since X' is $\mathbb{Q}$-factorial, there exists be a number $m \in \mathbb{N}^+$ such that mD' is actually Cartier.

As a first step towards showing triviality of a sufficiently high multiple, we prove numerical triviality of D'.

Lemma 7.8. *Setting as above, then the $\mathbb{Q}$-Cartier divisor D' is numerically trivial.*

Proof. We aim to apply Proposition 4.2 in order to conclude that D' is numerically trivial. As a first step in this direction, recall from Lemma 7.6 that $\mu_h(\mathscr{E}) \cdot h = 0$. Consequently, we have the following equality of intersection numbers of $\mathbb{Q}$-Cartier divisors,

$$0 = D' \cdot \phi^*(H_1) \cdots \phi^*(H_{n-1}) = -D' \cdot \phi^*(H_1) \cdots \phi^*(H_{n-1}). \tag{7.8.1}$$

Secondly, observe that the inclusion $\mathscr{E}' \hookrightarrow \mathscr{T}_{X'}$ yields an inclusion $\det \mathscr{E}' \hookrightarrow \wedge^{[p]} \mathscr{T}_{X'}$. Dualising, we obtain a non-zero morphism

$$\Omega^{[p]}_{X'} \to (\det \mathscr{E}')^* \cong \mathscr{O}_{X'}(-D').$$

Since K_X is trivial and since X has only canonical singularities, X is not uniruled and Proposition 5.6 therefore implies that the $\mathbb{Q}$-Cartier divisor $-D'$ is pseudoeffective on X'. Due to Equation (7.8.1), Proposition 4.2 applies to show that $-D'$ and hence D' is numerically trivial indeed. Q.E.D.

Next up, we conclude from numerical triviality that high reflexive powers of $\det \mathscr{E}$ and $\det \mathscr{E}'$ are trivial.

Corollary 7.9. *Setting as above. Then there exists a number $M \in \mathbb{N}^+$ such that $(\det \mathscr{E}')^{[M]} \cong \mathscr{O}_{X'}$ and $(\det \mathscr{E})^{[M]} \cong \mathscr{O}_X$.*

Proof. We are going to use the assumption that $\widetilde{q}(X) = 0$, which implies that

$$0 = q(X') = h^1\big(X', \mathscr{O}_{X'}\big),$$

hence the Picard group of X' is discrete. The subgroup $\mathrm{Pic}^0(X') \subsetneq \mathrm{Pic}(X')$ of invertible sheaves with numerically trivial Chern class is therefore finite. It follows that there exists a positive natural number k such that $(\det \mathscr{E}')^{[km]} \cong \mathscr{O}_{X'}$. Since the reflexive sheaves $\mathscr{O}_X \cong \phi_*\big((\det \mathscr{E}')^{[km]}\big)$ and $(\det \mathscr{E})^{[km]}$ agree in codimension one, they agree everywhere, and $(\det \mathscr{E})^{[km]}$ is likewise trivial. Q.E.D.

7.A.4. *Step 4: Constructing the splitting on a cover of X'* Since $\det \mathscr{E}'$ is torsion, there exists an index-one cover $g : \widetilde{X}' \to X'$ for $\det \mathscr{E}'$, see for example [KM98, 2.52]. This is a finite morphism from a normal variety, étale in codimension one, such that

$$g^{[*]} \det \mathscr{E}' \cong \mathscr{O}_{\widetilde{X}'}. \tag{7.9.1}$$

Recall from [KM98, 5.20] that $\widetilde{X}'$ has trivial canonical bundle and at worst canonical singularities. Set $\widetilde{\mathscr{E}}' := g^{[*]}\mathscr{E}'$. We aim to construct a splitting of $\mathscr{T}_{\widetilde{X}'}$ using the existence of complementary subsheaves shown in Corollary 6.12 on page 87. The following claim guarantees that the assumptions made in Corollary 6.12 are satisfied in our context.

Claim 7.10. The inclusion $\mathscr{E}' \hookrightarrow \mathscr{T}_{X'}$ induces an inclusion $\widetilde{\mathscr{E}}' \hookrightarrow \mathscr{T}_{\widetilde{X}'}$. With the inclusion understood, $\widetilde{\mathscr{E}}'$ is a saturated subsheaf of $\mathscr{T}_{\widetilde{X}'}$

Proof. Since $\phi \circ g : \widetilde{X}' \to X$ is étale in codimension one, the reflexive sheaf $\widetilde{\mathscr{E}}'$ injects into $\mathscr{T}_{\widetilde{X}'}$ outside of a small set, and is a saturated subsheaf there. Since both $\widetilde{\mathscr{E}}'$ and its saturation in $\mathscr{T}_{\widetilde{X}'}$ are reflexive and agree in codimension one, it follows that $\widetilde{\mathscr{E}}'$ actually is isomorphic to its saturation, as claimed. Q.E.D.

With Claim 7.10 in place, Corollary 6.12 asserts the existence of a sheaf $\widetilde{\mathscr{F}}'$ with trivial determinant such that $\mathscr{T}_{\widetilde{X}'} \cong \widetilde{\mathscr{E}}' \oplus \widetilde{\mathscr{F}}'$. As we have seen in Claim 7.7, this concludes the proof of Theorem 7.1. Q.E.D.

7.B. Integrability of direct summands

In this section, we show that the individual summands in the decomposition stated in Corollary 7.3 are integrable; that is, they define foliations.

Theorem 7.11 (Integrability of direct summands). *Let X be a normal projective variety with at worst canonical singularities. Assume* *that $\omega_X \cong \mathscr{O}_X$. Let $\mathscr{T}_X \cong \bigoplus \mathscr{E}_i$ be a decomposition into reflexive sheaves with trivial determinants. Then all $\mathscr{E}_i$ are integrable.*

Proof. We follow the arguments of [Hör07]. Without loss of generality we assume that $\mathscr{T}_X \cong \mathscr{E}_1 \oplus \mathscr{E}_2$, that is, we assume that $\mathscr{T}_X$ can be decomposed into two summands. We will show that $\mathscr{E}_2$ is integrable. The integrability of $\mathscr{E}_1$ then follows for symmetry reasons. Let r_1 be the rank of $\mathscr{E}_1$, and consider the trivialisable sheaf $\mathscr{L}_1 := \det \mathscr{E}_1$. Since $\mathscr{E}_1^*$ is a direct summand of Ω^1_X, the reflexive sheaf $\mathscr{L}_1 \otimes \Omega^{[r_1]}_X$ has a trivial direct summand. Let $\theta \in H^0\big(X,\, \mathscr{L}_1 \otimes \Omega^{[r_1]}_X\big)$ be the corresponding nowhere vanishing $\mathscr{L}_1$-valued differential form and let $\pi : \widetilde{X} \to X$ be a resolution of singularities. By the Extension Theorem 2.4, the reflexive differential form θ pulls back to a non-trivial section $\widetilde{\theta} \in H^0\big(\widetilde{X},\, \pi^*(\mathscr{L}_1) \otimes \Omega^{r_1}_{\widetilde{X}}\big)$.

At general points of $\widetilde{X}$, where π is isomorphic, the sheaf $\pi^*\mathscr{E}_2$ coincides with the degeneracy sheaf $S_{\widetilde{\theta}}$ of $\widetilde{\theta}$, that is, the sheaf of vector fields $\vec{v}$ such that the contraction

$$i_{\vec{v}}(\widetilde{\theta}) = \widetilde{\theta}(\vec{v}, \cdot) \in H^0\big(\widetilde{X},\, \pi^*(\mathscr{L}_1) \otimes \Omega^{r_1 - 1}_{\widetilde{X}}\big)$$

vanishes[1]. In this setting, it follows from [Dem02, Main Thm.] that $S_{\widetilde{\theta}}$ is integrable. As a consequence, we obtain that $\mathscr{E}_2$ is integrable at general points of X. Since $\mathscr{E}_2$ is a saturated subsheaf of $\mathscr{T}_X$, it follows that it is integrable everywhere. Q.E.D.

7.C. Proof of Theorem 1.3

We maintain notation and assumptions of Theorem 1.3. Corollary 3.6 implies the existence of an Abelian variety A and of a projective variety X' with at worst canonical singularities, with trivial canonical bundle and $\widetilde{q}(X') = 0$, together with a finite cover $A \times X' \to X$, étale in codimension one. Property (1.3.3) stated in Theorem 1.3 is hence fulfilled for any cover of the form $A \times \widetilde{X} \to A \times X'$, where $\widetilde{X} \to X'$ is a finite cover, étale in codimension one. The existence of such a cover $\widetilde{X} \to X$ and of a decomposition of $\mathscr{T}_{\widetilde{X}}$ satisfying Properties (1.3.1) and (1.3.2) follows by combining Corollary 7.3 and Theorem 7.11. In summary, this finishes the proof of Theorem 1.3. Q.E.D.

Remark 7.12. The decomposition theorem of Beauville-Bogomolov holds for compact Kähler manifolds. Therefore, we should expect a singular version in the non-algebraic context as well. In particular, Theorem 1.3 should hold for Kähler varieties. There are however two main

[1] Degneracy subsheaves are introduced and discussed in more detail in Section 8.B.2 below.

ingredients in our argument which are not yet available in the Kähler context: the Extension Theorem 2.4 and the pseudoeffectivity result Proposition 5.6.

§8. Towards a structure theory

If X is any projective manifold with Kodaira dimension zero, $\kappa(X) = 0$, standard conjectures of minimal model theory predict the existence of a birational contraction[2] map $\lambda : X \dashrightarrow X_\lambda$, where X_λ has terminal singularities and numerically trivial canonical divisor. Generalising the Beauville-Bogomolov Decomposition Theorem 1.1, it is widely expected that X_λ admits a finite cover, étale in codimension one, which can be birationally decomposed into a product

$$T \times \prod X_j,$$

where T is a torus and the X_j are singular versions of Calabi-Yau manifolds and irreducible symplectic manifold, which cannot be decomposed further. Such a decomposition result would clearly be a central pillar to any structure theory for varieties with Kodaira dimension zero. The main result of the present paper, Theorem 1.3, is a first step in this direction.

Section 8.A discusses the remaining problems of turning the decomposition found in the tangent sheaf into a decomposition of the variety. Section 8.B gives a conjectural description of the irreducible pieces coming out of the decomposition, discussing singular analogues of Calabi-Yau and irreducible holomorphic-symplectic varieties, and proving the conjectured description in low dimensions. Finally, fundamental groups of varieties with trivial canonical class, which are crucial for our understanding of this class of varieties, are discussed in the concluding Section 8.C.

Remark 8.1. Corollary 3.6 and Theorem 1.3 allow to restrict our attention to varieties with vanishing augmented irregularity. For most of the present Section 8, we will therefore only consider varieties X with $\widetilde{q}(X) = 0$.

8.A. Decomposing varieties with trivial canonical bundle

In technically correct terms, the setup of our discussion is now summarised as follows.

[2]Following standard use, we call a birational map a *contraction map* if its inverse does not contract any divisors.

Setup 8.2. Let X be a normal $\mathbb{Q}$-factorial projective variety with canonical singularities such that K_X is torsion and $\widetilde{q}(X) = 0$. By Theorem 1.3, there exists a finite cover $f : \widetilde{X} \to X$, étale in codimension one, such that $\omega_X = \mathscr{O}_X$ and such that there exists a decomposition

$$\mathscr{T}_{\widetilde{X}} = \bigoplus \mathscr{E}_i$$

of $\mathscr{T}_{\widetilde{X}}$ into strongly stable integrable reflexive subsheaves.

In view of the desired decomposition of the variety $\widetilde{X}$, this naturally leads to the following problems.

Problem 8.3 (Algebraicity of leaves). *In Setup 8.2, show that the leaves of the foliations $\mathscr{E}_i$ are algebraic, perhaps after passing to another cover.*

Problem 8.4 (Decomposition of the variety). *In the setup of Problem 8.3, show that the algebraicity of the leaves leads to a birational decomposition of $\widetilde{X}$, perhaps after passing to another cover. More precisely, show that there is a birational morphism*

$$g : \widetilde{X} \dashrightarrow \prod X_i,$$

isomorphic outside of a small set $V \subset \widetilde{X}$, such that the following holds.

(8.4.1) *The varieties X_j are smooth, projective with $\kappa(X_j) = 0$ for all j.*

(8.4.2) *If p_j denotes the composition of g with j^{th} projection $\Pi X_i \to X_j$, then $p_j^*(\mathscr{T}_{Y_j}) = \mathscr{E}_j$ over $X \setminus V$ for all j.*

Remark 8.5. Once it is known that the leaves of $\mathscr{E}_j$ are algebraic, one easily obtains rational maps $X \dashrightarrow Y_j$ to smooth projective varieties such that $\mathscr{E}_j = \mathscr{T}_{X/Y_j}$ generically. The main problem is now to show that the equality $\mathscr{E}_j = \mathscr{T}_{X/Y_j}$ holds everywhere, and that $\kappa(Y_j) = 0$.

A solution to Problem 8.4 is not the yet desired final outcome of our decomposition strategy for X: since K_X is (numerically) trivial one clearly aims for a decomposition into varieties with trivial canonical class. Assuming that the minimal model program works for varieties of Kodaira dimension zero, each X_j may be replaced by a minimal model X'_j. As a consequence we would obtain a birational map $g' : X \dashrightarrow \Pi X'_i$. If the singularities of X' are not only canonical but terminal, it follows from [Kaw08] that g' is isomorphic in codimension one and decomposes into a finite sequence of flops. One might hope that each terminal variety with numerically trivial canonical class decomposes into terminal varieties with trivial canonical class and strongly stable tangent bundle, after performing a finite cover, étale in codimension one, and after performing a finite number of flops.

8.B. Classifying the strongly stable pieces: Calabi-Yau and irreducible holomorphic-symplectic varieties

We start with after a short discussion of the notion of strong stability in Section 8.B.1, showing by way of example that varieties with strongly stable tangent sheaves are the "right" objects when building a structure theory for spaces of Kodaira dimension zero. The remainder of the present Section 8.B discusses these spaces in detail.

Sections 8.B.2 and 8.B.3 relate stability properties of the tangent bundle to non-degeneracy of differential forms, and discuss implications for the exterior algebra of reflexive forms. We apply these results in the concluding Section 8.B.4 to show that singular varieties with strongly stable tangent sheaf are in a very strong sense natural analogues of Calabi-Yau and irreducible holomorphic-symplectic manifolds, at least in dimension up to five. There is ample evidence to conjecture that this description holds in general, for strongly stable varieties of arbitrary dimension.

8.B.1. *Strong stability versus stability* At first sight, it seems tempting to consider varieties with stable tangent bundle as the building blocks of varieties with semistable tangent sheaf, such as varieties with trivial canonical bundle. However, the following example shows that *strong stability* is indeed the correct notion in our setup.

Example 8.6 (A variety with stable, but not strongly stable tangent sheaf). Let Z be a projective K3-surface, let $\widetilde{X} := Z \times Z$ with projections $p_1, p_2 : \widetilde{X} \to Z$, and let $\phi \in \operatorname{Aut}_{\mathscr{O}}(\widetilde{X})$ be the automorphism which interchanges the two factors. The quotient $X := \widetilde{X}/\langle\phi\rangle = \operatorname{Sym}^2(Z)$ is then a projective holomorphic-symplectic variety with trivial canonical bundle and rational Gorenstein singularities. The quotient map $\pi : \widetilde{X} \to X$ is finite and étale in codimension one. Let h be any ample polarisation on X. The tangent sheaf $\mathscr{T}_X$ of X is obviously not strongly stable.

However, we claim that $\mathscr{T}_X$ is h-stable. Indeed, suppose that there exists a non-trivial h-stable subsheaf $0 \subsetneq \mathscr{S} \subsetneq \mathscr{T}_X$ with slope zero that destabilises $\mathscr{T}_X$. Then, the reflexive pull-back $\widetilde{\mathscr{S}} := \pi^{[*]}(\mathscr{S})$ is $\pi^*(h)$-polystable, see [HL97, Lem. 3.2.3], and injects into $\mathscr{T}_{\widetilde{X}} = p_1^*(\mathscr{T}_Z) \oplus p_2^*(\mathscr{T}_Z)$. Since neither of the two sheaves $p_j^*\mathscr{T}_Z$ is stable under the action of ϕ, clearly $\widetilde{\mathscr{S}}$ is not one of these. More is true: looking at the maps to the two summands of $\mathscr{T}_{\widetilde{X}}$ and using that morphisms between stable sheaves with the same slope are either trivial or isomorphic, we see that

$\mathscr{S}$ has to be stable of rank two, and isomorphic to both $p_1^*\mathscr{T}_Z$ and $p_2^*\mathscr{T}_Z$. This is absurd, as restriction to p_1-fibers shows that the sheaves $p_1^*\mathscr{T}_Z$ and $p_2^*\mathscr{T}_Z$ are in fact not isomorphic.

8.B.2. *Non-degeneracy of differential forms* If X is a canonical variety with numerically trivial canonical class, stability of the tangent bundle has strong implications for the geometry of differential forms X. This section is concerned with degeneracy properties. Conjectural consequences for the structure of the exterior algebra of forms are discussed in the subsequent Section 8.B.3.

Non-degeneracy of differential forms will be measured using the following definition.

Definition 8.7 (Contraction of a reflexive form, degeneracy subsheaf). Let X be a normal complex variety, let $0 < q \leq \dim X$ be any number and $\sigma \in H^0\big(X, \Omega_X^{[q]}\big)$ any reflexive form. The *contraction map of* σ is the unique sheaf morphism

$$i_\sigma : \mathscr{T}_X \to \Omega_X^{[q-1]}$$

whose restriction to X_{reg} is given by $\vec{u} \mapsto \sigma(\vec{u}, \cdot)$. Let $S_\sigma := \ker(i_\sigma)$ be the kernel of i_σ. We call $S_\sigma \subseteq \mathscr{T}_X$ the *degeneracy subsheaf* of the reflexive form σ. If $S_\sigma = 0$, we say that σ is *generically non-degenerate.*

The main result of the present section asserts that in our setup, forms never degenerate. This can be seen as first evidence for the conjectural classification of the stable pieces into "Calabi-Yau" and "irreducible holomorphic-symplectic" which we discuss later in Section 8.B.4 below.

Proposition 8.8 (Non-degeneracy of forms on canonical varieties with stable $\mathscr{T}_X$). *Let X be a normal n-dimensional projective variety X having at worst canonical singularities, $n > 1$. Assume that the canonical divisor K_X is numerically trivial, and that the tangent sheaf $\mathscr{T}_X$ is stable with respect to some ample polarisation. If σ is any non-zero reflexive form on X, then σ is generically non-degenerate, in the sense of Definition 8.7.*

Proof. We argue by contradiction and assume that there exists a reflexive q-form σ whose degeneracy subsheaf does not vanish, $S_\sigma \neq 0$. Consider the exact sequence

$$0 \to S_\sigma \to \mathscr{T}_X \xrightarrow{i(\sigma)} \underbrace{\mathrm{Image}\, i(\sigma)}_{=:\mathscr{E} \subseteq \Omega_X^{[q-1]}} \to 0. \tag{8.8.1}$$

Recalling from Proposition 5.7 that $\mathscr{T}_X$ is stable with respect to any ample polarisation, we choose an ample Cartier divisor H on X, a sufficiently large number m, and let $(D_j)_{1\le j\le n-1} \in |mH|$ be general elements. Consider the corresponding general complete intersection curve $C := D_1 \cap \cdots \cap D_{n-1} \subsetneq X$, which avoids the singular locus of X.

Since K_X is torsion, the Kodaira-dimension of X is zero, $\kappa(X) = 0$. As X has only canonical singularities, this implies that X is not covered by rational curves. Miyaoka's Generic Semipositivity Theorem 5.1 therefore asserts that the vector bundle $\mathscr{T}_X|_C \cong \big(\Omega_X^{[n-1]} \otimes \omega_X^*\big)\big|_C$ is nef. This has two consequences in our setup. On the one hand, since $\mathscr{E}$ is a quotient of $\mathscr{T}_X$, it follows that $\mathscr{E}|_C$ and $\det \mathscr{E}|_C$ are nef. On the other hand, since $\mathscr{E} \subseteq \Omega_X^{[q-1]}$ by definition, its dual $\mathscr{E}^*|_C$ is a quotient of $\wedge^{q-1}\mathscr{T}_X|_C$, and it follows that $\mathscr{E}^*|_C$ and $\det \mathscr{E}^*|_C$ are likewise nef. Consequently, we obtain $\det \mathscr{E}|_C \equiv 0$. The exact Sequence (8.8.1) then implies that S_σ destabilises $\mathscr{T}_X$. This contradicts the assumed stability of $\mathscr{T}_X$, and finishes the proof of Proposition 8.8. Q.E.D.

Corollary 8.9 (Reflexive two-forms on canonical varieties with stable $\mathscr{T}_X$, I)**.** *In the setup of Proposition 8.8,* $h^0\big(X, \Omega_X^{[2]}\big) \le 1$.

Proof. We argue by contradiction and assume that there are two linearly independent forms $\sigma_1, \sigma_2 \in H^0\big(\widetilde{X}, \Omega_{\widetilde{X}}^{[2]}\big)$. Since both forms are non-degenerate by Proposition 8.8, they induce linearly independent isomorphisms $\phi_\bullet : \mathscr{T}_{\widetilde{X}} \to \Omega_{\widetilde{X}}^{[1]}$. The composition $\phi_1^{-1} \circ \phi_2$ is thus a non-trivial automorphism of $\mathscr{T}_{\widetilde{X}}$. We obtain that the stable sheaf $\mathscr{T}_{\widetilde{X}}$ is not simple, contradicting [HL97, Cor. 1.2.8] and thereby finishing the proof of Corollary 8.9. Q.E.D.

Corollary 8.10 (Reflexive two-forms on canonical varieties with stable $\mathscr{T}_X$, II)**.** *In the setup of Proposition 8.8, if there exists a non-trivial reflexive two-form* $\sigma \in H^0\big(X, \Omega_X^{[2]}\big)$*, then* σ *is a complex-symplectic form on the smooth part of* X*. In particular,* $\dim X$ *is even,* ω_X *is trivial, and* X *has only rational Gorenstein singularities.*

Proof. Proposition 8.8 implies that the non-degeneracy subsheaf S_σ vanishes. For general points $x \in X_{\mathrm{reg}}$, this implies that $\sigma|_x$ is a non-degenerate, and hence symplectic, form on the vector space $T_X|_x$. This already shows that the dimension of X is even, say $\dim X = 2k$. If $\tau \in H^0\big(X, \omega_X\big)$ is the section induced by $\wedge^k\sigma$, then τ does not vanish at x.

To prove that σ is a complex-symplectic form on the smooth part of X, we need to show that non-degeneracy holds at arbitrary points of X_{reg}. If not, there exists a point $y \in X_{\mathrm{reg}}$ such that $\sigma|_y$ is a degenerate

2-form on the vector space $T_X|_y$. The form τ will therefore vanish at y, showing that K_X can be represented by a non-trivial, effective $\mathbb{Q}$-Cartier divisor, contradicting the assumption that K_X is numerically trivial.

The remaining assertions of Corollary 8.10 follow immediately. Q.E.D.

8.B.3. *Exterior algebras of differential forms on the strongly stable pieces* The algebra of differential forms on irreducible holomorphic-symplectic and Calabi-Yau manifolds has a rather simple structure cf. [Bea83, Props. 1 and 4]. In order to characterise the strongly stable pieces in the singular case one would need a similar description which we formulate as the following problem.

Problem 8.11 (Forms on varieties with strongly stable tangent bundle). *Let X be a normal projective variety of dimension $n > 1$ with $\omega_X \cong \mathscr{O}_X$, having at worst canonical singularities. Assume that the tangent sheaf $\mathscr{T}_X$ is strongly stable. Then show that the following holds.*

(8.11.1) *For all odd numbers $q \neq n$, we have $H^0(\widetilde{X}, \Omega^{[q]}_{\widetilde{X}}) = 0$ for all finite covers $f : \widetilde{X} \to X$, étale in codimension one.*

(8.11.2) *If there exists a finite cover $g : X' \to X$, étale in codimension one, and an even number $0 < q < n$ such that $H^0(X', \Omega^{[q]}_{X'}) \neq 0$, then there exists a reflexive 2-form $\sigma' \in H^0(X', \Omega^{[2]}_{X'})$, symplectic on the smooth locus X'_{reg}, such that for any finite cover $f : \widetilde{X} \to X'$, étale in codimension one, the exterior algebra of global reflexive forms on $\widetilde{X}$ is generated by $f^*(\sigma')$. In other words,*

$$\bigoplus_p H^0\Big(\widetilde{X}, \Omega^{[p]}_{\widetilde{X}}\Big) = \mathbb{C}\big[f^*(\sigma)\big].$$

Remark 8.12. Notice that the assumptions on the strong stability of $\mathscr{T}_X$ and on the dimension of X automatically imply $\widetilde{q}(X) = 0$.

As we will discuss in more detail in the subsequent Section 8.B.4, a positive solution to Problem 8.11 leads to a characterisation of canonical varieties with trivial canonical class and strongly stable tangent bundle as singular analogues of Calabi-Yau or irreducible holomorphic symplectic manifolds. There are a number of cases where Problem 8.11 can be solved. We conclude Section 8.B.3 with two propositions that provide evidence by discussing the case where X is smooth, or of dimension ≤ 5, respectively.

Proposition 8.13. *The claims of Problem 8.11 hold if X is smooth.*

Proof. Note that on a smooth variety X the sheaves Ω^p_X and $\Omega^{[p]}_X$ coincide and that any finite cover of X that is étale in codimension one is actually étale by purity of the branch locus.

Let now X be a smooth projective variety of dimension n with $\omega_X \cong \mathscr{O}_X$. Assume that the tangent bundle $\mathscr{T}_X$ is strongly stable. As noticed in Remark 8.12, this implies that $\widetilde{q}(X) = 0$. Consequently, the fundamental group of X is finite by [Bea83, Thm. 2(2)]. Let $\widehat{X} \to X$ be the universal cover. Since $\mathscr{T}_X$ is strongly stable, $\mathscr{T}_{\widehat{X}}$ is stable with respect to any polarisation, and $\widehat{X}$ is hence irreducible in the sense of the Beauville-Bogomolov decomposition Theorem 1.1. As $\widetilde{q}(X) = 0$, the manifold $\widehat{X}$ is therefore either Calabi-Yau or irreducible holomorphic-symplectic.

In order to show (8.11.1), pulling back forms from any étale cover $\widetilde{X} \to X$ to the universal cover $\widehat{X}$ if necessary, it suffices to note that both in the Calabi–Yau and in the irreducible holomorphic-symplectic case, $\widehat{X}$ does not support differential forms of odd degree $p < n$ by [Bea83, Props. 1 and 3].

To show (8.11.2), let $X' \to X$ be any étale cover, and assume that there exists a non-vanishing form such that that $\sigma' \in H^0\big(X', \Omega^{[q]}_{X'}\big)$ for some even number $0 < q < n$. Pulling back σ' to the universal cover $\widehat{X}$, we see that $\widehat{X}$ cannot be Calabi–Yau and is therefore irreducible holomorphic-symplectic, say with symplectic form $\widehat{\sigma}$. Consequently, [Bea83, Prop. 3] implies that the algebra of differential forms on $\widehat{X}$ is generated by $\widehat{\sigma}$. Hence, in order to establish the claim it therefore suffices to show that $\widehat{X}$ is biholomorphic to X and therefore also to X'. In other words, we need to show that X is already simply-connected. This is done in Lemma 8.14 below. Q.E.D.

We are grateful to Keiji Oguiso for pointing us towards [OS11] and for explaining the following observation to us.

Lemma 8.14. *Let X be a projective manifold whose universal cover is an irreducible holomorphic-symplectic manifold. If the canonical bundle of X is trivial, $\omega_X \cong \mathscr{O}_X$, then X is simply-connected, and therefore itself irreducible holomorphic-symplectic.*

Proof. The assumptions on X imply that X is an *Enriques manifold* in the sense of Oguiso and Schröer [OS11], see also [BNWS11]. Since the canonical bundle of X is trivial, and since the universal cover of X is irreducible holomorphic-symplectic, the fundamental group of X is finite, cf. [Bea83, Thm. 2(2)]. Let d denote the degree of the universal covering map $\widehat{X} \to X$, and set $\dim X = \dim \widehat{X} = n = 2k$. It then follows from [OS11, Prop. 2.4] that $d \mid (k+1)$. Moreover, since X is assumed to have trivial canonical bundle, [OS11, Prop. 2.6] implies that additionally $d \mid k$. Consequently, we have $d = 1$, which proves the claim. Q.E.D.

Proposition 8.15. *The claims of Problem 8.11 hold if* $\dim X \leq 5$.

Proof. Let X be a projective variety of dimension greater than one, having at worst canonical singularities. Assume that X has a trivial canonical bundle, $\omega_X \cong \mathscr{O}_X$, and a strongly stable tangent sheaf $\mathscr{T}_X$. Again we have $\widetilde{q}(X) = 0$, since $\mathscr{T}_X$ is strongly stable. Fix a finite cover $\widetilde{X} \to X$, étale in codimension one.

If $\dim X = 3$, then Corollary 6.11 immediately implies that $h^0\big(\widetilde{X}, \Omega^{[1]}_{\widetilde{X}}\big) = h^0\big(\widetilde{X}, \Omega^{[2]}_{\widetilde{X}}\big) = 0$. Conditions (8.11.1) and (8.11.2) of Problem 8.11 are therefore satisfied.

Now assume that $\dim X = 4$. In this setting, Corollary 6.11 gives that $h^0\big(\widetilde{X}, \Omega^{[1]}_{\widetilde{X}}\big) = h^0\big(\widetilde{X}, \Omega^{[3]}_{\widetilde{X}}\big) = 0$. The claims of Problem 8.11 thus follow from Corollary 8.10 and from the fact that $h^0\big(\widetilde{X}, \Omega^{[2]}_{\widetilde{X}}\big) \leq 1$, as shown in Corollary 8.9.

It remains to consider the case where $\dim X = 5$, where Corollary 6.11 asserts that $h^0\big(\widetilde{X}, \Omega^{[1]}_{\widetilde{X}}\big) = h^0\big(\widetilde{X}, \Omega^{[4]}_{\widetilde{X}}\big) = 0$. The claims of Problem 8.11 will follow once we show that $h^0\big(\widetilde{X}, \Omega^{[2]}_{\widetilde{X}}\big)$ and $h^0\big(\widetilde{X}, \Omega^{[3]}_{\widetilde{X}}\big)$ vanish as well. For that, recall from item (6.10.2) of Corollary 6.10 that there exists a non-trivial 3-form on $\widetilde{X}$ if and only if there exists a non-trivial reflexive 2-form on $\widetilde{X}$. However, by Corollary 8.10 any non-trivial 2-form would be non-degenerate, forcing $\dim X$ to be even, a contradiction. Q.E.D.

8.B.4. *Calabi-Yau and holomorphic-symplectic varieties* The following definition is motivated by the description of the exterior algebra of Calabi-Yau manifolds and irreducible holomorphic-symplectic manifolds, [Bea83, Props. 1 and 4], and by the discussion of Problem 8.11 in the previous section.

Definition 8.16 (Calabi-Yau and symplectic varieties in the singular case). Let X be a normal projective variety with $\omega_X \cong \mathscr{O}_X$, having at worst canonical singularities.

(8.16.1) We call X *Calabi-Yau* if $H^0\big(\widetilde{X}, \Omega^{[q]}_{\widetilde{X}}\big) = 0$ for all numbers $0 < q < \dim X$ and all finite covers $\widetilde{X} \to X$, étale in codimension one.

(8.16.2) We call X *irreducible holomorphic-symplectic* if there exists a reflexive 2-form $\sigma \in H^0\big(X, \Omega^{[2]}_X\big)$ such that σ is everywhere non-degenerate on X_{reg}, and such that for all finite covers $f : \widetilde{X} \to X$, étale in codimension one, the exterior algebra of global reflexive forms is generated by $f^*(\sigma)$.

Remark 8.17 (Augmented irregularity of Calabi-Yau and symplectic varieties). If X is Calabi-Yau or irreducible holomorphic-symplectic in the sense of Definition 8.16, it follows immediately that the augmented irregularity of X vanishes, $\widetilde{q}(X) = 0$.

Remark 8.18 (Definition (8.16.1) for "Calabi-Yau" in the smooth case). By [Bea83, Sect. 3, Prop. 2] the conditions spelled out in (8.16.1) are in the smooth case equivalent to the existence of a Kähler metric with holonomy $SU(m)$. If X is smooth and Calabi-Yau in the sense of Definition 8.16, then X is not necessarily simply-connected, but may have finite fundamental group. If we assume additionally that $\dim X$ is even, then a simple computation with holomorphic Euler characteristics shows that X is in fact simply-connected, cf. [Bea83, Prop. 2 and Rem.].

Remark 8.19 (Definition (8.16.2) for "irreducible symplectic" in the smooth case). If X is smooth and irreducible holomorphic-symplectic in the sense of Definition 8.16, then X is simply-connected. In fact, even without the condition on the algebra of differential forms on étale covers, if X is a holomorphic-symplectic manifold of complex dimension $2n$ such that

$$H^{k,0}(X) \cong \begin{cases} \mathbb{C} & \text{if } k \text{ is even} \\ 0 & \text{if } k \text{ is odd,} \end{cases}$$

then X is simply-connected, that is, X is an irreducible holomorphic-symplectic manifold, see [HNW11, Prop. A.1].

Assuming Problem 8.11 can be solved, the following two propositions provide a classification of the strongly stable pieces in the conjectural version of the Beauville-Bogomolov decomposition for the singular case.

Proposition 8.20 (Characterisation of strongly stable pieces, I). *Let X be Calabi-Yau or irreducible holomorphic-symplectic in the sense of Definition 8.16. Then $\mathscr{T}_X$ is strongly stable in the sense of Definition 7.2.*

Proof. Let X be Calabi-Yau or irreducible symplectic. We argue by contradiction and assume that there exists a finite cover $g : \widetilde{X} \to X$, étale in codimension one, and ample Cartier divisors $\widetilde{H}_1, \dots \widetilde{H}_{n-1}$ on $\widetilde{X}$ such that the tangent sheaf $\mathscr{T}_{\widetilde{X}}$ is not stable with respect to the $\widetilde{H}_i$. In this setting, Theorem 7.1 asserts that there exists a further finite cover $h : \widehat{X} \to \widetilde{X}$ and a proper decomposition

$$\mathscr{T}_{\widehat{X}} \cong \mathscr{E} \oplus \mathscr{F}. \tag{8.20.1}$$

with $\det \mathscr{E} \cong \mathscr{O}_{\widehat{X}}$. Setting $r := \operatorname{rank} \mathscr{E}$, the splitting (8.20.1) immediately gives an embedding $\mathscr{O}_{\widehat{X}} \cong \det \mathscr{E}^* \hookrightarrow \Omega^{[r]}_{\widehat{X}}$, and an associated form $\tau \in H^0\big(X, \Omega^{[r]}_{\widehat{X}}\big)$. Since $0 < r < \dim X$, it follows that X cannot be Calabi-Yau.

Since $\mathscr{F}$ is contained in the degeneracy subsheaf S_τ, as introduced in Definition 8.7, it is clear that τ cannot be a wedge-power of the pull-back of any symplectic form on X. This rules out that X is irreducible holomorphic-symplectic in the sense of Definition 8.16. We obtain a contradiction, which finishes the proof of Proposition 8.20. Q.E.D.

A positive solution to Problem 8.11 would immediately give a partial converse to Proposition 8.20.

Proposition 8.21 (Characterisation of strongly stable pieces, II). *Let X be a normal projective variety with $\omega_X = \mathscr{O}_X$ having at worst canonical singularities. Assume that $\mathscr{T}_X$ is strongly stable. If the assertions of Problem 8.11 hold, then either*

(8.21.1) *the semistable sheaf $\wedge^{[2]} \mathscr{T}_X$ is strongly stable, and X is Calabi-Yau, or*

(8.21.2) *there exists a finite cover $\widetilde{X} \to X$, étale in codimension one, such that the sheaf $\wedge^{[2]} \mathscr{T}_{\widetilde{X}}$ is not $\widetilde{H}$-stable for some polarisation $\widetilde{H}$ on $\widetilde{X}$, and $\widetilde{X}$ is irreducible holomorphic-symplectic.* *Q.E.D.*

Remark 8.22. In the second case of the previous proposition one would of course rather like X itself to be irreducible holomorphic-symplectic. This is in fact true if X is additionally assumed to be smooth: If X is not Calabi–Yau, then the universal cover $\widetilde{X}$ of X is irreducible holomorphic-symplectic. Since additionally the canonical bundle ω_X is assumed to be trivial, Lemma 8.14 implies that X itself is irreducible holomorphic-symplectic.

8.C. Fundamental groups of varieties with trivial canonical class

A Kähler manifold X with trivial canonical class and vanishing augmented irregularity $\widetilde{q}(X)$ has finite fundamental group, see [Bea83, Thm. 1]. We believe that the same should hold for projective varieties, in our singular setting. We show that this is true, at least under the assumption that $\chi(X, \mathscr{O}_X) \neq 0$.

Proposition 8.23 (Fundamental groups of canonical varieties with $K_X \equiv 0$, I). *Let X be a normal projective variety with at worst canonical* *singularities. If K_X is torsion and if $\chi(X, \mathscr{O}_X) \neq 0$, then $\pi_1(X)$ is finite, of cardinality*

$$|\pi_1(X)| \leq \frac{2^{n-1}}{|\chi(X, \mathscr{O}_X)|}.$$

Remark 8.24. If X is smooth and K_X is torsion, then then classical Beauville-Bogomolov Decomposition Theorem 1.1 together with Proposition 8.23 shows that $\chi(X, \mathscr{O}_X) \neq 0$ implies $\widetilde{q}(X) = 0$.

Proof of Proposition 8.23. Set $n := \dim X$. Let $\pi : \widetilde{X} \to X$ be a strong resolution of singularities. Recalling from [KM98, Thm. 5.22] that X has rational singularities, we obtain that $\chi(\widetilde{X}, \mathscr{O}_{\widetilde{X}}) = \chi(X, \mathscr{O}_X) \neq 0$. Consider the invariant

$$\kappa^+\big(\widetilde{X}\big) := \max\Big\{\kappa\big(\det \mathscr{F}\big) \mid \mathscr{F} \subseteq \Omega^p_{\widetilde{X}} \text{ a coherent subsheaf, for some } p\Big\}.$$

We are going to show that $\kappa^+\big(\widetilde{X}\big) = 0$. Using the assumption that $\chi(\widetilde{X}, \mathscr{O}_{\widetilde{X}}) \neq 0$, Campana has then shown in [Cam95, Cor. 5.3] that $\pi_1(\widetilde{X})$ is finite, of cardinality at most $2^{n-1} \cdot |\chi(\widetilde{X}, \mathscr{O}_{\widetilde{X}})|^{-1}$. Since the natural map $\pi_1\big(\widetilde{X}\big) \to \pi_1(X)$ is isomorphic by [Tak03, Thm. 1.1], this implies that $\pi_1(X)$ is likewise finite of the same cardinality.

So let $0 \leq p \leq n$ be any number and let $\mathscr{F} \subseteq \Omega^p_{\widetilde{X}}$ be a coherent subsheaf. As a subsheaf of a torsion-free sheaf, $\mathscr{F}$ is itself torsion-free, and therefore locally free in codimension one. Next, let $C \subset X$ be a general complete intersection curve. Recall that the strong resolution map π is isomorphic along C, and denote the preimage curve by $\widetilde{C} := \pi^{-1}(C)$. The restricted sheaves $\Omega^p_{\widetilde{X}}|_{\widetilde{C}}$ and $\mathscr{F}|_{\widetilde{C}}$ are then both locally free.

Since K_X is torsion, the Kodaira-dimension of X is zero, $\kappa(X) = 0$. As X has only canonical singularities, this shows that X is not covered by rational curves. Miyaoka's Generic Semipositivity Theorem 5.1 therefore implies that $\Omega^q_{\widetilde{X}}|_{\widetilde{C}}$ is nef for all q. Better still, we have $\deg \Omega^n_{\widetilde{X}}|_{\widetilde{C}} = 0$, so that

$$\big(\Omega^p_{\widetilde{X}}|_{\widetilde{C}}\big)^* \cong \wedge^p \mathscr{T}_X|_{\widetilde{C}} \cong \operatorname{Hom}\Big(\Omega^n_{\widetilde{X}}|_{\widetilde{C}}, \Omega^{n-p}_{\widetilde{X}}|_{\widetilde{C}}\Big) \cong \big(\Omega^n_{\widetilde{X}}|_{\widetilde{C}}\big)^* \otimes \Omega^{n-p}_{\widetilde{X}}|_{\widetilde{C}}$$

is likewise as a nef vector bundle on the curve $\widetilde{C}$. Its quotient $\mathscr{F}^*|_{\widetilde{C}}$ is then nef as well. In summary, we obtain that $c_1(\mathscr{F}) \cdot \widetilde{C} \leq 0$. Since the curves $\widetilde{C}$ are moving, this implies $\kappa(\det \mathscr{F}) \leq 0$, and therefore $\kappa^+\big(\widetilde{X}\big) \leq 0$. Since $\kappa\big(\widetilde{X}\big) = 0$, we obtain $\kappa^+\big(\widetilde{X}\big) = 0$, as claimed. This finishes the proof of Proposition 8.23. Q.E.D.

Corollary 8.25 (Fundamental groups of canonical varieties with $K_X \equiv 0$, II)**.** *Let X be a normal projective variety with at worst canonical singularities. Assume that* $\dim X \leq 4$*, and that the canonical divisor K_X is numerically trivial. Then $\pi_1(X)$ is almost Abelian, that is, $\pi_1(X)$ contains an Abelian subgroup of finite index.*

Proof. Recall from [Kol95, 4.17.3] that the statement of Corollary 8.25 is well-known if $\dim X \leq 3$. We will therefore assume for the remainder of the proof that X is of dimension four.

Let $f : \widetilde{X} \to X$ be the index-one cover associated with K_X. As we have noted before, f is étale in codimension one, $\widetilde{X}$ has canonical singularities, and $\omega_{\widetilde{X}} = \mathscr{O}_{\widetilde{X}}$, cf. [KM98, 5.19 and 5.20]. The image of the natural map $\pi_1(\widetilde{X}) \to \pi_1(X)$ has finite index in $\pi_1(X)$, cf. [Kol95, Prop. 2.10(2)] and [Cam91, Prop. 1.3]. Replacing X by $\widetilde{X}$, if necessary, we may therefore assume without loss of generality that ω_X is trivial. Passing to a further cover, Corollary 3.6 even allows to assume that X is of the form $X = A \times Z$, where A is an Abelian variety, and Y is normal projective variety with at worst canonical singularities, with trivial canonical class and vanishing augmented irregularity, $\omega_Z \cong \mathscr{O}_Z$ and $\widetilde{q}(Z) = 0$.

If $\dim Z \leq 3$, then [Kol95, 4.17.3] asserts that $\pi_1(Z)$ is almost Abelian. Since $\pi_1(A)$ is Abelian, this finishes the proof.

It remains to consider that case where $\dim Z = 4$, that is, where $X = Z$ and $\widetilde{q}(Z) = 0$. In this case, we finish proof by showing that the fundamental group of X is finite. Recall from Corollary 6.11 that X does not carry any reflexive 1-form or 3-forms. Using Proposition 6.9 to relate $H^p\big(X, \mathscr{O}_X\big)$ with the space of reflexive p-forms we see that $\chi(X, \mathcal{O}_X) > 0$, and we conclude by Proposition 8.23 that $\pi_1(X)$ is finite, thus finishing the proof of Corollary 8.25. Q.E.D.

Remark 8.26 (Fundamental groups of smooth 4-folds with $\kappa = 0$). If X is a smooth projective 4-fold with $\kappa(X) = 0$ admitting a good minimal model X', then $\pi_1(X) = \pi_1(X')$ by [Tak03, Thm. 1.1] or [Kol93, Thm. 7.8.1]. Consequently, $\pi_1(X)$ is almost Abelian.

Assuming that the claims of Problem 8.11 hold, the following corollary complements the results obtained in Section 6, and in particular the results of Corollary 6.10.

Corollary 8.27 (Fundamental groups of even-dim. X with $\mathscr{T}_X$ strongly stable)**.** *Let X be a normal projective variety with $\omega_X \cong \mathscr{O}_X$ having at worst canonical singularities. Suppose furthermore that* $\dim X$ *is even and that $\mathscr{T}_X$ is strongly stable. If Problem 8.11 has a positive solution, then $\chi(X, \mathscr{O}_X) > 0$ and $\pi_1(X)$ is finite.* *Q.E.D.*

References

[BCHM10] Caucher Birkar, Paolo Cascini, Christopher D. Hacon, and James McKernan, *Existence of minimal models for varieties of log general type*, Journal of the AMS **23** (2010), 405–468, DOI:10.1090/S0894-0347-09-00649-3. MR 2601039 (2011f:14023)

[Bea83] Arnaud Beauville, *Variétés Kähleriennes dont la première classe de Chern est nulle*, J. Differential Geom. **18** (1983), no. 4, 755–782 (1984). MR 730926 (86c:32030)

[BNWS11] Samuel Boissière, Marc Nieper-Wißkirchen, and Alessandra Sarti, *Higher dimensional Enriques varieties and automorphisms of generalized Kummer varieties*, J. Math. Pures Appl. (9). **95** (2011), no. 5. 553–563. DOI:10.1016/j.matpur.2010.12.003. MR 2786223

[Bog74] Fedor A. Bogomolov, *Kähler manifolds with trivial canonical class*, Izv. Akad. Nauk SSSR Ser. Mat. **38** (1974), 11–21. MR 0338459 (49 #3223)

[Cam91] Frédéric Campana, *On twistor spaces of the class $\mathscr{C}$*, J. Differential Geom. **33** (1991), no. 2, 541–549. MR 1094468 (92g:32059)

[Cam95] ———, *Fundamental group and positivity of cotangent bundles of compact Kähler manifolds*, J. Algebraic Geom. **4** (1995), no. 3, 487–502. MR 1325789 (96f:32054)

[CP11] Frédéric Campana and Thomas Peternell, *Geometric stability of the cotangent bundle and the universal cover of a projective manifold*, Bull. Soc. Math. France **139** (2011), no. 1, 41–74.

[Dem02] Jean-Pierre Demailly, *On the Frobenius integrability of certain holomorphic p-forms*, Complex geometry (Göttingen, 2000), Springer, Berlin, 2002, pp. 93–98. MR 1922099 (2003f:32029)

[Dem09] ———, *Complex Analytic and Differential Geometry*, September 2009, OpenContent Book.
URL: http://www-fourier.ujf-grenoble.fr/~demailly/books.html.

[Fle84] Hubert Flenner, *Restrictions of semistable bundles on projective varieties*, Comment. Math. Helv. **59** (1984), no. 4, 635–650. MR 780080 (86m:14014)

[GKK10] Daniel Greb, Stefan Kebekus, and Sándor J Kovács, *Extension theorems for differential forms, and Bogomolov-Sommese vanishing on log canonical varieties*, Compos. Math. **146** (2010), 193–219, DOI:10.1112/S0010437X09004321. MR 2581247 (2011c:14054). A slightly extended version is available as arXiv:0808.3647.

[GKKP11] Daniel Greb, Stefan Kebekus, Sándor Kovács, and Thomas Peternell, *Differential forms on log canonical spaces*, Publ. Math. IHES. **114** (2011), no. 1, 87–169. DOI:10.1007/s10240-011-0036-0. MR 2854859. An extended version with additional graphics is available as arXiv:1003.2913.

[HL97] Daniel Huybrechts and Manfred Lehn, *The geometry of moduli spaces of sheaves*, Aspects of Mathematics, E31, Friedr. Vieweg & Sohn, Braunschweig, 1997. MR 1450870 (98g:14012)

[HNW11] Daniel Huybrechts and Marc Nieper-Wisskirchen, *Remarks on derived equivalences of Ricci-flat manifolds*, Math. Z. **267** (2011), no. 3-4, 939–963, DOI:10.1007/s00209-009-0655-z. MR 2776067

[Hör07] Andreas Höring, *Uniruled varieties with split tangent bundle*, Math. Z. **256** (2007), no. 3, 465–479, DOI:10.1007/s00209-006-0072-5. MR 2299565 (2008f:32024)

[Kaw85] Yujiro Kawamata, *Minimal models and the Kodaira dimension of algebraic fiber spaces*, J. Reine Angew. Math. **363** (1985), 1–46, DOI:10.1515/crll.1985.363.1. MR 814013 (87a:14013)

[Kaw08] ______, *Flops connect minimal models*, Publ. Res. Inst. Math. Sci. **44** (2008), no. 2, 419–423, DOI:10.2977/prims/1210167332. MR 2426353 (2009d:14011)

[Keb13] Stefan Kebekus, *Differential forms on singular spaces, the minimal model program, and hyperbolicity of moduli stacks*, in: Handbook of Moduli, Volume II, editors Gavril Farkas and Ian Morrison. Number 26 in Advanced Lectures in Mathematics, International Press, Beijing, 2013. Also available as arXiv:1107.4239.

[Kle66] Steven L. Kleiman, *Toward a numerical theory of ampleness*, Ann. of Math. (2) **84** (1966), 293–344. MR 0206009 (34 #5834)

[KM98] János Kollár and Shigefumi Mori, *Birational geometry of algebraic varieties*, Cambridge Tracts in Mathematics, vol. 134, Cambridge University Press, Cambridge, 1998, With the collaboration of C. H. Clemens and A. Corti, Translated from the 1998 Japanese original. MR 2000b:14018

[Kob87] Shoshichi Kobayashi, *Differential geometry of complex vector bundles*, Publications of the Mathematical Society of Japan, vol. 15, Iwanami Shoten and Princeton University Press, Princeton, NJ, 1987, Kanô Memorial Lectures, 5. MR 909698 (89e:53100)

[Kol93] János Kollár, *Shafarevich maps and plurigenera of algebraic varieties*, Invent. Math. **113** (1993), no. 1, 177–215, DOI:10.1007/BF01244307. MR 1223229 (94m:14018)

[Kol95] ______, *Shafarevich maps and automorphic forms*, M. B. Porter Lectures, Princeton University Press, Princeton, NJ, 1995. MR 1341589 (96i:14016)

[Miy87a] Yoichi Miyaoka, *The Chern classes and Kodaira dimension of a minimal variety*, Algebraic geometry, Sendai, 1985, Adv. Stud. Pure Math., vol. 10, North-Holland, Amsterdam, 1987, pp. 449–476. MR 89k:14022

[Miy87b] ______, *Deformations of a morphism along a foliation and applications*, Algebraic geometry, Bowdoin, 1985 (Brunswick, Maine, 1985), Proc. Sympos. Pure Math., vol. 46, Amer. Math. Soc., Providence, RI, 1987, pp. 245–268. MR 927960 (89e:14011)

[MR82] Vikram B. Mehta and Annamalai Ramanathan, *Semistable sheaves on projective varieties and their restriction to curves*, Math. Ann. **258** (1981/82), no. 3, 213–224. MR 83f:14013

[OS11] Keiji Oguiso and Stefan Schröer, *Enriques manifolds*, J. Reine Angew. Math. **661** (2011), 215–235, DOI:10.1515/CRELLE.2011.077.

[OSS80] Christian Okonek, Michael Schneider, and Heinz Spindler, *Vector bundles on complex projective spaces*, Progress in Mathematics, vol. 3, Birkhäuser Boston, Mass., 1980. MR 561910 (81b:14001)

[Pet94] Thomas Peternell, *Minimal varieties with trivial canonical classes. I*, Math. Z. **217** (1994), no. 3, 377–405, DOI:10.1007/BF02571950. MR 1306667 (96f:14041)

[Rei87] Miles Reid, *Young person's guide to canonical singularities*, Algebraic geometry, Bowdoin, 1985 (Brunswick, Maine, 1985), Proc. Sympos. Pure Math., vol. 46, Amer. Math. Soc., Providence, RI, 1987, pp. 345–414. MR 927963 (89b:14016)

[Ros68] Hugo Rossi, *Picard variety of an isolated singular point*, Rice Univ. Studies **54** (1968), no. 4, 63–73. MR 0244517 (39 #5831)

[Tak03] Shigeharu Takayama, *Local simple connectedness of resolutions of log-terminal singularities*, Internat. J. Math. **14** (2003), no. 8, 825–836, DOI:10.1142/S0129167X0300196X. MR 2013147 (2004m:14023)

Daniel Greb, Essener Seminar für Algebraische Geometrie und Arithmetik, Fakultät für Mathematik, Universität Duisburg-Essen, 45117 Essen, Germany
E-mail address: daniel.greb@uni-due.de
URL: http://www.esaga.uni-due.de/daniel.greb/

Stefan Kebekus, Mathematisches Institut, Albert-Ludwigs-Universität Freiburg, Eckerstraße 1, 79104 Freiburg im Breisgau, Germany
E-mail address: stefan.kebekus@math.uni-freiburg.de
URL: http://home.mathematik.uni-freiburg.de/kebekus

Thomas Peternell, Mathematisches Institut, Universität Bayreuth, 95440 Bayreuth, Germany
E-mail address: thomas.peternell@uni-bayreuth.de
URL: http://www.komplexe-analysis.uni-bayreuth.de/de/team/Peternell_Thomas

Advanced Studies in Pure Mathematics 70, 2016
Minimal Models and Extremal Rays (Kyoto, 2011)
pp. 115–214

Resolution of singularities of an idealistic filtration in dimension 3 after Benito-Villamayor

Hiraku Kawanoue and Kenji Matsuki

Dedicated to Prof. Shigefumi Mori on the occasion of his 60-th birthday

Abstract.

We establish an algorithm for resolution of singularities of an idealistic filtration in dimension 3 (a local version) in positive characteristic, incorporating the method recently developed by Benito-Villamayor into our framework. Although (a global version of) our algorithm only implies embedded resolution of surfaces in a smooth ambient space of dimension 3, a classical result known before, we introduce some new invariant which effectively measures how much the singularities are improved in the process of our algorithm and which strictly decreases after each blow up. This is in contrast to the well-known Abhyankar-Moh pathology of the increase of the residual order under blow up and the phenomenon of the "Kangaroo" points observed by Hauser.

Contents

Received March 23, 2012.
Revised May 19, 2012; August 3, 2012.
2010 *Mathematics Subject Classification.* 14E15.
Key words and phrases. resolution of singularities, IFP.

§1. Outline of the paper

The goals of this paper are two-fold. The first goal is to present the general mechanism of resolution of singularities (a local version) in the framework of the Idealistic Filtration Program in positive characteristic. The classical algorithm in characteristic zero works by induction on dimension based upon the notion of a hypersurface of maximal contact. Our algorithm in positive characteristic works by induction on the invariant "σ" based upon the notion of a leading generator system (cf. [18]). Roughly speaking, the general mechanism splits into two parts; the first part is to reduce the problem in the general case to the one in the so-called "monomial case", and the second part is to solve the problem in the monomial case. The first part of the general mechanism works in arbitrary dimension. The second part is quite subtle and difficult in positive characteristic (while it is easy in the classical setting in characteristic zero). The second goal is to establish the algorithm in dimension 3, by actually solving the problem of resolution of singularities in the monomial case. We incorporate the method recently developed by Benito-Villamayor [5] into our framework. We introduce some new invariant, which effectively measures how much the singularities are improved in the process of our algorithm and which strictly decreases after each blow up. This is in clear contrast to the well-known Abhyankar-Moh pathology of the increase of the residual order (cf. [21]) and the phenomenon of the "Kangaroo" points observed by Hauser (cf. [16]) and others. We note that the algorithm by Benito-Villamayor (cf. [3][4][5]), which works by induction on dimension based upon the notion of a generic projection, is different from our algorithm, and that even the setting of the monomial case is different from ours by definition. It is something of a surprise that we can share the "same" method when our and their approaches are different. Establishing the algorithm in dimension 4 or above remains as an open problem.

We remark that (a global version, which will be published elsewhere, of) our algorithm in dimension 3 only yields embedded resolution of surfaces in a nonsingular ambient 3-fold over an algebraically closed field of any characteristic. This is a classical result known for more than 50 years since the time of Abhyankar, Hironaka and others (cf. [1][12][15][17]). The front line of research, thanks to the works of Abhyankar, Cossart-Piltant, and Cutkosky (cf. [1][9][10][11][12]), goes way beyond, establishing resolution of singularities of a 3-fold over an algebraically closed field of positive characteristic. We believe, however, that the method of our paper should provide a first step toward the open problem of embedded resolution of singularities of 3-folds in a nonsingular ambient 4-fold in positive characteristic, and also in higher dimensional cases.

The outline of the paper goes as follows.

After §1, which describes the outline of the paper, we give a brief overview of the contents of the paper in §2. In §3, we present a quick review on the algorithm in characteristic zero. Our algorithm in positive characteristic is modeled upon the classical algorithm in characteristic zero. The review is given in such a way that the reader can see the similarities and differences between our algorithm and the classical algorithm through an easy and accessible comparison. In §4, we present the general mechanism of our algorithm for resolution of singularities in positive characteristic. In §5, we present a solution to the problem of resolution of singularities in the monomial case in dimension 3, thus completing our algorithm as a whole in dimension 3.

We assume, throughout the entire paper, that the base field is an algebraically closed field $k = \overline{k}$ of characteristic zero $\mathrm{char}(k) = 0$ or positive characteristic $\mathrm{char}(k) = p > 0$. Therefore, there is no danger in *not* distinguishing the two notions "smooth over k" and "regular" (meaning that every local ring of a variety is a regular local ring), and in using the word "nonsingular" as a synonym. In the case where the base field k is perfect, our algorithm over its algebraic closure $\overline{k}$ is invariant under the action of the Galois group $\mathrm{Gal}(\overline{k}/k)$ and hence descends to the algorithm over the original base field k. The case where the base field is not perfect will be investigated elsewhere.

Acknowledgments: We would like to express our hearty thanks to Profs. A. Benito and O. Villamayor for explaining their results to us with great patience. We owe §5 of this paper to their ideas, which inspired us to come up with further improvements. We would also like to thank Profs. R. Blanco, A. Bravo, S. Encinas, and Dr. M.L. García for their warm hospitality during our stay in Madrid, Spain, in May of 2011. The first author would like to thank Prof. Hauser in Vienna, Austria, and Dr.

Bernd Schober in Regensburg, Germany. The conversation with them led to a crucial development of the paper. The second author would like to thank Prof. Y. Lee for the hospitality during his stay in Seoul, Korea, and Prof. Y. Kawamata at Tokyo University, Japan, for the consistent and warm encouragement. He would also like to thank Profs. D. Arapura, W. Heinzer, J. Lipman, T.T. Moh, B. Ulrich, J. Włodarczyk at Purdue University in Indiana, U.S.A., for their advice and many invaluable comments. Our indebtedness to the people at RIMS in Kyoto, Japan, is more than a word could express. The deep appreciation goes to Profs. S. Mori and S. Mukai, who have been supporting us both mathematically and personally throughout our entire project toward resolution of singularities in positive characteristic. Prof. N. Nakayama's help was essential for the completion of our paper. We would like to thank the referee for giving us many helpful comments to improve the paper. The first author is partially supported by the Grant-in-Aid for Young Scientists (B) and by the Sumitomo Foundation. The second author is partially supported by the NSA grant H98230-10-1-0172.

§2. Overview

The problem of resolution of singularities in its simplest form is stated as follows:

Problem 1 (Resolution of singularities). *Given an algebraic variety X over k, find a proper birational map $X \stackrel{\pi}{\leftarrow} \widetilde{X}$ from a nonsingular variety $\widetilde{X}$.*

The above problem is reduced to the following problem of embedded resolution of singularities, if our solution to the latter is *functorial* in the sense that it is stable under the pull-back by smooth morphisms.

Problem 2 (Embedded resolution of singularities). *Given an algebraic variety X, embedded as a closed subvariety in a smooth ambient variety W over k, i.e., $X \subset W$, find a sequence starting from $(X_0 \subset W_0) = (X \subset W)$*

$$(X_0 \subset W_0) \leftarrow \cdots \leftarrow (X_i \subset W_i) \stackrel{\pi_{i+1}}{\leftarrow} (X_{i+1} \subset W_{i+1}) \leftarrow \cdots \leftarrow (X_l \subset W_l),$$

where $W_i \leftarrow W_{i+1}$ is a blow up with smooth center $C_i \subset W_i$ which does not contain the strict transform X_i of X_0 in year i , i.e., $C_i \not\supset X_i$, and which is transversal to the exceptional divisor D_i, i.e., $C_i \pitchfork D_i$ (maybe contained in D_i), such that the last strict transform X_l is nonsingular and transversal to D_l.

Note that in the above formulation we do *not* require that the center is contained in the singular locus of the strict transform $C_i \subset \mathrm{Sing}(X_i)$ or even that the center is contained in the strict transform $C_i \subset X_i$.

Note also that we say C_i is transversal to D_i, denoted by $C_i \pitchfork D_i$, if at any closed point $P \in W_i$ there exist a regular system of parameters $(x_1, \ldots, x_d)$ at P, taken from $\mathfrak{m}_P \subset \mathcal{O}_{W,P}$, and subsets $A, B \subset \{1, \ldots, d = \dim W_i\}$ such that $C_i = \bigcap_{\alpha \in A}\{x_\alpha = 0\}$ and $D_i = \bigcup_{\beta \in B}\{x_\beta = 0\}$ in a neighborhood of P.

The implication "Solution to Problem 2 (in a functorial way) $\Rightarrow$ Solution to Problem 1" can be seen as follows: Given an algebraic variety X, decompose it into the union of affine open subvarieties $X = \bigcup_{\lambda \in \Lambda} X_\lambda$ with embeddings $X_\lambda \subset W_\lambda = \mathbb{A}^{n_\lambda}$. Take embedded resolutions of singularities $(X_\lambda \subset W_\lambda) \leftarrow (\widetilde{X_\lambda} \subset \widetilde{W_\lambda})$. We have only to see that the $\widetilde{X_\lambda}$'s patch together by the *functoriality* to obtain resolution of singularities $X = \bigcup_{\lambda \in \Lambda} X_\lambda \leftarrow \widetilde{X} = \bigcup_{\lambda \in \Lambda} \widetilde{X_\lambda}$. For the detail of the proof, we refer the reader to [23].

How can we solve Problem 2 (in a functorial way so that we can solve Problem 1 also) ? We would like to use "induction" on dimension. However, the inductive scheme to approach Problem 2 is not clear as stated, at least not obvious. We will present in §3 the reformulation by Hironaka where in characteristic zero the inductive scheme on dimension is more transparent and used classically, and in §4 another reformulation in our framework where in positive characteristic the inductive scheme on the invariant "σ" emerges.

We give an overview of the contents of our paper in the following.

§3 is devoted to a quick review on the algorithm in characteristic zero.

In §3.1, we give the precise statement of the reformulation by Hironaka (following the language used by Villamayor). In short, the reformulation turns the problem of embedded resolution of singularities into a game of reducing the order of an ideal on a nonsingular ambient variety. We start from a triplet of data $(W, (\mathcal{I}, a), E)$, where W is a nonsingular variety over k, $(\mathcal{I}, a)$ is a pair consisting of a *nonzero* coherent ideal sheaf $\mathcal{I}$ on W and a fixed positive integer $a \in \mathbb{Z}_{>0}$, and where E is a simple normal crossing divisor on W, called a boundary (divisor). We define its singular locus to be $\mathrm{Sing}(\mathcal{I}, a) := \{P \in W \mid \mathrm{ord}_P(\mathcal{I}) \geq a\}$. Then we are required to find a sequence of transformations (see §3.1 for the precise definition of a transformation) starting from $(W_0, (\mathcal{I}_0, a), E_0) = (W, (\mathcal{I}, a), E)$

$$\begin{aligned}(W_0, (\mathcal{I}_0, a), E_0) \leftarrow \cdots \leftarrow (W_i, (\mathcal{I}_i, a), E_i) &\overset{\pi_{i+1}}{\leftarrow} (W_{i+1}, (\mathcal{I}_{i+1}, a), E_{i+1}) \\ &\leftarrow \cdots \leftarrow (W_l, (\mathcal{I}_l, a), E_l)\end{aligned}$$

such that $\mathrm{Sing}(\mathcal{I}_l, a) = \emptyset$. That is to say, the order of the last ideal $\mathcal{I}_l$ is everywhere below a. We call such a sequence "resolution of singularities for $(W, (\mathcal{I}, a), E)$".

In this reformulation, the inductive scheme of the problem can be stated, though naive, simply as follows: Given a triplet of data $(W, (\mathcal{I}, a), E)$, find another triplet of data $(H, (\mathcal{J}, b), F)$ with $H \underset{\text{closed}}{\subset} W$ and $\dim H = \dim W - 1$ such that constructing resolution of singularities for $(W, (\mathcal{I}, a), E)$ is equivalent to constructing one for $(H, (\mathcal{J}, b), F)$, a property symbolically denoted by

$$(W, (\mathcal{I}, a), E) \underset{\text{equivalent}}{\sim} (H, (\mathcal{J}, b), F).$$

In §3.2, we discuss the inductive scheme on dimension of the algorithm in characteristic zero. It starts with the following key inductive lemma: Given $(W, (\mathcal{I}, a), E)$ assumed to be under a certain condition $(\star)$, the lemma constructs $(H, (\mathcal{J}, b), F)$ with $H \underset{\text{closed}}{\subset} W$ and $\dim H = \dim W - 1$ (locally around a fixed point $P \in W$) which satisfies one of the following two.

(i) The ideal $\mathcal{J}$ is a zero sheaf, i.e., $\mathcal{J} \equiv 0$: In this case, we take the transformation with center $H = \mathrm{Sing}(\mathcal{I}, a)$

$$(W, (\mathcal{I}, a), E) \leftarrow (\widetilde{W}, (\widetilde{\mathcal{I}}, a), \widetilde{E}).$$

After the transformation, we have $\mathrm{Sing}(\widetilde{\mathcal{I}}, a) = \emptyset$ and hence resolution of singularities is achieved.

(ii) The ideal $\mathcal{J}$ is *not* a zero sheaf, i.e., $\mathcal{J} \not\equiv 0$: In this case, we have

$$(W, (\mathcal{I}, a), E) \underset{\text{equivalent}}{\sim} (H, (\mathcal{J}, b), F).$$

Therefore, by constructing resolution of singularities for $(H, (\mathcal{J}, b), F)$ by induction on dimension, we achieve resolution of singularities for $(W, (\mathcal{I}, a), E)$.

The hypersurface "H" in the key inductive lemma is called *a hypersurface of maximal contact.*

There are three shortcomings of the above key inductive lemma when we look at the goal of establishing the algorithm in characteristic zero:

① we have to impose condition $(\star)$ on $(W, (\mathcal{I}, a), E)$,
② the construction of $(H, (\mathcal{J}, b), F)$ is only local, and
③ the invariant "ord" may increase after a transformation, even though our ultimate goal is to reduce its value to be below the fixed level a.

In order to overcome these shortcomings, we adopt the following mechanism.

(1) We introduce a pair of invariants $(\text{w-ord}, s)$ and its associated triplet of data $(W, (\mathcal{K}, \kappa), G)$, called the modification of the original triplet $(W, (\mathcal{I}, a), E)$, having the following properties. (We remark that, when we want to emphasize the dimension of the ambient variety, we add "dim" to the pair of invariants as the first factor, making the pair into a triplet $(\dim, \text{w-ord}, s)$.)

- The maximum locus of the pair of invariants coincides with the singular locus of the modification, i.e.,

$$\text{MaxLocus}(\text{w-ord}, s) = \text{Sing}(\mathcal{K}, \kappa).$$

Moreover, after each transformation, the value of the pair $(\text{w-ord}, s)$ never increases, and the locus where the value of the pair takes the same maximum value as the original one coincides with the singular locus of the transformation of the modification. (Note that the transformations of $(W, (\mathcal{K}, \kappa), G)$ induce those of $(W, (\mathcal{I}, a), E)$.) This means that resolution of singularities for $(W, (\mathcal{K}, \kappa), G)$ implies the strict decrease of the (maximum) value of the pair $(\text{w-ord}, s)$.

- Even though the original triplet $(W, (\mathcal{I}, a), E)$ may not satisfy condition $(\star)$, the modification $(W, (\mathcal{K}, \kappa), G)$ does.

(2) We apply the key inductive lemma to $(W, (\mathcal{K}, \kappa), G)$. We find a triplet $(H, (\mathcal{J}, b), F)$ with $H \underset{\text{closed}}{\subset} W$ and $\dim H = \dim W - 1$ such that we are either in Case (i), or in Case (ii) where resolution of singularities for $(H, (\mathcal{J}, b), F)$ implies the one for $(W, (\mathcal{K}, \kappa), G)$.

(3) In Case (i), by a single blow up with center H, we achieve resolution of singularities for $(W, (\mathcal{K}, \kappa), G)$. In Case (ii), by induction on dimension, we achieve resolution of singularities for $(H, (\mathcal{J}, b), F)$, hence for $(W, (\mathcal{K}, \kappa), G)$. In both cases, we have the strict decrease of the (maximum) value of the pair $(\text{w-ord}, s)$.

(4) Repeatedly decreasing the value of the pair $(\text{w-ord}, s)$ this way, we reach the case where $\text{w-ord} = 0$. The condition $\text{w-ord} = 0$ is equivalent to saying that the ideal is generated by some monomial of the defining equations of the components in the boundary divisor E. Thus we call the case where $\text{w-ord} = 0$ *the monomial case*.

(5) Finally, we have only to construct resolution of singularities in the monomial case, which can be done easily in characteristic zero.

The description of the inductive scheme above is only local, and hence we have overcome shortcomings ① and ③ only so far. The way we overcome shortcoming ② is discussed in §3.5 via the strand of invariants woven in §3.3.

In §3.3, we present the inductive scheme explained in §3.2 in terms of weaving the strand of invariants "$\mathrm{inv}_{\mathrm{classic}}$". The strand "$\mathrm{inv}_{\mathrm{classic}}$" consists of the units of the form $(\dim H^j, \text{w-ord}^j, s^j)$, the dimension of the hypersurface of maximal contact H^j followed by the pair as described in §3.1 computed from the triplet $(H^j, (\mathcal{J}^j, b^j), F^j)$ at the j-th stage, and ends either with $(\dim H^m, \infty)$ or with $(\dim H^m, 0, \Gamma)$ depending on whether the last triplet $(H^m, (\mathcal{J}^m, b^m), F^m)$ is in Case (i) of the key inductive lemma or it is in the monomial case. That is to say, "$\mathrm{inv}_{\mathrm{classic}}$" takes the following form

$$\text{“}\mathrm{inv}_{\mathrm{classic}}\text{”} = (\dim H^0 = \dim W, \text{w-ord}^0, s^0) \cdots (\dim H^j, \text{w-ord}^j, s^j)$$
$$\cdots (\dim H^{m-1}, \text{w-ord}^{m-1}, s^{m-1}) \begin{cases} (\dim H^m, \infty), \text{ or} \\ (\dim H^m, 0, \Gamma). \end{cases}$$

Note that we do not include the invariant s in the last unit.

We choose the center of blow up to be H^m when $\text{w-ord}^m = \infty$ according to Case (i) of the key inductive lemma, or the maximum locus of the invariant Γ on H^m when $\text{w-ord}^m = 0$ according to the procedure in the monomial case (which is discussed in §3.4). This is equivalent to choosing the center of blow up to be the maximum locus of "$\mathrm{inv}_{\mathrm{classic}}$". This leads to the decrease of the (maximum) value of "$\mathrm{inv}_{\mathrm{classic}}$". Showing that the (maximum) value of "$\mathrm{inv}_{\mathrm{classic}}$" cannot decrease infinitely many times, we achieve resolution of singularities for $(W, (\mathcal{I}, a), E)$.

In §3.4, we discuss the procedure of constructing resolution of singularities in the monomial case, the only remaining task to complete the classical algorithm. In characteristic zero, this can be done easily and purely from the combinatorial data obtained by looking at the monomial in consideration, manifested as the invariant Γ.

The strand "$\mathrm{inv}_{\mathrm{classic}}$" a priori depends on the choice of the hypersurfaces of maximal contact we take in the process of weaving, and it is a priori only locally defined. However, the strand "$\mathrm{inv}_{\mathrm{classic}}$" is actually independent of the choice, and hence it is globally well-defined. This can be shown classically by the so-called Hironaka's trick, or more recently by inserting Włodarczyk's "homogenization" or the first author's "differential saturation" into the construction of the modification. Therefore, the process of resolution of singularities, where we take the center of blow up to be the maximum locus of "$\text{inv}_{\text{classic}}$", is also globally well-defined. This is how we overcome shortcoming ② of the key inductive lemma, accomplishing the globalization of the algorithm in §3.5.

This completes the overview of §3.

§4 is devoted to describing the general mechanism of our algorithm in positive characteristic, which is closely modeled upon the algorithm in characteristic zero explained in §3. (We remark that our algorithm is also valid in characteristic zero.)

In §4.1, we give the statement of a further reformulation (of Problem 2), which allows us to present the inductive structure on the invariant "σ". We start from a triplet of data $(W, \mathcal{R}, E)$, where we replace the pair $(\mathcal{I}, a)$ in the classical triplet $(W, (\mathcal{I}, a), E)$ with $\mathcal{R}$. Here $\mathcal{R} = \bigoplus_{a \in \mathbb{Z}_{\geq 0}} (\mathcal{I}_a, a)$ represents an idealistic filtration of i.f.g. type (short for "integrally and finitely generated type"), i.e., a finitely generated and graded (by the nonnegative integer $a \in \mathbb{Z}_{\geq 0}$ called the level of the ideal $\mathcal{I}_a$ in the first factor) $\mathcal{O}_W$-algebra satisfying the condition $\mathcal{O}_W = \mathcal{I}_0 \supset \mathcal{I}_1 \supset \mathcal{I}_2 \supset \cdots \supset \mathcal{I}_a \supset \cdots$. We define its singular locus to be $\text{Sing}(\mathcal{R}) := \{P \in W \mid \text{ord}_P(\mathcal{I}_a) \geq a \quad \forall a \in \mathbb{Z}_{\geq 0}\}$. Then we are required to find a sequence of transformations (see §4.1 for the precise definition of a transformation) starting from $(W_0, \mathcal{R}_0, E_0) = (W, \mathcal{R}, E)$

$$
\begin{aligned}
(W_0, \mathcal{R}_0, E_0) \leftarrow \cdots \leftarrow (W_i, \mathcal{R}_i, E_i) &\overset{\pi_{i+1}}{\leftarrow} (W_{i+1}, \mathcal{R}_{i+1}, E_{i+1}) \\
&\leftarrow \cdots \leftarrow (W_l, \mathcal{R}_l, E_l)
\end{aligned}
$$

such that $\text{Sing}(\mathcal{R}_l, a) = \emptyset$. We call such a sequence "resolution of singularities for $(W, \mathcal{R}, E)$". So far it is perfectly parallel to the story in characteristic zero, and there is nothing unique to the case in positive characteristic. In fact, resolution of singularities for $(W, (\mathcal{I}, a), E)$ is equivalent to resolution of singularities for $(W, \mathcal{R}, E)$ where the idealistic filtration of i.f.g. type is given by the formula $\mathcal{R} = \bigoplus_{n \in \mathbb{Z}_{\geq 0}} (\mathcal{I}^{\lceil n/a \rceil}, n)$. However, the reformulation allows us to introduce the invariant σ, which plays the key role in our inductive scheme.

In §4.2, we discuss the inductive scheme of our algorithm on the invariant σ in positive characteristic. Here, unlike in characteristic zero, we have no key inductive lemma. In fact, the statement of the key inductive lemma fails to hold in positive characteristic as is demonstrated by an example due to R. Narasimhan. That is to say, there is *no* smooth Hypersurface of Maximal Contact (HMC for short) in general. However, we can still introduce the notion of *a Leading Generator System* (LGS for short), which should be considered as a collective substitute in positive characteristic for the notion of an HMC in characteristic zero. Our basic strategy is to follow the construction of the algorithm in characteristic zero, replacing an HMC (leading to the induction on dimension) with an LGS (leading to the induction on the invariant σ). The description of the new inductive scheme goes as follows.

(1) We introduce a triplet of invariants $(\sigma, \widetilde{\mu}, s)$.

If $(\sigma, \widetilde{\mu}, s) = (\sigma, \infty, 0)$ or $(\sigma, 0, 0)$, then we do not construct the modification $(W', \mathcal{R}', E')$.

- In case $(\sigma, \widetilde{\mu}, s) = (\sigma, \infty, 0)$, we blow up with center $C = \mathrm{Sing}(\mathcal{R})$. The nonsingularity of C is guaranteed by the Nonsingularity Principle (cf. [18] [19]), while the transversality of C to the boundary E is guaranteed by the invariant $s = 0$. After the blow up, the singular locus becomes empty, and resolution of singularities for $(W, \mathcal{R}, E)$ is achieved.
- In case $(\sigma, \widetilde{\mu}, s) = (\sigma, 0, 0)$, we are in the monomial case by definition, and we go to Step (5).

If $(\sigma, \widetilde{\mu}, s) \neq (\sigma, \infty, 0)$ or $(\sigma, 0, 0)$, then we construct its associated triplet of data $(W', \mathcal{R}', E')$, called the modification of the original triplet $(W, \mathcal{R}, E)$, having the following properties. (Actually the ambient space for the modification stays the same, i.e., $W' = W$.)

- Resolution of singularities for $(W', \mathcal{R}', E')$ implies the strict decrease of the (maximum) value of the triplet $(\sigma, \widetilde{\mu}, s)$. Note that the value of the triplet never increases after transformations.
- Either the value of σ strictly decreases, or the value of σ stays the same but the number of the components in the boundary strictly drops. In either case, we have $(\sigma, \#E) > (\sigma', \#E')$.

(2) There is *no* key inductive lemma in our new setting.

(3) When $(\sigma, \widetilde{\mu}, s) \neq (\sigma, \infty, 0)$ or $(\sigma, 0, 0)$ in (1), we achieve resolution of singularities for $(W', \mathcal{R}', E')$ by induction on $(\sigma, \#E)$, which implies the strict decrease of the (maximum) value of the triplet $(\sigma, \widetilde{\mu}, s)$.

(4) Repeatedly decreasing the value of the triplet $(\sigma, \widetilde{\mu}, s)$, we reach the case where $(\sigma, \widetilde{\mu}, s) = (\sigma, \infty, 0)$ or $(\sigma, 0, 0)$, the monomial case.

(5) Finally we have only to construct resolution of singularities in the monomial case in order to achieve resolution of singularities of the original triplet $(W, \mathcal{R}, E)$. However, the problem of resolution of singularities in the monomial case in positive characteristic is quite subtle and very difficult. We only provide a solution in dimension 3 in §5.

In §4.3, we present the inductive scheme explained in §4.2 in terms of weaving the strand of invariants "$\mathrm{inv}_{\mathrm{new}}$". The strand "$\mathrm{inv}_{\mathrm{new}}$" consists of the units of the form $(\sigma^j, \widetilde{\mu}^j, s^j)$ computed from the triplet $(W^j, \mathcal{R}^j, E^j)$ at the j-th stage, and ends either with $(\sigma^m, \infty, 0)$ or with

$(\sigma^m, 0, 0)$. That is to say, "inv_{new}" takes the following form

$$\text{“}\mathrm{inv}_{\mathrm{new}}\text{”} = (\sigma^0, \widetilde{\mu}^0, s^0) \cdots (\sigma^j, \widetilde{\mu}^j, s^j) \cdots (\sigma^{m-1}, \widetilde{\mu}^{m-1}, s^{m-1}) \begin{cases} (\sigma^m, \infty, 0), & \text{or} \\ (\sigma^m, 0, 0). \end{cases}$$

Note that we *do* compute and include the invariant s in the last unit, in contrast to the weaving of "$\text{inv}_{\text{classic}}$".

Resolution of singularities of the last m-th triplet $(W^m, \mathcal{R}^m, E^m)$ can be achieved by taking the transformation with center $\text{Sing}(\mathcal{R}^m)$ when $(\sigma^m, \widetilde{\mu}^m, s^m) = (\sigma^m, \infty, 0)$ or by resolution of singularities in the monomial case when $(\sigma^m, \widetilde{\mu}^m, s^m) = (\sigma^m, 0, 0)$. This leads to the decrease of the value of "inv_{new}". Showing that the value of "inv_{new}" cannot decrease infinitely many times, we (should) achieve resolution of singularities for $(W, \mathcal{R}, E)$ (assuming that the problem of resolution of singularities in the monomial case is solved).

There is one important remark to make. In §4.2 we do *not* claim that the maximum locus of the triplet $(\sigma, \widetilde{\mu}, s)$ coincides with the singular locus $\text{Sing}(\mathcal{R}')$ of the modified idealistic filtration, and in §4.3 we do *not* claim that the strand "inv_{new}" is a global invariant whose maximum locus gives the globally well-defined center of blow up for the algorithm. The discussion of our new inductive scheme and the weaving of the new strand of invariants are, therefore, only local. (See Remark 3 for the precise meaning of "local".) In fact, "inv_{new}" as presented is independent of the choice of the LGS's we take in the process of weaving, just like "$\text{inv}_{\text{classic}}$" is independent of the choice of the HMC's we take in the process of weaving in characteristic zero. Nevertheless, the gap between $\text{MaxLocus}(\sigma, \widetilde{\mu}, s)$ and $\text{Sing}(\mathcal{R}')$ may occur when $\widetilde{\mu} = 1$, and hence that the maximum locus of "inv_{new}" does not provide the global center of blow up as it is. This calamity is unique to our setting, and never existed in the classical setting. We can fix this calamity by making certain adjustments to the strand, the description of which is rather technical and will be published elsewhere. Therefore, in this paper, we restrict ourselves to the local description, which, we believe, still captures the essence of the inductive scheme on the invariant σ.

In §4.4, we mention briefly the reason why resolution of singularities in the monomial case is difficult in positive characteristic, while it is easy in characteristic zero.

This completes the overview of §4.

§5 is devoted to the detailed discussion of resolution of singularities in the monomial case in dimension 3, incorporating the method recently developed by Benito-Villamayor into our framework with some improvements. In fact, we introduce a new invariant, which strictly drops after each blow up and hence effectively shows the termination of the procedure. This is the most subtle and difficult part of this paper.

The invariant τ represents the number of the elements in the LGS, which takes the value $0, 1, 2, 3$ in dimension 3 and which never decreases under transformations. When $\tau = 0, 2, 3$, resolution of singularities in the monomial case is rather easy. Therefore, we focus our attention to the case where $\tau = 1$ in §5.

We are thus in the situation where (analytically) we have a unique element (h, p^e) in the LGS and where h is of the following form via Weierstrass Preparation Theorem with respect to a regular system of parameters (x, y, z) at $P \in W$, taken from $\widehat{\mathfrak{m}_P} \subset \widehat{\mathcal{O}_{W,P}}$,

$$h = z^{p^e} + a_1 z^{p^e} + \cdots + a_{p^e} \quad \text{with} \quad a_i \in k[[x, y]]$$

satisfying $\operatorname{ord}_P(a_i) > i$ for $i = 1, \ldots, p^e$. We also have a monomial

$$(x^\alpha y^\beta, a) \in \mathcal{R}$$

of the defining equations x and y of the components of the boundary divisor E (actually E_{young}) such that every element $(f, \lambda) \in \mathcal{R}$ is divisible by $(x^\alpha y^\beta)^{\lambda/a}$ modulo h, a consequence of the condition $\widetilde{\mu} = 0$.

A naive idea for resolution of singularities in the monomial case may be stated as follows: Carry out the algorithm for resolution of singularities of $(x^\alpha y^\beta, a)$ on the hypersurface $\{z = 0\}$.

A bad news is that the above naive idea does not work for the reason that $(z, 1) \notin \mathcal{R}$ in general, which has the following bad implications:

(1) Even though we see that the coefficients a_i for $i = 1, \ldots, p^e - 1$ are under control (in the sense that a_i is divisible by $(x^\alpha y^\beta)^{i/a}$), the constant term a_{p^e} is not well controlled. (The idealistic filtration $\mathcal{R}$ in the monomial case is *not* differentially saturated but only saturated for $\{\partial^n / \partial z^n \mid n \in \mathbb{Z}_{\geq 0}\}$ in general. However, this is good enough to control the coefficients a_i for $i = 1, \ldots, p^e - 1$.) This leads to the calamity that a candidate for the center determined by the naive idea may not even be contained in the singular locus.

(2) The hypersurface $\{z = 0\}$ may not be of maximal contact. That is to say, after a blow up, its strict transform may no longer "contact" (contain) the singular locus at all.

In §5.1, we introduce the process of "cleaning" in order to eliminate the "mess" described in the bad news above. The idea of "cleaning" can be seen already in the work of Abhyankar and Hironaka, in the definition of the "residual order". Here we follow the process refined by Benito-Villamayor. After cleaning, the invariant

$$\mathrm{H}(P) := \min\left\{\frac{1}{p^e}\mathrm{ord}_P(a_{p^e}),\ \mu(P) = \frac{\alpha+\beta}{a}\right\}$$

is independent of the choice of h or a regular system of parameters (x, y, z). The invariant H is well-defined not only at a closed point $P \in W$ but also at the generic point $\xi_{\{x=0\}}$ (resp. $\xi_{\{y=0\}}$) of a component $\{x = 0\}$ (resp. $\{y = 0\}$) of the boundary divisor E.

In §5.2, we give the description of the procedure of resolution of singularities, depending on the description of the singular locus. Since we are in the monomial case, we have $\mathrm{Sing}(\mathcal{R}) \subset \{x = 0\} \cup \{y = 0\}$. The description of the singular locus restricted to $\{x = 0\}$, locally around P, is given as follows according to the value of the invariant H:

$$\mathrm{Sing}(\mathcal{R}) \cap \{x = 0\} = \begin{cases} \{z = x = 0\} & \text{if} \quad \mathrm{H}(\xi_{\{x=0\}}) \geq 1 \\ P & \text{if} \quad \mathrm{H}(\xi_{\{x=0\}}) < 1. \end{cases}$$

That is to say, by looking at the invariant H, we can tell if the singular locus has dimension 1, where it is a nonsingular curve, or has dimension zero, where it is an isolated point. We have a similar description of the singular locus restricted to $\{y = 0\}$.

The procedure goes as follows:

Step 1. Check if $\dim \mathrm{Sing}(\mathcal{R}) = 1$. If yes, then blow up the 1-dimensional components one by one. Since the invariant H strictly decreases for the component of the boundary divisor involved in the blow up, this step comes to an end after finitely many times with the dimension of the singular locus dropping to 0.

Step 2. Once $\dim \mathrm{Sing}(\mathcal{R}) = 0$, blow up the isolated points in the singular locus.

Step 3. Go back to Step 1.

Repeat these steps.

§5.3 is devoted to showing termination of the procedure described in §5.2. We closely follow the beautiful and delicate argument recently developed by Benito-Villamayor, which analyzes the behavior of the monomial, the invariant H, and the newton polygon of the constant term a_{p^e} under blow ups. Their argument is an extension of the classical ideas of Abhyankar and Hironaka, but highly refined taking into consideration the condition that we are in the monomial case. Benito-Villamayor also uses "a proof by contradiction" in some part of their argument for termination of the procedure. That is to say, they derive a contradiction, assuming the existence of an infinite sequence of the procedure. Therefore, their argument is not effective. They also use some "stratification" of the configuration of the boundary divisors. We introduce a new and explicit invariant, which makes the termination argument effective and which allows us to eliminate the use of "stratification" from our argument.

This completes the overview of §5, and hence the overview of the entire paper.

§3. A quick review on the algorithm in characteristic zero

The goal of this section is to give a quick review on the algorithm in characteristic zero, upon which our algorithm in positive characteristic is closely modeled. For the overview of this section, we refer the reader to §2.

3.1. Reformulation of the problem by Hironaka

First, we present the reformulation of the problem by Hironaka. The form of the presentation we use here is due to Villamayor.

Problem 3 (Hironaka's reformulation). *Suppose we are given the triplet of data $(W, (\mathcal{I}, a), E)$, where W is a nonsingular variety over k, $(\mathcal{I}, a)$ is a pair consisting of a coherent ideal sheaf $\mathcal{I}$ on W and a positive integer $a \in \mathbb{Z}_{>0}$, and E is a simple normal crossing divisor on W.*

We define its singular locus to be

$$\mathrm{Sing}(\mathcal{I}, a) := \{P \in W \mid \mathrm{ord}_P(\mathcal{I}) \geq a\}.$$

We only consider a nonzero *ideal sheaf $\mathcal{I}$ for the resolution problem.*

Then construct a sequence of transformations starting from $(W_0, (\mathcal{I}_0, a), E_0) = (W, (\mathcal{I}, a), E)$

$$\begin{aligned}(W_0, (\mathcal{I}_0, a), E_0) \leftarrow \cdots \leftarrow (W_i, (\mathcal{I}_i, a), E_i) &\stackrel{\pi_{i+1}}{\leftarrow} (W_{i+1}, (\mathcal{I}_{i+1}, a), E_{i+1}) \\ &\leftarrow \cdots \leftarrow (W_l, (\mathcal{I}_l, a), E_l)\end{aligned}$$

such that $\mathrm{Sing}(\mathcal{I}_l, a) = \emptyset$. *We call such a sequence "resolution of singularities for $(W, (\mathcal{I}, a), E)$".*

We note that the transformation

$$(W_i, (\mathcal{I}_i, a), E_i) \stackrel{\pi_{i+1}}{\leftarrow} (W_{i+1}, (\mathcal{I}_{i+1}, a), E_{i+1})$$

is required to satisfy the following conditions:

(1) $W_i \overset{\pi_{i+1}}{\leftarrow} W_{i+1}$ *is a blow up with smooth center* $C_i \subset W_i$,
(2) $C_i \subset \mathrm{Sing}(\mathcal{I}_i, a)$, *and* C_i *is transversal to* E_i *(maybe contained in* E_i*), i.e.,* $C_i \pitchfork E_i$,
(3) $\mathcal{I}_{i+1} = \mathcal{I}(\pi_{i+1}^{-1}(C_i))^{-a} \cdot \pi_{i+1}^{-1}(\mathcal{I}_i)\mathcal{O}_{W_{i+1}}$,
(4) $E_{i+1} = E_i \cup \pi_{i+1}^{-1}(C_i)$.

Lemma 1. *A solution to Hironaka's reformulation provides a solution to the problem of embedded resolution of singularities.*

Proof. Given $X \subset W$, set $(W, (\mathcal{I}, a), E) = (W, (\mathcal{I}_X, 1), \emptyset)$. Then $\mathrm{Sing}(\mathcal{I}_0, a) = \mathrm{Sing}(\mathcal{I}_X, 1) = X = X_0$. Take resolution of singularities for $(W, (\mathcal{I}, a), E)$. Observe that, if X_i, the strict transform of X in year i, is an irreducible component of $\mathrm{Sing}(\mathcal{I}_i, a)$ and if X_i is not an irreducible component of the center C_i in year i (and X_j has not been an irreducible component of C_j in year j for $0 \leq j < i$), then X_{i+1} is an irreducible component of $\mathrm{Sing}(\mathcal{I}_{i+1}, a)$ in year $(i+1)$. Since $\mathrm{Sing}(\mathcal{I}_l, a) = \emptyset$, we conclude that X_m must be an irreducible component of C_m for some $m < l$. Since the center C_m is nonsingular by requirement, so is X_m. Therefore, the truncation of the sequence up to year m provides a sequence for embedded resolution. (The other requirements for embedded resolution as stated in Problem 2 follow automatically from the construction.) Q.E.D.

The inductive scheme for solving Hironaka's reformulation can be simply stated in the following naive form.

Naive Inductive Scheme: *Given a triplet* $(W, (\mathcal{I}, a), E)$, *find another triplet* $(H, (\mathcal{J}, b), F)$ *with* $H \underset{closed}{\subset} W$ *and* $\dim H = \dim W - 1$ *such that the problem of constructing resolution of singularities for* $(W, (\mathcal{I}, a), E)$ *is equivalent to constructing one for* $(H, (\mathcal{J}, b), F)$, *i.e.,*

$$(W, (\mathcal{I}, a), E) \underset{\text{equivalent}}{\sim} (H, (\mathcal{J}, b), F).$$

As it is, the above inductive scheme is too naive to hold in general. In §3.2, we first state the key inductive lemma, which realizes the naive inductive scheme under a certain extra condition called $(\star)$, and then discuss how to turn it into the real inductive scheme, which works in general without the extra condition.

3.2. Inductive scheme on dimension

Lemma 2 (Key Inductive Lemma). *Given* $(W, (\mathcal{I}, a), E)$ *and a closed point* $P \in \mathrm{Sing}(\mathcal{I}, a)$, *suppose that the following condition* $(\star)$ *is satisfied:*

$$(\star) \quad \begin{cases} \mathrm{ord}_P(\mathcal{I}) & = & a \\ E & = & \emptyset. \end{cases}$$

Then there exists $(H, (\mathcal{J}, b), F)$ *with* $H \underset{closed}{\subset} W$ *and* $\dim H = \dim W - 1$, *in a neighborhood of* P, *which satisfies one of the following two.*

(i) *The ideal* $\mathcal{J}$ *is a zero sheaf, i.e.,* $\mathcal{J} \equiv 0$*: In this case, we take the transformation with center* $H = \mathrm{Sing}(\mathcal{I}, a)$

$$(W, (\mathcal{I}, a), E) \leftarrow (\widetilde{W}, (\widetilde{\mathcal{I}}, a), \widetilde{E}).$$

After the transformation, we have $\mathrm{Sing}(\widetilde{\mathcal{I}}, a) = \emptyset$ *and hence we achieve resolution of singularities for* $(W, (\mathcal{I}, a), E)$ *(in a neighborhood of* P*).*

(ii) *The ideal* $\mathcal{J}$ *is* not *a zero sheaf, i.e.,* $\mathcal{J} \not\equiv 0$*: In this case, we have*

$$(W, (\mathcal{I}, a), E) \underset{\text{equivalent}}{\sim} (H, (\mathcal{J}, b), F).$$

Therefore, constructing resolution of singularities for $(H, (\mathcal{J}, b), F)$ *by induction on dimension, we achieve resolution of singularities for* $(W, (\mathcal{I}, a), E)$ *(in a neighborhood of* P*).*

Proof. Take $f \in \mathcal{I}_P$ such that $\mathrm{ord}_P(f) = a$. Then there exists a differential operator δ of $\deg \delta = a - 1$ such that $\mathrm{ord}_P(\delta f) = 1$. We note that this is exactly the place where we use the "characteristic zero" condition.

We have only to set

$$\begin{cases} H &= \{h = 0\} \quad \text{where} \quad h = \delta f, \\ (\mathcal{J}, b) &= (\mathrm{Coeff}(\mathcal{I})|_H, a!), \\ F &= E|_H = \emptyset, \end{cases}$$

where the "coefficient ideal", denoted by $\mathrm{Coeff}(\mathcal{I})$, is defined by the formula

$$\mathrm{Coeff}(\mathcal{I}) := \sum_{j=0}^{a-1} \{Diff^j(\mathcal{I})\}^{a!/(a-j)}$$

with $Diff^j(\mathcal{I})$ being the sheaf characterized at the stalk for a point $Q \in W$ by

$$Diff^j(\mathcal{I})_Q = \{\theta(g) \mid g \in \mathcal{I}_Q,\ \theta : \text{a differential operator of } \deg(\theta) \leq j\}.$$

We note that the condition $E = \emptyset$ is only used to guarantee that H intersects E transversally. The condition on the boundary divisor can be weakened (and E can be non-empty) to the following: we can find such $H = \{h = 0\}$, with $h \in Diff^{a-1}(\mathcal{I})_P$ and $\mathrm{ord}_P(h) = 1$, that is transversal to E. We then call the condition $(\star)_{\text{weakened}}$. Q.E.D.

Remark 1. We remark that the statement of the key inductive lemma fails to hold in positive characteristic, as the following example by R. Narasimhan shows:

Consider $(W, (\mathcal{I}, a), E)$ defined by

$$\begin{cases} W & = & \mathbb{A}^4 = \operatorname{Spec} k[x, y, z, w] \quad \text{with} \quad \operatorname{char}(k) = 2, \\ (\mathcal{I}, a) & = & ((f), 2) \quad \text{with} \quad f = w^2 + x^3y + y^3z + z^7x, \\ E & = & \emptyset. \end{cases}$$

Note that the triplet $(W, (\mathcal{I}, a), E)$ satisfies the condition $(\star)$ at the origin $\mathbb{O}$.

Then we have a curve C, parametrized by t, sitting inside of the singular locus

$$C = \{x = t^{15}, y = t^{19}, z = t^7, w = t^{32}\} \subset \operatorname{Sing}(\mathcal{I}, a).$$

Observe that the curve C has full embedding dimension at the origin $\mathbb{O}$, i.e.,

$$\text{embedding-}\dim_{\mathbb{O}} C = 4 = \dim W.$$

Therefore, there exists *no* smooth hypersurface H which contains the singular locus $\operatorname{Sing}(\mathcal{I}, a)$, and hence there exists *no* smooth hypersurface of maximal contact.

We list the shortcomings of the key inductive lemma toward establishing the algorithm for resolution of singularities for $(W, (\mathcal{I}, a), E)$ in general.

List of shortcomings of the key inductive lemma

① We have to impose condition $(\star)$ on $(W, (\mathcal{I}, a), E)$.

② We construct $(H, (\mathcal{J}, b), F)$ only locally, and hence the resolution process is only locally constructed by induction.

③ The invariant "ord" may strictly increase under transformations, even though our ultimate goal is to reduce the invariant "ord" to be below the fixed level a.

Now we describe the mechanism which overcomes all of the shortcomings above in one stroke, turning the key inductive lemma into the real inductive structure in characteristic zero.

Mechanism to overcome the shortcomings in the list

<u>Outline of the mechanism</u>

(1) Given $(W, (\mathcal{I}, a), E)$ (Precisely speaking, the triplet sits in the middle of the sequence, say in year "i", for resolution of singularities. However, we omit the subscript "$(\ )_i$" indicating the year for simplicity

of the notation.), we introduce a pair of invariants $(\text{w-ord}, s)$ and its associated triplet of data $(W, (\mathcal{K}, \kappa), G)$, called the modification of the original triplet, having the following properties.

- The maximum locus of the pair $(\text{w-ord}, s)$, which is an upper semi-continuous function, coincides with the singular locus $\operatorname{Sing}(\mathcal{K}, \kappa)$ of the modification, i.e.,

$$\operatorname{MaxLocus}(\text{w-ord}, s) = \operatorname{Sing}(\mathcal{K}, \kappa).$$

Moreover, after each transformation, the value of the pair $(\text{w-ord}, s)$ never increases, and the locus where the value of the pair takes the same maximum value as the original one coincides with the singular locus of the transformation of the modification. (Note that the transformations of $(W, (\mathcal{K}, \kappa), G)$ induce those of $(W, (\mathcal{I}, a), E)$.) This means that resolution of singularities for $(W, (\mathcal{K}, \kappa), G)$ implies the strict decrease of the (maximum) value of the pair $(\text{w-ord}, s)$.

- Even though the original triplet $(W, (\mathcal{I}, a), E)$ may not satisfy condition $(\star)$ (or condition $(\star)_{\text{weakened}}$), the modification $(W, (\mathcal{K}, \kappa), G)$ satisfies $(\star)_{\text{weakened}}$.

(2) We apply the key inductive lemma to $(W, (\mathcal{K}, \kappa), G)$. We find a triplet $(H, (\mathcal{J}, b), F)$ with $H \underset{\text{closed}}{\subset} W$ and $\dim H = \dim W - 1$ such that we are either in Case (i), or in Case (ii) where resolution of singularities for $(H, (\mathcal{J}, b), F)$ implies the one for $(W, (\mathcal{K}, \kappa), G)$. (We note that $G = E_{\text{new}}$ may not be empty. However, we can still find a hypersurface of maximal contact H which is transversal to $G = E_{\text{new}}$. Therefore, the triplet $(H, (\mathcal{J}, b), F)$ with $F = G|_H$ works.)

(3) In Case (i), by a simple blow up with center H, we accomplish resolution of singularities for $(W, (\mathcal{K}, \kappa), G)$. In Case (ii), by induction on dimension, we achieve resolution of singularities for $(H, (\mathcal{J}, b), F)$, hence for $(W, (\mathcal{K}, \kappa), G)$. In both cases, we have the strict decrease of the (maximum) value of the pair $(\text{w-ord}, s)$.

(4) By repeating the above procedures (1), (2), (3) and decreasing the value of the pair $(\text{w-ord}, s)$, we reach the stage where $\text{w-ord} = 0$, i.e., $\overline{\mathcal{I}} = \mathcal{O}_W$. This means that we are in the monomial case, where the ideal $\mathcal{I}$ is generated by some monomial of the defining equations of the components in $E_{\text{young}} \subset E$.

(5) Finally, we have only to construct resolution of singularities for the triplet in the monomial case, which can be done easily in characteristic zero.

We give the detailed and explicit description of the pair $(\text{w-ord}, s)$, the triplets $(W, (\mathcal{K}, \kappa), G)$ and $(H, (\mathcal{J}, b), F)$ in the following:

Description of the pair (w-ord, s)

∘ w-ord: It is the so-called (normalized) weak order. It is the order of $\overline{\mathcal{I}}$ (divided by the level a), where $\overline{\mathcal{I}}$ is obtained from $\mathcal{I}$ by dividing it as much as possible by the defining equations of the components in E_{young}. The symbol E_{young} refers to the union of the exceptional divisors created after the process of resolution of singularities began. By abuse of notation, the symbol E_{young} (or E) sometimes refers to the set having its components as its elements.

∘ s: It is the number of the components in $E_{\text{old}} = E \setminus E_{\text{new}}$. The symbol E_{new} refers to the union of the exceptional divisors created after the time when the current value of w-ord first started.

Description of the triplet $(W, (\mathcal{K}, \kappa), G)$

$$W = W, \quad (\mathcal{K}, \kappa) = \text{Bdry}\,(\text{Comp}(\mathcal{I}, a)), \quad \text{and} \quad G = E \setminus E_{\text{old}} = E_{\text{new}},$$

where

- Comp($\mathcal{I}, a$) is either the transformation of the one in the previous year if w-ord stays the same, or

$$(\mathcal{I}^M + \overline{\mathcal{I}}^a, M \cdot a) \quad \text{with} \quad M = \text{w-ord} \cdot a$$

if w-ord strictly decreases, and where

- Bdry (Comp($\mathcal{I}, a$)) is either the transformation of the one in the previous year if (w-ord, s) stays the same, or

$$(\mathcal{C} + (\sum_{D \in E_{\text{old}}} \mathcal{I}(D))^c, c) \quad \text{where} \quad (\mathcal{C}, c) = \text{Comp}(\mathcal{I}, a),$$

if (w-ord, s) strictly decreases.

We note that we have $E_{\text{new}} \subset E_{\text{young}}$ and that they may not be equal in general. We also note that, if the value of the pair (w-ord, s) stays the same as in the previous year, then $(\mathcal{K}, \kappa)$ is the transformation of the one in the previous year. We remark that the symbols "Comp" and "Bdry" represent the "Companion" modification and the "Boundary" modification, respectively.

Description of the triplet $(H, (\mathcal{J}, b), F)$

Case: The value of the pair (w-ord, s) *stays the same as in the previous year.*

In this case, we simply take $(H, (\mathcal{J}, b), F)$ to be the transformation of the one in the previous year under blow up.

Case: The value of the pair (w-ord, s) *strictly decreases.*

In this case, we construct $(H, (\mathcal{J}, b), F)$ as follows.

$$\begin{cases} H & = & \text{the strict transform of } H_{i_{\text{old}}}, \\ (\mathcal{J}, b) & = & (\text{Coeff}(\mathcal{K})|_H, \kappa!), \\ F & = & G|_H, \end{cases}$$

where $H_{i_{\text{old}}}$ is taken in the following way: We go back to the year $\iota := i_{\text{old}}$ when the current value of w-ord first started. Let $(\mathcal{C}_\iota, c_\iota) = \text{Comp}(\mathcal{I}_\iota, a)$ be the companion modification constructed in year ι. We take $f_\iota \in (\mathcal{C}_\iota)_{P_\iota}$ and a differential operator δ_ι of $\deg \delta_\iota = c_\iota - 1$ such that $\text{ord}_{P_\iota}(f_\iota) = c_\iota$ and $\text{ord}_{P_\iota}(\delta_\iota f_\iota) = 1$. We set $H_{i_{\text{old}}} = H_\iota = \{\delta_\iota f_\iota = 0\}$.

This completes the discussion of the mechanism to achieve resolution of singularities for $(W, (\mathcal{I}, a), E)$ in general. We note, however, that we only overcome shortcomings ① and ③ on the list, since the description so far is only local. We discuss in §3.5 how to overcome shortcoming ② and how to globalize the procedure via the strand of invariants woven in the next section.

3.3. Weaving of the classical strand of invariants "$\text{inv}_{\text{classic}}$"

In §3.3, we interpret the inductive scheme explained in §3.2 in terms of weaving the strand of invariants "$\text{inv}_{\text{classic}}$", whose maximum locus (with respect to the lexicographical order) determines the center of blow up for the algorithm for resolution of singularities in characteristic zero.

We weave the strand of invariants "$\text{inv}_{\text{classic}}$" consisting of the units of the form $(\dim H^j, \text{w-ord}^j, s^j)$, computed from the modifications $(H^j, (\mathcal{J}^j, b^j), F^j)$ constructed simultaneously along the weaving process. We note that we are adding "$\dim H$" to the pair (w-ord, s) as the first factor of the unit, in order to emphasize the role of the dimension in the inductive scheme.

Weaving Process

We describe the weaving process inductively.

Note that constructing a sequence for resolution of singularities by blowing up is referred to as "proceeding in the vertical direction" passing from one year to the next, indicated by the subscript "i", while weaving the strand and constructing the modifications passing from one stage to the next, indicated by the superscript "j", is referred to as "proceeding in the horizontal direction" staying in a fixed year.

Suppose we have already woven the strands and constructed the modifications up to year $(i-1)$.

Now we are in year i (looking at the neighborhood of a closed point $P_i \in W_i$).

We start with $(W_i, (\mathcal{I}_i, a), E_i) = (H_i^0, (\mathcal{J}_i^0, b_i^0), F_i^0)$, just renaming the transformation $(W_i, (\mathcal{I}_i, a), E_i)$ in year i of the resolution sequence as the 0-th stage modification $(H_i^0, (\mathcal{J}_i^0, b_i^0), F_i^0)$ in year i.

Suppose that we have already woven the strand up to the $(j-1)$-th unit

$$(\mathrm{inv}_{\mathrm{classic}})_i^{\leq j-1} = (\dim H_i^0, \text{w-ord}_i^0, s_i^0)(\dim H_i^1, \text{w-ord}_i^1, s_i^1) \cdots (\dim H_i^{j-1}, \text{w-ord}_i^{j-1}, s_i^{j-1})$$

and that we have also constructed the modifications up to the j-th one

$$(H_i^0, (\mathcal{J}_i^0, b_i^0), F_i^0), \cdots, (H_i^{j-1}, (\mathcal{J}_i^{j-1}, b_i^{j-1}), F_i^{j-1}), (H_i^j, (\mathcal{J}_i^j, b_i^j), F_i^j).$$

Our task is to compute the j-th unit $(\dim H_i^j, \text{w-ord}_i^j, s_i^j)$ and construct the $(j+1)$-th modification $(H_i^{j+1}, (\mathcal{J}_i^{j+1}, b_i^{j+1}), F_i^{j+1})$ (unless the weaving process is over at the j-th stage).

<u>Computation of the j-th unit $(\dim H_i^j, \text{w-ord}_i^j, s_i^j)$</u>

○ $\dim H_i^j$: We just remark that we insert this first factor in characteristic zero only to emphasize the role of the dimension, which corresponds to the role of the invariant σ in our algorithm in positive characteristic.

○ w-ord_i^j: We compute the second factor as follows.

$$\text{w-ord}_i^j = \begin{cases} \infty & \text{if} \quad \mathcal{J}_i^j \equiv 0, \\ \mathrm{ord}(\overline{\mathcal{J}_i^j})/b_i^j & \text{if} \quad \mathcal{J}_i^j \not\equiv 0, \end{cases}$$

where the ideal $\overline{\mathcal{J}_i^j}$ is obtained from $\mathcal{J}_i^j$ by dividing the latter as much as possible by the defining ideals of the components in $(F_i^j)_{\mathrm{young}}$, i.e.,

$$\overline{\mathcal{J}_i^j} = \left(\prod_{D \in (F_i^j)_{\mathrm{young}}} \mathcal{I}(D)^{-\mathrm{ord}_{\eta(D)}(\mathcal{J}_i^j)} \right) \cdot \mathcal{J}_i^j$$

where $\eta(D)$ is the generic point of D and where $(F_i^j)_{\mathrm{young}} (\subset F_i^j)$ is the union of the exceptional divisors created after the year when the value $(\mathrm{inv}_{\mathrm{classic}})_i^{\leq j-1}(\dim H_i^j)$ first started.

We note that, if $\text{w-ord}_i^j = \infty$ or 0, we declare that the $(j = m)$-th unit is the last one, and we stop the weaving process at the m-th stage in year i. When $\text{w-ord}_i^j = \infty$, the third factor is *not* included. When $\text{w-ord}_i^j = 0$, we are in the monomial case and we insert the invariant Γ as the third factor instead of the invariant s.

$\circ\ s_i^j$: It is the number of the components in $(F_i^j)_{\text{old}} = F_i^j \setminus (F_i^j)_{\text{new}}$, where $(F_i^j)_{\text{new}} (\subset F_i^j)$ is the union of the exceptional divisors created after the year when the value $(\text{inv}_{\text{classic}})_i^{\leq j-1}(\dim H_i^j, \text{w-ord}_i^j)$ first started. We note that the third factor s is *only* included if $\text{w-ord}_i^j \neq \infty$ or 0.

At the end, the weaving process of the strand comes to an end in a fixed year i, with $(\text{inv}_{\text{classic}})_i$ taking the following form

$$(\text{inv}_{\text{classic}})_i = (\dim H_i^0, \text{w-ord}_i^0, s_i^0) \cdots (\dim H_i^j, \text{w-ord}_i^j, s_i^j)$$
$$\cdots (\dim H_i^{m-1}, \text{w-ord}_i^{m-1}, s_i^{m-1}) \begin{cases} (\dim H_i^m, \text{w-ord}_i^m = \infty), \text{ or} \\ (\dim H_i^m, \text{w-ord}_i^m = 0, \Gamma). \end{cases}$$

Termination in the horizontal direction

We note that termination of the weaving process in the horizontal direction is a consequence of the fact that going from the j-th unit to the $(j+1)$-th unit we have $\dim H_i^j > \dim H_i^{j+1}$ and that the dimension obviously satisfies the descending chain condition.

Construction of the (j+1)-th modification $(H_i^{j+1}, (\mathcal{J}_i^{j+1}, b_i^{j+1}), F_i^{j+1})$

We note that we construct the $(j+1)$-th modification only when $\text{w-ord}_i^j \neq \infty$ or 0.

Case: $(\text{inv}_{\text{classic}})_i^{\leq j} = (\text{inv}_{\text{classic}})_{i-1}^j$.

In this case, we simply take $(H_i^{j+1}, (\mathcal{J}_i^{j+1}, b_i^{j+1}), F_i^{j+1})$ to be the transformation of $(H_{i-1}^{j+1}, (\mathcal{J}_{i-1}^{j+1}, b_{i-1}^{j+1}), F_{i-1}^{j+1})$ under the blow up.

Case: $(\text{inv}_{\text{classic}})_i^{\leq j} < (\text{inv}_{\text{classic}})_{i-1}^j$.

In this case, we follow the construction described in the mechanism discussed in §3.2.

Starting from $(H_i^j, (\mathcal{J}_i^j, b_i^j), F_i^j)$, we firstly construct $(H_i^j, (\mathcal{K}_i^j, \kappa_i^j), G_i^j)$, whose description is given below.

$$H_i^j = H_i^j, \ (\mathcal{K}_i^j, \kappa_i^j) = \text{Bdry}(\mathcal{C}_i^j, c_i^j), \text{ and } G_i^j = F_i^j \setminus (F_i^j)_{\text{old}} = (F_i^j)_{\text{new}},$$

where

$$(\mathcal{C}_i^j, c_i^j) = \text{Comp}(\mathcal{J}_i^j, b_i^j), \quad \text{Bdry}(\mathcal{C}_i^j, c_i^j) = (\mathcal{C}_i^j + (\sum_{D \subset (F_i^j)_{\text{old}}} \mathcal{I}(D))^{c_i^j}, c_i^j),$$

and, denoting $(\text{inv}_{\text{classic}})_i^{\leq j-1}(\dim H_i^j, \text{w-ord}_i^j)$ by α_i^j, we set $\text{Comp}(\mathcal{J}_i^j, b_i^j)$ to be either the transformation of $\text{Comp}(\mathcal{J}_{i-1}^j, b_{i-1}^j)$ if $\alpha_i^j = \alpha_{i-1}^j$, or

$$(\mathcal{J}_i^{j\,M_i^j} + \overline{\mathcal{J}_i^j}^{b_i^j}, M_i^j \cdot b_i^j) \quad \text{with} \quad M_i^j = \text{w-ord}_i^j \cdot b_i^j$$

if $\alpha_i^j < \alpha_{i-1}^j$.

Then we secondly construct $(H_i^{j+1}, (\mathcal{J}_i^{j+1}, b_i^{j+1}), F_i^{j+1})$ as follows.

$$\begin{cases} H_i^{j+1} & = \text{the strict transform of } H_{i_{\text{old}}^j}^{j+1}, \\ (\mathcal{J}_i^{j+1}, b_i^{j+1}) & = (\text{Coeff}(\mathcal{K}_i^j)|_{H_i^{j+1}}, (\kappa_i^j)!) \\ F_i^{j+1} & = G_i^j|_{H_i^{j+1}} = (F_i^j)_{\text{new}}|_{H_i^{j+1}}, \end{cases}$$

where $H_{i_{\text{old}}^j}^{j+1}$ is taken in the following way: We go back to the year $\iota := i_{\text{old}}^j$ when the current value of $\text{inv}_i^{\leq j}(\dim H_i^j, \text{w-ord}_i^j)$ first started. Let $(\mathcal{C}_\iota^j, c_\iota^j) = \text{Comp}(\mathcal{I}_\iota^j, b_\iota^j)$ be the companion modification constructed in year ι. We take $f_\iota^j \in (\mathcal{C}_\iota^j)_{P_\iota}$ and a differential operator δ_ι^j of $\deg \delta_\iota^j = c_\iota^j - 1$ such that $\text{ord}_{P_\iota}(f_\iota^j) = c_\iota^j$ and $\text{ord}_{P_\iota}(\delta_\iota^j f_\iota^j) = 1$. We set $H_{i_{\text{old}}^j}^{j+1} = H_\iota^{j+1} = \{\delta_\iota^j f_\iota^j = 0\}$.

Summary of the algorithm in $\text{char}(k) = 0$ **in terms of "**$\text{inv}_{\text{classic}}$**"**

We start with $(W, (\mathcal{I}, a), E) = (W_0, (\mathcal{I}_0, a), E_0)$.

Suppose we have already constructed the resolution sequence up to year i

$$(W, (\mathcal{I}, a), E) = (W_0, (\mathcal{I}_0, a), E_0) \leftarrow \cdots \leftarrow (W_i, (\mathcal{I}_i, a), E_i).$$

We weave the strand of invariants in year i described as above

$$\begin{aligned} (\text{inv}_{\text{classic}})_i &= (\dim H_i^0, \text{w-ord}_i^0, s_i^0) \cdots (\dim H_i^j, \text{w-ord}_i^j, s_i^j) \\ &\cdots (\dim H_i^{m-1}, \text{w-ord}_i^{m-1}, s_i^{m-1}) \begin{cases} (\dim H_i^m, \text{w-ord}_i^m = \infty), \text{ or} \\ (\dim H_i^m, \text{w-ord}_i^m = 0, \Gamma). \end{cases} \end{aligned}$$

There are two cases according to the form of the last unit:

Case: $\text{w-ord}_i^m = \infty$.

In this case, we take the center of blow up in year i for the transformation to be the last hypersurface of maximal contact H_i^m.

Case: $\text{w-ord}_i^m = 0$.

In this case, we follow the procedure specified for resolution of singularities in the monomial case in §3.4, and take the center of blow up in year i for the transformation to be the maximum locus of the invariant Γ on H_i^m.

In both cases, the center of blow up coincides with the maximum locus of the strand $(\text{inv}_{\text{classic}})_i$.

Termination in the vertical direction

The value of the strand "$\mathrm{inv}_{\mathrm{classic}}$" never increases after the blow up described as above. By construction, the maximum locus of $(\mathrm{inv}_{\mathrm{classic}})^{\leq j}$ coincides with the singular locus of $(H^{j+1}, (\mathcal{J}^{j+1}, b^{j+1}), F^{j+1})$. Therefore, resolution of singularities for $(H^{j+1}, (\mathcal{J}^{j+1}, b^{j+1}), F^{j+1})$ implies the strict decrease of $(\mathrm{inv}_{\mathrm{classic}})^{\leq j}$. In particular, in the first case, $(\mathrm{inv}_{\mathrm{classic}})^{\leq m-1}$ strictly decreases. In the second case, either $(\mathrm{inv}_{\mathrm{classic}})^{\leq m-1}$ strictly decreases, or while $(\mathrm{inv}_{\mathrm{classic}})^{\leq m-1}$ may remain the same (and hence so does $(\mathrm{inv}_{\mathrm{classic}})^{\leq m-1}(\dim H_i^m, \text{w-ord}_i^m = 0)$), the invariant Γ strictly decreases. After all, we conclude that the value of the strand strictly decreases after each blow up, i.e., we have $(\mathrm{inv}_{\mathrm{classic}})_i > (\mathrm{inv}_{\mathrm{classic}})_{i+1}$. Now we claim that the value of the strand "$\mathrm{inv}_{\mathrm{classic}}$" cannot decrease infinitely many times. In fact, suppose by induction we have shown that the value of $(\mathrm{inv}_{\mathrm{classic}})^{\leq t-1}$ cannot decrease infinitely many times. Then after some year, the value of $(\mathrm{inv}_{\mathrm{classic}})^{\leq t-1}$ stabilizes. This in turn implies that the value b^t, the second factor of the pair in the t-th modification, stays the same, say b. Now the value of w-ord^t, having the fixed denominator b, cannot decrease infinitely many times, and neither can the value s^t being the nonnegative integer. Therefore, we conclude that the value of $(\mathrm{inv}_{\mathrm{classic}})^{\leq t}$ cannot decrease infinitely many times. As the value of t increases, the value of the dimension decreases by one. Since obviously the value of the dimension satisfies the descending chain condition, the increase of the value of t stops after finitely many times. Finally, therefore, we conclude that the value of the strand "$\mathrm{inv}_{\mathrm{classic}}$" cannot decrease infinitely many times.

Therefore, the algorithm terminates after finitely many years, achieving resolution of singularities for $(W, (\mathcal{I}, a), E)$.

3.4. The monomial case in characteristic zero

The purpose of §3.4 is to discuss how to construct resolution of singularities for $(W, (\mathcal{I}, a), E)$ which is in the monomial case. (Precisely speaking, the triplet sits in the middle of the sequence, say in year "i", for resolution of singularities. However, we omit the subscript "$(\ )_i$" indicating the year for simplicity of the notation.)

Recall that, in the monomial case, $\mathcal{I}$ is a monomial of the ideals defining the components of $E_{\mathrm{young}} = \bigcup_{t=1}^{e} D_t \subset E$ (See §3.3 for the definition of E_{young}.), i.e.,

$$\mathcal{I} = \prod_{t=1}^{e} \mathcal{I}(D_t)^{c_t} \quad \text{with} \quad c_t \in \mathbb{Z}_{\geq 0}.$$

Invariant "Γ"

Definition 1 (Invariant "Γ"). Let the situation be as above. We define the invariant $\Gamma = (\Gamma_1, \Gamma_2, \Gamma_3)$ at $P \in \mathrm{Sing}(\mathcal{I}, a)$ in the following way:

$$\begin{aligned}
\Gamma_1 &= -\min\{n \mid \exists\, t_1, \ldots, t_n \text{ s.t. } c_{t_1} + \cdots + c_{t_n} \geq a, P \in D_{t_1} \cap \cdots \cap D_{t_n}\},\\
\Gamma_2 &= \max\{(c_{t_1} + \cdots + c_{t_n})/a \mid n = -\Gamma_1, P \in D_{t_1} \cap \cdots \cap D_{t_n}\},\\
\Gamma_3 &= \max\left\{(t_1, \ldots, t_n) \mid \begin{array}{l} n = -\Gamma_1,\ (c_{t_1} + \cdots + c_{t_n})/a = \Gamma_2,\\ P \in D_{t_1} \cap \cdots \cap D_{t_n},\ t_1 < \cdots < t_n \end{array}\right\}.
\end{aligned}$$

It is immediate to see the following properties of the invariant "Γ".

(1) The invariant Γ is an upper semi-continuous function.

(2) The maximum locus MaxLocus(Γ) is nonsingular, since it is the intersection of some components in $E_{\text{young}} \subset E$, a simple normal crossing divisor.

Procedure and its termination

Now take the transformation with center $C = \mathrm{MaxLocus}(\Gamma)$

$$(W, (\mathcal{I}, a), E) \stackrel{\pi}{\leftarrow} (W', (\mathcal{I}', a), E')$$

where $E' = E \cup \pi^{-1}(C)$ and $E'_{\text{young}} = E_{\text{young}} \cup \pi^{-1}(C) = \bigcup_{t=1}^{e+1} D_t$ with $D_{e+1} = \pi^{-1}(C)$. Then it is easy to see that $(W', (\mathcal{I}', a), E')$ is again in the monomial case and that the invariant Γ strictly decreases, i.e.,

$$\Gamma > \Gamma'.$$

Since the value of Γ cannot decrease infinitely many times, this procedure must terminate after finitely many years, achieving resolution of singularities for $(W, (\mathcal{I}, a), E)$ in the monomial case.

This completes the discussion on how to construct resolution of singularities for $(W, (\mathcal{I}, a), E)$ in the monomial case.

3.5. Globalization

The strand "$\mathrm{inv}_{\text{classic}}$" a priori depends on the choice of the hypersurfaces of maximal contact we take in the process of weaving, and it is a priori only locally defined. However, the strand "$\mathrm{inv}_{\text{classic}}$" is actually independent of the choice, and hence it is globally well-defined. This can be shown classically by the so-called Hironaka's trick, or more recently by incorporating Włodarczyk's "homogenization" (cf.[23]) or the first author's "differential saturation" (cf.[18]) in the construction of the modification. Therefore, the process of resolution of singularities, where we take the center of blow up to be the maximum locus of "$\mathrm{inv}_{\mathrm{classic}}$", is also globally well-defined. This is how we overcome shortcoming (2) of the key inductive lemma.

This finishes the quick review on the classical algorithm in characteristic zero.

§4. Our algorithm in positive characteristic

The goal of this section is to discuss the general mechanism of our algorithm in positive characteristic, which is modeled closely upon the classical algorithm in characteristic zero reviewed in §3.

4.1. Reformulation of the problem in our setting

First, we present the reformulation of the problem in our setting.

Problem 4 (Reformulation in terms of an idealistic filtration (cf. [18] [19])). *Suppose we are given the triplet of data $(W, \mathcal{R}, E)$ such that W is a nonsingular variety over k, $\mathcal{R} = \bigoplus_{a \in \mathbb{Z}_{\geq 0}} (\mathcal{I}_a, a)$ is an idealistic filtration of i.f.g. type (resp. of i.g. type), i.e., a finitely generated graded $\mathcal{O}_W$-algebra (resp. a graded $\mathcal{O}_W$-algebra) satisfying the condition $\mathcal{O}_W = \mathcal{I}_0 \supset \mathcal{I}_1 \supset \mathcal{I}_2 \cdots \supset \mathcal{I}_a \supset \cdots$, where "$a$" in the second factor specifies the "level" of the ideal $\mathcal{I}_a$ in the first factor, and E is a simple normal crossing divisor on W.*

We define its singular locus to be

$$\mathrm{Sing}(\mathcal{R}) := \{P \in W \mid \mathrm{ord}_P(\mathcal{I}_a) \geq a \quad \forall a \in \mathbb{Z}_{\geq 0}\}.$$

(We note that we only consider $\mathcal{R}$ with $\mathcal{I}_1 \neq 0$ for the resolution problem.)

Then construct a sequence of transformations starting from $(W_0, \mathcal{R}_0, E_0) = (W, \mathcal{R}, E)$

$$\begin{aligned}(W_0, \mathcal{R}_0, E_0) \leftarrow \cdots \leftarrow (W_i, \mathcal{R}_i, E_i) &\stackrel{\pi_{i+1}}{\leftarrow} (W_{i+1}, \mathcal{R}_{i+1}, E_{i+1}) \\ &\leftarrow \cdots \leftarrow (W_l, \mathcal{R}_l, E_l)\end{aligned}$$

such that $\mathrm{Sing}(\mathcal{R}_l) = \emptyset$.

We call such a sequence "resolution of singularities for $(W, \mathcal{R}, E)$".

We note that the transformation

$$(W_i, \mathcal{R}_i, E_i) \stackrel{\pi_{i+1}}{\leftarrow} (W_{i+1}, \mathcal{R}_{i+1}, E_{i+1})$$

is required to satisfy the following conditions:

(1) *$W_i \stackrel{\pi_{i+1}}{\leftarrow} W_{i+1}$ is a blow up with smooth center $C_i \subset W_i$,*

(2) $C_i \subset \mathrm{Sing}(\mathcal{R}_i)$, *and* C_i *is transversal to* E_i *(maybe contained in* E_i*), i.e.,* $C \pitchfork E_i$,

(3) $\mathcal{R}_{i+1} = \mathcal{G}(\bigcup_{a \in \mathbb{Z}_{>0}} (\mathcal{J}_{a,i+1}, a))$, *where*

$$\mathcal{J}_{a,i+1} = \mathcal{I}(\pi_{i+1}^{-1}(C_i))^{-a} \cdot \pi_{i+1}^{-1}(\mathcal{I}_{a,i})\mathcal{O}_{W_{i+1}} \quad \textit{for} \quad a \in \mathbb{Z}_{\geq 0},$$

i.e., $\mathcal{R}_{i+1}$ *is the smallest idealistic filtration of i.f.g. type containing* $(\mathcal{J}_{a,i+1}, a)$ *for all* $a \in \mathbb{Z}_{\geq 0}$ *(We note that* $\mathcal{I}_{a,i+1} \supset \mathcal{J}_{a,i+1}$ *but they may not be equal in general.),*

(4) $E_{i+1} = E_i \cup \pi_{i+1}^{-1}(C_i)$.

Remark 2 (Local version). Problem 4 is the "global" version of the problem of resolution of singularities for the triplet of data $(W, \mathcal{R}, E)$. In the following, we formulate its local version: Starting from a closed point $P \in \mathrm{Sing}(\mathcal{R}) \subset W$ and its neighborhood, we have a sequence of closed points and their neighborhoods

$$P_0 \in \mathrm{Sing}(\mathcal{R}_0) \subset W_0 \leftarrow P_1 \in \mathrm{Sing}(\mathcal{R}_1) \subset W_1 \leftarrow \cdots \leftarrow P_i \in \mathrm{Sing}(\mathcal{R}_i) \subset W_i$$

in the resolution sequence, where $W = W_0$, $\mathcal{R} = \mathcal{R}_0$ and $P = P_0$. After we choose the center $P_i \in C_i \subset \mathrm{Sing}(\mathcal{R}_i) \subset W_i$ and take the transformation $W_i \overset{\pi_{i+1}}{\leftarrow} W_{i+1}$ to extend the resolution sequence, the "devil" tries to choose a closed point $P_{i+1} \in \pi_i^{-1}(P_i) \cap \mathrm{Sing}(\mathcal{R}_{i+1}) \subset W_i$. If $\pi_i^{-1}(P_i) \cap \mathrm{Sing}(\mathcal{R}_{i+1}) = \emptyset$, then the devil loses. Our task is to provide a prescription on how to choose the center so that, no matter how the devil makes his choice, he will end up losing. That is to say, the prescription should guarantee that we ultimately reach year $i = l - 1$ so that, with the choice of center C_{l-1}, after the blow up we have $\pi_l^{-1}(P_{l-1}) \cap \mathrm{Sing}(\mathcal{R}_l) = \emptyset$.

Our algorithm discussed in this paper is exclusively for this local version of the problem of resolution of singularities for the triplet of data $(W, \mathcal{R}, E)$. The adjustments we have to make to our algorithm in order to solve the global version of the problem will be published elsewhere.

The notion of "the differential saturation" $\mathcal{DR}$ (of an idealistic filtration $\mathcal{R}$ of i.f.g. type) plays an important role in our algorithm.

Definition 2. Let $\mathcal{R}$ be an idealistic filtration of i.f.g. type. We define its differential saturation $\mathcal{DR}$ (at the level of the stalk for a point $P \in W$) as follows:

$$\mathcal{DR}_P := \mathcal{G}\left(\{(\delta f, \max\{a - \deg \delta, 0\}) \mid (f, a) \in \mathcal{R}_P,\ \delta\text{: a diff. op.}\}\right),$$

where the symbol $\mathcal{G}(S)$ denotes "an idealistic filtration of i.g. type generated by the set S", i.e., the smallest idealistic filtration of i.g. type containing S.

Remark 3.

(1) It follows immediately from the generalized product rule (cf. [18]) that $\mathcal{DR}$ is again an idealistic filtration of i.f.g. type, $\mathcal{R} \subset \mathcal{DR}$, and that it is differentially saturated, i.e., $\mathcal{D}(\mathcal{DR}) = \mathcal{DR}$.

(2) The problem of constructing resolution of singularities for $(W, \mathcal{R}, E)$ is equivalent to the one for $(W, \mathcal{DR}, E)$, i.e.,

$$(W, \mathcal{R}, E) \underset{\text{equivalent to}}{\sim} (W, \mathcal{DR}, E).$$

4.2. Inductive scheme on the invariant "σ"

The classical algorithm in characteristic zero works by induction on dimension, based upon the notion of a *smooth* hypersurface of maximal contact, as reviewed in §3. Narasimhan's example (cf. Remark 1 in §3) tells us, however, that there is no hope of finding a *smooth* hypersurface of maximal contact in positive characteristic. The following proposition gives rise to the notion of "a leading generator system" (called an LGS for short), which we consider as a collective substitute in positive characteristic for the notion of a hypersurface of maximal contact (called an HMC for short) in characteristic zero. Our algorithm in positive characteristic works by induction on the invariant "σ", based upon the notion of an LGS. Roughly speaking, introducing the notion of an LGS corresponds to considering *singular* hypersurfaces of maximal contact.

Definition of the invariant "σ"

Proposition 1 (cf. [18]). *Let $\mathcal{R} = \bigoplus_{a \in \mathbb{Z}_{\geq 0}} (\mathcal{I}_a, a)$ be an idealistic filtration of i.f.g. type. Assume that $\mathcal{R}$ is differentially saturated, i.e., $\mathcal{R} = \mathcal{DR}$. Fix a closed point $P \in \mathrm{Sing}(\mathcal{R}) \subset W$.*

Consider the leading algebra $L_P(\mathcal{R})$

$$L_P(\mathcal{R}) := \bigoplus_{a \in \mathbb{Z}_{\geq 0}} \{f \bmod \mathfrak{m}_P^{a+1} \mid (f, a) \in \mathcal{R}_P, f \in \mathfrak{m}_P^a\} \subset \bigoplus_{a \in \mathbb{Z}_{\geq 0}} \mathfrak{m}_P^a / \mathfrak{m}_P^{a+1}.$$

Then there exists a regular system of parameters $(x_1, \ldots, x_t, x_{t+1}, \ldots, x_d)$ at P such that the leading algebra takes the following form:

Case: $\mathrm{char}(k) = 0$

$$L_P(\mathcal{R}) = k[x_1, \ldots, x_t] \subset k[x_1, \ldots, x_t, x_{t+1}, \ldots, x_d] = \bigoplus_{a \in \mathbb{Z}_{\geq 0}} \mathfrak{m}_P^a / \mathfrak{m}_P^{a+1}.$$

Moreover, we observe the following: if we take an element $(h_i, 1) \in \mathcal{R}_P$ *with* $h_i \equiv x_i \bmod \mathfrak{m}_P^2$ $(i = 1, \ldots, t)$, *then the hypersurface* $\{h_i = 0\}$ *is a hypersurface of maximal contact in the classical sense.*

Case: $\mathrm{char}(k) = p > 0$

$$L_P(\mathcal{R}) = k[x_1^{p^{e_1}}, \ldots, x_t^{p^{e_t}}] \subset k[x_1, \ldots, x_t, x_{t+1}, \ldots, x_d] = \bigoplus_{a \in \mathbb{Z}_{\geq 0}} \mathfrak{m}_P^a / \mathfrak{m}_P^{a+1}$$

for some $0 \leq e_1 \leq \cdots \leq e_t$.

Remark 4. The former *Case*: $\mathrm{char}(k) = 0$ in the above proposition can be regarded as a special case of the latter *Case*: $\mathrm{char}(k) = p > 0$, by formally setting $p = \infty$, $0 = e_1 = \cdots = e_t$ and $\infty^0 = 1$, where all the x^{p^e}-terms with $e > 0$ become "invisible" as p goes to ∞.

Definition 3 (Leading Generator System (cf. [18] [19])). Let the situation be as in the proposition above. Take a subset $\mathbb{H} = \{(h_\alpha, p^{e_\alpha})\}_{\alpha=1}^t \subset \mathcal{R}_P$ with

$$h_\alpha \equiv x_\alpha^{p^{e_\alpha}} \bmod \mathfrak{m}_P^{p^{e_\alpha}+1} \quad \text{for} \quad \alpha = 1, \ldots, t.$$

We say that $\mathbb{H}$ is a leading generator system for $\mathcal{R}_P$ (called an LGS for short). A leading generator system for $\widehat{\mathcal{R}_P}$ is defined similarly.

Definition 4 (Invariant "σ" and "τ" (cf. [18] [19])). Let the situation be as in the proposition above. Then the invariants σ and τ are defined by the following formulas

$$\sigma(P) := (a_n)_{n \in \mathbb{Z}_{\geq 0}} \quad \text{where} \quad a_n = d - \#\{e_\alpha \mid e_\alpha \leq n\}$$

where the value set of the invariant σ is given the lexicographical order, and

$$\tau(P) := \# \text{ of the elements in an LGS} = \#\mathbb{H} = t.$$

(We note that the invariants σ and τ are independent of the choice of a regular system of parameters or an LGS, and that it does not matter whether we compute them at the algebraic level or at the analytic level.) The moral here is that the more e_α's we have at the lower level, the smaller the value of the invariant σ is and hence we consider the better the LGS $\mathbb{H}$ is.

Basic strategy to establish our algorithm in positive characteristic: *Follow the construction of the algorithm in* $\mathrm{char}(k) = 0$, *replacing the notion of an HMC to use the induction on dimension by the notion of an LGS to use the induction on the invariant* σ.

Mechanism of the inductive scheme on the invariant σ

Given $(W, \mathcal{R}, E)$ (Precisely speaking, the triplet sits in the middle of the sequence, say in year "i", for resolution of singularities. However, we omit the subscript "$(\ )_i$" indicating the year for simplicity of the notation), we introduce a triplet of invariants $(\sigma, \widetilde{\mu}, s)$ and its associated triplet of data $(W', \mathcal{R}', E')$. Together they form the following mechanism to realize the inductive scheme on the invariant σ (We note that there is *no* Key Inductive Lemma in our setting):

<u>Outline of the mechanism</u>

(1) We introduce a triplet of invariants $(\sigma, \widetilde{\mu}, s)$.
If $(\sigma, \widetilde{\mu}, s) = (\sigma, \infty, 0)$ or $(\sigma, 0, 0)$, then we do not construct the modification $(W', \mathcal{R}', E')$.

- In case $(\sigma, \widetilde{\mu}, s) = (\sigma, \infty, 0)$, we blow up with center $C = \mathrm{Sing}(\mathcal{R})$. The nonsingularity of C is guaranteed by the Nonsingularity Principle (cf. [18] [19]), while the transversality of C to the boundary E is guaranteed by the invariant $s = 0$. After the blow up, the singular locus becomes empty, and resolution of singularities for $(W, \mathcal{R}, E)$ is achieved.
- In case $(\sigma, \widetilde{\mu}, s) = (\sigma, 0, 0)$, we are in the monomial case by definition, and we go to Step (5).

If $(\sigma, \widetilde{\mu}, s) \neq (\sigma, \infty, 0)$ or $(\sigma, 0, 0)$, then we construct its associated triplet of data $(W', \mathcal{R}', E')$, called the modification of the original triplet $(W, \mathcal{R}, E)$, having the following properties. (Actually the ambient space for the modification stays the same, i.e., $W' = W$.)

- Resolution of singularities for $(W', \mathcal{R}', E')$ implies the strict decrease of the (maximum) value of the triplet $(\sigma, \widetilde{\mu}, s)$. Note that the value of the triplet never increases after transformations.
- Either the value of σ strictly decrease, or the value of σ stays the same but the number of the components in the boundary strictly drops. In either case, we have $(\sigma, \#E) > (\sigma', \#E')$.

(2) There is *no* key inductive lemma in our new setting.

(3) When $(\sigma, \widetilde{\mu}, s) \neq (\sigma, \infty, 0)$ or $(\sigma, 0, 0)$ in (1), we achieve resolution of singularities for $(W', \mathcal{R}', E')$ by induction on $(\sigma, \#E)$, which implies the strict decrease of the (maximum) value of the triplet $(\sigma, \widetilde{\mu}, s)$.

(4) Repeatedly decreasing the value of the triplet $(\sigma, \widetilde{\mu}, s)$, we reach the case where $(\sigma, \widetilde{\mu}, s) = (\sigma, \infty, 0)$ or $(\sigma, 0, 0)$, the monomial case.

(5) Finally we have only to construct resolution of singularities in the monomial case in order to achieve resolution of singularities of the original triplet $(W, \mathcal{R}, E)$. However, the problem of resolution of singularities in the monomial case in positive characteristic is quite subtle and very difficult. We only provide a solution in dimension 3 in §5.

Warning: Even though the invariant σ never increases after blow ups chosen in our algorithm, the number of exceptional divisors $\#E$ and hence the pair $(\sigma, \#E)$ may increase. Therefore, the description of "by induction on $(\sigma, \#E)$" in (3) above is slightly imprecise. The precise mechanism of the induction is manifested as the weaving of the new strand of invariants "inv_{new}". See §4.3 for details.

Description of the triplet of invariants $(\sigma, \widetilde{\mu}, s)$

$\circ$ σ: It is the invariant σ associated to the differential saturation $\mathcal{DR}$ of the idealistic filtration of i.f.g. type $\mathcal{R}$ (cf. Definitions 2, 3, and 4).

$\circ$ $\widetilde{\mu}$: It is the (normalized) weak order modulo an LGS of the idealistic filtration $\mathcal{R}$, with respect to E_{young}.

$\circ$ s: It is the number of the components in $E_{\text{aged}} = E \setminus E_{\text{young}}$.

We explain how to compute the invariant $\widetilde{\mu}$ more in detail: Let $\mathbb{H}$ be the LGS chosen. Given $f \in \widehat{\mathcal{O}_{W,P}}$, let $f = \sum c_{f,B} H^B$ be its power series expansion with respect to the LGS (and its associated regular system of parameters) (cf. [19]). Then we define

$$\begin{aligned}
\mu_P(\mathcal{R}) &= \inf\left\{\frac{1}{a}\mathrm{ord}_P(c_{f,\mathbb{O}}) \mid (f,a) \in \mathcal{R}_P, a > 0\right\} \\
&= \inf\left\{\frac{1}{a}\mathrm{ord}_P(c_{f,\mathbb{O}}) \mid (f,a) \in \widehat{\mathcal{R}_P}, a > 0\right\}, \\
\mu_{P,D}(\mathcal{R}) &= \inf\left\{\frac{1}{a}\mathrm{ord}_{\xi_D}(c_{f,\mathbb{O}}) \mid (f,a) \in \mathcal{R}_P, a > 0\right\} \\
&= \inf\left\{\frac{1}{a}\mathrm{ord}_{\xi_D}(c_{f,\mathbb{O}}) \mid (f,a) \in \widehat{\mathcal{R}_P}, a > 0\right\},
\end{aligned}$$

where ξ_D is the generic point of a component D in E_{young}. (For the definition of E_{young}, see (iii) of the remark below.) Now we define the invariant $\widetilde{\mu}$ by the following formula

$$\widetilde{\mu} = \mu_P(\mathcal{R}) - \sum_{D \in E_{\text{young}}} \mu_{P,D}(\mathcal{R}).$$

It is straightforward to see via the formal coefficient lemma that $\widetilde{\mu}$ is independent of the choice of the LGS (and its associated regular system of parameters) and that $\widetilde{\mu}$ is a nonnegative rational number, since our idealistic filtration is of i.f.g. type (cf. [18][19]).

We make the following remarks on the technical but important points about the LGS (and its associated regular system of parameters), the idealistic filtration of i.f.g. type $\mathcal{R}$, and $E_{\text{young}} \subset E$ used in the computation above:

(i) The idealistic filtration of i.f.g.type $\mathcal{R}$ used in the computation depends on the history of the behavior of the invariant σ.

Case: *The value of σ remains the same as the one in the previous year.*

In this case, we keep $\mathcal{R}$ as it is, which is the transformation of the one in the previous year, even though we compute the invariant σ using the differential saturation $\mathcal{DR}$. We take our LGS (a priori only in $\mathcal{DR}$) to be the transformation of the one in the previous year, which hence sits inside of $\mathcal{R}$.

Case: *The value of σ is strictly less than the one in the previous year.*

In this case, we replace the original $\mathcal{R}$ with its differential saturation. We take our LGS from this replaced $\mathcal{R} = \mathcal{DR}$, which is differentially saturated, and compute $\widetilde{\mu}$ accordingly. We remark that, in this case, $E_{\text{young}} = \emptyset$ and hence that $\widetilde{\mu} = \mu_P(\mathcal{R})$.

We note that, in year 0, we also replace the original $\mathcal{R}$ with its differential saturation (cf. Remark 3 (2)).

(ii) The LGS $\mathbb{H} = \{(h_\alpha, p^{e_\alpha})\}_{\alpha=1}^t \subset \widehat{\mathcal{R}_P}$ and its associated regular system of parameters $X = (x_1, \ldots, x_t, x_{t+1}, \ldots, x_d) \subset \widehat{\mathcal{O}_{W,P}}$ are taken in such a way that they satisfy the condition ($\heartsuit$) consisting of the three requirements described below:

Condition ($\heartsuit$)

- $h_\alpha \equiv x_\alpha^{p^{e_\alpha}} \bmod \widehat{\mathfrak{m}}^{p^{e_\alpha}+1}$ for $\alpha = 1, \ldots, t$,
- the idealistic filtration of i.f.g. type $\widehat{\mathcal{R}_P}$ is saturated for $\{\partial^n/\partial x_\alpha{}^n \mid n \in \mathbb{Z}_{\geq 0}, \alpha = 1, \ldots, t\}$, and
- the defining equations for the components of E_{young}, which are transversal to the LGS, form a part of the regular system of parameters, i.e., $\{x_D \mid D \in E_{\text{young}}\} \subset \{x_{t+1}, \ldots, x_d\}$.

(We remark that it is easy to find such $H \subset \mathcal{R}$ and $X \subset \mathcal{O}_{W,P}$ that satisfy all the requirements but the second one in Condition ($\heartsuit$).)

We recall the following formal coefficient lemma (cf. [19]), where the assumption is slightly weaker than the original one in the sense that it does not require the idealistic filtration is $\mathcal{D}$-saturated. However, the same conclusion is valid with the same proof.

Formal Coefficient Lemma

Let $\mathcal{R}$ be an idealistic filtration of i.f.g. type. Take a subset $\mathbb{H} = \{(h_\alpha, p^{e_\alpha})\}_{\alpha=1}^t \subset \widehat{\mathcal{R}_P}$ (not necessarily an LGS) and a regular system of parameters $X = (x_1, \ldots, x_t, x_{t+1}, \ldots, x_d) \subset \widehat{\mathcal{O}_{W,P}}$, satisfying the first two requirements in condition ($\heartsuit$). Then, for $(f, a) \in \widehat{\mathcal{R}_P}$, we have

$$(c_{f,B}, \max\{a - |[B]|, 0\}) \in \widehat{\mathcal{R}_P}$$

for any B, and in particular,

$$(c_{f,\mathbb{O}}, a) \in \widehat{\mathcal{R}_P},$$

where $f = \sum c_{f,B} H^B$ is the power series expansion of f with respect to $\mathbb{H}$ and X (cf. [19]).

(iii) The symbol E_{young} refers to the union of the exceptional divisors created after the time when the current value of σ first started. Therefore, by construction, E_{young} is transversal to the LGS. We only use E_{young} in our algorithm, in contrast to the classical algorithm where we have to use both E_{young} and E_{new} (cf. §3).

<u>Description of the triplet $(W', \mathcal{R}', E')$</u>

$$W' = W, \quad \mathcal{R}' = \text{Bdry}\left(\text{Comp}(\mathcal{R})\right), \quad \text{and} \quad E' = E \setminus E_{\text{aged}} = E_{\text{young}},$$

where

- $\text{Comp}(\mathcal{R})$ is either the transformation of the one in the previous year if $(\sigma, \widetilde{\mu})$ stays the same, or the one constructed below if $(\sigma, \widetilde{\mu})$ strictly decreases, and
- $\text{Bdry}\left(\text{Comp}(\mathcal{R})\right)$ is either the transformation of the one in the previous year if $(\sigma, \widetilde{\mu}, s)$ stays the same, or

$$\mathcal{G}(\text{Comp}(\mathcal{R}) \cup \{(x_D, 1) \mid D \in E_{\text{aged}}\})$$

if $(\sigma, \widetilde{\mu}, s)$ strictly decreases, where $\mathcal{G}(S)$ is the idealistic filtration of i.g. type generated by the set S, i.e., the smallest idealistic filtration of i.g. type containing S, and x_D is the defining equation of a component $D \in E_{\text{aged}}$.

We note that, if the value of the triplet $(\sigma, \widetilde{\mu}, s)$ stays the same as in the previous year, then $(W', \mathcal{R}', E')$ is the transformation of the one in the previous year. We remark that the symbols "Comp" and "Bdry" represent the "Companion" modification and the "Boundary" modification, respectively.

<u>Construction of $\text{Comp}(\mathcal{R})$</u>

We describe the construction of the companion modification Comp $(\mathcal{R})$, first at the analytic level, following closely the construction in the classical setting, and then at the algebraic level, showing that the companion modification at the analytic level "descends" to the one at the algebraic level via the argument of "étale descent".

Construction at the analytic level

First, we take an LGS $\mathbb{H} \subset \widehat{\mathcal{R}_P}$ and its associated regular system of parameters $X \subset \widehat{\mathcal{O}_{W,P}}$ satisfying the condition (♡).

We set

$$\mathbb{M}_X = \prod_{D \in E_{\text{young}}} x_D^{\mu_{P,D}(\mathcal{R})}.$$

Recall that $\{x_D \mid D \in E_{\text{young}}\} \subset X$.

Fix a common multiple $L \in \mathbb{Z}_{>0}$ of the denominators of $\widetilde{\mu}, \mu_P(\mathcal{R})$, and $\{\mu_{P,D} \mid D \in E_{\text{young}}\}$. Set

$$\Xi = \widehat{\mathcal{O}_{W,P}} \otimes_k k[x_{t+1}^{\pm\frac{1}{L}}, \ldots, x_d^{\pm\frac{1}{L}}].$$

We consider the following notion of an idealistic filtration $\mathcal{Q}$ in the generalized sense:

$$\mathcal{Q} = \bigoplus_{n \in \mathbb{Z}_{\geq 0}} \left(\mathcal{Q}_{\frac{n}{L}}, \frac{n}{L}\right) \subset \bigoplus_{n \in \mathbb{Z}_{\geq 0}} \left(\Xi, \frac{n}{L}\right)$$

is a graded $\widehat{\mathcal{O}_{W,P}}$-algebra, where

- the grading is given by $\{n/L \mid n \in \mathbb{Z}_{\geq 0}\}$, and it is specified as the level in the second factor,
- $\mathcal{Q}_{n/L} \subset \Xi$ is an $\widehat{\mathcal{O}_{W,P}}$-submodule with the $\widehat{\mathcal{O}_{W,P}}$-module structure induced by the left multiplication on Ξ (We emphasize that the tensor "$\otimes$" is over k.) satisfying the condition $\mathcal{Q}_{0/L} \supset \mathcal{Q}_{1/L} \supset \mathcal{Q}_{2/L} \supset \cdots$,
- the algebra structure is given by the addition and multiplication on Ξ, while the $\widehat{\mathcal{O}_{W,P}}$-algebra structure is given by the left multiplication on the first factor of Ξ,
- the differential operators act on the first factor, i.e., for a differential operator δ of $\deg \delta$ and

$$q = \left(\sum f \otimes g, n/L\right) \in (\Xi, n/L)$$

with $f \in \widehat{\mathcal{O}_{W,P}}$ and $g \in k[x_{t+1}^{\pm\frac{1}{L}}, \ldots, x_d^{\pm\frac{1}{L}}]$, we set

$$\delta q = \left(\sum \delta f \otimes g, \max\{n/L - \deg \delta, 0\}\right).$$

We construct $\widehat{\mathrm{Comp}(\mathcal{R})}_{\mathbb{H},X}$ in the following manner:

Step 1. We take the idealistic filtration $\mathcal{Q}_1$ in the generalized sense generated by $\{(f \otimes 1, a) \mid (f, a) \in \mathcal{R}_P\}$ and $\{(c_{f,\mathbb{O}} \otimes (\mathbb{M}_X^{-1})^a, \widetilde{\mu} \cdot a) \mid (f, a) \in \mathcal{R}_P\}$, i.e.,

$$\mathcal{Q}_1 = \mathcal{G}\left(\left\{(f \otimes 1, a),\ \left(c_{f,\mathbb{O}} \otimes (\mathbb{M}_X^{-1})^a, \widetilde{\mu} \cdot a\right) \mid (f, a) \in \mathcal{R}_P\right\}\right),$$

where $c_{f,\mathbb{O}}$ is the constant term of the power series expansion $f = \sum c_{f,B} H^B$ with respect to $\mathbb{H}$ and X.

Step 2. We take the idealistic filtration $\mathcal{Q}_2$ in the generalized sense to be the $\mathcal{D}_{E_{\text{young}}}$-saturation of the idealistic filtration $\mathcal{Q}_1$ in the generalized sense, i.e.,

$$\mathcal{Q}_2 = \mathcal{D}_{E_{\text{young}}}(\mathcal{Q}_1),$$

where $\mathcal{D}_{E_{\text{young}}}$ represents the logarithmic differentials with respect to the simple normal crossing divisor E_{young} (cf. [18]).

Step 3. We take the integral level part $\mathcal{P} = \text{ILP}(\mathcal{Q}_2)$ of the idealistic filtration $\mathcal{Q}_2$ in the generalized sense. That is to say, $\mathcal{P} = \bigoplus_{a \in \mathbb{Z}_{\geq 0}} (\mathcal{P}_a, a)$ is a graded $\widehat{\mathcal{O}_{W,P}}$-algebra where, for $a \in \mathbb{Z}_{\geq 0}$, we set $\mathcal{P}_a = (\mathcal{Q}_2)_{n/L}$ with $a = n/L$.

Step 4. By taking the "round up" and contraction of $\mathcal{P}$, we obtain the usual idealistic filtration of i.f.g. type $\mathcal{C} = \bigoplus_{a \in \mathbb{Z}_{\geq 0}} (\mathcal{C}_a, a)$ where we set

$$\mathcal{C}_a = \text{RUC}(\mathcal{P}_a) \quad \text{for} \quad a \in \mathbb{Z}_{\geq 0}.$$

Note that the "round up" and contraction map $\text{RUC} : \mathcal{P}_a \to \widehat{\mathcal{O}_{W,P}}$ is given by $\text{RUC}\left(\sum f \otimes (\mathbb{M}_X^{-1})^l\right) = \sum f \cdot \lceil (\mathbb{M}_X^{-1})^l \rceil$ for $l \in \mathbb{Z}_{\geq 0}$, where

$$\lceil (\mathbb{M}_X^{-1})^l \rceil = \prod_{D \in E_{\text{young}}} x_D^{\lceil -\mu_{P,D}(\mathcal{R}) \cdot l \rceil}.$$

Since we have $\text{ord}_{\xi_D}(c_{f,\mathbb{O}}) \geq \mu_{P,D}(\mathcal{R}) \cdot a$ for $(f, a) \in \mathcal{R}_P$ and $D \in E_{\text{young}}$ by definition, and since a logarithmic differential operator with respect to E_{young} does not decrease the power of x_D for $D \in E_{\text{young}}$, we conclude that, though the image of the RUC map is only in $\widehat{\mathcal{O}_{W,P}}[x_{t+1}^{\pm 1}, \ldots, x_d^{\pm 1}]$ a priori, it actually lies within $\widehat{\mathcal{O}_{W,P}}[x_{t+1}, \ldots, x_d] = \widehat{\mathcal{O}_{W,P}}$, i.e.,

$$\mathcal{C}_a = \text{RUC}(\mathcal{P}_a) \subset \widehat{\mathcal{O}_{W,P}} \quad \text{for} \quad a \in \mathbb{Z}_{\geq 0}.$$

Step 5. We set $\mathcal{C} = \widehat{\text{Comp}(\mathcal{R})}_{\mathbb{H},X}$.

Remark 5.

(1) We remark that $\mathcal{C}$ is finitely generated. In fact, if $\{(f_\lambda, a_\lambda) \mid \lambda \in \Lambda\}$ with $\#\Lambda < \infty$ is a set of (local) generators for $\mathcal{R}$, then by setting

$$\begin{aligned}\mathcal{Q}_{1,\Lambda} = \mathcal{G}\,(\{(f_\lambda \otimes 1, a_\lambda) \mid (f_\lambda, a_\lambda) \in \mathcal{R}_P, \lambda \in \Lambda\} \\ \cup \{\left(c_{f_\lambda,\mathbb{O}} \otimes (\mathbb{M}_X^{-1})^{a_\lambda}, \widetilde{\mu} \cdot a_\lambda\right) \mid (f, a_\lambda) \in \mathcal{R}_P, \lambda \in \Lambda\}),\end{aligned}$$

we see by the formal coefficient lemma applied to the case of an idealistic filtration in the generalized sense (the same proof works as in the usual case (cf. [19])) that $\mathcal{Q}_1 \subset \mathcal{D}_{E_{\text{young}}}(\mathcal{Q}_{1,\Lambda})$ while obviously we have $\mathcal{Q}_{1,\Lambda} \subset \mathcal{Q}_1$, and hence that

$$\mathcal{D}_{E_{\text{young}}}(\mathcal{Q}_1) = \mathcal{D}_{E_{\text{young}}}(\mathcal{Q}_{1,\Lambda}).$$

It follows easily from this that $\mathcal{C}$ is finitely generated.

(2) In Step 1, we take the set $\{\left(c_{f,\mathbb{O}} \otimes (\mathbb{M}_X^{-1})^a, \widetilde{\mu} \cdot a\right) \mid (f, a) \in \mathcal{R}_P\}$ as a part of the generators, where the elements (f, a) belong to the idealistic filtration of i.f.g. type $\mathcal{R}_P$ at the algebraic level. Even if we replace this with $\{\left(c_{f,\mathbb{O}} \otimes (\mathbb{M}_X^{-1})^a, \widetilde{\mu} \cdot a\right) \mid (f, a) \in \widehat{\mathcal{R}_P}\}$, where the elements (f, a) belong to the idealistic filtration of i.f.g. type $\widehat{\mathcal{R}_P}$ at the analytic (completion) level, the resulting $\mathcal{C}$ will not change.

In fact, this can be seen as follows: Let $\mathcal{C}_{\text{an}}$ be the one obtained by using the analytic one. (Note that $\mathcal{Q}_{1,\text{an}}$ and $\mathcal{Q}_{2,\text{an}} = \mathcal{D}_{E_{\text{young}}}(\mathcal{Q}_{1,\text{an}})$ are defined similarly, to be used in the proof of Lemma 3.) Since

$$\begin{aligned}\{\left(c_{f,\mathbb{O}} \otimes (\mathbb{M}_X^{-1})^a, \widetilde{\mu} \cdot a\right) \mid (f, a) \in \mathcal{R}_P\} \\ \subset \{\left(c_{f,\mathbb{O}} \otimes (\mathbb{M}_X^{-1})^a, \widetilde{\mu} \cdot a\right) \mid (f, a) \in \widehat{\mathcal{R}_P}\},\end{aligned}$$

it is obvious that we have $\mathcal{C} \subset \mathcal{C}_{\text{an}}$. On the other hand, take $(f, a) \in \widehat{\mathcal{R}_P}$ with the power series expansion $f = \sum c_{f,B} H^B$ with respect to $\mathbb{H}$ and X, and $c_{f,\mathbb{O}}$ its constant term. Then by the formal coefficient lemma, we conclude $(c_{f,\mathbb{O}}, a) \in \widehat{\mathcal{R}_P} = \mathcal{R}_P \otimes_{\mathcal{O}_{W,P}} \widehat{\mathcal{O}_{W,P}}$. This means that there exits a finite set $\{(f_\lambda, a) \in \mathcal{R}_P \mid \lambda \in \Lambda\}$ such that $c_{f,\mathbb{O}} = \sum r_\lambda f_\lambda$ for some $r_\lambda \in \widehat{\mathcal{O}_{W,P}}$.

Set $g = \sum r_\lambda c_{f_\lambda,\mathbb{O}}$ and $h = \sum r_\lambda (f_\lambda - c_{f_\lambda,\mathbb{O}})$ with $c_{f,\mathbb{O}} = g + h$. Then since $c_{h,\mathbb{O}} = 0$, we conclude $c_{f,\mathbb{O}} = c_{g,\mathbb{O}} + c_{h,\mathbb{O}} = c_{g,\mathbb{O}}$. Now since $\{\left(c_{f_\lambda,\mathbb{O}} \otimes (\mathbb{M}_X^{-1})^a, \widetilde{\mu} \cdot a\right)\} \subset \mathcal{Q}_1$ and hence $\left(g \otimes (\mathbb{M}_X^{-1})^a, \widetilde{\mu} \cdot a\right) \in \mathcal{Q}_1$, we conclude, by the formal coefficient lemma applied to the case of an idealistic filtration in the generalized sense, that

$$\left(c_{f,\mathbb{O}} \otimes (\mathbb{M}_X^{-1})^a, \widetilde{\mu} \cdot a\right) = \left(c_{g,\mathbb{O}} \otimes (\mathbb{M}_X^{-1})^a, \widetilde{\mu} \cdot a\right) \in \mathcal{D}_{E_{\text{young}}}(\mathcal{Q}_1) = \mathcal{Q}_2,$$

which implies $\mathcal{C}_{\mathrm{an}} \subset \mathcal{C}$.

Therefore, we conclude $\mathcal{C} = \mathcal{C}_{\mathrm{an}}$.

Lemma 3. *The companion modification* $\widehat{\mathrm{Comp}(\mathcal{R})}_{\mathbb{H},X}$ *constructed at the analytic level as above is independent of the choice of an LGS* $\mathbb{H}$ *and its associated regular system of parameters* X *satisfying the condition* ($\heartsuit$). *That is to say, if* $\mathbb{H}'$ *and* X' *are another LGS and its associated regular system of parameters satisfying the condition* ($\heartsuit$), *then we have*

$$\widehat{\mathrm{Comp}(\mathcal{R})}_{\mathbb{H},X} = \widehat{\mathrm{Comp}(\mathcal{R})}_{\mathbb{H}',X'}.$$

We write, therefore,

$$\widehat{\mathrm{Comp}(\mathcal{R})} = \widehat{\mathrm{Comp}(\mathcal{R})}_{\mathbb{H},X}$$

omitting the reference to the LGS and its associated regular system of parameters used in the construction.

We call $\widehat{\mathrm{Comp}(\mathcal{R})}$ the companion modification at the analytic level.

Proof. We consider the following two cases.

Case: *For any* $e \in \mathbb{Z}_{\geq 0}$ *and* $(h_\alpha, p^{e_\alpha}) \in \mathbb{H}$ *with* $e_\alpha = e$, *the element* h_α *is a linear combination of*

$$\{(h'_\beta)^{p^{e-e'_\beta}} \mid e'_\beta \leq e\}.$$

There is no condition on X *or* X' *other than their being associated to* $\mathbb{H}$ *and* $\mathbb{H}'$, *respectively.*

In this case, we claim $\widehat{\mathrm{Comp}(\mathcal{R})}_{\mathbb{H}',X'} \subset \widehat{\mathrm{Comp}(\mathcal{R})}_{\mathbb{H},X}$.

Take $(f,a) \in \mathcal{R}_P$. Let $f = \sum c_{f,B} H$ be the power series expansion of f with respect to $\mathbb{H}$ and X, with its constant term $c_{f,\mathbb{O}}$. We have $\left(c_{f,\mathbb{O}} \otimes (\mathbb{M}_X^{-1})^a, \widetilde{\mu} \cdot a\right) \in \mathcal{Q}_1$. Set $g = c_{f,\mathbb{O}}$ and $h = f - c_{f,\mathbb{O}}$. Let $g = \sum c'_{g,B} H'^B$ and $h = \sum c'_{h,B} H'^B$ be the power series expansions of g and h, respectively, with respect to $\mathbb{H}'$ and X'. Now we conclude, by the formal coefficient lemma applied to the case of an idealistic filtration in the generalized sense, that

$$\left(c'_{g,\mathbb{O}} \otimes (\mathbb{M}_X^{-1})^a, \widetilde{\mu} \cdot a\right) \in \mathcal{D}_{E_{\mathrm{young}}}(\mathcal{Q}_{1,\mathrm{an}}) = \mathcal{Q}_{2,\mathrm{an}}.$$

On the other hand, the assumption of this case implies that $c'_{h,\mathbb{O}} = 0$ and hence that

$$c'_{g,\mathbb{O}} = c'_{g,\mathbb{O}} + c'_{h,\mathbb{O}} = c'_{g+h,\mathbb{O}} = c'_{f,\mathbb{O}},$$

the constant term of the power series expansion $f = \sum c'_{f,B} H'^B$ with respect to $\mathbb{H}'$ and X'. Thus, we have

$$\left(c'_{f,\mathbb{O}} \otimes (\mathbb{M}_X^{-1})^a, \widetilde{\mu} \cdot a\right) \in \mathcal{D}_{E_{\text{young}}}(\mathcal{Q}_{1,\text{an}}) = \mathcal{Q}_{2,\text{an}}.$$

This implies by Remark 5 (2)

$$\widehat{\text{Comp}(\mathcal{R})}_{\mathbb{H}',X'} \subset \mathcal{C}_{\text{an}} = \widehat{\text{Comp}(\mathcal{R})}_{\mathbb{H},X}.$$

(We note that the generators we choose in Step 1 of the construction for $\widehat{\text{Comp}(\mathcal{R})}_{\mathbb{H}',X'}$ are of the form $(c'_{f,\mathbb{O}} \otimes (\mathbb{M}_{X'}^{-1})^a, \widetilde{\mu} \cdot a)$ (not $\otimes(\mathbb{M}_X^{-1})^a$), which are sitting inside of

$$\widehat{\mathcal{O}_{W,P}} \otimes_k k[x'_{t+1}{}^{\pm\frac{1}{L}}, \ldots, x'_d{}^{\pm\frac{1}{L}}] \quad (\text{not of } \Xi = \widehat{\mathcal{O}_{W,P}} \otimes_k k[x_{t+1}^{\pm\frac{1}{L}}, \ldots, x_d^{\pm\frac{1}{L}}]).$$

However, these differences only contribute to the multiplication of units after Step 3 and Step 4, and hence do not matter for us to conclude the inclusion above.)

Case: $X = X'$.

In this case, we also claim $\widehat{\text{Comp}(\mathcal{R})}_{\mathbb{H}',X'} \subset \widehat{\text{Comp}(\mathcal{R})}_{\mathbb{H},X}$.

Take $(f, a) \in \mathcal{R}_P$. Let $f = \sum c'_{f,B} H'$ be the power series expansion of f with respect to $\mathbb{H}'$ and X', with its constant term $c'_{f,\mathbb{O}}$. By the formal coefficient lemma, we have $(c'_{f,\mathbb{O}}, a) \in \widehat{\mathcal{R}_P}$. Set $g' = c'_{f,\mathbb{O}} \in \widehat{\mathcal{O}_{W,P}}$. Let $g' = \sum c_{g',B} H^B$ be the power series expansion of g' with respect to H and X with its constant term $c_{g',\mathbb{O}}$. Then the assumption of $X = X'$ implies $g' = c_{g',\mathbb{O}}$. Therefore, we conclude that

$$\left(c'_{f,\mathbb{O}} \otimes (\mathbb{M}_X^{-1})^a, \widetilde{\mu} \cdot a\right) \in \left\{\left(c_{h,\mathbb{O}} \otimes (\mathbb{M}_X^{-1})^a, \widetilde{\mu} \cdot a\right) \mid (h, a) \in \widehat{\mathcal{R}_P}\right\}$$

and hence by Remark 5 (2) (cf. the note at the end of the previous case) that

$$\widehat{\text{Comp}(\mathcal{R})}_{\mathbb{H}',X'} \subset \mathcal{C}_{\text{an}} = \mathcal{C} = \widehat{\text{Comp}(\mathcal{R})}_{\mathbb{H},X}.$$

Observe that, given an LGS $\mathbb{H}$ and its associated regular system of parameters X satisfying the condition (♡), we can reach another LGS $\mathbb{H}'$ and its associated regular system of parameters X' satisfying the condition (♡) by a transformation described in the former case followed by another transformation described in the latter case. Therefore, by the above analysis, we have $\widehat{\text{Comp}(\mathcal{R})}_{\mathbb{H}',X'} \subset \widehat{\text{Comp}(\mathcal{R})}_{\mathbb{H},X}$. Reversing the role of $\mathbb{H}$ and X with that of $\mathbb{H}'$ and X', we then have $\widehat{\text{Comp}(\mathcal{R})}_{\mathbb{H},X} \subset \widehat{\text{Comp}(\mathcal{R})}_{\mathbb{H}',X'}$.

Finally we conclude $\widehat{\text{Comp}(\mathcal{R})}_{\mathbb{H},X} = \widehat{\text{Comp}(\mathcal{R})}_{\mathbb{H}',X'}$. Q.E.D.

Construction at the algebraic level

Proposition 2. *There exists an idealistic filtration of i.f.g. type* $\mathrm{Comp}(\mathcal{R})$ *at the algebraic level (i.e., over* $\mathcal{O}_{W,P}$*) such that its completion coincides with the companion modification* $\widehat{\mathrm{Comp}(\mathcal{R})}$ *at the analytic level, i.e.,*

$$\{\mathrm{Comp}(\mathcal{R})\}\widehat{\ } := \mathrm{Comp}(\mathcal{R}) \otimes_{\mathcal{O}_{W,P}} \widehat{\mathcal{O}_{W,P}} = \widehat{\mathrm{Comp}(\mathcal{R})}.$$

We call $\mathrm{Comp}(\mathcal{R})$ the companion modification at the algebraic level.

Proof. Step 1. Descent to the Henselization level.

We first note that the ingredients that we used to construct the companion modification at the analytic level;

(i) the LGS $\mathbb{H}$ and its associated regular system of parameters X satisfying the condition $(\heartsuit)$,

(ii) the constant term $c_{f,\mathbb{O}}$ of the power series expansion $f = \sum c_{f,B}H^B$ for $(f,a) \in \mathcal{R}_P$ with respect to $\mathbb{H}$ and X (which shows up in the form of $\left(c_{f,\mathbb{O}} \otimes (\mathbb{M}_X^{-1})^a, \widetilde{\mu} \cdot a\right)$ in the construction of the companion modification),

actually can be taken at the Henselization level (i.e., they can be taken from the Henselization $(\mathcal{O}_{W,P})^h$ of $\mathcal{O}_{W,P}$).

In fact, (i) at the Henselization level is a consequence of the classical Weierstrass Preparation Theorem and Weierstrass Division Theorem (cf. the proof of Proposition 4 (1)). (Alternatively, (i) at the Henselization level can be seen using the same argument as the one used to see (ii) at the Henselization level.)

We see (ii) at the Henselization level as follows: Set $R = \mathcal{O}_{W,P}$, R^h its Henselization, and $\widehat{R}$ its completion. By replacing R with some local ring of an étale cover of Spec R, we may assume that the LGS $\mathbb{H}$ and the regular system of parameters X satisfying the condition $(\heartsuit)$ are taken from R. Set $A = k[x_{t+1}, \ldots, x_d]_{(x_{t+1},\ldots,x_d)}$, A^h its Henselization, and $\widehat{A}$ its completion. By looking at the power series expansion with respect to $\mathbb{H}$ and its associated regular system of parameters X (cf. [19]), we have

$$\phi : \widehat{R}/(h_1, \ldots, h_t) \xrightarrow{\sim} \sum_K \widehat{A} X^K$$

where on the right hand side the subscript $K = (k_1, \ldots, k_t, k_{t+1}, \ldots, k_d) \in \mathbb{Z}^d_{\geq 0}$ for the summation varies in the finite range

$$\begin{cases} 0 \leq k_\alpha \leq p^{e_\alpha} - 1 & \text{for} \quad \alpha = 1, \ldots, t \\ k_\alpha = 0 & \text{for} \quad \alpha = t+1, \ldots, d. \end{cases}$$

Take an element $f \in R \subset \widehat{R}$. What we want to show is $c_{f,\mathbb{O}} \in R^h$.

It suffices to show

$$(\star) \quad \phi(f \bmod (h_1, \ldots, h_t)) = c_{f,\mathbb{O}} \in \sum_K A^h X^K.$$

We take the coordinate ring S of an affine open neighborhood $P \in \operatorname{Spec} S \subset W$ such that $f \in S$ and $\{h_\alpha\}_{\alpha=1}^t, X \subset S$. We denote by $\mathfrak{m}_{S,P}$ the maximal ideal of S corresponding to the point P. Note that the ideal $(h_1, \ldots, h_t, x_{t+1}, \ldots, x_d)R$ is $\mathfrak{m}_{S,P}R = \mathfrak{m}_P = (x_1, \ldots, x_t, x_{t+1}, \ldots, x_d)$-primary (in R). By shrinking $\operatorname{Spec} S$ if necessary, we may assume that the only prime ideal containing the ideal $(h_1, \ldots, h_t, x_{t+1}, \ldots, x_d)$ is $\mathfrak{m}_{S,P}$.

We regard R, R^h, S, A, A^h as subrings of $\widehat{R}$, i.e., $R, R^h, S, A, A^h \subset \widehat{R}$, and we consider the subring $SA^h \subset \widehat{R}$ generated by $S, A^h \subset \widehat{R}$. By abuse of notation, we denote the image of the natural projection $SA^h \subset \widehat{R} \to \widehat{R}/(h_1, \ldots, h_t)$ by $SA^h/(h_1, \ldots, h_t)$.

We claim

$$(\star\star) \quad \phi(SA^h/(h_1, \ldots, h_t)) = \sum_K A^h X^K,$$

which clearly implies the assertion $(\star)$.

In the following, we omit the isomorphism ϕ from the left hand side in order to ease the notation. Therefore, the claim $(\star\star)$ is expressed as an equality

$$SA^h/(h_1, \ldots, h_t) = \sum_K A^h X^K.$$

Obviously, we have

$$SA^h/(h_1, \ldots, h_t) \supset \sum_K A^h X^K.$$

Our goal is to show the equality after taking $\otimes_{A^h} \widehat{A}$, i.e.,

$$\text{L.H.S.} = SA^h/(h_1, \ldots, h_t) \otimes_{A^h} \widehat{A} = \sum_K A^h X^K \otimes_{A^h} \widehat{A} = \text{R.H.S.},$$

which, since $\widehat{A}$ is faithfully flat over A^h, implies the original equality above before taking $\otimes_{A^h} \widehat{A}$, i.e., $(\star\star)$.

- Analysis of R.H.S.

We have

$$\text{R.H.S.} = \sum_K A^h X^K \otimes_{A^h} \widehat{A} = \sum_K \widehat{A} X^K.$$

- Analysis of L.H.S.

In order to analyze L.H.S., we look at the morphism θ: $\mathrm{Spec}(SA^h/(h_1, \ldots, h_t)) \to \mathrm{Spec}(A^h)$. Observe that θ is quasi-finite, i.e.,

(a) it is of finite type, and
(b) it has finite fibers.

The assertion (a) is immediate, since S is finitely generated over k. In order to see the assertion (b), first look at the morphism $\mathrm{Spec}(SA/(h_1, \ldots, h_t)) \to \mathrm{Spec}(A)$. The fiber of this morphism over the origin downstairs is the origin upstairs, since the ideal $(h_1, \ldots, h_t, x_{t+1}, \ldots, x_d)$ is $(x_1, \ldots, x_t, x_{t+1}, \ldots, x_d)$-primary. Now by what is called "Zariski's Main Theorem by Grothendieck" (cf. [22]), we conclude that the morphism has finite fibers. From this it follows easily that the morphism $\theta : \mathrm{Spec}(SA^h/(h_1, \ldots, h_t)) \to \mathrm{Spec}(A^h)$ has finite fibers.

Therefore, since A^h is Henselian, we conclude that the morphism θ is finite, i.e., $SA^h/(h_1, \ldots, h_t)$ is a finite A^h-module. This implies that $SA^h/(h_1, \ldots, h_t) \otimes_{A^h} \widehat{A}$ is the $(x_{t+1}, \ldots, x_d)$-adic completion of the A^h-module $SA^h/(h_1, \ldots, h_t)$. The latter coincides with the $(h_1, \ldots, h_t, x_{t+1}, \ldots, x_d)$-adic completion of $SA^h/(h_1, \ldots, h_t)$ viewed as an SA^h-module. Since the ideal $(h_1, \ldots, h_t, x_{t+1}, \ldots, x_d)$ is $(x_1, \ldots, x_t, x_{t+1}, \ldots, x_d)$-primary, the $(h_1, \ldots, h_t, x_{t+1}, \ldots, x_d)$-adic completion coincides with the $(x_1, \ldots, x_t, x_{t+1}, \ldots, x_d)$-adic completion. Observing that the $(x_1, \ldots, x_t, x_{t+1}, \ldots, x_d)$-adic completion of SA^h is $\widehat{R}$, we see that the $(x_1, \ldots, x_t, x_{t+1}, \ldots, x_d)$-adic completion of $SA^h/(h_1, \ldots, h_t)$ is $\widehat{R}/(h_1, \ldots, h_t)$. Summarizing and remembering the convention of expressing the isomorphism ϕ as an equality, we conclude

$$\text{L.H.S.} = SA^h/(h_1, \ldots, h_t) \otimes_{A^h} \widehat{A} = \widehat{R}/(h_1, \ldots, h_t) = \sum_K \widehat{A} X^K.$$

Therefore, we have

$$\text{L.H.S.} = \sum_K \widehat{A} X^K = \text{R.H.S.}.$$

This completes the argument for Step 1.

Therefore, we conclude that, for each LGS $\mathbb{H}$ and its associated regular system of parameters X satisfying the condition ($\heartsuit$), taken at the Henselization level, we have an idealistic filtration of i.f.g. type $\mathrm{Comp}(\mathcal{R})_{\mathbb{H},X}$ at the Henselization level (i.e., all the ideals are those of $(\mathcal{O}_{W,P})^h$) such that $\widehat{\mathrm{Comp}(\mathcal{R})} = \widehat{\mathrm{Comp}(\mathcal{R})_{\mathbb{H},X}} = \widehat{\{\mathrm{Comp}(\mathcal{R})_{\mathbb{H},X}\}}$.

Step 2. Descent to the algebraic level.

Let $U = \text{Spec } \mathcal{O}_{W,P}$. Noting that the Henselization is the direct limit of the local rings at the closed points over P on the étale covers of U, we can take a collection of étale covers $\pi_\lambda : U_\lambda \to U$ and idealistic filtrations of i.f.g. type $\text{Comp}(\mathcal{R})_\lambda$ over U_λ such that, for $Q \in \pi_\lambda^{-1}(P)$, we have $\{\text{Comp}(\mathcal{R})_{\lambda,Q}\}^h = \text{Comp}(\mathcal{R})_{\mathbb{H},X}$ for some $\mathbb{H}$ and X described as in Step 1. This implies

$$\begin{aligned}\{\text{Comp}(\mathcal{R})_{\lambda,Q}\}\hat{} &= \{\{\text{Comp}(\mathcal{R})_{\lambda,Q}\}^h\}\hat{} = \{\text{Comp}(\mathcal{R})_{\mathbb{H},X}\}\hat{} \\ &= \widehat{\text{Comp}(\mathcal{R})}_{\mathbb{H},X} = \widehat{\text{Comp}(\mathcal{R})}.\end{aligned}$$

That is to say, the completion of $\text{Comp}(\mathcal{R})_{\lambda,Q}$ canonically coincides with the companion modification at the analytic level $\widehat{\text{Comp}(\mathcal{R})}$. This in turn implies that, over $U_\lambda \cap U_\mu = U_\lambda \times_U U_\mu$, we have

$$\text{Comp}(\mathcal{R})_\lambda|_{U_\lambda \cap U_\mu} \overset{\phi_{\lambda\mu}}{=} \text{Comp}(\mathcal{R})_\mu|_{U_\lambda \cap U_\mu},$$

and this identification $\phi_{\lambda\mu}$ is canonical (and hence the collection of these identifications automatically satisfies the cocycle condition $\phi_{\lambda\mu} \circ \phi_{\mu\nu} = \phi_{\lambda\nu}$). Now it is a consequence of the general étale descent argument (cf. [14][20]) that there exists an idealistic filtration of i.f.g. type $\text{Comp}(\mathcal{R})$ such that $\pi_\lambda^*(\text{Comp}(\mathcal{R})) = \text{Comp}(\mathcal{R})_\lambda$, and hence that $\{\text{Comp}(\mathcal{R})\}\hat{} = \widehat{\text{Comp}(\mathcal{R})}$.

This completes the proof of Proposition 2. Q.E.D.

Detailed discussion of the mechanism

Note that constructing the resolution sequence by blowing up is referred to as "proceeding in the vertical direction", where the process is numbered by "year", while constructing the triplet of invariants and its associated modification in a fixed year, is referred to as "proceeding in the horizontal direction", where the process is numbered by "stage". See §4.3 for more details.

Proceeding in the horizontal direction

Proposition 3. *The value of the pair $(\sigma, \#E)$ strictly decreases as we proceed in the horizontal direction from $(W, \mathcal{R}, E)$ to $(W', \mathcal{R}', E')$, i.e., we have*

$$(\sigma, \#E) > (\sigma', \#E').$$

Proof. We analyze the assertion in the following two cases. Note that, since $\mathcal{R} \subset \mathcal{R}'$, we have $\sigma \geq \sigma'$ in both cases.

Case: $\widetilde{\mu} \neq 0$ *or* ∞.

In this case, we claim $\sigma > \sigma'$. We take the LGS $\mathbb{H} = \{(h_\alpha, p^{e_\alpha})\}_{\alpha=1}^{t}$ and its associated regular system of parameters $X = (x_1, \ldots, x_t, x_{t+1}, \ldots, x_d)$ satisfying the condition ($\heartsuit$) as given in "Description of the triplet of invariants $(\sigma, \widetilde{\mu}, s)$". Since $\mathcal{R}$ is an idealistic filtration of i.f.g. type, there exists $(f, a) \in \mathcal{R}_P$ such that $\mu_P(\mathcal{R}) = \mathrm{ord}_P(c_{f,\mathbb{O}})/a$, where $f = \sum c_{f,B} H^B$ is the power series expansion of f with respect to $\mathbb{H}$ and X. Set

$$\mathbb{M} = \prod_{D \in E_{\text{young}}} x_D^{\mu_{P,D}(\mathcal{R})}.$$

Subcase: $\mu_{P,D}(\mathcal{R}) \cdot a \in \mathbb{Z}_{\geq 0} \quad \forall D \in E_{\text{young}}$.

In this subcase, we have $\mathbb{M}^a \in \widehat{\mathcal{O}_{W,P}}$ and $c_{f,\mathbb{O}} \cdot (\mathbb{M}^{-1})^a = c_{f,\mathbb{O}} \cdot (\mathbb{M}^a)^{-1} \in \widehat{\mathcal{O}_{W,P}}$ by definition. Moreover, by construction of the companion modification, we have $\left(c_{f,\mathbb{O}} \cdot (\mathbb{M}^a)^{-1}, \widetilde{\mu} \cdot a\right) \in \widehat{\mathrm{Comp}(\mathcal{R})} \subset \widehat{\mathcal{R}'}$. Note that

$$\widetilde{\mu} \cdot a = \left(\mu_P(\mathcal{R}) - \sum_{D \subset E_{\text{young}}} \mu_{P,D}(\mathcal{R})\right) \cdot a = \mathrm{ord}_P\left(c_{f,\mathbb{O}} \cdot (\mathbb{M}^{-1})^a\right).$$

Note also that, by the characterization of the power series expansion with respect to $\mathbb{H}$ and X, we have

$$c_{f,\mathbb{O}} = \sum_{K \in (\mathbb{Z}_{\geq 0})^d} b_{f,\mathbb{O},K} X^K$$

with $b_{f,\mathbb{O},K} \in k[[x_{t+1}, \ldots, x_d]]$ and with $K = (k_1, \ldots, k_t, k_{t+1}, \ldots, k_d)$ varying in the range satisfying the condition

$$\begin{cases} 0 \leq k_\alpha \leq p^{e_\alpha} - 1 & \text{for} \quad \alpha = 1, \ldots, t \\ k_\alpha = 0 & \text{for} \quad \alpha = t+1, \ldots, d. \end{cases}$$

Therefore, we conclude

$$\mathrm{In}\left(c_{f,\mathbb{O}} \cdot (\mathbb{M}^{-1})^a\right) = c_{f,\mathbb{O}} \cdot (\mathbb{M}^{-1})^a \bmod \widehat{\mathfrak{m}}^{\widetilde{\mu}\cdot a+1} = \sum_{K \in (\mathbb{Z}_{\geq 0})^d} a_{f,\mathbb{O},K} X^K$$

with $a_{f,\mathbb{O},K} \in k[x_{t+1}, \ldots, x_d]$. Since $L_P(\mathcal{R}) = k[x_1^{p^{e_1}}, \ldots, x_t^{p^{e_t}}]$, we conclude

$$\mathrm{In}\left(c_{f,\mathbb{O}} \cdot (\mathbb{M}^{-1})^a\right) \notin L_P(\mathcal{R}),$$

while by definition

$$\mathrm{In}\left(c_{f,\mathbb{O}} \cdot (\mathbb{M}^{-1})^a\right) \in L_P(\widehat{\mathcal{R}'}) = L_P(\mathcal{R}').$$

Therefore, we conclude $L_P(\mathcal{R}) \subsetneq L_P(\widehat{\mathcal{R}'})$ and hence $\sigma > \sigma'$.

Subcase: $\mu_{P,D}(\mathcal{R}) \cdot a \notin \mathbb{Z}_{\geq 0}$ *for some* $D \in E_{\text{young}}$.

In this case, take $l \in \mathbb{Z}_{>0}$ such that

$$l \cdot \mu_{P,D}(\mathcal{R}) \cdot a \in \mathbb{Z}_{\geq 0} \quad \forall D \in E_{\text{young}}.$$

Then we have $(\mathbb{M}^a)^l \in \mathcal{O}_{W,P}$ and $\left\{c_{f,\mathbb{O}} \cdot (\mathbb{M}^{-1})^a)\right\}^l \in \widehat{\mathcal{O}_{W,P}}$ by definition. Moreover, by construction of the companion modification, we have $(\left\{c_{f,\mathbb{O}} \cdot (\mathbb{M}^{-1})^a\right\}^l, l \cdot \widetilde{\mu} \cdot a) \in \widehat{\text{Comp}(\mathcal{R})} \subset \widehat{\mathcal{R}'}$. Note that

$$l \cdot \widetilde{\mu} \cdot a = l \cdot (\mu_P(\mathcal{R}) - \sum_{D \subset E_{\text{young}}} \mu_{P,D}(\mathcal{R})) \cdot a = \text{ord}_P\left(\left\{c_{f,\mathbb{O}} \cdot (\mathbb{M}^{-1})^a\right\}^l\right).$$

Note also that, since $\text{ord}_{\xi_D}(c_{f,\mathbb{O}}) \geq \mu_{P,D}(\mathcal{R}) \cdot a$ by definition, since $\text{ord}_{\xi_D}(c_{f,\mathbb{O}}) \in \mathbb{Z}_{\geq 0}$ and since $\mu_{P,D}(\mathcal{R}) \cdot a \notin \mathbb{Z}_{\geq 0}$ by the subcase assumption, we conclude

$$\text{ord}_{\xi_D}(c_{f,\mathbb{O}}) > \mu_{P,D}(\mathcal{R}) \cdot a.$$

This implies, $\left\{c_{f,\mathbb{O}} \cdot (\mathbb{M}^{-1})^a\right\}^l$ is divisible by x_D, and hence so is

$$\text{In}\left(\left\{c_{f,\mathbb{O}} \cdot (\mathbb{M}^{-1})^a\right\}^l\right) = \left\{c_{f,\mathbb{O}} \cdot (\mathbb{M}^{-1})^a\right\}^l \mod \widehat{\mathfrak{m}}^{l \cdot \widetilde{\mu} \cdot a + 1}.$$

Since $L_P(\mathcal{R}) = k[x_1^{p^{e_1}}, \ldots, x_t^{p^{e_t}}]$ and since $x_D \in \{x_{t+1}, \ldots, x_d\}$, we conclude

$$\text{In}\left(\left\{c_{f,\mathbb{O}} \cdot (\mathbb{M}^{-1})^a\right\}^l\right) \notin L_P(\mathcal{R}),$$

while by definition

$$\text{In}\left(\left\{c_{f,\mathbb{O}} \cdot (\mathbb{M}^{-1})^a\right\}^l\right) \in L_P(\widehat{\mathcal{R}'}) = L_P(\mathcal{R}').$$

Therefore, we conclude $L_P(\mathcal{R}) \subsetneq L_P(\widehat{\mathcal{R}'})$ and hence $\sigma > \sigma'$.

Case: $\widetilde{\mu} = 0$ *or* ∞.

In this case, we claim $(\sigma, \#E) > (\sigma', \#E')$. Observe $s \neq 0$. (If $s = 0$, then $(\sigma, \widetilde{\mu}, s) = (\sigma, 0, 0)$ or $(\sigma, \infty, 0)$, and hence we do not construct the modification $(W', \mathcal{R}', E')$.) Therefore, by definition, there exists a divisor $D \in E_{\text{aged}}$ containing P. Thus, we have $\#E > \#(E \setminus E_{\text{aged}}) = \#E'$. (Note that "#" represents the number of the components passing through P.) Q.E.D.

Proceeding in the vertical direction

Proposition 4. *Let* $(W, \mathcal{R}, E) \stackrel{\pi}{\leftarrow} (\widetilde{W}, \widetilde{\mathcal{R}}, \widetilde{E})$ *be a transformation in the resolution sequence (i.e.,* $(W_i, \mathcal{R}_i, E_i) \stackrel{\pi_{i+1}}{\leftarrow} (W_{i+1}, \mathcal{R}_{i+1}, E_{i+1})$ *in the sequence described in Problem 4 (cf. Remark 2)). Take a point* $\widetilde{P} \in \pi^{-1}(P) \cap \mathrm{Sing}(\widetilde{\mathcal{R}})$*. Then the value of the triplet* $(\sigma, \widetilde{\mu}, s)$ *does not increase as we proceed in the vertical direction, i.e., we have*

$$(\sigma, \widetilde{\mu}, s) \geq (\widetilde{\sigma}, \widetilde{\widetilde{\mu}}, \widetilde{s}) \quad (i.e., (\sigma_i, \widetilde{\mu}_i, s_i) \geq (\sigma_{i+1}, \widetilde{\mu}_{i+1}, s_{i+1})).$$

More precisely, we have the following:

(1) *The invariant* σ *does not increase, i.e.,* $\sigma \geq \widetilde{\sigma}$*. When* $\sigma = \widetilde{\sigma}$*, the transformation of the LGS for* $\widehat{\mathcal{R}_P}$ *is an LGS for* $\widehat{\widetilde{\mathcal{R}}_{\widetilde{P}}}$*. Moreover, the following property is preserved going from* $(W, \mathcal{R}, E)$ *to* $(\widetilde{W}, \widetilde{\mathcal{R}}, \widetilde{E})$*: We can choose an LGS* $\mathbb{H} = \{(h_\alpha, p^{e_\alpha})\}_{\alpha=1}^t \subset \widehat{\mathcal{R}_P}$ *and a regular system of parameters* $X = (x_1, \ldots, x_t, x_{t+1}, \ldots, x_d)$*, taken from* $\widehat{\mathcal{O}_{W,P}}$*, satisfying the condition* ($\heartsuit$) *below;*

- $h_\alpha \equiv x_\alpha^{p^{e_\alpha}} \bmod \widehat{\mathfrak{m}}^{p^{e_\alpha}+1}$ *for* $\alpha = 1, \ldots, t$,
- *the idealistic filtration of i.f.g. type* $\widehat{\mathcal{R}_P}$ *is saturated for* $\{\partial^n / \partial x_\alpha{}^n \mid n \in \mathbb{Z}_{\geq 0}, \alpha = 1, \ldots, t\}$*, and*
- *the defining equations for the components of* E_{young}*, which are transversal to the LGS, form a part of the regular system of parameters, i.e.,* $\{x_D \mid D \in E_{\mathrm{young}}\} \subset \{x_{t+1}, \ldots, x_d\}$.

(2) *When* $\sigma = \widetilde{\sigma}$*, the value of the invariant* $\widetilde{\mu}$ *does not increase, i.e.,* $(\sigma, \widetilde{\mu}) \geq (\sigma, \widetilde{\widetilde{\mu}}) = (\widetilde{\sigma}, \widetilde{\widetilde{\mu}})$.

(3) *When* $\sigma = \widetilde{\sigma}$*, the value of the invariant* s *does not increase, i.e.,* $s \geq \widetilde{s}$*, and hence combined with (2) we have*

$$(\sigma, \widetilde{\mu}, s) \geq (\sigma, \widetilde{\widetilde{\mu}}, \widetilde{s}) = (\widetilde{\sigma}, \widetilde{\widetilde{\mu}}, \widetilde{s}).$$

Proof. (1) Since $C \subset \mathrm{Sing}(\mathcal{R})$ and since C is nonsingular, we conclude that there exists a regular system of parameters $(y_1, \ldots, y_r, y_{r+1}, \ldots, y_d) \subset \mathcal{O}_{W,P}$ such that

- $C = \{y_1 = \cdots = y_r = 0\}$, and
- we have for $\alpha = 1, \ldots, t$

$$\begin{cases} h_\alpha & \equiv & x_\alpha^{p^{e_\alpha}} \bmod \widehat{\mathfrak{m}_P}^{p^{e_\alpha}+1}, \text{ and} \\ x_\alpha & \equiv & \sum_{\beta=1}^r c_{\alpha,\beta} y_\beta \bmod \widehat{\mathfrak{m}_P}^2 \text{ for some } c_{\alpha,\beta} \in k. \end{cases}$$

By replacing $(y_1, \ldots, y_r, y_{r+1}, \ldots, y_d)$ with some linear transformation and then by replacing $(x_1, \ldots, x_t, x_{t+1}, \ldots, x_d)$ with $(y_1, \ldots, y_r, y_{r+1}, \ldots, y_d)$, we may assume that we have a regular system of parameters $(x_1, \ldots, x_t, x_{t+1}, \ldots, x_d) \subset \widehat{\mathcal{O}_{W,P}}$ such that

- $C = \{x_1 = \cdots = x_r = 0\}$ $(t \leq r)$, and
- $h_\alpha \equiv x_\alpha^{p^{e_\alpha}} \mod \widehat{\mathfrak{m}_P}^{p^{e_\alpha}+1}$ for $\alpha = 1, \ldots, t$.

Observe that, if $t = r$, then after the blow up, for any point $\widetilde{P}$ over each x_α-chart $(\alpha = 1, \ldots, t)$ we have $\mathrm{ord}_{\widetilde{P}}(\widetilde{h_\alpha}) = 0 < p^{e_\alpha}$ where $\widetilde{h_\alpha} = h_\alpha / x_\alpha^{p^{e_\alpha}}$, and hence $\pi^{-1}(P) \cap \mathrm{Sing}(\mathcal{R}) = \emptyset$. Therefore, we may assume $t < r$ and that our point $\widetilde{P} \in \pi^{-1}(P) \cap \mathrm{Sing}(\widetilde{\mathcal{R}})$ is in the x_β-chart for some $(t < \beta \leq r)$ with the regular system of parameters $(\widetilde{x}_1, \ldots, \widetilde{x}_t, \widetilde{x}_{t+1}, \ldots, \widetilde{x}_r, \widetilde{x}_{r+1}, \ldots, \widetilde{x}_d)$ where

$$\widetilde{x}_\alpha = \begin{cases} x_\alpha / x_\beta & \text{for} \quad 1 \leq \alpha \leq r, \alpha \neq \beta \\ x_\beta & \text{for} \quad \alpha = \beta, \\ x_\alpha & \text{for} \quad r+1 \leq \alpha \leq d. \end{cases}$$

(For the indices $t+1 \leq \alpha \leq r, \alpha \neq \beta$, we may have to replace x_α with $x_\alpha - c_\alpha x_\beta$ for some $c_\alpha \in k$ if necessary.)

Now we look at the transformation $\widetilde{h_\alpha} = h_\alpha / x_\beta^{p^{e_\alpha}}$ of

$$h_\alpha = \sum_{K=(k_1,\ldots,k_d)\in(\mathbb{Z}_{\geq 0})^d} c_{\alpha,K} X^K \quad \text{with} \quad c_{\alpha,K} \in k$$

for $\alpha = 1, \ldots, t$. We compute

$$\widetilde{h_\alpha} = \frac{h_\alpha}{x_\beta^{p^{e_\alpha}}} = \sum_{K=(k_1,\ldots,k_d)\in(\mathbb{Z}_{\geq 0})^d} c_{\alpha,K} \frac{X^K}{x_\beta^{p^{e_\alpha}}} = \sum_{\widetilde{K}=(\widetilde{k}_1,\ldots,\widetilde{k}_d)\in(\mathbb{Z}_{\geq 0})^d} c_{\alpha,K} \widetilde{X}^{\widetilde{K}}$$

where $\widetilde{k}_u = k_u$ for $u \neq \beta$ and $\widetilde{k}_\beta = \sum_{u=1}^r k_u - p^{e_\alpha}$. Therefore, we have

$$\sum_{1 \leq u \leq r, u \neq \beta} k_u \geq p^{e_\alpha} \Rightarrow \deg_{\widetilde{X}}(\widetilde{X}^{\widetilde{K}}) \geq \deg_X(X^K).$$

Therefore, the only terms $\widetilde{X}^{\widetilde{K}}$ with $\deg_{\widetilde{X}}(\widetilde{X}^{\widetilde{K}}) = p^{e_\alpha} < \deg_X(X^K)$ that can possibly appear in $\widetilde{h_\alpha}$ have to contain one of $\widetilde{x}_\beta, \widetilde{x}_{r+1}, \ldots, \widetilde{x}_d$.

Looking at $\widetilde{h_1}, \widetilde{h_2}, \ldots, \widetilde{h_t}$ in the ascending order, we conclude that

- either $\sigma > \widetilde{\sigma}$,
- or $\sigma = \widetilde{\sigma}$, and we have

$$\widetilde{h_\alpha} \equiv \widetilde{x}_\alpha^{p^{e_\alpha}} + c_{\alpha,\beta}\widetilde{x}_\beta^{p^{e_\alpha}} + \sum_{u=r+1}^{d} c_{\alpha,u}\widetilde{x}_u^{p^{e_\alpha}} \bmod \widehat{\mathfrak{m}_P}^{p^{e_\alpha}+1}$$

with $\{c_{\alpha,\beta}, c_{\alpha,r+1}, \ldots, c_{\alpha,d}\} \subset k$ for $\alpha = 1, \ldots, t$, and hence $\widetilde{\mathbb{H}} = \{(\widetilde{h_\alpha}, p^{e_\alpha})\}_{\alpha=1}^t$ is an LGS for $\widetilde{\mathcal{R}}$.

This completes the proof of the first part of (1).

Next we look at the "Moreover" part of (1).

We show the existence of such an LGS $\mathbb{H} = \{(h_\alpha, p^{e_\alpha})\}_{\alpha=1}^t$ and its associated regular system of parameters $X = (x_1, \ldots, x_t, x_{t+1}, \ldots, x_d)$ that satisfy the condition ($\heartsuit$) by induction on the year .

When we are in the year when the value of σ first started (i.e., the value of σ is strictly less than the one in the previous year), we replace the original $\mathcal{R}$ with its differential saturation (cf. the technical but important points in the description of the triplet of invariants $(\sigma, \widetilde{\mu}, s)$ (i)). Thus we may assume $\mathcal{R}$ is $\mathcal{D}$-saturated. Moreover, we have $E_{\text{young}} = \emptyset$. Therefore, we have only to take an LGS $\mathbb{H}$ and X such that $h_\alpha \equiv x_\alpha^{p^{e_\alpha}} \bmod \widehat{\mathfrak{m}_P}^{p^{e_\alpha}+1}$ for $\alpha = 1, \ldots, t$. The remaining requirements in the condition ($\heartsuit$) are then automatically satisfied.

Now we assume that in the current year we have such an LGS $\mathbb{H}$ and its associated regular system of parameters X satisfying the condition ($\heartsuit$). We show that, assuming $\sigma = \widetilde{\sigma}$, even after the transformation we have such an LGS $\widetilde{\mathbb{H}}$ and $\widetilde{X}$ that satisfy the condition ($\heartsuit$).

Step 1. We modify our X so that the LGS $\mathbb{H}$ and X are still associated, satisfy the condition ($\heartsuit$) as before, and now satisfy the extra condition that the center is defined by $C = \{x_1 = \cdots = x_r = 0\}$ $(t \leq r)$.

(a) We take another regular system of parameters $Y = (y_1, \ldots, y_r, y_{r+1}, \ldots, y_d)$ such that the center is defined by $C = \{y_1 = \cdots = y_r = 0\}$. Note that, if $C \subset D$ for any component $D \in E_{\text{young}}$, then we include x_D in Y. Note that such x_D is included in X.

Then since $C \subset \operatorname{Sing}(\mathcal{R})$, we conclude that

$$x_\alpha \equiv \sum_{\beta=1}^{r} c_{\alpha,\beta} y_\beta \bmod \widehat{\mathfrak{m}_P}^2 \quad \text{for some } c_{\alpha,\beta} \in k$$

for $\alpha = 1, \ldots, t$. Therefore, by taking a suitable linear transformation among $\{y_1, \ldots, y_r\}$, we may assume

$$x_\alpha \equiv y_\alpha \bmod \widehat{\mathfrak{m}_P}^2 \quad \text{for} \quad \alpha = 1, \ldots, t.$$

It is straightforward to see that, since $\mathbb{H}$ and $X = (x_1, \ldots, x_t, x_{t+1}, \ldots, x_d)$ satisfy the condition ($\heartsuit$), so does $\mathbb{H}$ and $(y_1, \ldots, y_t, x_{t+1}, \ldots, x_d)$. By replacing X with $(y_1, \ldots, y_t, x_{t+1}, \ldots, x_d)$, we may assume that $\mathbb{H}$ and X satisfy the condition ($\heartsuit$), and $(x_1, \ldots, x_t, y_{t+1}, \ldots, y_r, y_{r+1}, \ldots, y_d)$ is a regular system of parameters with $C = \{x_1 = \cdots = x_t = y_{t+1} = \cdots = y_r = 0\}$.

(b) We look at $\{y_\beta \mid t+1 \leq \beta \leq r\}$. By applying the Weierstrass Division Theorem consecutively (and replacing the original y_β after multiplying some unit if necessary) or by simply looking at the power series expansion with respect to X, we write

$$y_\beta = \sum_{\alpha=1}^{t} q_{\beta,\alpha} x_\alpha + g(x_{t+1}, \ldots, x_d) \quad \text{for} \quad \beta = t+1, \ldots, r$$

with $q_{\beta,\alpha} \in \widehat{\mathcal{O}_{W,P}}$ and $g(x_{t+1}, \ldots, x_d) \in k[[x_{t+1}, \ldots, x_d]]$. Set

$$y'_\beta = y_\beta - \sum_{\alpha=1}^{t} q_{\beta,\alpha} x_\alpha \quad \text{for} \quad \beta = t+1, \ldots, r.$$

Choose $\{x_{r+1}, \ldots, x_d\} \subset \{x_{t+1}, \ldots, x_d\}$ (after renumbering of $x_{t+1}, \ldots, x_d$ if necessary and keeping the condition $\{x_D \mid D \in E_{\text{young}}\} \subset \{x_{t+1}, \ldots, x_d\}$) such that $(x_1, \ldots, x_t, y'_{t+1}, \ldots, y'_r, x_{r+1}, \ldots, x_d)$ is a regular system of parameters. Now replace X with $(x_1, \ldots, x_t, y'_{t+1}, \ldots, y'_r, x_{r+1}, \ldots, x_d)$. Then it is straightforward to see that, since $\mathbb{H}$ and the previous X satisfy the condition ($\heartsuit$), so do $\mathbb{H}$ and the new X. Now by construction, the center C is defined by $C = \{x_1 = \cdots = x_r = 0\}$ $(t \leq r)$.

Step 2. Analysis after blow up.

As in the proof of the first part, we may assume $t < r$ and that our point $\widetilde{P} \in \pi^{-1}(P) \cap \mathrm{Sing}(\widetilde{\mathcal{R}})$ is in the x_β-chart for some $(t < \beta \leq r)$ with the regular system of coordinates $\widetilde{X} = (\widetilde{x}_1, \ldots, \widetilde{x}_t, \widetilde{x}_{t+1}, \ldots, \widetilde{x}_r, \widetilde{x}_{r+1}, \ldots, \widetilde{x}_d)$ where

$$\widetilde{x}_\alpha = \begin{cases} x_\alpha / x_\beta & \text{for} \quad 1 \leq \alpha \leq r,\ \alpha \neq \beta, \\ x_\beta & \text{for} \quad \alpha = \beta, \\ x_\alpha & \text{for} \quad r+1 \leq \alpha \leq d. \end{cases}$$

(For the indices $t+1 \leq \alpha \leq r, \alpha \neq \beta$, we may have to replace x_α with $x_\alpha - c_\alpha x_\beta$ for some $c_\alpha \in k$ if necessary.)

It is straightforward to see that the above $\widetilde{X}$ satisfies the following two conditions;

• the idealistic filtration of i.f.g. type $\widehat{\widetilde{\mathcal{R}}_{\widetilde{P}}}$, which is (the completion of) the transformation of $\widehat{\mathcal{R}_P}$, is saturated for $\{\partial^n/\partial \widetilde{x}_\alpha^n \mid n \in \mathbb{Z}_{\geq 0}, \alpha = 1, \ldots, t\}$, and

• the defining equations for the $\widetilde{E}_{\text{young}}$ form a part of the regular system of parameters, i.e., $\{\widetilde{x}_{\widetilde{D}} \mid \widetilde{D} \in \widetilde{E}_{\text{young}}\} \subset \{\widetilde{x}_{t+1}, \ldots, \widetilde{x}_d\}$.

We have only to replace $\widetilde{X}$ with $(\widetilde{x}'_1, \ldots, \widetilde{x}'_t, \widetilde{x}_{t+1}, \ldots, \widetilde{x}_d)$ in order to satisfy the remaining condition

• $\widetilde{h_\alpha} \equiv \widetilde{x}_\alpha^{p^{e_i}} \bmod \widehat{\mathfrak{m}_{\widetilde{P}}}^{p^{e_\alpha}+1}$ for $\alpha = 1, \ldots, t$,

while keeping the other two requirements as above, where

$$\widetilde{x}'_\alpha = \widetilde{x}_\alpha + c_{\alpha,\beta}^{1/p^{e_\alpha}} \widetilde{x}_\beta + \sum_{u=r+1}^{d} c_{\alpha,u}^{1/p^{e_\alpha}} \widetilde{x}_u \quad \text{for} \quad \alpha = 1, \ldots, t$$

(using the same notation as in the proof of the first part).

Then the LGS $\widetilde{\mathbb{H}}$ and $\widetilde{X}$ satisfy the condition (♡).

This finishes the proof of "Moreover" part of (1).

(2) We use the same notation used in the proof of (1). We take an LGS $\mathbb{H}$ and its associated regular system of parameters X, satisfying the condition (♡) and the extra condition that the center is defined by $C = \{x_1 = \cdots = x_r = 0\}$ (see Step 1 in the proof of "Moreover" part of (1)). When $\sigma = \widetilde{\sigma}$, we make the following two observations.

• The transformation $\widetilde{\mathbb{H}}$ is an LGS of $\widetilde{\mathcal{R}}$.

• For $(f, a) \in \widehat{\mathcal{R}_P}$, let $\sum c_{f,B} H^B$ be the power series expansion of f with respect to $\mathbb{H}$ and its associated regular system of parameters $(x_1, \ldots, x_t, x_{t+1}, \ldots, x_d)$, with the "constant term" being $c_{f,\mathbb{O}}$. Look at its transformation $(\widetilde{f}, a)$ with $\widetilde{f} = f/x_\beta^a$. The constant term $c_{\widetilde{f},\mathbb{O}}$ of the transformation $\widetilde{f}$ with respect to $\widetilde{H}$ and its associated regular system of parameters $(\widetilde{x}'_1, \ldots, \widetilde{x}'_t, \widetilde{x}_{t+1}, \ldots, \widetilde{x}_d)$ is the transformation $\widetilde{c_{f,\mathbb{O}}} = c_{f,\mathbb{O}}/x_\beta^a$, where

$$\widetilde{x}'_\alpha = \widetilde{x}_\alpha + c_{\alpha,\beta}^{1/p^{e_\alpha}} \widetilde{x}_\beta + \sum_{u=r+1}^{d} c_{\alpha,u}^{1/p^{e_\alpha}} \widetilde{x}_u \quad \text{for} \quad \alpha = 1, \ldots, t.$$

Now the inequality $\widetilde{\mu} \geq \widetilde{\widetilde{\mu}}$ follows from these two observations and the condition $C \subset \operatorname{Sing}(\operatorname{Comp}(\mathcal{R}))$ in our new setting by the same argument as the one used to show that the invariant w-ord does not increase under transformation in the classical setting.

(3) This follows immediately from the fact that the center C of blow up for the transformation π is nonsingular and transversal to the

boundary E (hence to E_{aged}), and from the fact that, since $\sigma = \widetilde{\sigma}$, the aged part $\widetilde{E}_{\text{aged}}$ of $\widetilde{E}$ is the strict transform of E_{aged} by definition.

Q.E.D.

4.3. Weaving of the new strand of invariants "inv_{new}"

In §4.3, we interpret the inductive scheme explained in §4.2 in terms of weaving the new strand of invariants "inv_{new}".

We weave the strand of invariants "inv_{new}" consisting of the units of the form $(\sigma^j, \widetilde{\mu}^j, s^j)$, computed from the modifications $(W^j, \mathcal{R}^j, E^j)$ constructed simultaneously along the weaving process.

Weaving Process

We describe the weaving process inductively.

Note that constructing a sequence for resolution of singularities by blowing up is referred to as "proceeding in the vertical direction" passing from one year to the next, indicated by the subscript "i", while weaving the strand and constructing the modifications passing from one stage to the next, indicated by the superscript "j", is referred to as "proceeding in the horizontal direction" staying in a fixed year.

Suppose we have already woven the strands and constructed the modifications up to year $(i-1)$.

Now we are in year i (looking at the neighborhood of a point $P_i \in \text{Sing}(\mathcal{R}_i) \subset W_i$ (cf. Remark 2)).

We start with $(W_i, \mathcal{R}_i, E_i) = (W_i^0, \mathcal{R}_i^0, E_i^0)$, just renaming the transformation $(W_i, \mathcal{R}_i, E_i)$ in year i of the resolution sequence as the 0-th stage modification $(W_i^0, \mathcal{R}_i^0, E_i^0)$ in year i.

Suppose that we have already woven the strand up to the $(j-1)$-th unit

$$(\text{inv}_{\text{new}})_i^{\leq j-1} = (\sigma_i^0, \widetilde{\mu}_i^0, s_i^0)(\sigma_i^1, \widetilde{\mu}_i^1, s_i^1) \cdots (\sigma_i^{j-1}, \widetilde{\mu}_i^{j-1}, s_i^{j-1})$$

and that we have also constructed the modifications up to the j-th one

$$(W_i^0, \mathcal{R}_i^0, E_i^0), (W_i^1, \mathcal{R}_i^1, E_i^1), \ldots, (W_i^{j-1}, \mathcal{R}_i^{j-1}, E_i^{j-1}), (W_i^j, \mathcal{R}_i^j, E_i^j).$$

Our task is to compute the j-th unit $(\sigma_i^j, \widetilde{\mu}_i^j, s_i^j)$ and construct the $(j+1)$-th modification $(W_i^{j+1}, \mathcal{R}_i^{j+1}, E_i^{j+1})$ (unless the weaving process is over at the j-th stage).

Computation of the j-th unit $(\sigma_i^j, \widetilde{\mu}_i^j, s_i^j)$

∘ σ_i^j: It is the invariant σ associated to the differential saturation $\mathcal{DR}_i^j$ of the idealistic filtration of i.f.g. type $\mathcal{R}_i^j$ (cf. Proposition 1).

$\circ\ \widetilde{\mu}_i^j$: It is the (normalized) weak order modulo LGS of the idealistic filtration of i.f.g. type $\mathcal{R}_i^j$ with respect to $(E_i^j)_{\text{young}}$. (For the definition of $(E_i^j)_{\text{young}} (\subset E_i^j)$, see the description of the invariant s_i^j below.

We note that, in order to compute $\widetilde{\mu}_i^j$,

$\begin{cases} \bullet \quad \text{we keep } \mathcal{R}_i^j \text{ as it is, which is the transformation of } \mathcal{R}_{i-1}^j, \\ \bullet \quad \text{we take the LGS to be the transformation of the one in the} \\ \text{previous year} \end{cases}$

if $(\text{inv}_{\text{new}})_i^{\leq j-1}(\sigma_i^j) = (\text{inv}_{\text{new}})_{i-1}^{\leq j-1}(\sigma_{i-1}^j)$,

$\begin{cases} \bullet \quad \text{we replace the original } \mathcal{R}_i^j \text{ with its differential saturation } \mathcal{DR}_i^j, \\ \bullet \quad \text{we take the LGS from the replaced } \mathcal{R}_i^j = \mathcal{DR}_i^j \end{cases}$

if $(\text{inv}_{\text{new}})_i^{\leq j-1}(\sigma_i^j) < (\text{inv}_{\text{new}})_{i-1}^{\leq j-1}(\sigma_{i-1}^j)$.

$\circ\ s_i^j$: It is the number of the components in $(E_i^j)_{\text{aged}} = E_i^j \setminus (E_i^j)_{\text{young}}$, where $(E_i^j)_{\text{young}} (\subset E_i^j)$ is the union of the exceptional divisors created after the year when the value $(\text{inv}_{\text{new}})_i^{\leq j-1}(\sigma_i^j)$ first started. We note that this third factor is included even if $\widetilde{\mu}_i^j = \infty$ or 0.

We note that, if $(\sigma_i^j, \widetilde{\mu}_i^j, s_i^j) = (\sigma_i^j, \infty, 0)$ or $(\sigma_i^j, 0, 0)$, then we declare that the $(j = m)$-th unit is the last one, and we stop the weaving process at the m-th stage in year i.

Thus the weaving process of the strand comes to an end in a fixed year i, with the strand $(\text{inv}_{\text{new}})_i$ taking the following form

$$(\text{inv}_{\text{new}})_i = (\sigma_i^0, \widetilde{\mu}_i^0, s_i^0)(\sigma_i^1, \widetilde{\mu}_i^1, s_i^1) \cdots (\sigma_i^j, \widetilde{\mu}_i^j, s_i^j) \cdots (\sigma_i^m, \widetilde{\mu}_i^m, s_i^m),$$

where $(\sigma_i^m, \widetilde{\mu}_i^m, s_i^m) = (\sigma_i^m, \infty, 0)$ or $(\sigma_i^m, 0, 0)$.

Termination in the horizontal direction

We note that termination of the weaving process in the horizontal direction is a consequence of the fact that going from the j-th unit to the $(j+1)$-th unit we have $(\sigma_i^j, \#E_i^j) > (\sigma_i^{j+1}, \#E_i^{j+1})$ (cf. Proposition 3) and that the value set of $(\sigma, \#E)$ satisfies the descending chain condition.

Construction of the $(j+1)$-th modification $(W_i^{j+1}, \mathcal{R}_i^{j+1}, E_i^{j+1})$

We note that we construct the $(j+1)$-th modification only when $(\sigma_i^j, \widetilde{\mu}_i^j, s_i^j) \neq (\sigma_i^j, \infty, 0)$ or $(\sigma_i^j, 0, 0)$.

We follow the construction described in the mechanism discussed in §4.2. Starting from $(W_i^j, \mathcal{R}_i^j, E_i^j)$, we construct $(W_i^{j+1}, \mathcal{R}_i^{j+1}, E_i^{j+1})$ as below:

$$W_i^{j+1} = W_i^j, \quad \mathcal{R}_i^{j+1} = \mathrm{Bdry}\left(\mathrm{Comp}(\mathcal{R}_i^j)\right), \quad \text{and}$$
$$E_i^{j+1} = E_i^j \setminus (E_i^j)_{\mathrm{aged}} = (E_i^j)_{\mathrm{young}},$$

where

• denoting $(\mathrm{inv}_{\mathrm{new}})_i^{\leq j-1}(\sigma_i^j, \widetilde{\mu}_i^j)$ by β_i^j, we set $\mathrm{Comp}(\mathcal{R}_i^j)$ to be either the transformation of $\mathrm{Comp}(\mathcal{R}_{i-1}^j)$ if $\beta_i = \beta_{i-1}$, or the one obtained by applying the construction in §4.2 if $\beta_i^j < \beta_{i-1}^j$, and

• denoting $(\mathrm{inv}_{\mathrm{new}})_i^{\leq j}$ by γ_i^j, we set $\mathrm{Bdry}(\mathrm{Comp}(\mathcal{R}_i^j))$ to be either the transformation of $\mathrm{Bdry}(\mathrm{Comp}(\mathcal{R}_{i-1}^j))$ if $\gamma_i^j = \gamma_{i-1}^j$, or

$$\mathcal{G}(\mathrm{Comp}(\mathcal{R}_i^j) \cup \{(x_D, 1) \mid D \subset (E_i^j)_{\mathrm{aged}}\})$$

if $\gamma_i^j < \gamma_{i-1}^j$.

We note that, if $\gamma_i^j = \gamma_{i-1}^j$, then $(W_i^{j+1}, \mathcal{R}_i^{j+1}, E_i^{j+1})$ is the transformation of $(W_{i-1}^{j+1}, \mathcal{R}_{i-1}^{j+1}, E_{i-1}^{j+1})$.

We also observe the following.

∗ The ambient space remains the same throughout a fixed year i, i.e.,

$$W_i = W_i^0 = W_i^1 = \cdots = W_i^j = W_i^{j+1} = \cdots = W_i^{m-1} = W_i^m.$$

This is in clear contrast to the classical setting, where we take a consecutive sequence of the hypersurfaces of maximal contact (cf. §3.3)

$$W_i = H_i^0 \supset H_i^1 \supset \cdots \supset H_i^j \supset H_i^{j+1} \supset \cdots \supset H_i^{m-1} \supset H_i^m.$$

∗ The idealistic filtration of i.f.g. type gets enlarged (not necessarily strictly) under modification, i.e.,

$$\mathcal{R}_i = \mathcal{R}_i^0 \subset \mathcal{R}_i^1 \subset \cdots \subset \mathcal{R}_i^j \subset \mathcal{R}_i^{j+1} \cdots \subset \mathcal{R}_i^{m-1} \subset \mathcal{R}_i^m.$$

∗ The boundary divisor decreases (not necessarily strictly) under modification, i.e.,

$$E_i = E_i^0 \supset E_i^1 \supset \cdots \supset E_i^j \supset E_i^{j+1} \supset \cdots \supset E_i^{m-1} \supset E_i^m.$$

Summary of our algorithm in char(k) $= p > 0$ in terms of "$\mathrm{inv}_{\mathrm{new}}$"

We start with $(W, \mathcal{R}, E) = (W_0, \mathcal{R}_0, E_0)$.

Suppose we have already constructed the resolution sequence up to year i

$$(W, \mathcal{R}, E) = (W_0, \mathcal{R}_0, E_0) \leftarrow \cdots \leftarrow (W_i, \mathcal{R}_i, E_i).$$

We weave the strand of invariants in year i described as above

$$(\mathrm{inv}_{\mathrm{new}})_i = (\sigma_i^0, \widetilde{\mu}_i^0, s_i^0)(\sigma_i^1, \widetilde{\mu}_i^1, s_i^1) \cdots (\sigma_i^j, \widetilde{\mu}_i^j, s_i^j) \cdots (\sigma_i^m, \widetilde{\mu}_i^m, s_i^m),$$

where $(\sigma_i^m, \widetilde{\mu}_i^m, s_i^m) = (\sigma_i^m, \infty, 0)$ or $(\sigma_i^m, 0, 0)$.

Case: $(\sigma_i^m, \widetilde{\mu}_i^m, s_i^m) = (\sigma_i^m, \infty, 0)$.

In this case, we take the center of blow up in year i for the transformation to be the singular locus of the last modification $(W_i^m, \mathcal{R}_i^m, E_i^m)$, i.e., $\mathrm{Sing}(\mathcal{R}_i^m)$, which is easily seen to be nonsingular as follows. We go back to the year $\iota := i_{\mathrm{aged}}^m$ when the value of $(\mathrm{inv}_{\mathrm{new}})_i^{\leq m-1}(\sigma_i^m)$ first started. Observe that $\widetilde{\mu}_i^m = \infty$ implies $\widetilde{\mu}_\iota^m = \infty$. The Nonsingularity Principle (cf. [18] and [19]) applied to $\mathcal{R}_\iota^m = \mathcal{D}\mathcal{R}_\iota^m$ tells us that there exists a regular system of parameters $(x_1, \ldots, x_t, x_{t+1}, \ldots, x_d)$ at P_ι such that $\mathcal{R}_\iota^m = \mathcal{G}((x_1^{p^{e_1}}, p^{e_1}), \ldots, (x_t^{p^{e_t}}, p^{e_t}))$. Note that the center C_ι of blow up in year ι is contained in $\mathrm{Sing}(\mathcal{R}_\iota^m)$. From this it follows inductively that there exists a regular system of parameters $(x_{1,i'}, \ldots, x_{t,i'}, x_{t+1,i'}, \ldots, x_{d,i'})$ at $P_{i'} \in \mathrm{Sing}(\mathcal{R}_{i'}^m) \subset W_{i'}$ such that $\mathrm{Sing}(\mathcal{R}_{i'}^j) = \mathcal{G}((x_{1,i'}^{p^{e_1}}, p^{e_1}), \ldots, (x_{t,i'}^{p^{e_t}}, p^{e_t}))$ for $\iota \leq i' \leq i$. In particular, $\mathrm{Sing}(\mathcal{R}_i^m)$ is nonsingular. We note that $\mathrm{Sing}(\mathcal{R}_i^m) \subset \bigcap_{D \in E_i \setminus (E_i^m)_{\mathrm{young}}} D$ by construction of the boundary modifications, and that $\mathrm{Sing}(\mathcal{R}_i^m)$ is transversal to $(E_i^m)_{\mathrm{young}}$. This implies that the center $\mathrm{Sing}(\mathcal{R}_i^m)$ is transversal to E_i. After blow up, the singular locus of $(W_i^m, \mathcal{R}_i^m, E_i^m)$ disappears. Since resolution of singularities for $(W_i^m, \mathcal{R}_i^m, E_i^m)$ implies the strict decrease of the value of $(\mathrm{inv}_{\mathrm{new}})^{\leq m-1}$, we have $(\mathrm{inv}_{\mathrm{new}})_i^{\leq m-1} > (\mathrm{inv}_{\mathrm{new}})_{i+1}^{\leq m-1}$.

Case: $(\sigma_i^m, \widetilde{\mu}_i^m, s_i^m) = (\sigma_i^m, 0, 0)$.

In this case, by following the procedure specified for resolution of singularities in the monomial case, we achieve resolution of singularities of the m-th modification, which implies the strict decrease of the value of $(\mathrm{inv}_{\mathrm{new}})^{\leq m-1}$. (It is possible that in the middle of the procedure the value of the $(\mathrm{inv}_{\mathrm{new}})^{\leq m-1}$ (or $(\mathrm{inv}_{\mathrm{new}})^{\leq m-1}(\sigma^m)$) strictly decreases).

Note: In both cases above, we are using the fact that the value of "$\mathrm{inv}_{\mathrm{new}}$" never increases after each transformation in the resolution sequence, in order to derive the strict decrease of $(\mathrm{inv}_{\mathrm{new}})^{\leq m-1}$ (or $(\mathrm{inv}_{\mathrm{new}})^{\leq m-1}(\sigma^m)$).

Termination in the vertical direction

By looking at the conclusions of the two cases above at the end of the weaving process in the horizontal direction, we conclude that in some year i', the value of the strand strictly decreases, i.e., $(\mathrm{inv}_{\mathrm{new}})_i > (\mathrm{inv}_{\mathrm{new}})_{i'}$.

Now we claim that the value of the strand "$\mathrm{inv}_{\mathrm{new}}$" cannot decrease infinitely many times. In fact, suppose by induction we have shown that the value of $(\mathrm{inv}_{\mathrm{new}})^{\leq t-1}$ cannot decrease infinitely many times. Then after some year the value of $(\mathrm{inv}_{\mathrm{new}})^{\leq t-1}$ stabilizes. Since the value of the invariant σ satisfies the descending chain condition, after some year (say after year i'_{t-1}), the value of $(\mathrm{inv}_{\mathrm{new}})^{\leq t-1}(\sigma^t)$ stabilizes. Therefore, after year i'_{t-1}, we use the transformation of $\mathcal{R}^t_{i'_{t-1}}$ in order to compute the invariant $\widetilde{\mu}$. (See (i) in the technical but important points in the computation of the triplet $(\sigma, \widetilde{\mu}, s)$.) This implies that the denominator of the invariant $\widetilde{\mu}$ is bounded, and hence that the invariant $\widetilde{\mu}$ cannot decrease infinitely many times. Since the invariant s, being a nonnegative integer, cannot decrease infinitely many times, we conclude that $(\mathrm{inv}_{\mathrm{new}})^{\leq t-1}(\sigma^t, \widetilde{\mu}^t, s^t) = (\mathrm{inv}_{\mathrm{new}})^{\leq t}$ cannot decrease infinitely many times.

For each t, let year i_t be the time when the stabilization of $(\mathrm{inv}_{\mathrm{new}})^{\leq t}$ first starts, i.e.,

$$(\mathrm{inv}_{\mathrm{new}})^{\leq t}_{i_t - 1} > (\mathrm{inv}_{\mathrm{new}})^{\leq t}_{i_t} = (\mathrm{inv}_{\mathrm{new}})^{\leq t}_{i} \quad \text{for} \quad i \geq i_t.$$

Note that $\{i_t\}$ is a (not necessarily strictly) increasing sequence, i.e., $i_t \leq i_{t'}$ if $t \leq t'$. Let $\sigma_t = \sigma^t_{i_t}$ be the first factor of the t-th unit of $(\mathrm{inv}_{\mathrm{new}})^{\leq t}_{i_t}$. Note that $\{\sigma_t\}$ is a (not necessarily strictly) decreasing sequence. That is to say, we have $\sigma_t \geq \sigma_{t+1}$, which follows easily if we look at year i_{t+1} and see $\sigma_t = \sigma^t_{i_t} = \sigma^t_{i_{t+1}} \geq \sigma^{t+1}_{i_{t+1}} = \sigma_{t+1}$.

We claim that we have either $\sigma_t > \sigma_{t+1}$ or $\sigma_t = \sigma_{t+1} > \sigma_{t+2}$. This can be seen by the following reasoning.

- First look at the t-th unit in year i_t, and observe $(\sigma^t_{i_t}, \widetilde{\mu}^t_{i_t}, s^t_{i_t}) \neq (\sigma^t_{i_t}, \infty, 0)$ or $(\sigma_{i_t}, 0, 0)$. In fact, if $(\sigma^t_{i_t}, \widetilde{\mu}^t_{i_t}, s^t_{i_t}) = (\sigma^t_{i_t}, \infty, 0)$ or $(\sigma_{i_t}, 0, 0)$, then the weaving process is over at the t-th stage in year i_t. When $(\sigma^t_{i_t}, \widetilde{\mu}^t_{i_t}, s^t_{i_t}) = (\sigma^t_{i_t}, \infty, 0)$, by the single blow up with center $\mathrm{Sing}(\mathcal{R}^t_i)$, we accomplish resolution of singularities for $(W^t_{i_t}, \mathcal{R}^t_{i_t}, E^t_{i_t})$. This, however, implies the strict decrease of $(\mathrm{inv}_{\mathrm{new}})^{\leq t-1}$, contradicting its stability after year $i_{t-1} (\leq i_t)$. If $(\sigma^t_{i_t}, \widetilde{\mu}^t_{i_t}, s^t_{i_t}) = (\sigma_{i_t}, 0, 0)$, then by the procedure of resolution of singularities in the monomial case, we accomplish resolution of singularities for $(W^t_{i_t}, \mathcal{R}^t_{i_t}, E^t_{i_t})$, which implies the strict decrease of the value of $(\mathrm{inv}_{\mathrm{new}})^{\leq t-1}$. This also contradicts the stability of $(\mathrm{inv}_{\mathrm{new}})^{\leq t-1}$ after year $i_{t-1}(\leq i_t)$. (It is possible that in the middle of the procedure the value of the $(\mathrm{inv}_{\mathrm{new}})^{\leq t-1}$ (or $(\mathrm{inv}_{\mathrm{new}})^{\leq t-1}(\sigma^t)$) strictly decreases. In the former case, it would contradict the stability of $(\mathrm{inv}_{\mathrm{new}})^{\leq t-1}$ after year $i_{t-1}(\leq i_t)$. In the latter case, it would contradict the stability of $(\mathrm{inv}_{\mathrm{new}})^{\leq t}$ after year i_t.)

- If $\widetilde{\mu}^t_{i_t} \neq 0$ or ∞, then $\sigma^t_{i_t} > \sigma^{t+1}_{i_t}$ (cf. the proof of Proposition 3). Since $\sigma^{t+1}_{i_t} \geq \sigma^{t+1}_{i_{t+1}}$, we conclude $\sigma_t = \sigma^t_{i_t} > \sigma^{t+1}_{i_t} \geq \sigma^{t+1}_{i_{t+1}} = \sigma_{t+1}$.
- We consider the case where $\widetilde{\mu}^t_{i_t} = 0$. By the first observation, we have $s^t_{i_t} \neq 0$ and the weaving process continues onto the $(t+1)$-th unit in year i_t. We have $\sigma^t_{i_t} \geq \sigma^{t+1}_{i_t}$.

Case: $\sigma^t_{i_t} > \sigma^{t+1}_{i_t}$.

We have $\sigma_t = \sigma^t_{i_t} > \sigma^{t+1}_{i_t} \geq \sigma^{t+1}_{i_{t+1}} = \sigma_{t+1}$.

Case: $\sigma^t_{i_t} = \sigma^{t+1}_{i_t}$.

Since year i_t is the time when the value of $(\mathrm{inv}_{\mathrm{new}})^{\leq t}_{i_t}$ first started, we have $\left(E^{t+1}_{i_t}\right)_{\mathrm{young}} = \emptyset$. The idealistic filtration of i.f.g. type $\mathcal{R}^{t+1}_{i_t} \supset \mathcal{R}^t_{i_t}$ contains a monomial of the defining equations of $(E^t_{i_t})_{\mathrm{young}}$. This implies $\widetilde{\mu}^{t+1}_{i_t} \neq 0$ or ∞. We have $\sigma^{t+1}_{i_t} \geq \sigma^{t+1}_{i_{t+1}}$.

Subcase: $\sigma^{t+1}_{i_t} > \sigma^{t+1}_{i_{t+1}}$.

We have $\sigma_t = \sigma^t_{i_t} = \sigma^{t+1}_{i_t} > \sigma^{t+1}_{i_{t+1}} = \sigma_{t+1}$.

Subcase: $\sigma^{t+1}_{i_t} = \sigma^{t+1}_{i_{t+1}}$.

Since $\widetilde{\mu}^{t+1}_{i_t} \neq \infty$, we have $\widetilde{\mu}^{t+1}_{i_{t+1}} \neq \infty$.

Subsubcase: $\widetilde{\mu}^{t+1}_{i_{t+1}} \neq 0$.

We have $\sigma^{t+1}_{i_{t+1}} > \sigma^{t+2}_{i_{t+1}}$. This implies $\sigma_t = \sigma^t_{i_t} = \sigma^{t+1}_{i_t} = \sigma^{t+1}_{i_{t+1}} > \sigma^{t+2}_{i_{t+1}} \geq \sigma^{t+2}_{i_{t+2}} = \sigma_{t+2}$.

Subsubcase: $\widetilde{\mu}^{t+1}_{i_{t+1}} = 0$.

By the first observation, we have $s^{t+1}_{i_{t+1}} \neq 0$. That is to say, there is a component D of $(E^{t+1}_{i_{t+1}})_{\mathrm{aged}} \subset E^{t+1}_{i_{t+1}} = E^t_{i_{t+1}} \setminus (E^t_{i_{t+1}})_{\mathrm{aged}} = (E^t_{i_{t+1}})_{\mathrm{young}}$ passing through that point. $(E^t_{i_{t+1}})_{\mathrm{young}}$ is the union of the exceptional divisors created after the year when the value of $(\mathrm{inv}_{\mathrm{new}})^{\leq t-1}_{i_{t+1}}(\sigma^t_{i_{t+1}}) = (\mathrm{inv}_{\mathrm{new}})^{\leq t-1}_{i_t}(\sigma^t_{i_t})$ first started. Therefore, D is transversal to the LGS of $\mathcal{R}^{t+1}_{i_{t+1}}$, which is the transformation of the LGS of $\mathcal{R}^t_{i_t}$ since $\sigma^t_{i_t} = \sigma^{t+1}_{i_t} = \sigma^{t+1}_{i_{t+1}}$. Since $\mathcal{R}^{t+2}_{i_{t+1}}$ contains $(x_D, 1)$, where x_D is the defining equation of D, we conclude $\sigma^{t+1}_{i_{t+1}} > \sigma^{t+2}_{i_{t+1}}$. Therefore, we have $\sigma_t = \sigma^t_{i_t} = \sigma^{t+1}_{i_t} = \sigma^{t+1}_{i_{t+1}} > \sigma^{t+2}_{i_{t+1}} \geq \sigma^{t+2}_{i_{t+2}} = \sigma_{t+2}$.

• We consider the case where $\widetilde{\mu}^t_{i_t} = \infty$. By the first observation, we have $s^t_{i_t} \neq 0$ and the weaving process continues onto the $(t+1)$-th unit in year i_t. We have $\sigma^t_{i_t} \geq \sigma^{t+1}_{i_t}$.

Case: $\sigma^t_{i_t} > \sigma^{t+1}_{i_t}$.

We have $\sigma_t = \sigma^t_{i_t} > \sigma^{t+1}_{i_t} \geq \sigma^{t+1}_{i_{t+1}} = \sigma_{t+1}$.

Case: $\sigma^t_{i_t} = \sigma^{t+1}_{i_t}$.

Since year i_t is the time when the value of $(\mathrm{inv}_{\mathrm{new}})^{\leq t}_{i_t}$ first started, we have $\left(E^{t+1}_{i_t}\right)_{\mathrm{young}} = \emptyset$. This implies $\widetilde{\mu}^{t+1}_{i_t} \neq 0$. If $\widetilde{\mu}^{t+1}_{i_t} \neq \infty$, then we can carry the same argument as above ($\widetilde{\mu}^t_{i_t} = 0$ and *Case*: $\sigma^t_{i_t} = \sigma^{t+1}_{i_t}$) to conclude that either $\sigma_t > \sigma_{t+1}$ or $\sigma_t = \sigma_{t+1} > \sigma_{t+2}$. Therefore, we have only to consider the case where $\widetilde{\mu}^{t+1}_{i_t} = \infty$. We have $\sigma^{t+1}_{i_t} \geq \sigma^{t+1}_{i_{t+1}}$.

Subcase: $\sigma^{t+1}_{i_t} > \sigma^{t+1}_{i_{t+1}}$.

We have $\sigma_t = \sigma^t_{i_t} = \sigma^{t+1}_{i_t} > \sigma^{t+1}_{i_{t+1}} = \sigma_{t+1}$.

Subcase: $\sigma^{t+1}_{i_t} = \sigma^{t+1}_{i_{t+1}}$.

Since $\widetilde{\mu}^{t+1}_{i_t} = \infty$, we have $\widetilde{\mu}^{t+1}_{i_{t+1}} = \infty$. By the first observation, we have $s^{t+1}_{i_{t+1}} \neq 0$. That is to say, there is a component D of $(E^{t+1}_{i_{t+1}})_{\mathrm{aged}} \subset E^{t+1}_{i_{t+1}} = E^t_{i_{t+1}} \setminus (E^t_{i_{t+1}})_{\mathrm{aged}} = (E^t_{i_{t+1}})_{\mathrm{young}}$ passing through that point. $(E^t_{i_{t+1}})_{\mathrm{young}}$ is the union of the exceptional divisors created after the year when the value of $(\mathrm{inv}_{\mathrm{new}})^{\leq t-1}_{i_{t+1}}(\sigma^t_{i_{t+1}}) = (\mathrm{inv}_{\mathrm{new}})^{\leq t-1}_{i_t}(\sigma^t_{i_t})$ first started. Therefore, D is transversal to the LGS of $\mathcal{R}^{t+1}_{i_{t+1}}$, which is the transformation of the LGS of $\mathcal{R}^t_{i_t}$ since $\sigma^t_{i_t} = \sigma^{t+1}_{i_t} = \sigma^{t+1}_{i_{t+1}}$. Since $\mathcal{R}^{t+2}_{i_{t+1}}$ contains $(x_D, 1)$, where x_D is the defining equation of D, we conclude $\sigma^{t+1}_{i_{t+1}} > \sigma^{t+2}_{i_{t+1}}$. Therefore, we have $\sigma_t = \sigma^t_{i_t} = \sigma^{t+1}_{i_t} = \sigma^{t+1}_{i_{t+1}} > \sigma^{t+2}_{i_{t+1}} \geq \sigma^{t+2}_{i_{t+2}} = \sigma_{t+2}$.

This completes the reasoning for the claim that we have either $\sigma_t > \sigma_{t+1}$ or $\sigma_t = \sigma_{t+1} > \sigma_{t+2}$.

Now we finish the argument for termination in the vertical direction as follows.

Since the value of the invariant σ satisfies the descending chain condition, the increase of the value of t stops after finitely many times. Finally, therefore, we conclude that the value of the strand "$\mathrm{inv}_{\mathrm{new}}$" cannot decrease infinitely many times.

Therefore, the algorithm terminates after finitely many years, achieving resolution of singularities for $(W, \mathcal{R}, E)$.

4.4. Brief discussion on the monomial case in positive characteristic

Here in §4.4, we briefly discuss why the problem of resolution of singularities in the monomial case in positive characteristic is much more subtle and difficult than the one in characteristic zero.

Recall we say that the triplet $(W, \mathcal{R}, E)$ is in the monomial case (at $P \in \mathrm{Sing}(\mathcal{R}) \subset W$) in our setting if (and only if) the triplet of invariants takes the value $(\sigma, \widetilde{\mu}, s) = (\sigma, 0, 0)$. (Precisely speaking, the triplet $(W, \mathcal{R}, E)$ sits in the middle of constructing the sequence, say in year "i" and at stage "j", for resolution of singularities. However, we omit the subscript and superscript "$(\ )_i^j$", indicating the year and the stage, for simplicity of the notation.)

The description of the monomial case at the analytic level is given below.

SITUATION

∘ The condition $\widetilde{\mu} = 0$ is interpreted as follows:

We can choose a regular system of parameters $(x_1, \ldots, x_t, x_{t+1}, \ldots, x_d)$, taken from $\widehat{\mathcal{O}_{W,P}}$, such that

(1) the elements in the LGS $\mathbb{H} = \{(h_\alpha, p^{e_\alpha})\}_{\alpha=1}^t$ are of the form

$$h_\alpha = x_\alpha^{p^{e_\alpha}} + \text{higher terms} \quad \text{for} \quad \alpha = 1, \ldots, t,$$

(We sometimes call the higher terms in the above expression "the tail part".)

(2) there is a monomial $\mathbb{M} = \prod_{D \in E_{\text{young}}} x_D^{r_D}$ of the defining equations x_D of the components D in E_{young} with

$$(\mathbb{M}, a) \in \widehat{\mathcal{R}_P} \text{ for some } a \in \mathbb{Z}_{>0},$$

where $\{x_D \mid D \in E_{\text{young}}\} \subset \{x_\alpha \mid \alpha = t+1, \ldots, d\}$ and $\sum r_D > a$,

(3) the idealistic filtration of i.f.g. type $\widehat{\mathcal{R}_P}$ is saturated for $\{\partial^n / \partial {x_\alpha}^n \mid n \in \mathbb{Z}_{\geq 0}, \alpha = 1, \ldots, t\}$,

satisfying the following condition: for an arbitrary $(f, \lambda) \in \widehat{\mathcal{R}_P}$ with $f = \sum c_{f,B} H^B$ being the power series expansion with respect to the LGS $\mathbb{H}$ and the associated to the regular system of parameters $(x_1, \ldots, x_t, x_{t+1}, \ldots, x_d)$, we have $\mathbb{M}^{\lambda/a}$ dividing the constant term $c_{f,\mathbb{O}}$, i.e., $\mathbb{M}^{\lambda/a} | c_{f,\mathbb{O}}$. Note that, using the formal coefficient lemma, we see $(c_{f,\mathbb{O}}, \lambda) \in \widehat{\mathcal{R}_P}$.

∘ The condition $s = 0$ is of course equivalent to saying that there is no component of E_{aged} passing through P.

$\boxed{\text{char}(k) = 0}$

In fact, the above description $\boxed{\text{SITUATION}}$ of the monomial case is also valid when $\text{char}(k) = 0$, with all the elements of the LGS concentrated at level 1, i.e., $p^{e_\alpha} = 1$ for $\alpha = 1, \ldots, t$. Moreover, we can replace x_α with h_α so that we have $h_\alpha = x_\alpha$ for $\alpha = 1, \ldots, t$. Then there is *no* tail part. In order to construct resolution of singularities for $(W, \mathcal{R}, E)$, we have only to carry out the resolution process for $(V, ((\mathbb{M}), a)|_V, E|_V)$, which is the triplet in the monomial case in the classical setting, where $V = \{x_1 = \cdots = x_t = 0\}$ is a nonsingular subvariety inside of W. Note that, since $E_{\text{aged}} = \emptyset$ (in a neighborhood of P, because $s = 0$), the third factor $E|_V$ is a simple normal crossing divisor on V.

$\boxed{\text{char}(k) = p > 0}$

In contrast to the case in $\text{char}(k) = 0$, the elements in the LGS may not be concentrated at level 1 in general. Therefore, we usually have the tail parts for those h_α's at higher levels in $\text{char}(k) = p > 0$. The hypersurfaces defined by $\{h_\alpha = 0\}$ by those elements are singular hypersurfaces. Therefore, one is forced to analyze the monomial restricted to a singular subvariety (defined as the intersection of the singular hypersurfaces), or alternatively to analyze the combination of the monomial and the elements of the LGS with the tail parts included, while sticking to the original nonsingular ambient space. The latter is what we do in §5 of this paper in dimension 3.

In the simplest terms, NO or YES tail part is what makes the difference between the monomial case in $\text{char}(k) = 0$ and the one in $\text{char}(k) = p > 0$.

§5. Detailed discussion on the monomial case in dimension 3

The purpose of §5 is to discuss how to construct resolution of singularities (at the local level (cf. Remark 2)) in the monomial case. We refer the reader to §4.4 $\boxed{\text{SITUATION}}$ for the precise description of the monomial case.

5.1. Case analysis according to the invariant "τ"

In §5.1, we analyze the situation according to the value of the invariant τ.

Recall that the invariant τ is just the number of the elements in an LGS, and hence that, in dimension 3, it takes the value $\tau = 0, 1, 2, 3$.

It turns out that the analysis of the case $\tau = 0, 2$ or 3 is rather easy. We devote §5.2, §5.3, §5.4 to the analysis of the most subtle and difficult case $\tau = 1$.

Case: $\tau = 0$

In this case, there is no element in an LGS. We conclude that for an arbitrary $(f, \lambda) \in \widehat{\mathcal{R}_P}$ the monomial $\mathbb{M}^{\lambda/a}$ divides $c_{f,\mathbb{O}} = f$. Therefore, we can carry out the same algorithm for resolution of singularities of the triplet $(W, (\mathcal{I}, a), E) = (W, ((\mathbb{M}), a), E)$ in the monomial case in the classical setting in characteristic zero, using the invariant Γ discussed in §3.4. (Note that in the middle of the procedure the invariant σ may drop. If that happens, then we are no longer in the monomial case. In that case, we go through the mechanism described in §4.2 with the reduced new value of σ to reach the new monomial case.)

Case: $\tau = 1$

This case will be thoroughly discussed in §5.2, §5.3, §5.4.

Case: $\tau = 2$

In this case, we can choose a regular system of parameters (x, y, z), taken from $\widehat{\mathcal{O}_{W,P}}$, such that

(1) the two elements in the LGS $\mathbb{H} = \{(h_1, p^{e_1}), (h_2, p^{e_2})\}$ are of the form

$$\begin{cases} h_1 &= z^{p^{e_1}} + \text{higher terms} \\ h_2 &= y^{p^{e_2}} + \text{higher terms,} \end{cases}$$

(2) there is a monomial $\mathbb{M} = x^r$ of the defining equation x of the component $\{x = 0\} \in E_{\text{young}}$ with

$$(\mathbb{M}, a) = (x^r, a) \in \widehat{\mathcal{R}_P} \text{ for some } a \in \mathbb{Z}_{>0},$$

(3) the idealistic filtration $\widehat{\mathcal{R}_P}$ is saturated for $\{\partial^n/\partial z^n, \partial^m/\partial y^m \mid n, m \in \mathbb{Z}_{\geq 0}\}$.

Then it is easy to see that $\text{Sing}(\mathcal{R}) = P$ (in a neighborhood of P) and hence that the only possible transformation is the blow up with center P. After blow up, we see that the (possibly) non-empty singular locus lies only over the x-chart. We also see that the singular locus, if non-empty, consists of a single point $\widetilde{P} \in \widetilde{W}$ (in a neighborhood of the inverse image of P) with a regular system of parameters $(\widetilde{x}, \widetilde{y}, \widetilde{z}) = (x, y/x, z/x)$. The new LGS

$$\widetilde{\mathbb{H}} = \{(\widetilde{h_1}, p^{e_1}), (\widetilde{h_2}, p^{e_2})\} = \{(h_1/x^{p^{e_1}}, p^{e_1}), (h_2/x^{p^{e_2}}, p^{e_2})\}$$

are of the form

$$\begin{cases} \widetilde{h_1} & = & (\widetilde{z}')^{p^{e1}} + \text{higher terms} \\ \widetilde{h_2} & = & (\widetilde{y}')^{p^{e2}} + \text{higher terms} \end{cases}$$

where (cf. the proof of Proposition 4)

$$\begin{cases} \widetilde{z}' & = & \widetilde{z} + (c_z)^{1/p^{e1}}\widetilde{x} \\ \widetilde{y}' & = & \widetilde{y} + (c_y)^{1/p^{e2}}\widetilde{x} \end{cases} \quad \text{for some} \quad c_z, c_y \in k$$

(in case the invariant σ does not decrease). We calculate the new monomial to be

$$(\widetilde{x}^{r-a}, a).$$

Since the power of x in the above monomial cannot decrease infinitely many times, we achieve resolution of singularities after finitely many repetitions of this procedure. (Note that in the middle of the procedure the invariant σ may drop. If that happens, then we are no longer in the monomial case. In that case, we go through the mechanism described in §4.2 with the reduced new value of σ to reach the new monomial case.)

Case: $\tau = 3$

This case does not happen. In fact, suppose this case did happen. Then go back to the year $\iota := i_{\text{aged}}$ when the current value of σ first started. Since the current value of τ is equal to 3, so is the value of τ back in year ι. Take the blow up with center P_ι, which is the image of P in year ι. Then it is immediate to see that there is no singular locus (in a neighborhood of the inverse image of P_ι) after blow up. This implies in turn that $\text{Sing}(\mathcal{R}_\iota) = P_\iota$ (in a neighborhood of P_ι). The only possible transformation in the resolution sequence, therefore, is the blow up with center P_ι. After blow up, however, we already saw $\text{Sing}(\mathcal{R}_{\iota+1}) = \emptyset$ over P_ι. This is a contradiction, since $P_{\iota+1}$, which is the image of P in year $\iota + 1$, should be included in $\text{Sing}(\mathcal{R}_{\iota+1})$. (Note that we have $\iota < i$, since the value of $\widetilde{\mu}$ is never zero when the new value of σ starts in year ι and since the current value of $\widetilde{\mu}$ is zero being in the monomial case in year i.)

Focus on the case $\tau = 1$

In the following §5.2, §5.3, §5.4, we focus on, and restrict ourselves to, the case $\tau = 1$. We carry out the computation of the invariants at the analytic (completion) level, even though the centers of blow ups are chosen at the algebraic level, and hence all the procedures in the algorithm are carried out at the algebraic level.

We restate the $\boxed{\text{SITUATION}}$ described in §4.4 in a slightly refined form in our particular case $\tau = 1$, for the sole purpose of fixing the notation for §5.2, §5.3, §5.4.

$\boxed{\text{SITUATION}}$

We can choose a regular system of parameters (x, y, z), taken from $\widehat{\mathcal{O}_{W,P}}$, such that

(1) via the Weierstrass Preparation Theorem the unique element in the LGS $\mathbb{H} = \{(h, p^e)\}$ is of the form

$$h = z^{p^e} + a_1 z^{p^e-1} + a_2 z^{p^e-2} + \cdots + a_{p^e-1} z + a_{p^e}$$

with

$$a_i \in k[[x,y]] \quad \text{and} \quad \mathrm{ord}_P(a_i) > i \quad \text{for} \quad i = 1, \ldots, p^e,$$

(2) there is a monomial $\mathbb{M} = x^\alpha y^\beta$ of the defining equation(s) of the component(s) $H_x = \{x = 0\}$ (and possibly $H_y = \{y = 0\}$) in E_{young} with

$$(\mathbb{M}, a) = (x^\alpha y^\beta, a) \in \mathcal{R}_P \quad \text{and} \quad \alpha + \beta > a,$$

(We write $\mathbb{M}_u := x^{\alpha/a} y^{\beta/a}$ and call $\mathbb{M}_u$ the *usual* monomial.)

(3) the idealistic filtration $\widehat{\mathcal{R}_P}$ is saturated for $\{\partial^n/\partial z^n \mid n \in \mathbb{Z}_{\geq 0}\}$,

satisfying the following condition: for an arbitrary $(f, \lambda) \in \widehat{\mathcal{R}_P}$ with $f = \sum c_{f,B} H^B = \sum c_{f,b} h^b$ being the power series expansion with respect to the LGS $\mathbb{H} = \{(h, p^e)\}$ and its associated regular system of parameters (x, y, z), we have $\mathbb{M}^{\lambda/a}$ dividing the constant term $c_{f,\mathbb{O}} = c_{f,0}$, i.e., $\mathbb{M}^{\lambda/a} \mid c_{f,\mathbb{O}} = c_{f,0}$.

In particular, by looking at $\partial^n h/\partial z^n$ for $n = 1, \ldots, p^e - 1$, we see that the coefficient a_i is divisible by $(\mathbb{M}_u)^i$ for $i = 1, \ldots, p^e - 1$ (but maybe not for $i = p^e$). That is to say,

$$(\mathbb{M}_u)^i \mid a_i, \quad \text{i.e.,} \quad x^{\lceil i\alpha/a \rceil} y^{\lceil i\beta/a \rceil} \mid a_i \quad \text{for} \quad i = 1, \ldots, p^e - 1.$$

Throughout §5.2, §5.3, §5.4, we are under $\boxed{\text{SITUATION}}$ described as above.

5.2. "Cleaning" and the invariant "H" (in the case $\tau = 1$)

The purpose of this section is to introduce the invariant "H" through the process of "cleaning". We follow closely the argument developed by Benito-Villamayor [5], making some modifications to fit it into our own setting in the framework of the Idealistic Filtration Program.

Definition 5. We define the slope of h at P with respect to (x, y, z) by the formula

$$\mathrm{Slope}_{h,(x,y,z)}(P) = \min\left\{\frac{1}{p^e}\mathrm{ord}_P(a_{p^e}),\ \mu(P)\right\}.$$

Remark 6.

(i) Since we are in the monomial case and hence $\widetilde{\mu} = 0$ and since the monomial is $(\mathbb{M}, a) = (x^\alpha y^\beta, a)$, we compute

$$\mu(P) = \frac{1}{a}\mathrm{ord}_P(x^\alpha y^\beta) = \frac{\alpha+\beta}{a} = \mathrm{ord}_P(\mathbb{M}_u),$$

while

$$\mu(\xi_{H_x}) = \frac{\alpha}{a} \quad \text{and} \quad \mu(\xi_{H_y}) = \frac{\beta}{a}.$$

(ii) We have a "very good" control over the coefficients a_i except for the constant term a_{p^e}, in the sense that

$$(\mathbb{M}_u)^i = \left(x^{\alpha/a} y^{\beta/a}\right)^i \mid a_i \quad \text{for} \quad i = 1, \ldots, p^e - 1,$$

which implies

$$\frac{1}{i}\mathrm{ord}_P(a_i) \geq \mu(P) \quad \text{for} \quad i = 1, \ldots, p^e - 1.$$

Therefore, this control leads to the following observation

$$\begin{aligned}\mathrm{Slope}_{h,(x,y,z)}(P) &= \min\left\{\mu(P),\ \frac{1}{p^e}\mathrm{ord}_P(a_{p^e})\right\} \\ &= \min\left\{\mu(P),\ \frac{1}{i}\mathrm{ord}_P(a_i)\ ;\ i = 1, \ldots, p^e\right\}.\end{aligned}$$

Definition 6 (Well-adaptedness (cf. [5])). We say h is well-adapted at P with respect to (x, y, z) if one of the following two conditions holds:

A. $\mathrm{Slope}_{h,(x,y,z)}(P) = \mu(P)$.

B. $\mathrm{Slope}_{h,(x,y,z)}(P) = \mathrm{ord}_P(a_{p^e})/p^e < \mu(P)$ and the initial form $\mathrm{In}_P(a_{p^e})$ is *not* a p^e-th power.

Similarly, we say h is well-adapted at ξ_{H_x}, where ξ_{H_x} is the generic point of the hypersurface $H_x = \{x = 0\}$ in E_{young}, if one of the following two conditions holds:

A. $\mathrm{Slope}_{h,(x,y,z)}(\xi_{H_x}) = \mu(\xi_{H_x}) = \alpha/a$.

B. $\mathrm{Slope}_{h,(x,y,z)}(\xi_{H_x}) = \mathrm{ord}_{\xi_{H_x}}(a_{p^e})/p^e < \mu(\xi_{H_x})$ and the initial form $\mathrm{In}_{\xi_{H_x}}(a_{p^e})$ is *not* a p^e-th power.

The notion of h being well-adapted at ξ_{H_y}, where ξ_{H_y} is the generic point of the hypersurface $H_y = \{y = 0\}$ in E_{young}, is defined in an identical manner.

Note that if

$$a_{p^e} = \sum_{k+l \geq d} c_{kl} x^k y^l = x^r \{g(y) + x \cdot \omega(x, y)\}$$

where $c_{kl} \in k$, $0 \neq g(y) \in k[[y]]$, $\omega(x, y) \in k[[x, y]]$, $d = \mathrm{ord}_P(a_{p^e})$ and $r = \mathrm{ord}_{\xi_{H_x}}(a_{p^e})$, then

$$\mathrm{In}_P(a_{p^e}) = \sum_{k+l=d} c_{kl} x^k y^l \quad \text{and} \quad \mathrm{In}_{\xi_{H_x}}(a_{p^e}) = x^r g(y).$$

Proposition 5.

(1) *There exist an LGS $\mathbb{H} = \{(h, p^e)\}$ and a regular system of parameters (x, y, z), as described in* SITUATION *in §5.1, such that h is well-adapted at P, ξ_{H_x} and ξ_{H_y} simultaneously with respect to (x, y, z). Note that we require the property that the idealistic filtration $\mathcal{R}_P$ is $\left\{ \frac{\partial^n}{\partial z^n} \mid n \in \mathbb{Z}_{\geq 0} \right\}$-saturated with respect to the regular system of parameters (x, y, z) (cf.* SITUATION *(3)).*

(2) *If h is well-adapted at $* = P, \xi_{H_x}$ or ξ_{H_y} with respect to (x, y, z), then $\mathrm{Slope}_{h,(x,y,z)}(*)$ is independent of the choice of h and (x, y, z).*

Proof.

(1) We start with h and (x, y, z) as given in SITUATION in §5.1.

Step 1. Modify h and (x, y, z) to be well-adapted at P.

Suppose we are in Case A or Case B as described in Definition 6. Then h is already well-adapted at P with respect to (x, y, z) and there is no modification needed.

Therefore, we may assume that we are not in either Case A or Case B. That is to say, we have

$$\mathrm{Slope}_{h,(x,y,z)}(P) = \frac{1}{p^e} \mathrm{ord}_P(a_{p^e}) < \mu(P),$$

and

$$\mathrm{In}_P(a_{p^e}) = \sum_{k+l=d} c_{kl} x^k y^l, \quad \text{with } d = \mathrm{ord}_P(a_{p^e}), \text{ is a } p^e\text{-th power.}$$

Take

$$\{\mathrm{In}_P(a_{p^e})\}^{1/p^e} \in k[x, y],$$

and set

$$z' = z + \{\mathrm{In}_P(a_{p^e})\}^{1/p^e}, \quad \text{i.e.,} \quad z = z' - \{\mathrm{In}_P(a_{p^e})\}^{1/p^e}.$$

Plug this into

$$h = z^{p^e} + a_1 z^{p^e-1} + a_2 z^{p^e-2} + \cdots + a_{p^e-1} z + a_{p^e}$$

to obtain

$$h = z'^{p^e} + a'_1 z'^{p^e-1} + a'_2 z'^{p^e-2} + \cdots + a'_{p^e-1} z' + a'_{p^e}$$

with

$$a'_i \in k[[x,y]] \quad \text{for} \quad i = 1, \ldots, p^e - 1, p^e.$$

Since for $i = 1, \ldots, p^e - 1$, we have (cf. Remark 6 (ii))

$$\frac{1}{i}\mathrm{ord}_P(a_i) \geq \mu(P) > \frac{1}{p^e}\mathrm{ord}_P(a_{p^e}) = \mathrm{ord}_P\left(\{\mathrm{In}_P(a_{p^e})\}^{1/p^e}\right),$$

we conclude that a'_{p^e} is of the form

$$a'_{p^e} = a_{p^e} - \mathrm{In}_P(a_{p^e}) + \text{higher terms}$$

and hence that

$$\frac{1}{p^e}\mathrm{ord}_P(a'_{p^e}) > \frac{1}{p^e}\mathrm{ord}_P(a_{p^e}).$$

We go back to the starting point, replacing the original h and (x, y, z) by h' and $(x', y', z') = (x, y, z')$, with strictly increased $\mathrm{ord}_P(a'_{p^e})/p^e$. Since $\mu(P) < \infty$, we conclude that, after finitely many repetitions of this process, we have to come to the situation where we are in Case A or Case B, i.e., where h is well-adapted at P with respect to (x, y, z).

Step 2. Modify h and (x, y, z) further to be well-adapted at ξ_{H_x} without destroying the well-adaptedness at P.

Take h which is well-adapted at P with respect to (x, y, z), as obtained through Step 1.

Suppose we are in Case A or Case B as described in the second half of Definition 6. Then h is already well-adapted at ξ_{H_x} with respect to (x, y, z) and there is no modification needed.

Therefore, we may assume that we are not in either Case A or Case B. That is to say, we have

$$\mathrm{Slope}_{h,(x,y,z)}(\xi_{H_x}) = \frac{1}{p^e}\mathrm{ord}_{\xi_{H_x}}(a_{p^e}) < \mu(\xi_{H_x}),$$

and

$$\mathrm{In}_{\xi_{H_x}}(a_{p^e}) = x^r g(y) \text{ is a } p^e\text{-th power,}$$

where $a_{p^e} = x^r \{g(y) + x \cdot \omega(x,y)\}$ and $r = \mathrm{ord}_{\xi_{H_x}}(a_{p^e})$. Take

$$\left\{\mathrm{In}_{\xi_{H_x}}(a_{p^e})\right\}^{1/p^e} \in k[[y]][x],$$

and set

$$z' = z + \left\{\mathrm{In}_{\xi_{H_x}}(a_{p^e})\right\}^{1/p^e}, \quad \text{i.e.,} \quad z = z' - \left\{\mathrm{In}_{\xi_{H_x}}(a_{p^e})\right\}^{1/p^e}.$$

Then as in Step 1, we see

$$\frac{1}{p^e}\mathrm{ord}_{\xi_{H_x}}(a'_{p^e}) > \frac{1}{p^e}\mathrm{ord}_{\xi_{H_x}}(a_{p^e}).$$

We go back to the starting point, replacing the original h and (x,y,z) by h' and $(x',y',z') = (x,y,z')$, with strictly increased $\mathrm{ord}_{\xi_{H_x}}(a'_{p^e})/p^e$. Since $\mu(\xi_{H_x}) < \infty$, we conclude that, after finitely many repetitions of this process, we have to come to the situation where we are in Case A or Case B, i.e., where h is well-adapted at ξ_{H_x} with respect to (x,y,z).

The only issue here is to check, in the process, the property that h is well-adapted at P is preserved.

Case: $\mathrm{ord}_P(a_{p^e})/p^e \geq \mu(P)$.

In this case, we have

$$\mathrm{ord}_P\left(\left\{\mathrm{In}_{\xi_{H_x}}(a_{p^e})\right\}^{1/p^e}\right) \geq \frac{1}{p^e}\mathrm{ord}_P(a_{p^e}) \geq \mu(P)$$

and, for $i = 1, \ldots, p^e - 1$, we have (cf. Remark 6 (ii))

$$\frac{1}{i}\mathrm{ord}_P(a_i) \geq \mu(P).$$

Therefore, we conclude that

$$\frac{1}{p^e}\mathrm{ord}_P(a'_{p^e}) \geq \mu(P),$$

and hence that h' stays well-adapted at P with respect to (x',y',z').

Case: $\mathrm{ord}_P(a_{p^e})/p^e < \mu(P)$.

In this case, we have

$$\mathrm{ord}_P\left(\left\{\mathrm{In}_{\xi_{H_x}}(a_{p^e})\right\}^{1/p^e}\right) \geq \frac{1}{p^e}\mathrm{ord}_P(a_{p^e})$$

and, for $i = 1, \ldots, p^e - 1$, we have (cf. Remark 6 (ii))

$$\frac{1}{i}\mathrm{ord}_P(a_i) \geq \mu(P) > \frac{1}{p^e}\mathrm{ord}_P(a_{p^e}).$$

Hence, since $\mathrm{In}_P(a_{p^e})$ is not a p^e-th power and since the degree $d = \mathrm{ord}_p(a_{p^e})$-part of $\mathrm{In}_{\xi_{H_x}}(a_{p^e})$ is a p^e-th power, we have

$$\mathrm{In}_P(a'_{p^e}) = \mathrm{In}_P(a_{p^e}) - \left\{\text{the degree } d = \mathrm{ord}_p(a_{p^e})\text{-part of } \mathrm{In}_{\xi_{H_x}}(a_{p^e})\right\}.$$

Therefore, we conclude that

$$\mathrm{ord}_P(a'_{p^e}) = \mathrm{ord}_P\left(\mathrm{In}_P(a'_{p^e})\right) = \mathrm{ord}_P\left(\mathrm{In}_P(a_{p^e})\right) = \mathrm{ord}_P(a_{p^e}) < \mu(P)$$

and $\mathrm{In}_P(a'_{p^e})$ is not a p^e-th power, and hence that h' stays well-adapted at P with respect to (x', y', z').

Step 3. Modify h and (x, y, z) still further to be well-adapted at ξ_{H_y} without destroying the well-adaptedness at P and ξ_{H_x}.

The process of this step is almost identical to that of Step 2, and hence is left to the reader as an exercise.

We note that the requirement as described in SITUATION (3) is met, since the original (x, y, z) satisfies the property and since we only modify z by adding the elements in $k[[x, y]]$ throughout the process.

This finishes the proof of (1).

(2) We only give a proof for the case where $* = P$, since the proof for the case where $* = \xi_{H_x}$ is identical.

Take h which is well-adapted at P with respect to (x, y, z), with the property that the idealistic filtration $\mathcal{R}_P$ is saturated for $\{\partial^n/\partial z^n \mid n \in \mathbb{Z}_{\geq 0}\}$.

We set

$$\mathrm{H}(P) = \min\left\{\mu(P),\ \max\left\{\frac{1}{p^e}\mathrm{ord}_P(h'|_{Z'}) \;\middle|\; \begin{matrix} h', (x', y', z'), \\ Z' = \{z' = 0\} \end{matrix}\right\}\right\}$$

where, computing the above "max", we let h' and (x', y', z') vary among all such pairs consisting of the unique element in an LGS and a regular system of parameters that satisfy the condition

$$h' \equiv z'^{p^e} \bmod \widehat{\mathfrak{m}_P}^{p^e+1}.$$

It suffices to show

$$\mathrm{Slope}_{h,(x,y,z)}(P) = \mathrm{H}(P),$$

since the number $\mathrm{H}(P)$ is obviously independent of the choice of h and (x, y, z).

Observe

$$\mathrm{Slope}_{h,(x,y,z)}(P) = \min\left\{\frac{1}{p^e}\mathrm{ord}_P(a_{p^e})\ , \mu(P)\right\} \leq \mathrm{H}(P),$$

since

$$\frac{1}{p^e}\mathrm{ord}_P(h|_Z) = \frac{1}{p^e}\mathrm{ord}_P(a_{p^e})$$

and since h and (x, y, z) form such a pair consisting of the unique element in an LGS and a regular system of parameters that satisfy the condition

$$h \equiv z^{p^e} \bmod \widehat{\mathfrak{m}_P}^{p^e+1}.$$

Now we prove the inequality in the opposite direction

$$\mathrm{Slope}_{h,(x,y,z)}(P) = \min\left\{\frac{1}{p^e}\mathrm{ord}_P(a_{p^e}),\ \mu(P)\right\} \geq \mathrm{H}(P).$$

If $\mathrm{Slope}_{h,(x,y,z)}(P) = \mu(P)$, then the above inequality obviously holds. Therefore, we may assume that $\mathrm{Slope}_{h,(x,y,z)}(P) = \mathrm{ord}_P(a_{p^e})/p^e < \mu(P)$ and that $\mathrm{In}_P(a_{p^e})$ is not a p^e-th power.

Take an arbitrary pair h' and (x', y', z') as described in the definition of $H(P)$ above.

We claim that

$$\mathrm{ord}_P(h'|_{Z'}) = \mathrm{ord}_P(h|_{Z'}) \leq \mathrm{ord}_P(h|_Z) = \mathrm{ord}_P(a_{p^e}) < p^e\mu(P),$$

which implies the required inequality.

Let

$$h' = \sum c_B H^B = \sum_{b \in \mathbb{Z}_{\geq 0}} c_b h^b$$

be the power series expansion of h' with respect to the LGS $\mathbb{H} = \{(h, p^e)\}$ and its associated regular system of parameters (x, y, z).

Since

$$h' \equiv z'^{p^e} \equiv c \cdot z^{p^e} \bmod \widehat{\mathfrak{m}_P}^{p^e+1} \quad \text{for some} \quad c \in k^\times,$$

we conclude

$$h' = \sum_{b>0} c_b h^b + c_0 = u \cdot h + c_0$$

for some unit u in $\widehat{\mathcal{O}_{W,P}}$. Moreover, by the formal coefficient lemma, we have $(c_0, p^e) \in \widehat{\mathcal{R}_P}$. This implies that $(\mathbb{M}_u)^{p^e} \mid c_0$ and hence that $\mathrm{ord}_P(c_0|_{Z'}) \geq \mathrm{ord}_P(c_0) \geq p^e\mu(P)$.

Therefore, it suffices to prove

$$\mathrm{ord}_P(h|_{Z'}) \leq \mathrm{ord}_P(h|_Z)\,(= \mathrm{ord}_P(a_{p^e}) < p^e\mu(P))\,.$$

(Then $\mathrm{ord}_P(h|_{Z'}) = \mathrm{ord}_P((u \cdot h)|_{Z'}) = \mathrm{ord}_P((u \cdot h + c_0)|_{Z'}) = \mathrm{ord}_P(h'|_{Z'})$.)

Now by the Weierstrass Preparation Theorem, we have

$$z' = v \cdot (z + w)$$

for some unit v in $\widehat{\mathcal{O}_{W,P}}$ and $w \in k[[x,y]]$. Since $Z' = \{z' = 0\} = \{z + w = 0\}$, by replacing z' with $z + w$, we may assume that z' is of the form

$$z' = z + w, \quad \text{i.e.,} \quad z = z' - w \quad \text{with} \quad w \in k[[x,y]].$$

Plug this into

$$h = z^{p^e} + a_1 z^{p^e-1} + a_2 z^{p^e-2} + \cdots + a_{p^e-1} z + a_{p^e}$$

to obtain

$$h = z'^{p^e} + a'_1 z'^{p^e-1} + a'_2 z'^{p^e-2} + \cdots + a'_{p^e-1} z' + a'_{p^e}$$

with

$$a'_i \in k[[x,y]] \quad \text{for} \quad i = 1, \ldots, p^e - 1, p^e,$$

where

$$a'_{p^e} = (-w)^{p^e} + a_1(-w)^{p^e-1} + a_2(-w)^{p^e-2} + \cdots + a_{p^e-1}(-w) + a_{p^e}.$$

Observe that, since (cf. SITUATION in §5.1)

$$(\mathbb{M}_u)^i \mid a_i \quad \text{for} \quad i = 1, \ldots, p^e - 1,$$

we have

$$\frac{1}{i}\mathrm{ord}_P(a_i) \geq \mathrm{ord}_P(\mathbb{M}_u) = \mu(P) > \frac{1}{p^e}\mathrm{ord}_P(a_{p^e}) \quad \text{for} \quad 1 \leq i < p^e.$$

Case: $\mathrm{ord}_P(w) > \mathrm{ord}_P(a_{p^e})/p^e$.

In this case, we have

$$\mathrm{ord}_P(h|_{Z'}) = \mathrm{ord}_P(a'_{p^e}) = \mathrm{ord}_P(a_{p^e}) = \mathrm{ord}_P(h|_Z).$$

Case: $\mathrm{ord}_P(w) = \mathrm{ord}_P(a_{p^e})/p^e$.

In this case, since $\mathrm{In}_P(a_{p^e})$ is not a p^e-th power, we see via the observation above that

$$\mathrm{In}_P(a'_{p^e}) = \mathrm{In}_P((-w)^{p^e}) + \mathrm{In}_P(a_{p^e}) \neq 0$$

and

$$\mathrm{ord}_P(h|_{Z'}) = \mathrm{ord}_P(a'_{p^e}) = \mathrm{ord}_P(a_{p^e}) = \mathrm{ord}_P(h|_Z).$$

Case: $\mathrm{ord}_P(w) < \mathrm{ord}_P(a_{p^e})/p^e$.

In this case, we have

$$\begin{aligned}\mathrm{ord}_P(h|_{Z'}) &= \mathrm{ord}_P(a'_{p^e}) = \mathrm{ord}_P((-w)^{p^e}) \\ &= p^e \mathrm{ord}_p(w) < \mathrm{ord}_P(a_{p^e}) = \mathrm{ord}_P(h|_Z).\end{aligned}$$

Therefore, in all the cases above, we have

$$\mathrm{ord}_P(h|_{Z'}) \leq \mathrm{ord}_P(h|_Z).$$

This completes the proof of Proposition 5. Q.E.D.

Definition 7 (Invariant "H"). We define the invariant H by the following formula

$$\mathrm{H}(*) := \mathrm{Slope}_{h,(x,y,z)}(*)$$

where h is well-adapted at $* = P, \xi_{H_x}$, or ξ_{H_y} with respect to (x, y, z). By Proposition 5, the invariant H is independent of the choice of h and (x, y, z).

Definition 8 (the *tight* monomial). We define the *tight* monomial $\mathbb{M}_t$ by the formula

$$\mathbb{M}_t = x^{h_x} y^{h_y} \quad \text{where} \quad h_x = \mathrm{H}(\xi_{H_x}), h_y = \mathrm{H}(\xi_{H_y}).$$

Recall that the *usual* monomial $\mathbb{M}_u$ is defined by the formula

$$\mathbb{M}_u = x^{\alpha/a} y^{\beta/a} \quad \text{where} \quad \frac{\alpha}{a} = \mu(\xi_{H_x}), \ \frac{\beta}{a} = \mu(\xi_{H_y}).$$

Note that we have $\mathbb{M}_t \mid \mathbb{M}_u$, that is to say, we have $0 \leq h_x \leq \mu(\xi_{H_x})$ and $0 \leq h_y \leq \mu(\xi_{H_y})$, which follow easily from the definition.

5.3. Description of the procedure (in the case $\tau = 1$)

Analysis of the singular locus $\mathrm{Sing}(\mathcal{R})$ **at** P

First we analyze the singular locus $\mathrm{Sing}(\mathcal{R})$ of the idealistic filtration $\mathcal{R}$ of i.f.g. type at P.

Proposition 6. *We have the following description of the singular locus* $\mathrm{Sing}(\mathcal{R})$ *at* P*, denoted by* $\mathrm{Sing}(\mathcal{R})_P$*, according to the values of* $h_x = \mathrm{H}(\xi_{H_x})$ *and* $h_y = \mathrm{H}(\xi_{H_y})$*:*

$$\mathrm{Sing}(\mathcal{R})_P = \begin{cases} V(z,x) \cup V(z,y) & \textit{if} \quad h_x \geq 1 \quad \textit{and} \quad h_y \geq 1 \\ V(z,x) & \textit{if} \quad h_x \geq 1 \quad \textit{and} \quad h_y < 1 \\ V(z,y) & \textit{if} \quad h_x < 1 \quad \textit{and} \quad h_y \geq 1 \\ V(z,x,y) = P & \textit{if} \quad h_x < 1 \quad \textit{and} \quad h_y < 1, \end{cases}$$

where "V*" denotes the vanishing locus and where* (x,y,z) *is a regular system of parameters at* P *with respect to which* h *is well-adapted simultaneously at* P*,* ξ_{H_x}*, and* ξ_{H_y}*.*

Proof. Note first that, since $(\mathbb{M}, a) = (x^\alpha y^\beta, a) \in \widehat{\mathcal{R}_P}$ with $a \in \mathbb{Z}_{>0}$, we have

$$\mathrm{Sing}(\mathcal{R})_P \subset \{x = 0\} \cup \{y = 0\} = H_x \cup H_y.$$

Then the asserted description is a consequence of the following analysis of $\mathrm{Sing}(\mathcal{R})_P \cap H_x$ (and that of $\mathrm{Sing}(\mathcal{R})_P \cap H_y$, which is identical and hence omitted).

Case: $h_x \geq 1$.

In this case, we have

- $(h, p^e) \in \widehat{\mathcal{R}_P}$ with $h = z^{p^e} + a_1 z^{p^e-1} + \cdots + a_{p^e-1} z + a_{p^e}$ being well-adapted both at P and ξ_{H_x},
- $x \mid a_i$ for $i = 0, \ldots, p^e - 1$, since $\alpha/a \geq h_x \geq 1$ and since $\mathbb{M}_u \mid a_i$ for $i = 0, \ldots, p^e - 1$,
- $x \mid a_{p^e}$, since $h_x \geq 1$,
- $h = 0$ on $\mathrm{Sing}(\mathcal{R})_P$,

which imply

- $z = 0$ on $\mathrm{Sing}(\mathcal{R})_P \cap H_x$.

Therefore, we conclude

$$\mathrm{Sing}(\mathcal{R})_P \cap H_x \subset V(z,x).$$

On the other hand, for $Q \in V(z,x)$, we have

- $\mathrm{ord}_Q(a_i) \geq \mathrm{ord}_Q((\mathbb{M}_u)^i) \geq \alpha/a \cdot i \geq i$ for $i = 0, \ldots, p^e - 1$, since $\mathbb{M}_u \mid a_i$ for $i = 0, \ldots, p^e - 1$ and since $\alpha/a \geq h_x \geq 1$,
- $\mathrm{ord}_Q(a_{p^e}) \geq h_x \cdot p^e \geq p^e$, since $h_x \geq 1$,

which imply

- $\mathrm{ord}_Q(h) \geq p^e$.

Therefore, for any $(f, \lambda) \in \widehat{\mathcal{R}_P}$ with $f = \sum c_b h^b$ being the power series expansion with respect to the LGS $\mathbb{H} = \{(h, p^e)\}$ and its associated regular system of parameters (x, y, z), we have

$$\mathrm{ord}_Q(f) \geq \lambda$$

since

- $\mathrm{ord}_Q(c_b) \geq \mathrm{ord}_Q((\mathbb{M}_u)^{\lambda - p^e \cdot b}) \geq \alpha/a \cdot (\lambda - p^e \cdot b) \geq \lambda - p^e \cdot b$ for b with $\lambda - p^e \cdot b \geq 0$, since $(c_b, \lambda - p^e \cdot b) \in \widehat{\mathcal{R}_P}$, which follows from the formal coefficient lemma.

Therefore, we conclude

$$\mathrm{Sing}(\mathcal{R})_P \cap H_x \supset V(z, x),$$

and hence

$$\mathrm{Sing}(\mathcal{R})_P \cap H_x = V(z, x).$$

Case: $1 > h_x > 0$.

In this case, we have

- $(h, p^e) \in \widehat{\mathcal{R}_P}$ with $h = z^{p^e} + a_1 z^{p^e - 1} + \cdots + a_{p^e - 1} z + a_{p^e}$ being well-adapted both at P and ξ_{H_x},
- $x \mid a_i$ for $i = 0, \ldots, p^e - 1$, since $\alpha/a \geq h_x > 0$ and since $\mathbb{M}_u \mid a_i$ for $i = 0, \ldots, p^e - 1$,
- $x \mid a_{p^e}$, since $h_x > 0$,
- $h = 0$ on $\mathrm{Sing}(\mathcal{R})_P$,

which imply

- $z = 0$ on $\mathrm{Sing}(\mathcal{R})_P \cap H_x$.

Subcase: $h_x = \mu(\xi_x)$. In this subcase, the inequality $1 > h_x$ and the inclusion $(M, a) \in \widehat{\mathcal{R}_P}$ imply

$$\mathrm{Sing}(\mathcal{R})_P \cap H_x \subset V(z, x, y).$$

Subcase: $h_x = \mathrm{ord}_{\xi_x}(a_{p^e})/p^e$. In this subcase, we have $1 > h_x = r/p^e$ where $r = \mathrm{ord}_{\xi_{H_x}}(a_{p^e})$, and hence $a_{p^e} = x^r \cdot \gamma(x, y)$ where $\gamma(x, y)$ is not divisible by x. Therefore, we conclude

$$\mathrm{Sing}(\mathcal{R})_P \cap H_x \subset V(z, x) \cap \{\gamma(x, y) = 0\} = V(z, x, y).$$

Therefore, in both subcases, we have

$$\mathrm{Sing}(\mathcal{R})_P \cap H_x = V(z, x, y) = P.$$

Case: $h_x = 0$.

Subcase: $\alpha/a = \mu(\xi_{H_x}) = 0$. In this subcase, we have $\beta/a = \mu(\xi_{H_y}) \geq 1$. Therefore, we conclude

$\mathrm{Sing}(\mathcal{R})_P \cap H_x \subset V(z, x, y)$ and hence $\mathrm{Sing}(\mathcal{R})_P \cap H_x = V(z, x, y) = P$.

Subcase: $\alpha/a = \mu(\xi_{H_x}) > 0$. In this subcase, we have

- $x \mid a_i$ for $i = 0, \ldots, p^e - 1$, since $\alpha/a > 0$ and since $\mathbb{M}_u \mid a_i$ for $i = 0, \ldots, p^e - 1$.

Set $a_{p^e} = g(y) + x \cdot \omega(x, y)$. Then $g(y) \neq 0$ and $g(y) = \mathrm{In}_{\xi_{H_x}}(a_{p^e})$ is not a p^e-th power, since h is well-adapted at ξ_{H_x}. We also observe that, for $Q \in \mathrm{Sing}(\mathcal{R})_P \cap H_x$, we have

$$p^e \leq \mathrm{ord}_Q(h) \leq \mathrm{ord}_Q(h|_{H_x}) = \mathrm{ord}_Q(z^{p^e} + g(y)).$$

Therefore, there exists $e' < e$ such that

$$\frac{\partial^{p^{e'}}}{\partial y^{p^{e'}}}(z^{p^e} + g(y)) = \frac{\partial^{p^{e'}}}{\partial y^{p^{e'}}} g(y) \neq 0$$

and that

$$\mathrm{ord}_Q\left(\frac{\partial^{p^{e'}}}{\partial y^{p^{e'}}} g(y)\right) = \mathrm{ord}_Q\left(\frac{\partial^{p^{e'}}}{\partial y^{p^{e'}}}(z^{p^e} + g(y))\right) \geq p^e - p^{e'} > 0.$$

This implies $y = 0$ at Q. Therefore, we conclude

$$\mathrm{Sing}(\mathcal{R})_P \cap H_x \subset V(z, x, y), \text{ and hence } \mathrm{Sing}(\mathcal{R})_P \cap H_x = V(z, x, y) = P.$$

This completes the proof of Proposition 6. Q.E.D.

Description of the procedure for resolution of singularities

Now based upon the analysis of the support $\mathrm{Sing}(\mathcal{R})_P$, we give the following (local) description of the procedure (around the point P) for resolution of singularities in the monomial case with $\tau = 1$:

Step 1. Check if $\dim \mathrm{Sing}(\mathcal{R})_P = 1$.

If the answer is yes, then blow up the 1-dimensional components one by one. If there are two 1-dimensional components meeting at P, then we blow up the one associated to the boundary divisor with bigger H first. If the boundary divisors associated to the two 1-dimensional components have the same H, then we blow up first the one associated to the boundary divisor created later in the history. Since the invariant H strictly decreases under this procedure, this step comes to an end after finitely many times with the dimension of the singular locus dropping to 0.

If the answer is no, i.e., $\dim \mathrm{Sing}(\mathcal{R})_P = 0$, then go to Step 2.

Step 2. Once $\dim \mathrm{Sing}(\mathcal{R})_P = 0$, blow up the isolated point in the singular locus. Then go back to Step 1.

Repeat these steps.

(We note that, as long as the value of the invariant σ remains the same, we stay in the monomial case during the procedure above. We also note that in the middle of the procedure the invariant σ may drop. If the latter happens, we are no longer in the monomial case. In this case, we go through the mechanism described in §4.2 with the reduced new value of σ to reach the new monomial case.)

What remains to be shown is that the above procedure terminates after finitely many repetitions. This termination of the procedure is the main subject of §5.4.

5.4. Termination of the procedure (in the case $\tau = 1$)

Notion of a good/bad point (resp. hypersurface)

In order to analyze the termination of the procedure, we introduce the following notion of the point P being "good or bad" and the boundary divisor H_x (or H_y) being "good or bad".

Definition 9 ("good/bad" point (cf. [5])). We say P is a *good* (resp. a *bad*) point if $\mu(P) - \mathrm{H}(P) = 0$ (resp. > 0). Similarly, we say H_x is a *good* (resp. *bad*) hypersurface if $\mu(\xi_{H_x}) - \mathrm{H}(\xi_{H_x}) = 0$ (resp. > 0), where ξ_{H_x} is the generic point of the hypersurface H_x.

The notion of H_y being a good or bad hypersurface is defined in an identical manner.

Lemma 4. *Let $W \overset{\pi}{\leftarrow} \widetilde{W}$ be the blow up with center P, E_P the exceptional divisor, $\widetilde{\mathcal{R}}$ the transformation of the idealistic filtration of i.f.g. type $\mathcal{R}$. Then P is a good (resp. bad) point if and only if E_P is a good (resp. bad) hypersurface.*

Proof. Take h and a regular system of parameters as described in SITUATION in §5.1. We may further assume that h is well-adapted at P with respect to the regular system of parameters (x, y, z) (cf. Proposition 5). Take a point $\widetilde{P} \in \pi^{-1}(P) \cap \mathrm{Supp}(\widetilde{\mathcal{R}}) \subset \widetilde{W}$. Since the singular locus is empty over the z-chart, $\widetilde{P}$ should be either in the x-chart or in the y-chart. Say, $\widetilde{P}$ is in the x-chart with a regular system of parameters $(\widetilde{x}, \widetilde{y}, \widetilde{z}) = (x, y/x - c, z/x)$ for some $c \in k$. We also assume that the invariant σ stays the same, and hence (cf. Proposition 4) that $(\widetilde{h}, p^e)$ is the unique element in the LGS at $\widetilde{P}$ where

$$\widetilde{h} = \frac{\pi^*(h)}{x^{p^e}} = \widetilde{z}^{p^e} + \widetilde{a_1}\widetilde{z}^{p^e - 1} + \cdots + \widetilde{a_{p^e}} \quad \text{with} \quad \widetilde{a_i} = \frac{\pi^*(a_i)}{x^i} \in k[[\widetilde{x}, \widetilde{y}]].$$

We compute

$$\begin{aligned}\mathrm{Slope}_{\widetilde{h},(\widetilde{x},\widetilde{y},\widetilde{z})}(\xi_{E_P}) &= \min\left\{\frac{1}{p^e}\mathrm{ord}_{\xi_{E_P}}(\widetilde{a_{p^e}}),\ \mu(\xi_{E_P})\right\}\\ &= \min\left\{\frac{1}{p^e}\mathrm{ord}_P(a_{p^e})-1,\ \mu(P)-1\right\} = \min\left\{\frac{1}{p^e}\mathrm{ord}_P(a_{p^e}),\ \mu(P)\right\}-1\\ &= \mathrm{H}(P)-1.\end{aligned}$$

Case: *P is a good point, i.e.,* $\mathrm{H}(P)=\mu(P)$.

In this case, we have

$$\mathrm{Slope}_{\widetilde{h},(\widetilde{x},\widetilde{y},\widetilde{z})}(\xi_{E_P}) = \mathrm{H}(P)-1 = \mu(P)-1 = \mu(\xi_{E_P}).$$

Therefore, we conclude that E_P is a good hypersurface (with $\widetilde{h}$ well-adapted at ξ_{E_P} with respect to $(\widetilde{x},\widetilde{y},\widetilde{z})$ and $\mathrm{H}(\xi_{E_P})=\mu(\xi_{E_P})$).

Case: *P is a bad point, i.e.,* $\mathrm{H}(P)<\mu(P)$.

In this case, we have

$$\mathrm{Slope}_{\widetilde{h},(\widetilde{x},\widetilde{y},\widetilde{z})}(\xi_{E_P}) = \mathrm{H}(P)-1 < \mu(P)-1 = \mu(\xi_{E_P}).$$

Therefore, in order to see that E_P is a bad hypersurface, i.e., $\mathrm{H}(\xi_{E_P})<\mu(\xi_{E_P})$, we have only to show that $\widetilde{h}$ is well-adapted at ξ_{E_P} with respect to $(\widetilde{x},\widetilde{y},\widetilde{z})$, i.e., $\mathrm{In}_{\xi_{E_P}}(\widetilde{a_{p^e}})$ is not a p^e-th power, and hence that $\mathrm{Slope}_{\widetilde{h},(\widetilde{x},\widetilde{y},\widetilde{z})}(\xi_{E_P}) = H(\xi_{E_P})$.

Set $a_{p^e}=\sum_{k+l\geq d}c_{kl}x^ky^l$ where $d=\mathrm{ord}_P(a_{p^e})$. Then

$$\begin{aligned}\pi^*(a_{p^e}) &= \sum_{k+l\geq d}c_{kl}x^{k+l}\left(\frac{y}{x}\right)^l = x^d\left\{\sum_{k+l=d}c_{kl}\left(\frac{y}{x}\right)^l + x\cdot\Omega\left(x,\frac{y}{x}\right)\right\}\\ &= x^d\left\{\phi\left(\frac{y}{x}\right)+x\cdot\Omega\left(x,\frac{y}{x}\right)\right\},\end{aligned}$$

where $\phi(T)=\sum_{k+l=d}c_{kl}T^l$. Hence, we have

$$\begin{aligned}\widetilde{a_{p^e}} &= \frac{\pi^*(a_{p^e})}{x^{p^e}} = x^{d-p^e}\left\{\phi\left(\frac{y}{x}-c+c\right)+x\cdot\Omega\left(x,\frac{y}{x}-c+c\right)\right\}\\ &= \widetilde{x}^{d-p^e}\left\{\phi(\widetilde{y}+c)+\widetilde{x}\cdot\Omega(\widetilde{x},\widetilde{y}+c)\right\}.\end{aligned}$$

Therefore, we conclude

$$\begin{aligned} \mathrm{In}_P(a_{p^e}) = \sum_{k+l=d} c_{kl}x^k y^l = x^d \phi\left(\frac{y}{x}\right) &\text{ is a } p^e\text{-th power} \\ \Longleftrightarrow p^e \mid d \text{ and } \phi(T) &\text{ is a } p^e\text{-th power} \\ \Longleftrightarrow \mathrm{In}_{\xi_{E_P}}(\widetilde{a_{p^e}}) = \widetilde{x}^{d-p^e}\phi(\widetilde{y}+c) &\text{ is a } p^e\text{-th power.} \end{aligned}$$

Now since P is a bad point in this case, $\mathrm{In}_P(a_{p^e})$ is not a p^e-th power. Therefore, by the above equivalence, $\mathrm{In}_{\xi_{E_P}}(\widetilde{a_{p^e}})$ is not a p^e-th power, either.

This completes the proof of Lemma 4. Q.E.D.

Remark 7.

(1) The above proof is slightly sloppy in the sense that, after blow up, we may end up having $\mathrm{ord}_{\widetilde{P}}(\widetilde{a_{p^e}}) = p^e$ and $\mathrm{In}_{\widetilde{P}}(\widetilde{a_{p^e}}) = c\widetilde{x}^{p^e}$ for some $c \in k \setminus \{0\}$, even under the condition $\widetilde{P} \in \pi^{-1}(P) \cap \mathrm{Supp}(\widetilde{\mathcal{R}}) \subset \widetilde{W}$ and the assumption that the invariant σ stays the same (cf. the proof of Proposition 4 (1)). Then we would have to replace $(\widetilde{x}, \widetilde{y}, \widetilde{z})$ with $(\widetilde{x}, \widetilde{y}, \widetilde{z}' = \widetilde{z} + c^{1/p^e}\widetilde{x})$ to guarantee condition (1) in $\boxed{\text{SITUATION}}$ in §5.1. Accordingly, we have to analyze $\widetilde{a_{p^e}}'$. It is straightforward, however, to see that the same statement holds for $\widetilde{a_{p^e}}'$. The details are left to the reader as an exercise.

(2) As will be clear in the presentation that follows, especially in the way we classify the configurations and we define the new invariant "$\mathrm{inv}_{\mathrm{MON}}$", the focus of our proof centers around the analysis looking at whether the hypersurface of our concern is good/bad. The notion of a point being good/bad, though related to our analysis via Lemma 4, is somewhat auxiliary.

Configurations

Looking at the boundary divisor(s) in E_{young} at the point $P \in \mathrm{Sing}(\mathcal{R})$ and seeing whether they are good or bad, we come up with the following classification of the "configurations". Note that the pictures depict the configurations in a 2-dimensional manner, taking the intersection with the hypersurface $Z = \{z = 0\}$.

① The point P is only on one boundary divisor (in E_{young}), say H_x, which is good.

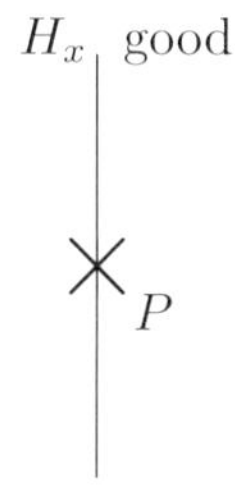

② The point P is at the intersection of two boundary divisors (in E_{young}), both of which are good.

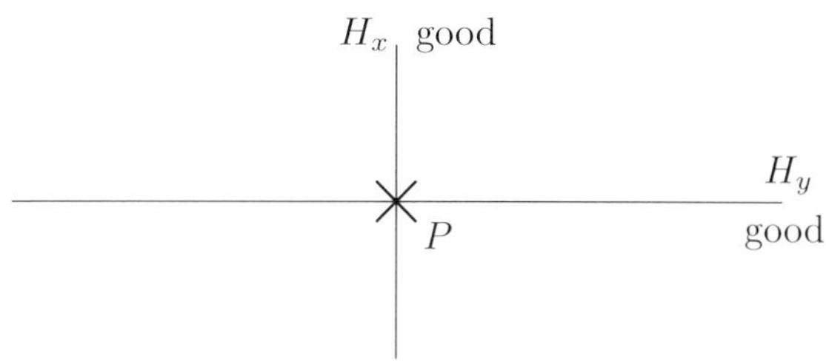

③ The point P is only on one boundary divisor (in E_{young}), say H_x, which is bad.

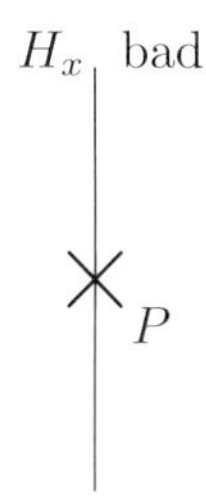

④ The point P is at the intersection of two boundary divisors (in E_{young}), one of which, say, H_x, is bad, while the other, say H_y, is good.

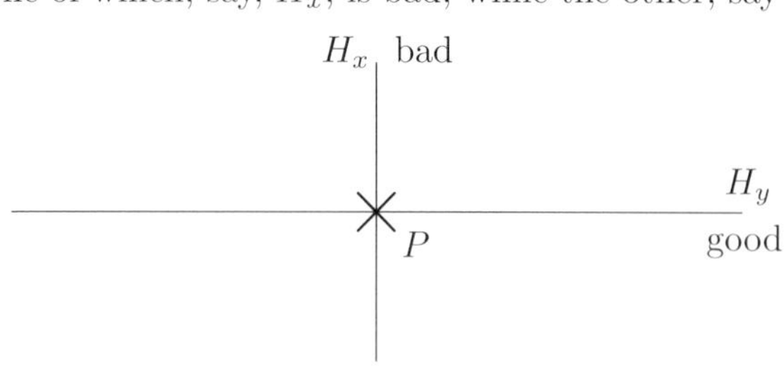

⑤ The point P is at the intersection of two boundary divisors (in E_{young}), say H_x and H_y, both of which are bad.

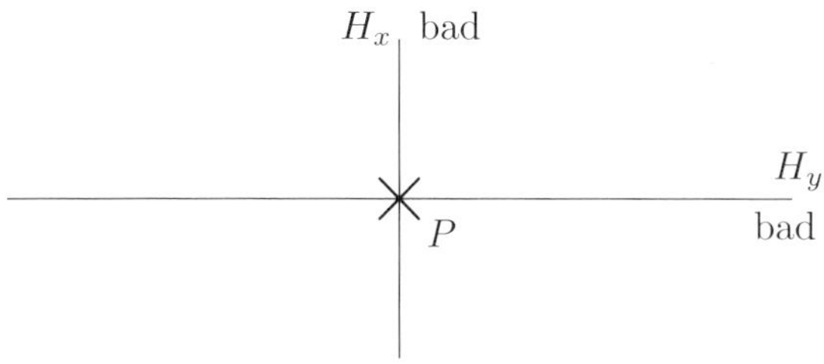

Basic strategy to show termination of the procedure

After the blow up $W \overset{\pi}{\leftarrow} \widetilde{W}$ specified by the procedure described in §5.3, we show that, at $\widetilde{P} \in \pi^{-1}(P) \in \widetilde{W}$, one of the following holds:

- $\widetilde{P} \notin \text{Sing}(\widetilde{R})$, i.e., $\text{Sing}(\widetilde{R}) = \emptyset$ in a neighborhood of $\widetilde{P}$,
- $\widetilde{P} \in \text{Sing}(\widetilde{R})$ and the invariant σ drops, or
- $\widetilde{P} \in \text{Sing}(\widetilde{R})$ and the invariant "$\text{inv}_{\text{MON}}(P)$" strictly decreases, i.e., $\text{inv}_{\text{MON}}(P) > \text{inv}_{\text{MON}}(\widetilde{P})$. We note that "$\text{inv}_{\text{MON}}(P)$" is a new invariant, which we introduce below, attached to the point $P \in \text{Sing}(\mathcal{R})$ in any one of the configurations ① through ⑤.

Since the invariant "inv_{MON}" cannot decrease infinitely many times, the procedure described in §5.3 terminates either with $\widetilde{P} \notin \text{Supp}(\widetilde{R})$ or with the drop of the invariant σ after finitely many repetitions.

This completes the description of the basic strategy.

Definition 10 (Invariant "inv_{MON}"). We define the invariant "$\text{inv}_{\text{MON}}(P)$" associated to a point $P \in \text{Sing}(\widetilde{R})$ in each of the configurations ① through ⑤ (cf. **Configurations**) as follows (Note that "MON" is short for "MONOMIAL".):

$$\text{inv}_{\text{MON}}(P) = \begin{cases} (0, 0, \mu_x) & \text{in configuration ①,} \\ (0, 0, \min\{\mu_x, \mu_y\}, \max\{\mu_x, \mu_y\}) & \text{in configuration ②,} \\ (\rho_x, 0, \mu_x) & \text{in configuration ③,} \\ (\min\{\rho_x, \mu_x\}, \max\{\rho_x, \mu_x\}) & \text{in configuration ④,} \\ (\min\{\rho_x, \rho_y\}, \max\{\rho_x, \rho_y\}) & \text{in configuration ⑤,} \end{cases}$$

where $\mu_x = \mu(\xi_{H_x}), \mu_y = \mu(\xi_{H_y})$ and the invariant ρ is defined as below to determine ρ_x, ρ_y.

Proposition 7. *Let P be the point in configuration ③, ④, or ⑤, and let H_x be a bad boundary divisor (in E_{young}). Suppose h is well-adapted at ξ_{H_x} with respect to (x, y, z) (satisfying the conditions as described in* $\boxed{SITUATION}$ *in §5.1 at the same time). Write*

$$a_{p^e} = x^r \{g(y) + x \cdot \omega(x, y)\} \quad \textit{where} \quad r = \mathrm{ord}_{\xi_{H_x}}(a_{p^e}).$$

Set

$$\begin{aligned}\rho_{h,(x,y,z),H_x}(P) &= \frac{1}{p^e}\{\text{res-ord}_P^{(p^e)}(\mathrm{In}_{\xi_{H_x}}(a_{p^e})) - \mathrm{ord}_{\xi_{H_x}}(a_{p^e})\} \\ &= \frac{1}{p^e}\{\text{res-ord}_P^{(p^e)}(x^r g(y)) - r\},\end{aligned}$$

i.e.,

$$\rho_{h,(x,y,z),H_x}(P) = \begin{cases} \mathrm{ord}_P(g(y))/p^e & \textit{in case } r \not\equiv 0 \textit{ mod } p^e \\ \text{res-ord}_P^{(p^e)}(g(y))/p^e & \textit{in case } r \equiv 0 \textit{ mod } p^e, \end{cases}$$

where res-ord$_P^{(p^e)}$ *is the smallest degree of the term which appears with a nonzero coefficient and which is not a p^e-th power.*

Then $\rho_{h,(x,y,z),H_x}(P)$ is independent of the choice of h and (x, y, z).

Proof. Note first that $r = \mathrm{H}(\xi_{H_x}) \cdot p^e$ is independent of the choice of h and (x, y, z) by Proposition 5 (2).

We set

$$\rho_x = \max\left\{ \frac{1}{p^e}\mathrm{ord}_P\left(\left\{(h'|_{Z'}) \cdot x'^{-r}\right\}|_{Z' \cap H_{x'}}\right) \;\middle|\; \begin{array}{c} h', (x', y', z'), \\ Z' = \{z' = 0\}, \\ H_{x'} = \{x' = 0\} \\ = H_x \end{array} \right\},$$

where, computing the above "max", we let h' and (x', y', z') vary among all such pairs consisting of the unique element in an LGS and a regular system of parameters that satisfy the condition

$$\begin{cases} h' & \equiv \quad z'^{p^e} \bmod \widehat{\mathfrak{m}_P}^{p^e+1}, \text{ and} \\ \mathrm{ord}_{\xi_{H_{x'}}}(h'|_{Z'}) & = \quad \mathrm{ord}_{\xi_{H_x}}(h'|_{Z'}) = r. \end{cases}$$

It suffices to show that

$$\rho_{h,(x,y,z),H_x}(P) = \rho_x,$$

as the number ρ_x is obviously independent of the choice of h and (x, y, z).

Firstly we claim that the inequality

$$\rho_{h,(x,y,z),H_x}(P) \leq \rho_x$$

holds. Note that h and (x,y,z) form such a pair satisfying the above condition described in the definition of ρ_x computing the "max", since h is well-adapted at ξ_{H_x} with respect to (x,y,z).

Case: $r \not\equiv 0 \bmod p^e$.

In this case, we have

$$\rho_{h,(x,y,z),H_x}(P) = \frac{1}{p^e}\mathrm{ord}_P(g(y)) = \frac{1}{p^e}\mathrm{ord}_P\left(\{(h|_Z)\cdot x^{-r}\}|_{Z\cap H_x}\right) \leq \rho_x$$

by the definition of ρ_x above taking the "max" among all such pairs.

Case: $r \equiv 0 \bmod p^e$.

In this case, we modify z in the following way.

Set $g(y) = \sum_{n\in\mathbb{Z}_{\geq 0}} b_n y^n$ with $b_n \in k$. We have

$$x^r\left\{\sum_{n<\text{res-ord}_P^{(p^e)}(g(y))} b_n y^n\right\} = w^{p^e} \quad \text{for some} \quad w \in k[x,y],$$

since L.H.S. is a p^e-th power by the case assumption $r \equiv 0 \bmod p^e$ and by the definition of $\text{res-ord}_P^{(p^e)}$. Then set $z' = z + w$, i.e., $z = z' - w$.

Plug this into

$$h = z^{p^e} + a_1 z^{p^e-1} + a_2 z^{p^e-2} + \cdots + a_{p^e-1}z + a_{p^e}$$

to obtain

$$h = z'^{p^e} + a'_1 z'^{p^e-1} + a'_2 z'^{p^e-2} + \cdots + a'_{p^e-1}z' + a'_{p^e}$$

with $a'_i \in k[[x,y]]$ for $i = 1,\ldots,p^e-1,p^e$, where

$$a'_{p^e} = (-w)^{p^e} + a_1(-w)^{p^e-1} + a_2(-w)^{p^e-2} + \cdots + a_{p^e}.$$

Observe that, since (cf. SITUATION in §5.1)

$$(\mathbb{M}_u)^i \mid a_i \quad \text{for} \quad i = 1,\ldots,p^e-1,$$

we have

$$\frac{1}{i}\mathrm{ord}_{\xi_{H_x}}(a_i) \geq \mathrm{ord}_{\xi_{H_x}}(\mathbb{M}_u) = \mu(\xi_{H_x}) > \frac{1}{p^e}\mathrm{ord}_{\xi_{H_x}}(a_{p^e}) = \frac{r}{p^e}$$

for $i = 1, \ldots, p^e - 1$, where the second strict inequality follows from the assumption that H_x is a bad boundary divisor.

Hence, we observe that a'_{p^e} is of the following form

$$a'_{p^e} = x^r \{g'(y) + x \cdot \omega'(x, y)\}$$

where

$$g'(y) = g(y) - w^{p^e} \cdot x^{-r} = \sum_{n \geq \text{res-ord}_P^{(p^e)}(g(y))} b_n y^n.$$

Therefore, we conclude that

$$\begin{aligned}\rho_{h,(x,y,z),H_x}(P) &= \frac{1}{p^e}\text{res-ord}_P^{(p^e)}(g(y)) = \frac{1}{p^e}\text{ord}_P(g'(y)) \\ &= \frac{1}{p^e}\text{ord}_P\left(\left\{(h|_{Z'}) \cdot x^{-r}\right\}|_{Z' \cap H_x}\right) \leq \rho_x.\end{aligned}$$

Secondly we prove the inequality in the opposite direction holds, i.e.,

$$\rho_{h,(x,y,z),H_x}(P) \geq \rho_x.$$

Take an arbitrary pair of h' and (x', y', z') satisfying the conditions described in the definition of ρ_x computing the "max".

We claim that

$$\rho_{h,(x,y,z),H_x}(P) \geq \frac{1}{p^e}\text{ord}_P\left(\left\{(h'|_{Z'}) \cdot x'^{-r}\right\}|_{Z' \cap H_{x'}}\right),$$

which implies the required inequality.

Let

$$h' = \sum c_B H^B = \sum_{b \in \mathbb{Z}_{\geq 0}} c_b h^b$$

be the power series expansion of h' with respect to the LGS $\mathbb{H} = \{(h, p^e)\}$ and its associated regular system of parameters (x, y, z).

Since

$$h' \equiv z'^{p^e} \equiv c \cdot z^{p^e} \bmod \widehat{\mathfrak{m}_P}^{p^e+1} \quad \text{for some} \quad c \in k^\times,$$

we conclude

$$h' = \sum_{b>0} c_b h^b + c_0 = u \cdot h + c_0$$

for some unit u in $\widehat{\mathcal{O}_{W,P}}$.

Moreover, by the formal coefficient lemma, we have $(c_0, p^e) \in \widehat{\mathcal{R}_P}$. This implies that $(\mathbb{M}_u)^{p^e} \mid c_0$, and hence that

$$\text{ord}_{\xi_{H_x}}(c_0|_{Z'}) \geq \text{ord}_{\xi_{H_x}}(c_0) \geq p^e \mu(\xi_{H_x}) > r.$$

Hence we have

$$\left\{(c_0|_{Z'}) \cdot x'^{-r}\right\}|_{Z'\cap H_{x'}} = \{(c_0|_{Z'}) \cdot x^{-r}\}|_{Z'\cap H_x} = 0,$$

which implies

$$\begin{aligned}\mathrm{ord}_P\Big(\left\{(h'|_{Z'}) \cdot x'^{-r}\right\}|_{Z'\cap H_{x'}}\Big) &= \mathrm{ord}_P\Big(\left\{((u\cdot h + c_0)|_{Z'}) \cdot x'^{-r}\right\}|_{Z'\cap H_{x'}}\Big)\\ = \mathrm{ord}_P\Big(\left\{(u \cdot h|_{Z'}) \cdot x'^{-r}\right\}|_{Z'\cap H_{x'}}\Big) &= \mathrm{ord}_P\Big(\left\{(h|_{Z'}) \cdot x'^{-r}\right\}|_{Z'\cap H_{x'}}\Big)\\ = \mathrm{ord}_P\left(\{(h|_{Z'}) \cdot x^{-r}\}|_{Z'\cap H_x}\right).\end{aligned}$$

Therefore, it suffices to prove

$$\rho_{h,(x,y,z),H_x}(P) \geq \frac{1}{p^e}\mathrm{ord}_P\left(\{(h|_{Z'}) \cdot x^{-r}\}|_{Z'\cap H_x}\right).$$

Now by the Weierstrass Preparation Theorem, we have

$$z' = v \cdot (z + w)$$

for some unit v in $\widehat{\mathcal{O}_{W,P}}$ and $w \in k[[x, y]]$. Since $Z' = \{z' = 0\} = \{z + w = 0\}$, by replacing z' with $z + w$, we may assume that z' is of the form

$$z' = z + w, \quad \text{i.e.,} \quad z = z' - w \quad \text{with} \quad w \in k[[x, y]].$$

Plug this into

$$h = z^{p^e} + a_1 z^{p^e-1} + a_2 z^{p^e-2} + \cdots + a_{p^e-1}z + a_{p^e}$$

to obtain

$$h = z'^{p^e} + a'_1 z'^{p^e-1} + a'_2 z'^{p^e-2} + \cdots + a'_{p^e-1}z' + a'_{p^e}$$

with $a'_i \in k[[x, y]]$ for $i = 1, \ldots, p^e - 1, p^e$, where

$$a'_{p^e} = (-w)^{p^e} + a_1(-w)^{p^e-1} + a_2(-w)^{p^e-2} + \cdots + a_{p^e}.$$

Observe that, since (cf. SITUATION in §5.1)

$$(\mathbb{M}_u)^i \mid a_i \quad \text{for} \quad i = 1, \ldots, p^e - 1,$$

we have

$$\frac{1}{i}\mathrm{ord}_{\xi_{H_x}}(a_i) \geq \mathrm{ord}_{\xi_{H_x}}(\mathbb{M}_u) = \mu(\xi_{H_x}) > \frac{1}{p^e}\mathrm{ord}_{\xi_{H_x}}(a_{p^e}) = \frac{r}{p^e}$$

for $i = 1, \ldots, p^e - 1$. We claim that

$$\mathrm{ord}_{\xi_{H_x}}(w) \geq \frac{r}{p^e}.$$

In fact, suppose $\mathrm{ord}_{\xi_{H_x}}(w) < r/p^e$. Then using the above observation we would have

$$\mathrm{ord}_{\xi_{H_x}}(h|_{Z'}) = \mathrm{ord}_{\xi_{H_x}}(a'_{p^e}) = \mathrm{ord}_{\xi_{H_x}}((-w)^{p^e}) < r.$$

By the equation $h' = u \cdot h + c_0$ and by the inequality $\mathrm{ord}_{\xi_{H_x}}(c_0|_{Z'}) > r$, this would also imply

$$\mathrm{ord}_{\xi_{H_x}}(h'|_{Z'}) = \mathrm{ord}_{\xi_{H_x}}(h|_{Z'}) < r,$$

which is against the choice of h' and (x', y', z') that we started with, satisfying the conditions described in the definition of ρ_x computing the "max".

Now we are at the stage to finish the argument to prove the inequality

$$\rho_{h,(x,y,z),H_x}(P) \geq \frac{1}{p^e}\mathrm{ord}_P\left(\left\{(h|_{Z'}) \cdot x^{-r}\right\}|_{Z' \cap H_x}\right).$$

Case: $r \not\equiv 0 \mod p^e$.

In this case, the claimed inequality $\mathrm{ord}_{\xi_{H_x}}(w) \geq r/p^e$ implies the strict inequality $\mathrm{ord}_{\xi_{H_x}}(w) > r/p^e$, since $\mathrm{ord}_{\xi_{H_x}}(w)$ is an integer. Together with the observation, we conclude that

$$\mathrm{In}_{\xi_{H_x}}(a'_{p^e}) = \mathrm{In}_{\xi_{H_x}}(a_{p^e}) = x^r g(y),$$

and hence that

$$\rho_{h,(x,y,z),H_x}(P) = \frac{1}{p^e}\mathrm{ord}_P(g(y)) = \frac{1}{p^e}\mathrm{ord}_P\left(\left\{(h|_{Z'}) \cdot x^{-r}\right\}|_{Z' \cap H_x}\right).$$

Case: $r \equiv 0 \mod p^e$.

In this case, set

$$w = x^{r/p^e}\{h(y) + x \cdot \theta(x,y)\} \quad \text{with} \quad h(y) \in k[[y]], \theta(x,y) \in k[[x,y]].$$

Together with the observation, we conclude that

$$\mathrm{In}_{\xi_{H_x}}(a'_{p^e}) = \mathrm{In}_{\xi_{H_x}}(a_{p^e}) - x^r h(y)^{p^e} = x^r\left\{g(y) - h(y)^{p^e}\right\},$$

and hence that

$$\begin{aligned}\rho_{h,(x,y,z),H_x}(P) &= \frac{1}{p^e}\text{res-ord}_P^{(p^e)}(g(y)) \geq \frac{1}{p^e}\mathrm{ord}_P\left(g(y) - h(y)^{p^e}\right) \\ &= \frac{1}{p^e}\mathrm{ord}_P\left(\left\{(h|_{Z'}) \cdot x^{-r}\right\}|_{Z' \cap H_x}\right).\end{aligned}$$

This completes the proof of Proposition 7. Q.E.D.

Definition 11 (Invariant "ρ"). Let the situation be as described in Proposition 7. We define the invariant ρ of H_x at P, denoted by ρ_x, by the formula

$$\rho_x = \rho_{h,(x,y,z),H_x}(P).$$

Invariant ρ_y is defined in an identical manner, in case H_y is a bad boundary divisor (in E_{young}) passing through P in configuration ⑤.

Study of the behavior of the invariants under blow up

Case: blow up with 1-dimensional center

Claim 1. *Let $W \stackrel{\pi}{\leftarrow} \widetilde{W}$ be the blow up with center being a 1-dimensional component $C = V(z,x)$ of* $\operatorname{Sing}(\mathcal{R})$ *(cf. Proposition 6), and $\widetilde{\mathcal{R}}$ the transformation of the idealistic filtration $\mathcal{R}$ of i.f.g. type. Then there is possibly only one point $\widetilde{P} \in \operatorname{Sing}(\widetilde{\mathcal{R}}) \cap \pi^{-1}(P) \subset \widetilde{W}$, lying in the x-chart, with the regular system of parameters $(\widetilde{x},\widetilde{y},\widetilde{z}) = (x,y,z/x)$. The behavior of the invariants under blow up is described as follows (in case the invariant σ does not drop):*

$$\begin{array}{llll} \mu_{\widetilde{x}} = \mu_x - 1, & h_{\widetilde{x}} = h_x - 1, & \textit{and, in case } H_x \textit{ is bad,} & \rho_{\widetilde{x}} = \rho_x. \\ \mu_{\widetilde{y}} = \mu_y, & h_{\widetilde{y}} = h_y, & \textit{and, in case } H_y \textit{ is bad,} & \rho_{\widetilde{y}} = \rho_y - 1. \end{array}$$

Proof. Write down

$$h = z^{p^e} + a_1 z^{p^e-1} + a_2 z^{p^e-2} + \cdots + a_{p^e-1} z + a_{p^e}$$

with

$$a_i \in k[[x,y]] \quad \text{and} \quad \operatorname{ord}_P(a_i) > i \quad \text{for} \quad i = 1,\ldots,p^e,$$

as described in | SITUATION | in §5.1. We may assume further that h is well-adapted at P, ξ_{H_x} and ξ_{H_y} simultaneously with respect to (x,y,z) (cf. Proposition 5).

It is straightforward to see that, if a point $\widetilde{P} \in \pi^{-1}(P) \subset \widetilde{W}$ lies in the z-chart, then $\operatorname{ord}_{\widetilde{P}}(\widetilde{h}) < p^e$, where $\widetilde{h} = \pi^*(h)/z^{p^e}$, and hence $\widetilde{P} \notin \operatorname{Sing}(\widetilde{\mathcal{R}})$. Therefore, there is possibly only one point $\widetilde{P} \in \operatorname{Sing}(\widetilde{\mathcal{R}}) \cap \pi^{-1}(P) \subset \widetilde{W}$, lying in the x-chart, with the regular system of parameters $(\widetilde{x},\widetilde{y},\widetilde{z}) = (x,y,z/x)$.

With respect to this regular system of parameters, we compute the transform of the monomial $(x^\alpha y^\beta, a)$ to be $(\widetilde{x}^{\alpha-a}\widetilde{y}^\beta, a)$. Therefore, we conclude

$$\mu_{\widetilde{x}} = \mu_x - 1, \quad \mu_{\widetilde{y}} = \mu_y.$$

Set

$$a_{p^e} = x^r \{g(y) + x \cdot \omega(x,y)\}$$

with

$$r = \mathrm{ord}_{\xi_{H_x}}(a_{p^e}), \quad 0 \neq g(y) \in k[[y]], \quad \text{and} \quad \omega(x,y) \in k[[x,y]].$$

After blow up, we compute

$$\widetilde{h} = \frac{\pi^*(h)}{x^{p^e}} = \widetilde{z}^{p^e} + \widetilde{a_1}\widetilde{z}^{p^e-1} + \widetilde{a_2}\widetilde{z}^{p^e-2} + \cdots + \widetilde{a_{p^e-1}}\widetilde{z} + \widetilde{a_{p^e}}$$

with

$$\widetilde{a_i} = \frac{\pi^*(a_i)}{x^i} \quad \text{for} \quad i = 1, \ldots, p^e.$$

In particular, we have

$$\widetilde{a_{p^e}} = \frac{\pi^*(a_{p^e})}{x^{p^e}} = x^{r-p^e}\{g(y) + x \cdot \omega(x,y)\} = \widetilde{x}^{r-p^e}\{g(\widetilde{y}) + \widetilde{x} \cdot \omega(\widetilde{x},\widetilde{y})\}.$$

(Note that we have $r \geq p^e$, since $h_x = r/p^e \geq 1$ (cf. Proposition 6).) Therefore, we compute

$$\begin{aligned}
\mathrm{Slope}_{\widetilde{h},(\widetilde{x},\widetilde{y},\widetilde{z})}(\xi_{H_{\widetilde{x}}}) &= \min\left\{\frac{1}{p^e}\mathrm{ord}_{\xi_{H_{\widetilde{x}}}}(\widetilde{a_{p^e}}),\ \mu(\xi_{H_{\widetilde{x}}})\right\} \\
&= \min\left\{\frac{r-p^e}{p^e}, \mu_{\widetilde{x}}\right\} = \min\left\{\frac{1}{p^e}\mathrm{ord}_{\xi_{H_x}}(a_{p^e}) - 1, \mu_x - 1\right\} \\
&= \min\left\{\frac{1}{p^e}\mathrm{ord}_{\xi_{H_x}}(a_{p^e}), \mu_x\right\} - 1 = \mathrm{Slope}_{h,(x,y,z)}(\xi_{H_x}) - 1.
\end{aligned}$$

Hence, if $\mathrm{Slope}_{h,(x,y,z)}(\xi_{H_x}) = \mu(\xi_{H_x})$, then

$$\mathrm{Slope}_{\widetilde{h},(\widetilde{x},\widetilde{y},\widetilde{z})}(\xi_{H_{\widetilde{x}}}) = \mu(\xi_{H_x}) - 1 = \mu(\xi_{H_{\widetilde{x}}}).$$

If $\mathrm{Slope}_{h,(x,y,z)}(\xi_{H_x}) < \mu(\xi_{H_x})$, then

$$\mathrm{Slope}_{\widetilde{h},(\widetilde{x},\widetilde{y},\widetilde{z})}(\xi_{H_{\widetilde{x}}}) = \mathrm{Slope}_{h,(x,y,z)}(\xi_{H_x}) - 1 < \mu(\xi_{H_{\widetilde{x}}}) - 1 = \mu(\xi_{H_{\widetilde{x}}})$$

and

$$\mathrm{In}_{\xi_{H_{\widetilde{x}}}}(\widetilde{a_{p^e}}) = \widetilde{x}^{r-p^e} g(\widetilde{y}) \text{ is not a } p^e\text{-th power},$$

since $\mathrm{In}_{\xi_{H_x}}(a_{p^e}) = x^r g(y)$ is not a p^e-th power. Therefore, $\widetilde{h}$ is well-adapted at $\xi_{H_{\widetilde{x}}}$ with respect to $(\widetilde{x},\widetilde{y},\widetilde{z})$.

Therefore, we conclude

$$h_{\widetilde{x}} = \mathrm{Slope}_{\widetilde{h},(\widetilde{x},\widetilde{y},\widetilde{z})}(\xi_{H_{\widetilde{x}}}) = \mathrm{Slope}_{h,(x,y,z)}(\xi_{H_x}) - 1 = h_x - 1.$$

In case H_x is bad, i.e., $h_x < \mu_x$, so is $H_{\widetilde{x}}$ since $h_{\widetilde{x}} = h_x - 1 < \mu_x - 1 = \mu_{\widetilde{x}}$. Moreover, we compute

$$\begin{aligned}\rho_{\widetilde{x}} = \rho_{\widetilde{h},(\widetilde{x},\widetilde{y},\widetilde{z}),H_{\widetilde{x}}}(\widetilde{P}) &= \begin{cases}\mathrm{ord}_{\widetilde{P}}\left(g(\widetilde{y})\right)/p^e & \text{if } r - p^e \not\equiv 0 \bmod p^e,\\ \text{res-ord}_{\widetilde{P}}^{(p^e)}\left(g(\widetilde{y})\right)/p^e & \text{if } r - p^e \equiv 0 \bmod p^e\end{cases}\\ &= \begin{cases}\mathrm{ord}_{P}\left(g(y)\right)/p^e & \text{if } r \not\equiv 0 \bmod p^e,\\ \text{res-ord}_{P}^{(p^e)}\left(g(y)\right)/p^e & \text{if } r \equiv 0 \bmod p^e\end{cases}\\ &= \rho_{h,(x,y,z),H_x}(P) = \rho_x.\end{aligned}$$

In summary, we have

$$h_{\widetilde{x}} = h_x - 1, \quad \rho_{\widetilde{x}} = \rho_x.$$

The proof for the formulas

$$h_{\widetilde{y}} = h_y, \quad \rho_{\widetilde{y}} = \rho_y - 1$$

is similar, and left to the reader as an exercise. Q.E.D.

Remark 8. The above proof is slightly sloppy in the sense that, after blow up, we may end up having $\mathrm{ord}_{\widetilde{P}}(\widetilde{a_{p^e}}) = p^e$ and $\mathrm{In}_{\widetilde{P}}(\widetilde{a_{p^e}}) = c\widetilde{x}^{p^e} + c'\widetilde{y}^{p^e}$ for some $(c, c') \in k^2 \setminus \{(0,0)\}$, even under the condition $\widetilde{P} \in \pi^{-1}(P) \cap \mathrm{Sing}(\widetilde{\mathcal{R}}) \subset \widetilde{W}$ and the assumption that the invariant σ stays the same (cf. the proof of Proposition 4 (1)). Then we have to replace $(\widetilde{x}, \widetilde{y}, \widetilde{z})$ with

$$(\widetilde{x}, \widetilde{y}, \widetilde{z}' = \widetilde{z} + c^{1/p^e}\widetilde{x} + c'^{1/p^e}\widetilde{y})$$

to guarantee condition (1) in SITUATION in §5.1. Accordingly, we have to analyze $\widetilde{a_{p^e}}'$. It is straightforward, however, to see that the same calculations hold with $\widetilde{a_{p^e}}'$. The details are left to the reader as an exercise.

Case: blow up with 0-dimensional center

Claim 2. *Let $W \xleftarrow{\pi} \widetilde{W}$ be the blow up with a 0-dimensional center $C = P = V(z,x,y) \in \mathrm{Sing}(\mathcal{R})$ (cf. Proposition 6 and the description of the procedure in §5.3), and $\widetilde{\mathcal{R}}$ the transformation of the idealistic filtration $\mathcal{R}$ of i.f.g. type. Set $Z = \{z = 0\}$. Then a point $\widetilde{P} \in \mathrm{Sing}(\widetilde{\mathcal{R}}) \cap \pi^{-1}(P) \subset \widetilde{W}$ must be on the strict transform Z' of Z, lying either in the x-chart or in the y-chart. Assume that the invariant σ stays the same, i.e., $\sigma(P) = \sigma(\widetilde{P})$. We make the following three observations regarding the behavior of the invariants under blow up. (We denote the strict* *transforms of H_x and H_y by H'_x and H'_y, and the exceptional divisor by E_P. Note that the pictures depict the configurations in a 2-dimensional manner, taking the intersection with the hypersurface Z before blow up and with its strict transform Z' after blow up.):*

(1) *The point P is in case ③, ④ or ⑤.*

(1.1) *Suppose $h_x < 1$. Look at the point $\widetilde{P} = E_P \cap H'_x \cap Z'$ in the y-chart with a regular system of parameters $(\widetilde{x}, \widetilde{y}, \widetilde{z}) = (x/y, y, z/y)$. Then the hypersurface $H'_x = H_{\widetilde{x}}$ is bad, and we have*

$$\rho_x > \rho_{\widetilde{x}}.$$

(1.2) *Suppose P is bad, and hence E_P is also bad (cf. Lemma 4). Look at a point $\widetilde{P} \in (E_P \setminus H'_x) \cap Z'$ in the x-chart with a regular system of parameters $(\widetilde{x} = e, \widetilde{y}, \widetilde{z}) = (x, y/x - c, z/x)$ for some $c \in k$. Then we have*

$$\rho_x \geq \rho_e.$$

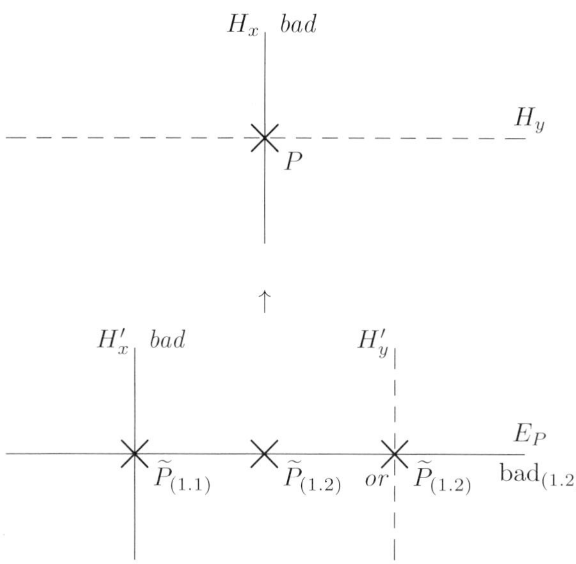

(2) *The point P is in case ④. Suppose P is bad, and hence E_P is also bad (cf. Lemma 4). Look at a point $\widetilde{P} \in (E_P \setminus H'_y) \cap Z'$ with a regular system of parameters $(\widetilde{x}, \widetilde{y} = e, \widetilde{z}) = (x/y - c, y, z/y)$ for some $c \in k$. Then we have*

$$\mu_x > \rho_e.$$

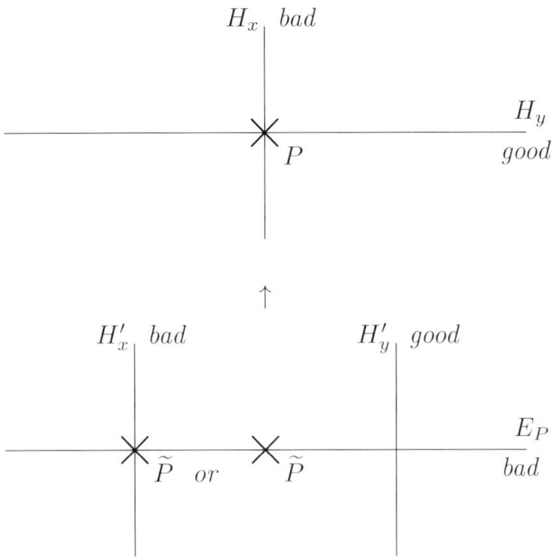

(3) *The point P is in case ⑤. Suppose P is good. Then we have*

$$\rho_x > \mu_y \quad \text{and} \quad \rho_y > \mu_x.$$

Look at the point $\widetilde{P} = E_P \cap H'_x \cap Z'$ in the y-chart with a regular system of parameters $(\widetilde{x}, \widetilde{y}, \widetilde{z}) = (x/y, y, z/y)$. Since $\mu_{\widetilde{x}} = \mu_x$, we have as a consequence

$$\rho_y > \mu_{\widetilde{x}}.$$

We draw a similar conclusion looking at the point $E_P \cap H'_y \cap Z'$ in the x-chart.

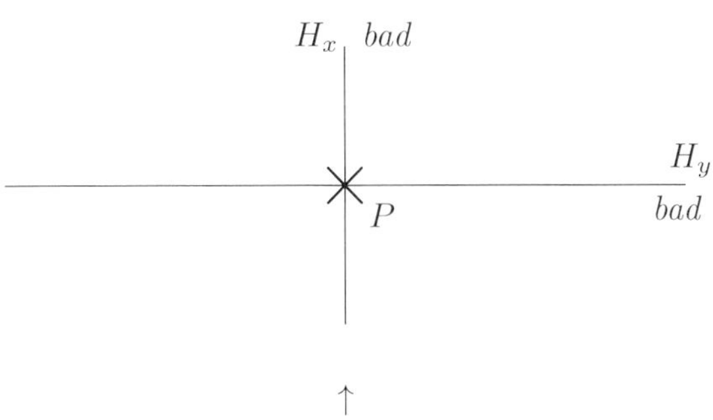

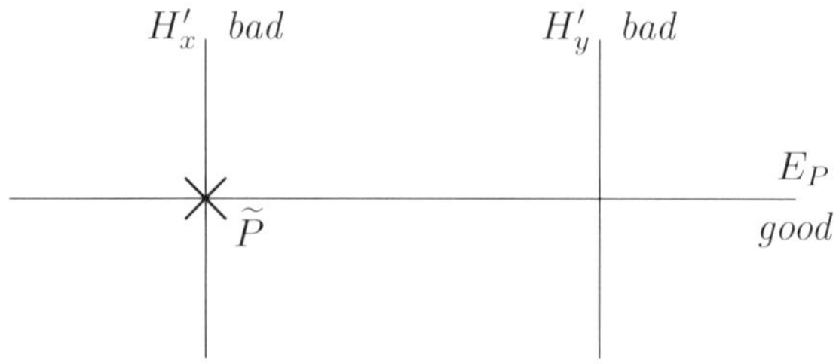

Proof. (1) (1.1) Take (as described in SITUATION in §5.1)

$$h = z^{p^e} + a_1 z^{p^e-1} + a_2 z^{p^e-2} + \cdots + a_{p^e-1} z + a_{p^e}$$

which is well-adapted at ξ_{H_x} with respect to (x, y, z) (cf. Proposition 5). Set

$$a_{p^e} = x^r \{g(y) + x \cdot \omega(x, y)\} \text{ with } 0 \neq g(y) \in k[[y]],\ \omega(x, y) \in k[[x, y]].$$

Then we have

$$h_x = \mathrm{H}(\xi_{H_x}) = \mathrm{Slope}_{h,(x,y,z)}(\xi_{H_x}) = \frac{1}{p^e}\mathrm{ord}_{\xi_{H_x}}(a_{p^e}) = \frac{r}{p^e} < \mu(\xi_{H_x}).$$

Since $h_x < 1$ by assumption, we have $r < p^e$. We compute

$$\widetilde{a_{p^e}} = \frac{a_{p^e}}{y^{p^e}} = \frac{(\widetilde{x}\widetilde{y})^r}{\widetilde{y}^{p^e}} \{g(\widetilde{y}) + \widetilde{x}\widetilde{y} \cdot \omega(\widetilde{x}\widetilde{y}, \widetilde{y})\} = \widetilde{x}^r \{\widetilde{g}(\widetilde{y}) + \widetilde{x} \cdot \widetilde{\omega}(\widetilde{x}, \widetilde{y})\}$$

where

$$\widetilde{g}(\widetilde{y}) = \widetilde{y}^{r-p^e} g(\widetilde{y}) \quad \text{and} \quad \widetilde{\omega}(\widetilde{x}, \widetilde{y}) = \widetilde{y}^{r-p^e+1}\omega(\widetilde{x}\widetilde{y}, \widetilde{y}).$$

We observe that $\widetilde{h} = h/y^{p^e}$ is well-adapted at $\xi_{H_{\widetilde{x}}}$ with respect to $(\widetilde{x}, \widetilde{y}, \widetilde{z})$, since

$$\mathrm{Slope}_{\widetilde{h},(\widetilde{x},\widetilde{y},\widetilde{z})}(\xi_{H_{\widetilde{x}}}) = \frac{r}{p^e} < \mu(\xi_{H_x}) = \mu(\xi_{H_{\widetilde{x}}})$$

and since

$$\mathrm{In}_{\xi_{H_{\widetilde{x}}}}(\widetilde{a_{p^e}}) = \widetilde{x}^r \widetilde{g}(\widetilde{y}) = \widetilde{x}^r \widetilde{y}^{r-p^e} g(\widetilde{y})$$

is not a p^e-th power, a fact which follows easily from the fact that $\mathrm{In}_{\xi_{H_x}}(a_{p^e}) = x^r g(y)$ is not a p^e-th power. Therefore, we conclude that

$$\begin{aligned}
\rho_x &= \rho_{h,(x,y,z),H_x}(P) \\
&= \begin{cases} \mathrm{ord}_P(g(y))/p^e & \text{in case } r \not\equiv 0 \bmod p^e, \\ \text{res-ord}_P^{(p^e)}(g(y))/p^e & \text{in case } r \equiv 0 \bmod p^e, \quad \text{i.e., } r = 0 \end{cases} \\
&> \begin{cases} \mathrm{ord}_{\widetilde{y}}\left(\widetilde{y}^{r-p^e} g(\widetilde{y})\right)/p^e & \text{in case } r \not\equiv 0 \bmod p^e, \\ \text{res-ord}_{\widetilde{y}}^{(p^e)}\left(\widetilde{y}^{-p^e} g(\widetilde{y})\right)/p^e & \text{in case } r \equiv 0 \bmod p^e, \quad \text{i.e., } r = 0 \end{cases} \\
&= \begin{cases} \mathrm{ord}_{\widetilde{y}}(\widetilde{g}(\widetilde{y}))/p^e & \text{in case } r \not\equiv 0 \bmod p^e, \\ \text{res-ord}_{\widetilde{y}}^{(p^e)}(\widetilde{g}(\widetilde{y}))/p^e & \text{in case } r \equiv 0 \bmod p^e, \quad \text{i.e. } r = 0 \end{cases} \\
&= \rho_{\widetilde{h},(\widetilde{x},\widetilde{y},\widetilde{z}),H_{\widetilde{x}}}(\widetilde{P}) = \rho_{\widetilde{x}}.
\end{aligned}$$

(Note that, even under the condition $\widetilde{P} \in \pi^{-1}(P) \cap \mathrm{Sing}(\widetilde{\mathcal{R}}) \subset \widetilde{W}$ and the assumption that the invariant σ stays the same, there is a possibility that we may end up having $\mathrm{ord}_{\widetilde{P}}(\widetilde{a_{p^e}}) = p^e$ and $\mathrm{In}_{\widetilde{P}}(\widetilde{a_{p^e}}) = c\widetilde{y}^{p^e}$ for some $c \in k \setminus \{0\}$. (In this case, we necessarily have $r = 0$.) Then we have to replace $(\widetilde{x}, \widetilde{y}, \widetilde{z})$ with $(\widetilde{x}, \widetilde{y}, \widetilde{z}' = \widetilde{z} + c^{1/p^e}\widetilde{y})$ to guarantee condition (1) in SITUATION. Accordingly, we have to analyze $\widetilde{a_{p^e}}'$. It is straightforward, however, to see that the same calculations hold with $\widetilde{a_{p^e}}'$.)

(1) (1.2) Take (as described in SITUATION in §5.1)

$$h = z^{p^e} + a_1 z^{p^e-1} + a_2 z^{p^e-2} + \cdots + a_{p^e-1}z + a_{p^e}$$

which is well-adapted at P and ξ_{H_x} simultaneously with respect to (x, y, z) (cf. Proposition 5).

Set

$$\begin{aligned}
a_{p^e} &= \sum_{k+l \geq d} c_{kl} x^k y^l & &\text{with } d = \mathrm{ord}_P(a_{p^e}) > p^e \\
&= x^r\{g(y) + x \cdot \omega(x, y)\} & &\text{with } r = \mathrm{ord}_{\xi_{H_x}}(a_{p^e}).
\end{aligned}$$

Then we compute

$$\begin{aligned}
\pi^*(a_{p^e}) &= \sum_{k+l \geq d} c_{kl} x^{k+l} \left(\frac{y}{x}\right)^l = x^d \left\{ \sum_{k+l=d} c_{kl} \left(\frac{y}{x}\right)^l + x \cdot \Omega\left(x, \frac{y}{x}\right) \right\} \\
&= x^d \left\{ \phi\left(\frac{y}{x}\right) + x \cdot \Omega\left(x, \frac{y}{x}\right) \right\} \quad \text{where} \quad \phi(T) = \sum_{k+l=d} c_{kl} T^l,
\end{aligned}$$

and hence

$$\begin{aligned}\widetilde{a_{p^e}} &= \frac{\pi^*(a_{p^e})}{x^{p^e}} = x^{d-p^e}\left\{\phi\left(\frac{y}{x} - c + c\right) + x \cdot \Omega\left(x, \frac{y}{x} - c + c\right)\right\} \\ &= \widetilde{x}^{d-p^e}\left\{\phi(\widetilde{y} + c) + \widetilde{x} \cdot \Omega(\widetilde{x}, \widetilde{y} + c)\right\}.\end{aligned}$$

Moreover, just as in the analysis of the "*Case: P is a bad point*" in Lemma 4, we see that $\widetilde{h}$ is well-adapted at ξ_{E_P} with respect to $(\widetilde{x}, \widetilde{y}, \widetilde{z})$, since

$$\begin{aligned}\mathrm{Slope}_{\widetilde{h},(\widetilde{x},\widetilde{y},\widetilde{z})}(\xi_{H_{\widetilde{x}}}) &= \frac{1}{p^e}\mathrm{ord}_{\xi_{H_{\widetilde{x}}}}(\widetilde{a_{p^e}}) = \frac{d-p^e}{p^e} \\ &= \frac{1}{p^e}\mathrm{ord}_P(a_{p^e}) - 1 \underset{\text{since } P \text{ is bad}}{<} \mu(P) - 1 = \mu(\xi_{H_{\widetilde{x}}}),\end{aligned}$$

and since $\mathrm{In}_{\xi_{H_{\widetilde{x}}}(\widetilde{a_{p^e}})} = \widetilde{x}^{d-p^e}\phi(\widetilde{y}+c)$ is not a p^e-th power, which is obvious if $d \not\equiv 0 \bmod p^e$ and which follows from the fact $\mathrm{In}_P(a_{p^e}) = \sum_{k+l=d} c_{kl}x^k y^l$ is not a p^e-th power and hence $\phi(T) = \sum_{k+l=d} c_{kl}T^l$ is not a p^e-th power if $d \equiv 0 \bmod p^e$.

Therefore, we conclude

$$\begin{aligned}\rho_{\widetilde{x}=e} &= \rho_{\widetilde{h},(\widetilde{x},\widetilde{y},\widetilde{z}),H_{\widetilde{x}}} \\ &= \begin{cases}\mathrm{ord}_{\widetilde{y}}\left(\phi(\widetilde{y}+c)\right)/p^e & \text{in case } r \not\equiv 0 \bmod p^e \\ \text{res-ord}^{(p^e)}_{\widetilde{y}}\left(\phi(\widetilde{y}+c)\right)/p^e & \text{in case } r \equiv 0 \bmod p^e\end{cases} \\ &\le \frac{1}{p^e}\deg\phi(\widetilde{y}+c) = \frac{1}{p^e}\deg\phi(y).\end{aligned}$$

That is to say, we have

$$(\star) \quad \rho_e \le \frac{1}{p^e}\deg\phi(y).$$

On the other hand, set

$$M = \begin{cases}\mathrm{ord}_y\left(g(y)\right) & \text{in case } r \not\equiv 0 \bmod p^e \\ \text{res-ord}^{(p^e)}_y\left(g(y)\right) & \text{in case } r \equiv 0 \bmod p^e\end{cases}$$

Then we conclude, for all those (k,l) with $k+l=d$ and $c_{kl} \neq 0$, that we have

$$k + l = d \le r + M, \quad \text{and} \quad k \ge r$$

and hence that

$$l = d - k \le d - r \le (r + M) - r = M.$$

This implies

$$(\star\star) \quad \deg \phi(y) = \deg \left(\sum_{k+l=d} c_{kl} y^l \right) \leq M.$$

From the inequalities $(\star)$ and $(\star\star)$, we finally conclude

$$\rho_x = \frac{M}{p^e} \geq \frac{1}{p^e} \deg(\phi(y)) \geq \rho_e.$$

(Note that, even under the condition $\widetilde{P} \in \pi^{-1}(P) \cap \operatorname{Sing}(\widetilde{\mathcal{R}}) \subset \widetilde{W}$ and the assumption that the invariant σ stays the same, there is a possibility that we may end up having $\operatorname{ord}_{\widetilde{P}}(\widetilde{a_{p^e}}) = p^e$ and $\operatorname{In}_{\widetilde{P}}(\widetilde{a_{p^e}}) = c\widetilde{x}^{p^e}$ for some $c \in k \setminus \{0\}$. Then we have to replace $(\widetilde{x}, \widetilde{y}, \widetilde{z})$ with $(\widetilde{x}, \widetilde{y}, \widetilde{z}' = \widetilde{z} + c^{1/p^e}\widetilde{x})$ to guarantee condition (1) in $\boxed{\text{SITUATION}}$ in §5.1. Accordingly, we have to analyze $\widetilde{a_{p^e}}'$. It is straightforward, however, to see that the same calculations hold with $\widetilde{a_{p^e}}'$.)

(2) Take (as described in $\boxed{\text{SITUATION}}$ in §5.1)

$$h = z^{p^e} + a_1 z^{p^e-1} + a_2 z^{p^e-2} + \cdots + a_{p^e-1} z + a_{p^e}$$

which is well-adapted at P and ξ_{H_y} simultaneously with respect to (x, y, z) (cf. Proposition 5). Set

$$a_{p^e} = \sum_{k+l \geq d} c_{kl} x^k y^l \quad \text{with} \quad d = \operatorname{ord}_P(a_{p^e}).$$

Then we see

$$\frac{d}{p^e} = \frac{1}{p^e} \operatorname{ord}_P(a_{p^e}) < \mu(P) \quad \text{and} \quad \operatorname{In}_P(a_{p^e}) \text{ is not a } p^e\text{-th power},$$

since P is bad and since h is well-adapted at P with respect to (x, y, z). Then we conclude, for all those (k, l) with $k + l = d$ and $c_{kl} \neq 0$, that we have

$$\frac{\alpha + \beta}{a} = \mu(P) > \frac{d}{p^e} = \frac{k + l}{p^e} = \frac{k}{p^e} + \frac{l}{p^e} \geq \frac{k}{p^e} + \frac{\beta}{a},$$

since

$$\frac{l}{p^e} \geq \frac{1}{p^e} \operatorname{ord}_{\xi_{H_y}}(a_{p^e}) \geq \mu(\xi_{H_y}) = \frac{\beta}{a},$$

where the second inequality follows from the assumption that H_y is good and that h is well-adapted at ξ_{H_y} with respect to (x, y, z). Therefore, we conclude

$$\mu_x = \mu(\xi_{H_x}) = \frac{\alpha}{a} > \frac{k}{p^e}.$$

On the other hand, we compute

$$\begin{aligned}\pi^*(a_{p^e}) &= \sum_{k+l \geq d} c_{kl} \left(\frac{x}{y}\right)^k y^{k+l} = y^d \left\{ \sum_{k+l=d} c_{kl} \left(\frac{x}{y}\right)^k + y \cdot \Omega\left(\frac{x}{y}, y\right)\right\} \\ &= y^d \left\{ \varphi\left(\frac{x}{y}\right) + y \cdot \Omega\left(\frac{x}{y}, y\right)\right\},\end{aligned}$$

where $\varphi(T) = \sum_{k+l=d} c_{kl} T^k$, and hence

$$\begin{aligned}\widetilde{a_{p^e}} &= \frac{\pi^*(a_{p^e})}{y^{p^e}} = y^{d-p^e} \left\{ \varphi\left(\frac{x}{y} - c + c\right) + y \cdot \Omega\left(\frac{x}{y} - c + c, y\right)\right\} \\ &= \widetilde{y}^{d-p^e} \left\{ \varphi(\widetilde{x} + c) + \widetilde{y} \cdot \Omega(\widetilde{x} + c, \widetilde{y})\right\}.\end{aligned}$$

Moreover, just as in the analysis of the "*Case: P is a bad point*" in Lemma 4, we see that $\widetilde{h}$ is well-adapted at ξ_{E_P} with respect to $(\widetilde{x}, \widetilde{y}, \widetilde{z})$, since

$$\begin{aligned}\mathrm{Slope}_{\widetilde{h},(\widetilde{x},\widetilde{y},\widetilde{z})}(\xi_{H_{\widetilde{y}}}) &= \frac{1}{p^e} \mathrm{ord}_{\xi_{H_{\widetilde{y}}}}(\widetilde{a_{p^e}}) = \frac{d - p^e}{p^e} \\ &= \frac{1}{p^e} \mathrm{ord}_P(a_{p^e}) - 1 \underset{\text{since } P \text{ is bad}}{<} \mu(P) - 1 = \mu(\xi_{H_{\widetilde{y}}}),\end{aligned}$$

and since $\mathrm{In}_{\xi_{H_{\widetilde{y}}}}(\widetilde{a_{p^e}}) = \widetilde{y}^{d-p^e} \varphi(\widetilde{x} + c)$ is not a p^e-th power, which is obvious if $d \not\equiv 0 \bmod p^e$ and which follows from the fact $\mathrm{In}_P(a_{p^e}) = \sum_{k+l=d} c_{kl} x^k y^l$ is not a p^e-th power and hence $\varphi(T) = \sum_{k+l=d} c_{kl} T^k$ is not a p^e-th power if $d \equiv 0 \bmod p^e$.

Therefore, we conclude

$$\rho_{\widetilde{y}=e} = \rho_{\widetilde{h},(\widetilde{x},\widetilde{y},\widetilde{z}),H_{\widetilde{y}}} = \begin{cases} \mathrm{ord}_{\widetilde{x}}\left(\varphi(\widetilde{x} + c)\right)/p^e & \text{if } d \not\equiv 0 \bmod p^e \\ \text{res-ord}^{(p^e)}_{\widetilde{x}}\left(\varphi(\widetilde{x} + c)\right)/p^e & \text{if } d \equiv 0 \bmod p^e \end{cases}$$

$$\leq \frac{1}{p^e} \deg \varphi(\widetilde{y} + c) = \frac{1}{p^e} \deg \varphi(y) = \frac{1}{p^e} \deg \left(\sum_{k+l=d} c_{kl} T^k \right) < \frac{\alpha}{a} = \mu_x.$$

(Note that, even under the condition $\widetilde{P} \in \pi^{-1}(P) \cap \mathrm{Sing}(\widetilde{\mathcal{R}}) \subset \widetilde{W}$ and the assumption that the invariant σ stays the same, there is a possibility that we may end up having $\mathrm{ord}_{\widetilde{P}}(\widetilde{a_{p^e}}) = p^e$ and $\mathrm{In}_{\widetilde{P}}(\widetilde{a_{p^e}}) = c\widetilde{y}^{p^e}$ for some

$c \in k \setminus \{0\}$. (In this case, we necessarily have $r = 0$.) Then we have to replace $(\widetilde{x}, \widetilde{y}, \widetilde{z})$ with $(\widetilde{x}, \widetilde{y}, \widetilde{z}' = \widetilde{z} + c^{1/p^e}\widetilde{y})$ to guarantee condition (1) in $\boxed{\text{SITUATION}}$ in §5.1. Accordingly, we have to analyze $\widetilde{a_{p^e}}'$. It is straightforward, however, to see that the same calculations hold with $\widetilde{a_{p^e}}'$.)

(3) Take (as described in $\boxed{\text{SITUATION}}$ in §5.1)

$$h = z^{p^e} + a_1 z^{p^e-1} + a_2 z^{p^e-2} + \cdots + a_{p^e-1} z + a_{p^e}$$

which is well-adapted at P and ξ_{H_x} simultaneously with respect to (x, y, z) (cf. Proposition 5).

Set

$$a_{p^e} = x^r \{g(y) + x \cdot \omega(x, y)\} \text{ with } 0 \neq g(y) \in k[[y]],\ \omega(x, y) \in k[[x, y]].$$

Set

$$M = \begin{cases} \mathrm{ord}_P(g(y)) & \text{in case } r \not\equiv 0 \bmod p^e, \\ \text{res-ord}_P^{(p^e)}(g(y)) & \text{in case } r \equiv 0 \bmod p^e \end{cases}$$

so that

$$\rho_x = \rho_{h,(x,y,z),H_x}(P) = \frac{M}{p^e}.$$

Now we compute

$$\begin{aligned} \frac{r + M}{p^e} \geq \frac{1}{p^e}\mathrm{ord}_P(a_{p^e}) & \underset{\text{since } P \text{ is good}}{\geq} \mu(P) = \frac{\alpha + \beta}{a} = \mu(\xi_{H_x}) + \frac{\beta}{a} \\ & \underset{\text{since } H_x \text{ is bad}}{>} \mathrm{H}(\xi_{H_x}) + \frac{\beta}{a} = \frac{r}{p^e} + \frac{\beta}{a}. \end{aligned}$$

Therefore, we conclude

$$\rho_x = \frac{M}{p^e} > \frac{\beta}{a} = \mu(\xi_{H_y}) = \mu_y.$$

The proof for the inequality $\rho_y > \mu_x$ is identical.

This completes the proof of Claim 2. Q.E.D.

Theorem 1. *Let $P \in \mathrm{Sing}(\mathcal{R}) \subset W$ be a point in the monomial case as described in $\boxed{\textit{SITUATION}}$ in §5.1. Let $W \overset{\pi}{\leftarrow} \widetilde{W}$ be the blow up with center C specified by the procedure described in §5.3, and $\widetilde{\mathcal{R}}$ the transformation of the idealistic filtration $\mathcal{R}$ of i.f.g. type.*

Then at $\widetilde{P} \in \pi^{-1}(P) \in \widetilde{W}$, one of the following holds:

- $\widetilde{P} \notin \mathrm{Sing}(\widetilde{R})$, *i.e.,* $\mathrm{Sing}(\widetilde{R}) = \emptyset$ *in a neighborhood of* $\widetilde{P}$,

- $\widetilde{P} \in \mathrm{Sing}(\widetilde{R})$ *and the invariant* σ *drops, or*
- $\widetilde{P} \in \mathrm{Sing}(\widetilde{R})$ *and the invariant "*$\mathrm{inv}_{\mathrm{MON}}(P)$*" strictly decreases, i.e.,* $\mathrm{inv}_{\mathrm{MON}}(P) > \mathrm{inv}_{\mathrm{MON}}(\widetilde{P})$.

Since the invariant "$\mathrm{inv}_{\mathrm{MON}}$*" cannot decrease infinitely many times, the procedure described in §5.3 terminates either with* $\widetilde{P} \notin \mathrm{Sing}(\widetilde{R})$ *or with the drop of the invariant* σ *after finitely many repetitions.*

That is to say, the basic strategy to show termination of the procedure in the monomial case with $\tau = 1$ *is established.*

Proof. We have only to show (cf. Proposition 4) that, at $\widetilde{P} \in \pi^{-1}(P) \in \widetilde{W}$, assuming $\widetilde{P} \in \mathrm{Sing}(\widetilde{R})$ and $\sigma(P) = \sigma(\widetilde{P})$, we have

$$\mathrm{inv}_{\mathrm{MON}}(P) > \mathrm{inv}_{\mathrm{MON}}(\widetilde{P}).$$

Case: $\dim C = 1$.

In this case, by Claim 1, it is easy to see that P and $\widetilde{P}$ are in the same configuration and

$$\mathrm{inv}_{\mathrm{MON}}(P) > \mathrm{inv}_{\mathrm{MON}}(\widetilde{P}).$$

Case: $\dim C = 0$, *i.e.*, $C = P$.

1. P is in configuration ① or ②

Since $P \in \mathrm{Sing}(\mathcal{R})$, we have $h_x = \mu_x \geq 1$ in configuration ①, and we do not choose a point center $C = P$ in this case (cf. Proposition 6). So we only have to consider configuration ②, where we have $\mu_x = h_x < 1$ and $\mu_y = h_y < 1$. This implies $\mu_e < \min\{\mu_x, \mu_y\}$. Moreover, it is easy to see that P is necessarily a good point, hence E_P is also good, and that $\widetilde{P}$ is in configuration ②, since we assume $\widetilde{P} \in \mathrm{Sing}(\widetilde{R})$. Now it is straightforward to see

$$\mathrm{inv}_{\mathrm{MON}}(P) > \mathrm{inv}_{\mathrm{MON}}(\widetilde{P}).$$

2. P is in configuration ③

(i) $\widetilde{P} = (E_P \setminus H'_x) \cap Z'$ *with* E_P *being bad, and hence* $\widetilde{P}$ *is in configuration* ③. In this subcase, we have $\rho_x \geq \rho_e$ by Claim 2 (1.2), while $\mu_x > \mu_x - 1 = \mu_e$. Therefore, we conclude

$$\mathrm{inv}_{\mathrm{MON}}(P) = (\rho_x, 0, \mu_x) > (\rho_e, 0, \mu_e) = \mathrm{inv}_{\mathrm{MON}}(\widetilde{P}).$$

(ii) $\widetilde{P} = E_P \cap H'_x \cap Z'$ *with* E_P *being good, and hence* $\widetilde{P}$ *is in configuration* ④. In this subcase, we have $\rho_x > \rho_{\widetilde{x}}$ by Claim 2 (1.1). Therefore, we conclude

$$\mathrm{inv}_{\mathrm{MON}}(P) = (\rho_x, 0, \mu_x) > (\min\{\rho_{\widetilde{x}}, \mu_{\widetilde{x}}\}, \max\{\rho_{\widetilde{x}}, \mu_{\widetilde{x}}\}) = \mathrm{inv}_{\mathrm{MON}}(\widetilde{P}).$$

(iii) $\widetilde{P} = E_P \cap H'_x \cap Z'$ *with* E_P *being bad, and hence* $\widetilde{P}$ *is in configuration* ⑤. In this case, we have $\rho_x > \rho_{\widetilde{x}}$ by Claim 2 (1.1). Therefore, we conclude

$$\mathrm{inv}_{\mathrm{MON}}(P) = (\rho_x, 0, \mu_x) > (\min\{\rho_{\widetilde{x}}, \rho_{\widetilde{y}}\}, \max\{\rho_{\widetilde{x}}, \rho_{\widetilde{y}}\}) = \mathrm{inv}_{\mathrm{MON}}(\widetilde{P}).$$

3. P is in configuration ④

(i) $\widetilde{P} = \left(E_P \setminus (H'_x \cup H'_y)\right) \cap Z'$ *with* E_P *being bad, and hence* $\widetilde{P}$ *is in configuration* ③. In this case, we have $\rho_x \geq \rho_e$ and $\mu_x > \rho_e (\geq 0)$ by Claim 2 (1.2) and (2). Therefore, we conclude

$$\mathrm{inv}_{\mathrm{MON}}(P) = (\min\{\rho_x, \mu_x\}, \max\{\rho_x, \mu_x\}) > (\rho_e, 0, \mu_e) = \mathrm{inv}_{\mathrm{MON}}(\widetilde{P}).$$

(ii) $\widetilde{P} = E_P \cap H'_x \cap Z'$ *with* E_P *being good, and hence* $\widetilde{P}$ *is in configuration* ④. In this case, we have $\rho_x > \rho_{\widetilde{x}}$ by Claim 2 (1.1), while $\mu_x = \mu_{\widetilde{x}}$. Therefore, we conclude

$$\begin{aligned}\mathrm{inv}_{\mathrm{MON}}(P) &= (\min\{\rho_x, \mu_x\}, \max\{\rho_x, \mu_x\}) \\ &> (\min\{\rho_{\widetilde{x}}, \mu_{\widetilde{x}}\}, \max\{\rho_{\widetilde{x}}, \mu_{\widetilde{x}}\}) = \mathrm{inv}_{\mathrm{MON}}(\widetilde{P}).\end{aligned}$$

(iii) $\widetilde{P} = E_P \cap H'_y \cap Z'$ *with* E_P *being bad, and hence* $\widetilde{P}$ *is in configuration* ④. In this case, we have $\rho_x \geq \rho_e$ by Claim 2 (1.2), while $\mu_x = \alpha/a > (\alpha + \beta)/a - 1 = \mu_e$ since $\beta/a = \mu_y = h_y < 1$. Therefore, we conclude

$$\begin{aligned}\mathrm{inv}_{\mathrm{MON}}(P) &= (\min\{\rho_x, \mu_x\}, \max\{\rho_x, \mu_x\}) \\ &> (\min\{\rho_e, \mu_e\}, \max\{\rho_e, \mu_e\}) = \mathrm{inv}_{\mathrm{MON}}(\widetilde{P}).\end{aligned}$$

(iv) $\widetilde{P} = E_P \cap H'_x \cap Z'$ *with* E_P *being bad, and hence* $\widetilde{P}$ *is in configuration* ⑤. In this case, we have $\rho_x > \rho_{\widetilde{x}}$ and $\mu_x > \rho_{\widetilde{y}=e}$ by Claim 2 (1.1) and (2). Therefore, we conclude

$$\begin{aligned}\mathrm{inv}_{\mathrm{MON}}(P) &= (\min\{\rho_x, \mu_x\}, \max\{\rho_x, \mu_x\}) \\ &> (\min\{\rho_{\widetilde{x}}, \rho_{\widetilde{y}}\}, \max\{\rho_{\widetilde{x}}, \rho_{\widetilde{y}}\}) = \mathrm{inv}_{\mathrm{MON}}(\widetilde{P}).\end{aligned}$$

4. P is in configuration ⑤

(i) $\widetilde{P} = \left(E_P \setminus (H'_x \cup H'_y)\right) \cap Z'$ *with* E_P *being bad, and hence* $\widetilde{P}$ *is in configuration* ③. In this case, we have $\rho_x \geq \rho_e$ and $\rho_y \geq \rho_e$ by Claim 2 (1.2). We also have $\rho_x, \rho_y > 0$ since $h_y, h_x < 1$ with $P \in \mathrm{Sing}(\mathcal{R})$. Therefore, we conclude

$$\mathrm{inv}_{\mathrm{MON}}(P) = (\min\{\rho_x, \rho_y\}, \max\{\rho_x, \rho_y\}) > (\rho_e, 0, \mu_e) = \mathrm{inv}_{\mathrm{MON}}(\widetilde{P}).$$

(ii) $\widetilde{P} = E_P \cap H'_x \cap Z'$ *with* E_P *being good, and hence* $\widetilde{P}$ *is in configuration* ④. In this case, we have $\rho_x > \rho_{\widetilde{x}}$ and $\rho_y > \mu_{\widetilde{x}}$ by Claim 2 (1.1) and (3). Therefore, we conclude

$$\begin{aligned}\mathrm{inv}_{\mathrm{MON}}(P) &= (\min\{\rho_x, \rho_y\}, \max\{\rho_x, \rho_y\}) \\ &> (\min\{\rho_{\widetilde{x}}, \mu_{\widetilde{x}}\}, \max\{\rho_{\widetilde{x}}, \mu_{\widetilde{x}}\}) = \mathrm{inv}_{\mathrm{MON}}(\widetilde{P}).\end{aligned}$$

(iii) $\widetilde{P} = E_P \cap H'_x \cap Z'$ *with* E_P *being bad, and hence* $\widetilde{P}$ *is in configuration* ⑤. In this case, we have $\rho_x > \rho_{\widetilde{x}}$ and $\rho_y \geq \rho_{\widetilde{y}=e}$ by Claim 2 (1.1) and (1.2). Therefore, we conclude

$$\begin{aligned}\mathrm{inv}_{\mathrm{MON}}(P) &= (\min\{\rho_x, \rho_y\}, \max\{\rho_x, \rho_y\}) \\ &> (\min\{\rho_{\widetilde{x}}, \rho_{\widetilde{y}}\}, \max\{\rho_{\widetilde{x}}, \rho_{\widetilde{y}}\}) = \mathrm{inv}_{\mathrm{MON}}(\widetilde{P}).\end{aligned}$$

This completes the proof of Theorem 1. Q.E.D.

This completes the detailed discussion of the monomial case in dimension 3, and hence completes the presentation of (the local version of) our algorithm in dimension 3.

We finish this paper by making a couple of remarks.

Remark 9 (**Invariant whose maximum locus determines the center of blow up in the monomial case?**). The invariant "$\mathrm{inv}_{\mathrm{MON}}$" is only used to show effectively the termination of the procedure (in the monomial case), while the choice of the center is dictated by the study of the dimension of the singular locus (cf. Proposition 6 and the description of the procedure in 5.3). Actually all the existing algorithms, including the one in [5], use the analysis of the dimension of the singular locus for the choice of the center.

It would be desirable to have an invariant (manifested as the invariant Γ in characteristic zero), which satisfies the following properties:

(1) it is upper semi-continuous,

(2) its maximum locus determines the nonsingular center of blow up for constructing the sequence of transformations for resolution of singularities, and

(3) it strictly drops after each blow up (over the center), and can not strictly decrease infinitely many times.

Such an invariant would not only show the termination effectively but also dictate the choice of the center.

Remark 10.

(1)(**Global version**) The presentation of our algorithm in this paper is restricted to the local version (cf. Remark 2). However, it is not so difficult, though technical and rather lengthy, to make some adjustments in order for us to turn the local version into the global version. The detailed discussion of these adjustments will be published elsewhere.

(2)(**Embedded resolution of singularities of a surface $X \subset W$ with** $\dim W = 3$) Consider a surface $X \subset W$ embedded in a nonsingular ambient space W of $\dim W = 3$. Then we can establish embedded resolution of singularities for $X \subset W$ (cf. Problem 2) in the following manner:

(i) The problem of embedded resolution of singularities for $X \subset W$ is reduced to the problem of resolution of singularities for the triplet $(W, (\mathcal{I}_X, 1), \emptyset)$ (cf. Lemma 1).

(ii) The problem of resolution of singularities for the triplet $(W, (\mathcal{I}_X, 1), \emptyset)$ is equivalent to the problem of resolution of singularities for $(W, \mathcal{R}, \emptyset)$ where the idealistic filtration of i.f.g.type is given by $\mathcal{R} = \mathcal{G}((\mathcal{I}_X, 1))$ (cf. §2. Overview).

(iii) The problem of resolution of singularities for $(W, \mathcal{R}, \emptyset)$ is solved by the global version of our algorithm in $\dim W = 3$.

(3)(**Embedded resolution of singularities of a surface $X \subset W$ with** $\dim W$ **arbitrary**) We can establish the global version of our algorithm for resolution of singularities of the triplet $(W, \mathcal{R}, E)$ with $\dim W$ arbitrary, as long as it satisfies the condition $\tau(P) \geq \dim W - 2$ for all $P \in W$. As an application, we can give an alternative proof for embedded resolution of a surface $X \subset W$ with $\dim W$ arbitrary, a theorem established by [4] and [11]. We only describe the outline of our alternative proof below (while the details will be published elsewhere):

(i) We consider the Hilbert-Samuel function HS over X, and its maximum value $M = \max_{x \in X}\{\mathrm{HS}(x)\}$.

(ii) Construct a covering $X = \bigcup_{\lambda \in \Lambda} U_\lambda$, and a pair $(\mathcal{I}_\lambda, a_\lambda)$ of an ideal $\mathcal{I}_\lambda$ over U_λ with level $a_\lambda \in \mathbb{Z}_{>0}$ for each $\lambda \in \Lambda$ (in étale topology using the result due to Aroca (cf. [13]) or in Zariski topology using the result by [2]) satisfying the following property:

(♠) the singular locus of the pair coincides with the maximum locus of the Hilbert-Samuel function, i.e.,

$$\mathrm{Sing}(\mathcal{I}_\lambda, a_\lambda) = \{P \in U_\lambda \mid \mathrm{HS}(P) = M\},$$

and this relation persists through *any* sequence of transformations (and smooth morphisms).

(iii) For each $\lambda \in \Lambda$, consider the idealistic filtration of i.f.g.type $\mathcal{R}'_\lambda = \mathcal{G}((\mathcal{I}_\lambda, a_\lambda))$. We take its $\mathcal{R}$adical-&-$\mathcal{D}$ifferential saturation $\mathcal{R}_\lambda := \mathcal{RD}(\mathcal{R}'_\lambda)$ (cf. [18]). Then by the result of [6] we see that the $\mathcal{R}_\lambda$'s patch together, i.e., $\mathcal{R}_\lambda|_{U_\lambda \cap U_\mu} = \mathcal{R}_\mu|_{U_\mu \cap U_\lambda}$, to give rise to the idealistic filtration $\mathcal{R}$ over W such that $\mathcal{R}|_{U_\lambda} = \mathcal{R}_\lambda$ for all $\lambda \in \Lambda$, that $\mathrm{Sing}(\mathcal{R}) = \{P \in W \mid \mathrm{HS}(P) = M\}$, and that this relation persists through any sequence of transformations (and smooth morphisms). Note that the $\mathcal{R}$-&-$\mathcal{D}$ saturation gives the "biggest" idealistic filtration of i.f.g.type with the property (♠).

(iv) Observe that, for this idealistic filtration $\mathcal{R}$, we have $\tau(P) \geq \dim W - 2$ for all $P \in W$ by the result in [7][8].

(v) Apply the global version of our algorithm to obtain resolution of singularities for $(W, \mathcal{R}, E_M = \emptyset)$, which implies the strict decrease of (the maximum value of) the Hilbert-Samuel function.

(vi) Repeat the procedure. (Note that in the middle of the repetition of the procedure, the boundary E_M, which is the union of all the exceptional divisors created so far, may not be empty.)

(vii) Since the value of the Hilbert-Samuel function can not strictly decrease infinitely many times, the procedure must come to an end after finitely many repetitions, providing embedded resolution of singularities for $X \subset W$.

References

[1] Shreeram S. Abhyankar. Resolution of singularities of embedded algebraic surfaces. *Pure and Applied Mathematics*, Vol. 24. Academic Press, New York, 1966.

[2] Edward Bierstone and Pierre Milman Canonical desingularization in characteristic zero by blowing-up the maximal strata of a local invariant *Inv. Math.*, 182(2):207–302, 1997.

[3] Angélica Benito and Orlando Villamayor U. Monoidal transforms and invariants of singularities in positive characteristic. *Compositio Math.*, to appear, 2013

[4] Angélica Benito and Orlando Villamayor U. On elimination of variables in the study of singularities in positive characteristic. *preprint*, `http://arXiv.org/abs/1103.3462`, 2011.

[5] Angélica Benito and Orlando Villamayor U. Techniques for the study of singularities with applications to resolution of 2-dimensional schemes. *Math. Ann.*, 353(3):1037–1068, 2012.

[6] Ana Bravo, M.L. García-Escamilla, and Orlando Villamayor U. On Rees algebras and invariants for singularities over perfect fields preprint, 2012

[7] Ana Bravo and Orlando Villamayor U. Singularities in positive characteristic, stratification and simplification of the singular locus *Advances in Mathematics* 224:1349 – 1418, 2010

[8] Ana Bravo and Orlando Villamayor U. Elimination algebras and inductive arguments in resolution of singularities. *Asian J. Math.* Vol. 15, No. 3:321 – 356, 2011

[9] Vincent Cossart and Olivier Piltant. Resolution of singularities of threefolds in positive characteristic. I. Reduction to local uniformization on Artin-Schreier and purely inseparable coverings. *J. Algebra*, 320(3):1051–1082, 2008.

[10] Vincent Cossart and Olivier Piltant. Resolution of singularities of threefolds in positive characteristic. II. *J. Algebra*, 321(7):1836–1976, 2009.

[11] Vincent Cossart, Uwe Jannsen and Shuji Saito. Canonical embedded and non-embedded resolution of singularities for excellent two-dimensional schemes. *preprint*, `http://arXiv.org/abs/0905.2191`, 2009.

[12] Steven Dale Cutkosky. Resolution of singularities for 3-folds in positive characteristic. *Amer. J. Math.*, 131(1):59–127, 2009.

[13] Santiago Encinas and Orlando Villamayor. A course on constructive desingularization and equivariance In *Resolution of singularities (Obergurgl, 1997)*, volume 181 of *Progr. Math.*, pages 147–227. Birkhäuser, Basel, 2000.

[14] Alexander Grothendieck. Revêtements Étales et Groupe Fondamental (SGA 1) *Lecture Notes in Math.* 224, Springer Verlag, 1971

[15] Herwig Hauser. Excellent surfaces and their taut resolution. In *Resolution of singularities (Obergurgl, 1997)*, volume 181 of *Progr. Math.*, pages 341–373. Birkhäuser, Basel, 2000.

[16] Herwig Hauser. On the problem of resolution of singularities in positive characteristic (or: a proof we are still waiting for). *Bull. Amer. Math. Soc. (N.S.)*, 47(1):1–30, 2010.

[17] Heisuke Hironaka. Desingularization of excellent surfaces. In *Resolution of surface singularities*, volume 1101 of *Lecture Notes in Mathematics*, pages 99–132. Springer-Verlag, Berlin, 1984. (Lectures at Bowdoin on 1967. Noted by B.M. Bennett).

[18] Hiraku Kawanoue. Toward resolution of singularities over a field of positive characteristic Part I. Foundation; the language of the idealistic filtration. *Publ. Res. Inst. Math. Sci.*, 43(3):819–909, 2007.

[19] Hiraku Kawanoue and Kenji Matsuki. Toward resolution of singularities over a field of positive characteristic (the idealistic filtration program) Part II. Basic invariants associated to the idealistic filtration and their properties. *Publ. Res. Inst. Math. Sci.*, 46(2):359–422, 2010.

[20] James S. Milne. *Étale cohomology*, volume 33 of *Princeton Mathematical Series*. Princeton University Press, Princeton, N.J., 1980.

[21] T. T. Moh. On a stability theorem for local uniformization in characteristic *p*. *Publ. Res. Inst. Math. Sci.*, 23(6):965–973, 1987.

[22] Michel Raynaud. *Anneaux locaux henséliens*. Lecture Notes in Mathematics, Vol. 169. Springer-Verlag, Berlin, 1970.

[23] Jarosław Włodarczyk. Simple Hironaka resolution in characteristic zero. *J. Amer. Math. Soc.*, 18(4):779–822 (electronic), 2005.

Hiraku Kawanoue
Research Institute for Mathematical Sciences
Kyoto University, Kyoto 606-8502
E-mail address: kawanoue@kurims.kyoto-u.ac.jp

Kenji Matsuki
Department of Mathematics, Purdue University
150 N. University Street, West Lafayette, IN 47907-2067
E-mail address: kmatsuki@math.purdue.edu

Advanced Studies in Pure Mathematics 70, 2016
Minimal Models and Extremal Rays (Kyoto, 2011)
pp. 215–245

Finite generation and geography of models

Anne-Sophie Kaloghiros, Alex Küronya and Vladimir Lazić

To Professor Shigefumi Mori on the occasion of his 60th birthday, with admiration

Abstract.

There are two main examples where a version of the Minimal Model Program can, at least conjecturally, be performed successfully: the first is the classical MMP associated to the canonical divisor, and the other is Mori Dream Spaces. In this paper we formulate a framework which generalises both of these examples. Starting from divisorial rings which are finitely generated, we determine precisely when we can run the MMP, and we show why finite generation alone is not sufficient to make the MMP work.

§1. Introduction

There are two classes of projective varieties whose birational geometry is particularly interesting and rich. The first family consists of varieties where the classical Minimal Model Program (MMP) can be performed successfully with the current techniques. The other class is

---

Received February 6, 2012.
Revised August 6, 2012.
2010 *Mathematics Subject Classification.* 14E30.
The first author was supported by the Engineering and Physical Sciences Research Council [grant number EP/H028811/1]. The second author was supported by the DFG-Forschergruppe 790 "Classification of Algebraic Surfaces and Compact Complex Manifolds", and by the OTKA Grants 77476 and 81203 of the Hungarian Academy of Sciences. The third author was supported by the DFG-Forschergruppe 790 "Classification of Algebraic Surfaces and Compact Complex Manifolds". We would like to thank P. Cascini, A.-M. Castravet, Y. Gongyo, K. Matsuki and the referee for many useful comments, and the following institutions for providing additional support and excellent working conditions: Albert-Ludwigs-Universität Freiburg, Imperial College London, University of Illinois at Chicago.

that of Mori Dream Spaces. We now know that, in both cases, the geometry of birational contractions from the varieties in question is entirely determined by suitable finitely generated divisorial rings.

More precisely, let X be a $\mathbb{Q}$-factorial projective variety that belongs to one of these two classes. Then, there are effective $\mathbb{Q}$-divisors $D_1, \dots, D_r$ strongly related to the geometry of X such that the multigraded divisorial ring

$$\mathfrak{R} = R(X; D_1, \dots, D_r) = \bigoplus_{(n_1,\dots,n_r)\in\mathbb{N}^r} H^0(X, n_1D_1 + \cdots + n_rD_r)$$

is finitely generated. In the first case, $\mathfrak{R}$ is an adjoint ring; in the second, it is a Cox ring. Then, for any divisor D in the span $\mathcal{S} = \sum \mathbb{R}_+ D_i$, finite generation implies the existence of a birational map $\varphi_D \colon X \dashrightarrow X_D$, where φ_D is a composition of elementary surgery operations that can be fully understood. Both X_D and $(\varphi_D)_*D$ have good properties: X_D is projective and $\mathbb{Q}$-factorial, and $(\varphi_D)_*D$ is semiample.

In addition, there is a decomposition of $\mathcal{S} = \bigcup \mathcal{S}_j$ into finitely many rational polyhedral cones, together with birational maps $\varphi_j \colon X \dashrightarrow X_j$, such that the pushforward under φ_j of every divisor in $\mathcal{S}_j$ is a nef divisor on X_j. In this paper, we say that these models $\varphi_j \colon X \dashrightarrow X_j$ are *optimal*; the precise definition is in Section 2. By analogy with the classical case, the map $\varphi_j \colon X \dashrightarrow X_j$ is called a *D-MMP*. After Shokurov, the decomposition of $\mathcal{S}$ above is called a *geography* of optimal models associated to $\mathfrak{R}$.

The goal of this paper is twofold. On the one hand, we want to put these two families of varieties under the same roof. That is to say, we want to identify the maximal class of varieties and divisors on them where a suitable MMP can be performed. On the other hand, we want to understand why this class is the right one, i.e. what the key ingredients that make the MMP work are.

Let D be a $\mathbb{Q}$-divisor on a variety X in one of the two families above. The D-MMP has two significant features, which we would like to extend to a more general setting:

(i) all varieties in the MMP are $\mathbb{Q}$-factorial,
(ii) the section ring $R(X, D)$ is preserved under the operations of the MMP.

Condition (ii) is by now well understood: contracting maps that preserve sections of D are D-nonpositive – we recall this definition in Section 2. Somewhat surprisingly, preserving $\mathbb{Q}$-factoriality is the main obstacle to extending the MMP to arbitrary varieties X and divisors D, even when the rings $R(X, D)$ are finitely generated; this is explained in Section 4.

In this work, we introduce the notion of *gen* divisors. We say that a $\mathbb{Q}$-divisor D on a $\mathbb{Q}$-factorial projective variety X is gen if every $\mathbb{Q}$-divisor in its numerical equivalence class has a finitely generated section ring. Ample divisors are examples of gen divisors. As we explain in Section 4, in the situations of interest to us, these form essentially the only source of examples: indeed, all gen divisors there come from ample divisors on the end products of some MMP. However, one should bear in mind that semiample divisors are not necessarily gen.

The main result of this paper is the following.

Theorem 1.1. *Let X be a projective $\mathbb{Q}$-factorial variety, let $D_1, \ldots, D_r$ be effective $\mathbb{Q}$-divisors on X, and assume that the numerical classes of D_i span $N^1(X)_{\mathbb{R}}$. Assume that the ring $R(X; D_1, \ldots, D_r)$ is finitely generated, that the cone $\sum \mathbb{R}_+ D_i$ contains an ample divisor, and that every divisor in the interior of this cone is gen.*

Then there is a finite decomposition

$$\sum \mathbb{R}_+ D_i = \coprod \mathcal{N}_i$$

into cones having the following properties:

(1) *each $\overline{\mathcal{N}_i}$ is a rational polyhedral cone,*
(2) *for each i, there exists a $\mathbb{Q}$-factorial projective variety X_i and a birational contraction $\varphi_i \colon X \dashrightarrow X_i$ such that φ_i is an optimal model for every divisor in $\mathcal{N}_i$.*

In fact, we prove a stronger result: we show that for any $\mathbb{Q}$-divisor $D \in \sum \mathbb{R}_+ D_i$, we can run a D-MMP which terminates, see Theorem 5.4 for the precise statement. The decomposition in Theorem 1.1 determines a geography of optimal models associated to $R(X; D_1, \ldots, D_r)$. The techniques used in this paper build on and extend those from [4].

Our work has been influenced by several lines of research. The original idea that geographies of various models are the right thing to look at is due to Shokurov [17], and the first unconditional results were proved in [1]. Similar decompositions were considered in the context of Mori Dream Spaces by Hu and Keel [8], and as we demonstrate here, these are closely related to the study of asymptotic valuations in [6]. Theorem 1.1 reproves and generalises some of the main results from these papers. We obtain in Corollary 5.6 the finiteness of models due to [1] by using the main theorem from [3]. Further, in Corollary 5.7 we prove a characterisation of Mori Dream Spaces in terms of the finite generation of their Cox rings due to [8] without using GIT techniques.

Along the way, we establish several results of independent interest. Theorem 3.8 shows that the nef and movable cones on Calabi-Yau varieties are locally rational polyhedral inside the big cone, and Lemma

3.11 gives a sufficient condition for the stable base loci of numerically equivalent divisors to coincide.

We spend a few words on the organisation of the paper. Section 2 sets the notation and gathers some preliminary results. Section 3 focuses on finite generation of divisorial rings, and contains a number of results which are either of independent interest, or are used later in the paper.

In Section 4, we show the existence of a decomposition $\sum \mathbb{R}_+ D_i = \coprod \mathcal{A}_i$ similar to that from Theorem 1.1, where all divisors in a given chamber $\mathcal{A}_i$ have a common *ample model*, see Theorem 4.2. We study the geography of ample models. The main drawback of this decomposition is that the corresponding models are not $\mathbb{Q}$-factorial in general. Moreover, we show in Example 4.8 that the conditions of Theorem 4.2 are not sufficient to ensure the existence of optimal models as in Theorem 1.1. We explain why the presence of gen divisors is essential to the proof of Theorem 1.1. However, we give a short proof that some of these models are indeed $\mathbb{Q}$-factorial in the case of adjoint divisors in Theorem 4.5.

In Section 5, we define what is meant by the MMP in our setting; it is easy to see that this generalises the classical MMP constructions. We then prove Theorem 5.4, which is a strengthening of Theorem 1.1. The main technical result is Theorem 5.2, and the presence of gen divisors is essential to its proof. We mention here that this reveals the philosophical role of the gen condition: it enables one to prove a version of the classical Basepoint free theorem, which is why we can then run the Minimal Model Program and preserve $\mathbb{Q}$-factoriality in the process. We end the paper with several corollaries that recover quickly some of the main results from [1] and [8].

§2. Preliminary results

Throughout this paper we work with varieties defined over $\mathbb{C}$. Unless otherwise stated, all varieties are projective and normal. We denote by $\mathbb{R}_+$ and $\mathbb{Q}_+$ the sets of non-negative real and rational numbers.

Convex geometry. Let $\mathcal{C} \subseteq \mathbb{R}^N$ be a convex set. A subset $F \subseteq \mathcal{C}$ is a *face* of $\mathcal{C}$ if it is convex, and whenever $tu + (1-t)v \in F$ for some $u, v \in \mathcal{C}$ and $0 < t < 1$, then $u, v \in F$.

The topological closure of a set $\mathcal{S} \subseteq \mathbb{R}^N$ is denoted by $\overline{\mathcal{S}}$. The boundary of a closed set $\mathcal{C} \subseteq \mathbb{R}^N$ is denoted by $\partial\mathcal{C}$.

A *rational polytope* in $\mathbb{R}^N$ is a compact set which is the convex hull of finitely many rational points in $\mathbb{R}^N$. A *rational polyhedral cone* in $\mathbb{R}^N$ is a convex cone spanned by finitely many rational vectors. The dimension of a cone in $\mathbb{R}^N$ is the dimension of the minimal $\mathbb{R}$-vector space containing it.

A finite rational polyhedral subdivision $\mathcal{C} = \bigcup \mathcal{C}_i$ of a rational polyhedral cone $\mathcal{C}$ is a *fan* if each face of $\mathcal{C}_i$ is also a cone in the decomposition, and the intersection of two cones in the decomposition is a face of each.

Divisors and line bundles. Let X be a normal projective variety and let $\mathbf{k} \in \{\mathbb{Z}, \mathbb{Q}, \mathbb{R}\}$. We denote by $\mathrm{Div}_{\mathbf{k}}(X)$ the group of $\mathbf{k}$-Cartier $\mathbf{k}$-divisors on X, and $\sim_{\mathbf{k}}$ and $\equiv$ denote the $\mathbf{k}$-linear and numerical equivalence of $\mathbb{R}$-divisors. If there is a morphism $X \to Y$ to another normal projective variety, numerical equivalence over Y is denoted by $\equiv_Y$. We denote $\mathrm{Pic}(X)_{\mathbf{k}} = \mathrm{Div}_{\mathbf{k}}(X)/\sim_{\mathbf{k}}$ and $N^1(X)_{\mathbf{k}} = \mathrm{Div}_{\mathbf{k}}(X)/\equiv$.

The ample, big, nef, effective, and pseudo-effective cones in $N^1(X)_{\mathbb{R}}$ are denoted by $\mathrm{Amp}(X)$, $\mathrm{Big}(X)$, $\mathrm{Nef}(X)$, $\mathrm{Eff}(X)$, and $\overline{\mathrm{Eff}}(X)$. The movable cone $\overline{\mathrm{Mov}}(X)$ is the closure of the cone in $N^1(X)_{\mathbb{R}}$ spanned by the classes of divisors whose base locus has codimension at least 2.

If X is a normal projective variety, and if D is an $\mathbb{R}$-divisor on X, then the group of global sections of D is

$$H^0(X, D) = \{f \in k(X) \mid \mathrm{div}\, f + D \geq 0\},$$

and the associated *section ring* is defined as

$$R(X, D) = \bigoplus_{m \in \mathbb{N}} H^0(X, mD).$$

A *pair* (X, Δ) consists of a normal projective variety X and an $\mathbb{R}$-divisor $\Delta \geq 0$ on X such that $K_X + \Delta$ is $\mathbb{R}$-Cartier. When (X, Δ) is a pair, $K_X + \Delta$ is an *adjoint divisor*. The pair (X, Δ) has *klt* (respectively *log canonical*) singularities if for every log resolution $f \colon Y \to X$, the divisor $K_Y - f^*(K_X + \Delta)$ has all coefficients > -1 (respectively ≥ -1).

A projective variety X is said to be of *Calabi-Yau type* if there exists a $\mathbb{Q}$-divisor $\Delta \geq 0$ such that (X, Δ) is klt and $K_X + \Delta \equiv 0$.

If X is a normal projective variety, and if D is an integral divisor on X, we denote by $\mathrm{Bs}\,|D|$ the base locus of D, whereas $\mathrm{Fix}\,|D|$ and $\mathrm{Mob}(D)$ denote the fixed and mobile parts of D. If D is an $\mathbb{R}$-divisor on X, we denote

$$|D|_{\mathbb{R}} = \{D' \geq 0 \mid D \sim_{\mathbb{R}} D'\} \qquad \text{and} \qquad \mathbf{B}(D) = \bigcap_{D' \in |D|_{\mathbb{R}}} \mathrm{Supp}\, D',$$

and we call $\mathbf{B}(D)$ the *stable base locus* of D. We set $\mathbf{B}(D) = X$ if $|D|_{\mathbb{R}} = \emptyset$.

As mentioned in the introduction, we pay special attention to two classes of varieties: varieties for which the classical MMP can be performed successfully on the one hand, and Mori Dream Spaces on the other. We recall some definitions related to their study.

Definition 2.1. Let X be a projective $\mathbb{Q}$-factorial variety, let $S_1, \dots, S_p$ be distinct prime divisors on X, denote $V = \sum_{i=1}^p \mathbb{R}S_i \subseteq \mathrm{Div}_{\mathbb{R}}(X)$, and let A be an ample $\mathbb{Q}$-divisor on X. We define

$$\mathcal{L}(V) = \{\Delta \in V \mid (X, \Delta) \text{ is log canonical}\},$$
$$\mathcal{E}_A(V) = \{\Delta \in \mathcal{L}(V) \mid |K_X + A + \Delta|_{\mathbb{R}} \neq \emptyset\}.$$

It is easy to check that $\mathcal{L}(V)$ is a rational polytope, cf. [1, Lemma 3.7.2]. On the other hand, the fact that $\mathcal{E}_A(V)$ is a rational polytope is much harder, see Corollary 3.6.

Definition 2.2. A projective $\mathbb{Q}$-factorial variety X is a Mori Dream Space if

(1) $\mathrm{Pic}(X)_{\mathbb{Q}} = N^1(X)_{\mathbb{Q}}$,
(2) $\mathrm{Nef}(X)$ is the affine hull of finitely many semiample line bundles, and
(3) there are finitely many birational maps $f_i\colon X \dashrightarrow X_i$ to projective $\mathbb{Q}$-factorial varieties X_i such that each f_i is an isomorphism in codimension 1, each X_i satisfies (2), and $\overline{\mathrm{Mov}}(X) = \bigcup f_i^*\big(\mathrm{Nef}(X_i)\big)$.

Models. Let X be a normal projective variety and let D be an $\mathbb{R}$-Cartier divisor on X. We adopt some of the definitions of *models* of D from [1]. The models defined below are significant because they provide correct generalisations of minimal and canonical models for divisors which are not necessarily adjoint.

Definition 2.3. Let $D \in \mathrm{Div}_{\mathbb{R}}(X)$ and let $\varphi\colon X \dashrightarrow Y$ be a contraction map to a normal projective variety Y such that $D' = \varphi_* D$ is $\mathbb{R}$-Cartier.

(1) The map φ is *D-nonpositive* (respectively *D-negative*) if it is birational, and for a common resolution $(p, q)\colon W \to X \times Y$, we can write
$$p^* D = q^* D' + E,$$
where $E \geq 0$ is q-exceptional (respectively $E \geq 0$ is q-exceptional and $\mathrm{Supp}\, E$ contains the strict transform of the φ-exceptional divisors).
(2) The map φ is an *optimal model* of D if φ is D-negative, Y is $\mathbb{Q}$-factorial and D' is nef.

(3) The map φ is a *semiample model* of D if φ is D-nonpositive and D' is semiample.
(4) The map φ is the *ample model* of D if there exist a birational contraction $f\colon X \dashrightarrow Z$ which is a semiample model of D, and a morphism with connected fibres $g\colon Z \to Y$ such that $\varphi = g \circ f$ and $f_*D = g^*A$, where A is an ample $\mathbb{R}$-divisor on Y.

Remark 2.4. (i) If the birational contraction $\varphi\colon X \dashrightarrow Y$ is D-nonpositive for some $D \in \operatorname{Div}_{\mathbb{Q}}(X)$, then $H^0(X,D) \simeq H^0(Y,\varphi_*D)$.
(ii) The ample model is unique up to isomorphism. Indeed, with the notation from Definition 2.3, we may assume that D is a $\mathbb{Q}$-divisor, and (i) shows that $R(X,D) \simeq R(Z,f_*D)$. This last ring is isomorphic to $R(Y,A)$, and therefore $Y \simeq \operatorname{Proj} R(X,D)$. Note that in Definition 2.3(4), when D is a $\mathbb{Q}$-divisor, the map g is the *semiample fibration* associated to D, see [14, Theorem 2.1.27].
(iii) In this paper we require that the ample model factors through a semiample model (compare with [1, Definition 3.6.5, Lemma 3.6.6(3)]).
(iv) When D is an adjoint divisor, then optimal, semiample and ample models are called *log terminal*, *good* and *log canonical* models, respectively.

We recall the following important result known as the Negativity lemma, see [10, Lemma 2.19]. This result and its corollary will be used in Sections 4 and 5.

Lemma 2.5. *Let $f\colon X \to Y$ be a proper birational morphism, where X is normal, and let E be an f-exceptional divisor on X. Assume that $E \equiv_Y H + D$, where H is f-nef and $D \geq 0$ has no common components with E. Then $E \leq 0$.*

Corollary 2.6. *Let $X \to Z$ and $Y \to Z$ be projective morphisms of normal projective varieties. Let $f\colon X \dashrightarrow Y$ be a birational contraction over Z, and let $(p,q)\colon W \to X \times Y$ be a resolution of f. Let D and D' be $\mathbb{R}$-Cartier divisors on X such that f_*D and f_*D' are $\mathbb{R}$-Cartier on Y, and assume that $D \equiv_Z D'$. Then*

$$p^*D - q^*f_*D = p^*D' - q^*f_*D'.$$

In particular, f is D-nonpositive (respectively D-negative) if and only if f is D'-nonpositive (respectively D'-negative).

Proof. This result is [1, Lemma 3.6.4]. The divisor $E = p^*(D - D') - q^* f_*(D - D')$ is q-exceptional since f is a contraction, and we have $E \equiv_Y 0$. We conclude by Lemma 2.5. Q.E.D.

Asymptotic valuations. A *geometric valuation* Γ over a normal variety X is a valuation on $k(X)$ given by the order of vanishing at the generic point of a prime divisor on some birational model $f\colon Y \to X$. If D is an $\mathbb{R}$-Cartier divisor on X, we abuse notation and write $\operatorname{mult}_\Gamma D$ to denote $\operatorname{mult}_\Gamma f^*D$.

The following definition is due to Nakayama.

Definition 2.7. Let X be a normal projective variety, let D be an $\mathbb{R}$-Cartier divisor such that $|D|_\mathbb{R} \neq \emptyset$, and let Γ be a geometric valuation over X. The *asymptotic order of vanishing* of D along Γ is

$$o_\Gamma(D) = \inf\{\operatorname{mult}_\Gamma D' \mid D' \in |D|_\mathbb{R}\}.$$

If D is a big divisor, we define

$$N_\sigma(D) = \textstyle\sum_\Gamma o_\Gamma(D) \cdot \Gamma \quad \text{and} \quad P_\sigma(D) = D - N_\sigma(D),$$

where the sum runs over all prime divisors Γ on X.

Remark 2.8. On a surface X, the construction above gives the classical Zariski decomposition: this is a unique decomposition $D = P_\sigma(D) + N_\sigma(D)$, where $P_\sigma(D)$ is nef, and $N_\sigma(D) = \sum \gamma_i \Gamma_i$ is an effective divisor such that $P_\sigma(D) \cdot \Gamma_i = 0$ for all i, and the matrix $(\Gamma_i \cdot \Gamma_j)$ is negative definite. We use this characterisation in Example 4.8.

Lemma 2.9. *Let X be a $\mathbb{Q}$-factorial projective variety, let D be a big $\mathbb{R}$-divisor, and let Γ be a prime divisor. Then $o_\Gamma(D)$ depends only on the numerical class of D. The function o_Γ is homogeneous of degree one, convex and continuous on* $\operatorname{Big}(X)$. *The formal sum $N_\sigma(D)$ is an $\mathbb{R}$-divisor, the divisor $P_\sigma(D)$ is movable, and for any $\mathbb{R}$-divisor $0 \leq F \leq N_\sigma(D)$ we have $N_\sigma(D - F) = N_\sigma(D) - F$. If $E \geq 0$ is an $\mathbb{R}$-divisor on X such that $D - E \in \overline{\operatorname{Mov}}(X)$, then $E \geq N_\sigma(D)$.*

Proof. See [15, §III.1]. Q.E.D.

In certain situations we have more information on the divisor $P_\sigma(D)$.

Lemma 2.10. *Let X be a $\mathbb{Q}$-factorial projective variety, and let D be a big $\mathbb{Q}$-divisor on X. Assume that the cone $\overline{\operatorname{Mov}}(X)$ is rational polyhedral.*

Then $P_\sigma(D)$ is a $\mathbb{Q}$-divisor, and $R(X, D)$ is finitely generated if and only if $R(X, P_\sigma(D))$ is finitely generated.

Proof. Let Γ_i be the components of $N_\sigma(D)$, and denote

$$\mathcal{H} = D - \sum \mathbb{R}_+\Gamma_i \quad \text{and} \quad \mathcal{G} = P_\sigma(D) - \sum \mathbb{R}_+\Gamma_i.$$

Then we have $\overline{\mathrm{Mov}}(X) \cap \mathcal{H} \subseteq \mathcal{G}$ by Lemma 2.9. Since $\overline{\mathrm{Mov}}(X) \cap \mathcal{H}$ is an intersection of finitely many rational half-spaces, and as $P_\sigma(D) \in \overline{\mathrm{Mov}}(X)$ is an extremal point of $\mathcal{G}$, we conclude that $P_\sigma(D)$ is a $\mathbb{Q}$-divisor.

For the second statement, we may assume that D is an integral divisor and that $|D| \neq \emptyset$, so the claim follows from $P_\sigma(mD) \geq \mathrm{Mob}(mD)$ for every positive integer m. Q.E.D.

The proof of the following lemma is analogous to that of [4, Lemma 5.2], and it will be used in Section 5 to ensure that a certain MMP terminates.

Lemma 2.11. *Let $f\colon X \dashrightarrow Y$ be a birational contraction between projective $\mathbb{Q}$-factorial varieties, and let $\mathcal{C} \subseteq \mathrm{Div}_{\mathbb{R}}(X)$ be a cone such that f is D-nonpositive for all $D \in \mathcal{C}$. Let Γ be a geometric valuation on $k(X)$.*

Then o_Γ is linear on $\mathcal{C}$ if and only if it is linear on the cone $f_*\mathcal{C} \subseteq \mathrm{Div}_{\mathbb{R}}(Y)$.

Proof. Let $(p,q)\colon W \to X \times Y$ be a resolution of f. Then for every $D \in \mathcal{C}$ we have $p^*D = q^*f_*D + E_D$, where $E_D \geq 0$ is a q-exceptional divisor. This implies that f_* restricts to an isomorphism between $|D|_{\mathbb{R}}$ and $|f_*D|_{\mathbb{R}}$. Denote

$$V_D = \{D_X - D \mid D_X \in |D|_{\mathbb{R}}\} \quad \text{and} \quad W_D = \{D_Y - f_*D \mid D_Y \in |f_*D|_{\mathbb{R}}\}.$$

By the above, we have the isomorphism $f_*|_{V_D}\colon V_D \simeq W_D$, and also $\mathrm{mult}_\Gamma P_X = \mathrm{mult}_\Gamma f_*P_X$ for every $P_X \in V_D$ by [4, Lemma 5.1(2)]. Therefore

$$\begin{aligned} o_\Gamma(D) - \mathrm{mult}_\Gamma D &= \inf_{P_X \in V_D} \mathrm{mult}_\Gamma P_X \\ &= \inf_{P_X \in V_D} \mathrm{mult}_\Gamma f_*P_X = o_\Gamma(f_*D) - \mathrm{mult}_\Gamma f_*D, \end{aligned}$$

hence the function $o_\Gamma(\cdot) - o_\Gamma\big(f_*(\cdot)\big)\colon \mathcal{C} \to \mathbb{R}$ is equal to the linear map $\mathrm{mult}_\Gamma(\cdot) - \mathrm{mult}_\Gamma f_*(\cdot)$. The lemma follows. Q.E.D.

§3. Around finite generation

In this section, we recall the definition of divisorial rings on a normal projective variety from [3, 4], and we give some examples that we use later – adjoint rings, and rings spanned by big divisors on varieties of

Calabi-Yau type. We then relate finite generation to asymptotic valuations and to properties of the stable base locus.

Divisorial rings. If X is a normal projective variety, and if $\mathcal{S} \subseteq \mathrm{Div}_{\mathbb{Q}}(X)$ is a finitely generated monoid, then

$$R(X,\mathcal{S}) = \bigoplus_{D\in\mathcal{S}} H^0(X,D)$$

is a *divisorial $\mathcal{S}$-graded ring*. If $\mathcal{C} \subseteq \mathrm{Div}_{\mathbb{R}}(X)$ is a rational polyhedral cone, then $\mathcal{S} = \mathcal{C} \cap \mathrm{Div}(X)$ is a finitely generated monoid by Gordan's lemma, and we define the ring $R(X,\mathcal{C})$ to be $R(X,\mathcal{S})$. We also use divisorial rings of the form

$$\mathfrak{R} = R(X;D_1,\dots,D_r) = \bigoplus_{(n_1,\dots,n_r)\in\mathbb{N}^r} H^0(X,n_1D_1+\cdots+n_rD_r),$$

where $D_1,\dots,D_r \in \mathrm{Div}_{\mathbb{Q}}(X)$. If D_i are adjoint divisors, the ring $\mathfrak{R}$ is an *adjoint ring*. The *support* of $\mathfrak{R}$ is the cone

$$\mathrm{Supp}\,\mathfrak{R} = \{D \in \textstyle\sum \mathbb{R}_+D_i \mid |D|_{\mathbb{R}} \neq \emptyset\} \subseteq \mathrm{Div}_{\mathbb{R}}(X),$$

and similarly for rings of the form $R(X,\mathcal{C})$.

If X is a $\mathbb{Q}$-factorial projective with $\mathrm{Pic}(X)_{\mathbb{Q}} = N^1(X)_{\mathbb{Q}}$, and if $D_1,\dots,D_r$ is a basis of $\mathrm{Pic}(X)_{\mathbb{Q}}$ such that $\overline{\mathrm{Eff}}(X) \subseteq \sum \mathbb{R}_+D_i$, then $R(X;D_1,\dots,D_r)$ is a *Cox ring* of X. The finite generation of this ring is independent of the choice of $D_1,\dots,D_r$.

Throughout the paper, we use several properties of finitely generated divisorial rings without explicit mention, see [3, §2.4] for details and background. The one we use most is recalled in the following lemma.

Lemma 3.1. *Let X be a normal projective variety, let $D_1,\dots,D_r$ be divisors in $\mathrm{Div}_{\mathbb{Q}}(X)$, and let $p_1,\dots,p_r$ be positive rational numbers.*

Then the ring $R(X;D_1,\dots,D_r)$ is finitely generated if and only if the ring $R(X;p_1D_1,\dots,p_rD_r)$ is finitely generated.

Relation to asymptotic valuations. Finite generation of a divisorial ring $\mathfrak{R}$ has important consequences on the behavior of the asymptotic order functions, and therefore on the convex geometry of $\mathrm{Supp}\,\mathfrak{R}$, as observed in [6].

Theorem 3.2. *Let X be a projective $\mathbb{Q}$-factorial variety, and let $\mathcal{C} \subseteq \mathrm{Div}_{\mathbb{R}}(X)$ be a rational polyhedral cone. Assume that the ring $\mathfrak{R} = R(X,\mathcal{C})$ is finitely generated. Then:*

(1) $\mathrm{Supp}\,\mathfrak{R}$ *is a rational polyhedral cone,*

(2) *if* $\operatorname{Supp}\mathfrak{R}$ *contains a big divisor, then all pseudo-effective divisors in* $\operatorname{Supp}\mathfrak{R}$ *are in fact effective,*

(3) *there is a finite rational polyhedral subdivision* $\operatorname{Supp}\mathfrak{R} = \bigcup \mathcal{C}_i$ *such that* o_Γ *is linear on* $\mathcal{C}_i$ *for every geometric valuation* Γ *over* X*, and the cones* $\mathcal{C}_i$ *form a fan,*

(4) *there is a positive integer* d *and a resolution* $f\colon \widetilde{X} \to X$ *such that* $\operatorname{Mob} f^*(dD)$ *is basepoint free for every* $D \in \operatorname{Supp}\mathfrak{R} \cap \operatorname{Div}(X)$*, and* $\operatorname{Mob} f^*(kdD) = k \operatorname{Mob} f^*(dD)$ *for every positive integer* k*.*

Proof. This is essentially [6, Theorem 4.1], see [4, Theorem 3.6]. Q.E.D.

Part (i) of the following lemma is [4, Lemma 3.8]. Part (ii) is a result of Zariski and Wilson, cf. [14, Theorem 2.3.15].

Lemma 3.3. *Let* X *be a normal projective variety and let* D *be a divisor in* $\operatorname{Div}_{\mathbb{Q}}(X)$*.*

(i) *If* $|D|_{\mathbb{Q}} \neq \emptyset$*, then* D *is semiample if and only if* $R(X, D)$ *is finitely generated and* $o_\Gamma(D) = 0$ *for all geometric valuations* Γ *over* X*.*

(ii) *If* D *is nef and big, then* D *is semiample if and only if* $R(X, D)$ *is finitely generated.*

Proof. If D is semiample, then some multiple of D is basepoint free, thus $R(X, D)$ is finitely generated by Lemma 3.1, and all $o_\Gamma(D) = 0$. Now, fix a point $x \in X$. If $R(X, D)$ is finitely generated and $o_x(D) = 0$, then $x \notin \mathbf{B}(D)$ by Theorem 3.2(4), which proves (i).

For (ii), let A be an ample divisor. Then $D + \varepsilon A$ is ample for any $\varepsilon > 0$, hence $o_\Gamma(D + \varepsilon A) = 0$ for any geometric valuation Γ over X. But then $o_\Gamma(D) = \lim_{\varepsilon \to 0} o_\Gamma(D + \varepsilon A) = 0$ by Lemma 2.9, so we conclude by (i). Q.E.D.

Corollary 3.4. *Let* X *be a normal projective variety and let* $D_1, \dots, D_r$ *be divisors in* $\operatorname{Div}_{\mathbb{Q}}(X)$*. Assume that the ring* $\mathfrak{R} = R(X; D_1, \dots, D_r)$ *is finitely generated, and let* $\operatorname{Supp}\mathfrak{R} = \bigcup_{i=1}^N \mathcal{C}_i$ *be a finite rational polyhedral subdivision as in Theorem 3.2(3). Denote by* $\pi\colon \operatorname{Div}_{\mathbb{R}}(X) \to N^1(X)_{\mathbb{R}}$ *the natural projection.*

Then there is a set $I_1 \subseteq \{1, \dots, N\}$ *such that*

$$\operatorname{Supp}\mathfrak{R} \cap \pi^{-1}\big(\overline{\operatorname{Mov}}(X)\big) = \bigcup_{i \in I_1} \mathcal{C}_i.$$

Assume further that $\operatorname{Supp}\mathfrak{R}$ *contains an ample divisor. Then there is a set* $I_2 \subseteq \{1,\ldots,N\}$ *such that the cone* $\operatorname{Supp}\mathfrak{R}\cap\pi^{-1}\big(\operatorname{Nef}(X)\big)$ *equals* $\bigcup_{i\in I_2}\mathcal{C}_i$*, and every element of this cone is semiample.*

Proof. For every prime divisor Γ on X denote $\mathcal{C}_\Gamma = \{D \in \operatorname{Supp}\mathfrak{R} \mid o_\Gamma(D) = 0\}$. If $\mathcal{C}_\Gamma$ intersects the interior of some $\mathcal{C}_\ell$, then $\mathcal{C}_\ell \subseteq \mathcal{C}_\Gamma$ since o_Γ is a linear non-negative function on $\mathcal{C}_\ell$. Therefore, there exists a set $I_\Gamma \subseteq \{1,\ldots,N\}$ such that $\mathcal{C}_\Gamma = \bigcup_{i\in I_\Gamma}\mathcal{C}_i$. Now the first claim follows since $\overline{\operatorname{Mov}}(X)$ is the intersection of all $\mathcal{C}_\Gamma$.

For the second claim, note that since $\operatorname{Supp}\mathfrak{R}\cap\pi^{-1}\big(\operatorname{Nef}(X)\big)$ is a cone of dimension $\dim\operatorname{Supp}\mathfrak{R}$, we can consider only maximal dimensional cones $\mathcal{C}_\ell$. Now, for every $\mathcal{C}_\ell$ whose interior contains an ample divisor, all asymptotic order functions o_Γ are identically zero on $\mathcal{C}_\ell$ similarly as above. Therefore, by Lemma 3.3, every element of $\mathcal{C}_\ell$ is semiample, and thus $\mathcal{C}_\ell \subseteq \operatorname{Supp}\mathfrak{R}\cap\pi^{-1}\big(\operatorname{Nef}(X)\big)$. The claim follows. Q.E.D.

Examples of finitely generated rings. The following is a small variation of the main result of [3], where it is proved by a self-contained argument avoiding the techniques of the MMP. It was first proved in the seminal paper [1] by MMP methods.

Theorem 3.5. *Let* X *be a* $\mathbb{Q}$*-factorial projective variety, and let* $\Delta_1,\ldots,\Delta_r$ *be* $\mathbb{Q}$*-divisors such that all pairs* (X,Δ_i) *are klt.*

(1) *If* $A_1,\ldots,A_r$ *are ample* $\mathbb{Q}$*-divisors, then the adjoint ring*

$$R(X;K_X+\Delta_1+A_1,\ldots,K_X+\Delta_r+A_r)$$

is finitely generated.

(2) *If* Δ_i *are big, then the adjoint ring*

$$R(X;K_X+\Delta_1,\ldots,K_X+\Delta_r)$$

is finitely generated.

Proof. See [4, Theorem 3.2]. Q.E.D.

Corollary 3.6. *Let* X *be a projective* $\mathbb{Q}$*-factorial variety, let* $S_1,\ldots,S_p$ *be distinct prime divisors on* X*, denote* $V=\sum_{i=1}^p\mathbb{R}S_i\subseteq\operatorname{Div}_{\mathbb{R}}(X)$*, and let* A *be an ample* $\mathbb{Q}$*-divisor on* X*. Let* $\mathcal{C}\subseteq\mathcal{L}(V)$ *be a rational polytope such that for every* $\Delta\in\mathcal{C}$*, the pair* (X,Δ) *is klt.*

Then the set $\mathcal{C}\cap\mathcal{E}_A(V)$ *is a rational polytope, and the ring*

$$R(X,\mathbb{R}_+(K_X+A+\mathcal{C}\cap\mathcal{E}_A(V)))$$

is finitely generated.

Proof. Let $B_1, \ldots, B_r$ be the vertices of $\mathcal{C}$. Then the ring $\mathfrak{R} = R(X; K_X + B_1 + A, \ldots, K_X + B_r + A)$ is finitely generated by Theorem 3.5, and we have $\operatorname{Supp}\mathfrak{R} = \mathbb{R}_+(K_X + A + \mathcal{C} \cap \mathcal{E}_A(V))$. Now the result follows from Theorem 3.2(1). Q.E.D.

The following corollary is well known.

Corollary 3.7. *Let X be a projective $\mathbb{Q}$-factorial variety of Calabi-Yau type, and let $B_1, \ldots, B_q$ be big $\mathbb{Q}$-divisors on X. Then the ring $R(X; B_1, \ldots, B_q)$ is finitely generated.*

Proof. Let $\Delta \geq 0$ be a $\mathbb{Q}$-divisor such that (X, Δ) is klt and $K_X + \Delta \equiv 0$, and write $B_i = A_i + E_i$, where each A_i is ample and $E_i \geq 0$. Let $\varepsilon > 0$ be a rational number such that all pairs $(X, \Delta + \varepsilon E_i)$ are klt, and denote $A'_i = \varepsilon B_i - (K_X + \Delta + \varepsilon E_i)$. Then each A'_i is ample since $A'_i \equiv \varepsilon A_i$, hence the adjoint ring

$$R(X; K_X + \Delta + \varepsilon E_1 + A'_1, \ldots, K_X + \Delta + \varepsilon E_q + A'_q) = R(X; \varepsilon B_1, \ldots, \varepsilon B_q)$$

is finitely generated by Theorem 3.5. Therefore $R(X; B_1, \ldots, B_q)$ is finitely generated by Lemma 3.1. Q.E.D.

Part (1) of the following theorem was proved in [11], while part (2) is a generalisation of the analogous result for 3-folds proved in [12].

Theorem 3.8. *Let X be a projective $\mathbb{Q}$-factorial variety of Calabi-Yau type.*

(1) The cone $\operatorname{Nef}(X)$ *is locally rational polyhedral in* $\operatorname{Big}(X)$*, and moreover, every element of* $\operatorname{Nef}(X) \cap \operatorname{Big}(X)$ *is semiample.*

(2) The cone $\overline{\operatorname{Mov}}(X)$ *is locally rational polyhedral in* $\operatorname{Big}(X)$*.*

Proof. We prove the result for $\overline{\operatorname{Mov}}(X)$; the case of $\operatorname{Nef}(X)$ is analogous.

Let V be a relatively compact subset of the boundary of $\overline{\operatorname{Mov}}(X) \cap \operatorname{Big}(X)$, and denote by $\pi\colon \operatorname{Div}_{\mathbb{R}}(X) \to N^1(X)_{\mathbb{R}}$ the natural projection. Then we can choose finitely many big $\mathbb{Q}$-divisors $B_1, \ldots, B_q$ such that $V \subseteq \pi(\sum_{i=1}^q \mathbb{R}_+ B_i)$. Corollary 3.7 implies that the ring $\mathfrak{R} = R(X; B_1, \ldots, B_q)$ is finitely generated, and hence $\pi^{-1}\big(\overline{\operatorname{Mov}}(X)\big) \cap \operatorname{Supp}\mathfrak{R}$ is a rational polyhedral cone by Corollary 3.4. But then V is contained in finitely many rational hyperplanes. Q.E.D.

Remark 3.9. The proof of Theorem 3.8 shows existence of a locally polyhedral decomposition of the big cone on a variety X of Calabi-Yau type, which comes from the behaviour of asymptotic valuations on this cone. This is a consequence of the finite generation of any divisorial ring which is supported on $\operatorname{Big}(X)$. On the other hand, it was shown in [2] that an analogous decomposition exists on any smooth surface, and it is not a consequence of finite generation. It is an interesting problem to establish in which contexts that result generalises to higher dimensions.

Relations to the stable base locus. As Example 3.10 shows, the stable base locus and finite generation of section rings are not, in general, numerical invariants. However, we prove in Lemma 3.11 that under some finite generation hypotheses, the stable base loci of numerically equivalent big divisors coincide.

Example 3.10. We recall [14, Example 10.3.3]. Let B be a smooth elliptic curve, and let A be an ample divisor of degree 1 on B. Let $X = \mathbb{P}(\mathcal{O}_B \oplus \mathcal{O}_B(A))$ be a projective bundle with the natural map $p\colon X \to B$. Let P_1 be a torsion divisor on B, let P_2 be a non-torsion degree 0 divisor on B, and consider $L_i = \mathcal{O}_X(1) \otimes p^*\mathcal{O}_B(P_i)$. Then L_1 and L_2 are numerically equivalent nef and big line bundles with $\emptyset = \mathbf{B}(L_1) \neq \mathbf{B}(L_2)$, and $R(X, L_1)$ is finitely generated while $R(X, L_2)$ is not by Lemma 3.3(2).

Lemma 3.11. *Let X be a $\mathbb{Q}$-factorial projective variety, and let D_1 and D_2 be big $\mathbb{Q}$-divisors such that $D_1 \equiv D_2$. Assume that the rings $R(X, D_i)$ are finitely generated, and consider the maps $\varphi_i\colon X \dashrightarrow \operatorname{Proj} R(X, D_i)$.*

Then we have $\mathbf{B}(D_1) = \mathbf{B}(D_2)$, and there is an isomorphism

$$\eta\colon \operatorname{Proj} R(X, D_1) \to \operatorname{Proj} R(X, D_2)$$

such that $\varphi_2 = \eta \circ \varphi_1$.

Proof. Since finite generation holds, we have $\mathbf{B}(D_i) = \{x \in X \mid o_x(D_i) > 0\}$, so the first claim follows immediately from Lemma 2.9.

For the second claim, by passing to a resolution and by Theorem 3.2, we may assume that there is a positive integer k such that $\operatorname{Mob}(kD_i)$ are basepoint free, and $\operatorname{Mob}(pkD_i) = p\operatorname{Mob}(kD_i)$ for all positive integers p. Note that then $P_\sigma(D_i) = \frac{1}{k}\operatorname{Mob}(kD_i)$, and that

$$P_\sigma(D_1) \equiv P_\sigma(D_2) \tag{1}$$

since $N_\sigma(D_1) = N_\sigma(D_2)$ by Lemma 2.9. Thus φ_i is given by the linear system $|kpP_\sigma(D_i)|$ for some $p \gg 0$. But then (1) shows that φ_1 and φ_2 contract the same curves, which implies the claim. Q.E.D.

§4. Geography of ample models

In this section we study the geography of ample models associated to a finitely generated divisorial ring $\mathfrak{R} = R(X; D_1, \ldots, D_r)$. More precisely, there is a decomposition $\operatorname{Supp}\mathfrak{R} = \coprod \mathcal{A}_i$ into finitely many chambers together with contracting maps $\varphi_i \colon X \dashrightarrow X_i$, such that φ_i is the ample model for every divisor in $\mathcal{A}_i$. We study these ample models in the special case of adjoint divisors; then, the varieties X_i are $\mathbb{Q}$-factorial when the numerical classes of the elements of $\mathcal{A}_i$ span $N^1(X)_\mathbb{R}$. This is a highly desirable feature which we would like to preserve in the general case. We then formally introduce the gen condition, and show – both by analysis and by example – that it is necessary in order to perform a Minimal Model Program in a more general setting.

We first recall the following important result [16, Proposition 1.2]. We follow closely the proof of [5, Lemma 7.10].

Lemma 4.1. *Let X be a smooth variety and let D be a big divisor on X. Assume that, for every positive integer m, the divisor $M_m = \operatorname{Mob}(mD)$ is basepoint free, that $M_m = mM_1$, and that $\operatorname{Fix}|D|$ has simple normal crossings. Let $\varphi \colon X \to Y$ be the semiample fibration associated to M_1.*

Then every component of $\operatorname{Fix}|D|$ is contracted by φ. In particular, we have $R(X, D) \simeq R(Y, \varphi_* D)$.

Proof. Denote $n = \dim X$. We may assume that φ is the morphism associated to M_1, and then $\mathcal{O}_X(M_1) = \varphi^*\mathcal{O}_Y(1)$ for a very ample line bundle $\mathcal{O}_Y(1)$ on Y. Let Γ be a component of $\operatorname{Fix}|D|$. We need to show that $h^0(\varphi(\Gamma), \mathcal{O}_{\varphi(\Gamma)}(m)) \leq O(m^{n-2})$.

Since $\mathcal{O}_X(M_m) = \varphi^*\mathcal{O}_Y(m)$ and the natural map $\mathcal{O}_{\varphi(\Gamma)} \to \varphi_*\mathcal{O}_\Gamma$ is injective, we have

$$(2)\quad h^0(\varphi(\Gamma), \mathcal{O}_{\varphi(\Gamma)}(m)) \leq h^0(\varphi(\Gamma), \mathcal{O}_Y(m) \otimes \varphi_*\mathcal{O}_\Gamma) = h^0(\Gamma, \mathcal{O}_\Gamma(M_m)).$$

Write $\Gamma_{|\Gamma} \sim G^+ - G^-$, where $G^+, G^- \geq 0$ are Cartier divisors on Γ. Consider the exact sequences

$$(3)\qquad 0 \to H^0(\Gamma, M_{m|\Gamma} - G^-) \to H^0(\Gamma, M_{m|\Gamma}) \to H^0(G^-, M_{m|G^-})$$

and

$$(4)\qquad \begin{aligned} H^0(X, M_m) &\to H^0(X, M_m + \Gamma) \\ &\to H^0(\Gamma, (M_m + \Gamma)_{|\Gamma}) \to H^1(X, M_m). \end{aligned}$$

Since $\operatorname{Fix}|mD| = m \operatorname{Fix}|D|$, the divisor Γ is a component of $\operatorname{Fix}|mD|$, hence the first map in (4) is an isomorphism and the last map in (4) is an injection. Therefore, from (2), (3) and (4) we have

$$\begin{aligned} h^0(\varphi(\Gamma), \mathcal{O}_{\varphi(\Gamma)}(m)) &\leq h^0(\Gamma, M_{m|\Gamma}) \\ &\leq h^0(\Gamma, M_{m|\Gamma} - G^-) + h^0(G^-, M_{m|G^-}) \\ &\leq h^0(\Gamma, (M_m + \Gamma)_{|\Gamma}) + h^0(G^-, M_{m|G^-}) \\ &\leq h^1(X, M_m) + h^0(G^-, M_{m|G^-}). \end{aligned}$$

As $h^0(G^-, M_{m|G^-}) \leq O(m^{n-2})$ for dimension reasons, it is enough to show that $h^1(X, M_m) \leq O(m^{n-2})$. To this end, consider the Leray spectral sequence

$$H^p(Y, R^{1-p}\varphi_*\mathcal{O}_X(M_m)) \Rightarrow H^1(X, \mathcal{O}_X(M_m)).$$

The terms $H^1(Y, \varphi_*\mathcal{O}_X(M_m)) = H^1(Y, \mathcal{O}_Y(m))$ vanish for $m \gg 0$ by Serre vanishing, so we need to prove

(5) $$h^0(Y, R^1\varphi_*\mathcal{O}_X(M_m)) \leq O(m^{n-2}).$$

Let $U \subseteq Y$ be the maximal open subset over which φ is an isomorphism. By [7, III.11.2], for each m the sheaf $R^1\varphi_*\mathcal{O}_X(M_m)$ is supported on the set $Y \setminus U$ of dimension at most $n-2$, hence

$$\chi(Y, R^1\varphi_*\mathcal{O}_X(M_m)) \leq O(m^{n-2}).$$

But by Serre vanishing again, all the higher cohomology groups of the sheaf $R^1\varphi_*\mathcal{O}_X(M_m)$ vanish for $m \gg 0$, and this implies (5). Q.E.D.

The following is the main result of this section – the geography of ample models.

Theorem 4.2. *Let X be a projective $\mathbb{Q}$-factorial variety, and let $\mathcal{C} \subseteq \mathrm{Div}_{\mathbb{R}}(X)$ be a rational polyhedral cone such that the ring $\mathfrak{R} = R(X, \mathcal{C})$ is finitely generated. Assume that* $\operatorname{Supp}\mathfrak{R}$ *contains a big divisor. Then there is a finite decomposition*

$$\operatorname{Supp}\mathfrak{R} = \coprod \mathcal{A}_i$$

into cones such that the following holds:

1. *each $\overline{\mathcal{A}_i}$ is a rational polyhedral cone,*
2. *for each i, there exists a normal projective variety X_i and a rational map $\varphi_i \colon X \dashrightarrow X_i$ such that φ_i is the ample model for every $D \in \mathcal{A}_i$,*

(3) *if $\mathcal{A}_j \subseteq \overline{\mathcal{A}_i}$, then there is a morphism $\varphi_{ij}\colon X_i \to X_j$ such that the diagram*

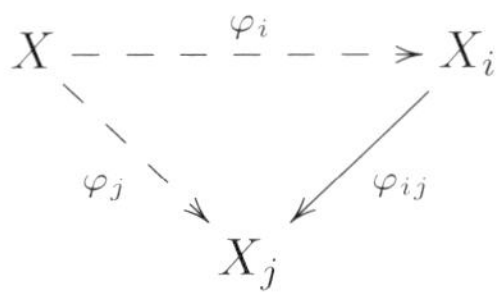

commutes.

(4) *if $\mathcal{A}_i$ contains a big divisor, then φ_i is a semiample model for every $D \in \overline{\mathcal{A}_i}$.*

Proof. Let $\operatorname{Supp}\mathfrak{R} = \bigcup \mathcal{C}_i$ be a finite rational polyhedral decomposition as in Theorem 3.2, and let $\mathcal{A}_i$ be the relative interior of $\mathcal{C}_i$ for each i. We show that this is the required decomposition.

Let $f\colon \widetilde{X} \to X$ be a resolution and let d be a positive integer as in Theorem 3.2. For each i, fix $D_i \in \mathcal{A}_i \cap \operatorname{Div}(X)$, and denote

$$M_i = \operatorname{Mob} f^*(dD_i) \quad \text{and} \quad F_i = \operatorname{Fix}|f^*(dD_i)|.$$

Then M_i is basepoint free, and let $\psi_i\colon \widetilde{X} \to X_i$ be the semiample fibration associated to M_i. Let $\varphi_i\colon X \dashrightarrow X_i$ be the induced map.

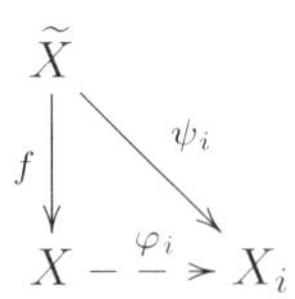

Claim 4.3. *Assume that $\mathcal{A}_j \subseteq \overline{\mathcal{A}_i}$, and let $C \subseteq \widetilde{X}$ be a curve such that $M_i \cdot C = 0$. Then $M_j \cdot C = 0$. In other words, all curves contracted by ψ_i are contracted by ψ_j.*

Indeed, since $\mathcal{A}_i$ is relatively open, there exist a divisor $D^\circ \in \mathcal{A}_i \cap \operatorname{Div}(X)$ and positive integers k_i, k_j, k° such that $k_iD_i = k^\circ D^\circ + k_jD_j$. By the definition of f and d, the divisor $M^\circ = \operatorname{Mob} f^*(dD^\circ)$ is basepoint free, and we have $k_iM_i = k^\circ M^\circ + k_jM_j$. In particular, if $C \subset \widetilde{X}$ is a curve such that $M_i \cdot C = 0$, then $M^\circ \cdot C = M_j \cdot C = 0$, which shows the claim.

The claim immediately implies $\varphi_j = \varphi_{ij} \circ \varphi_i$ for some morphism $\varphi_{ij}\colon X_i \to X_j$, which shows (3). In particular, when $i = j$ and since the divisors D_i are arbitrary, this shows that the definition of φ_i is independent of the choice of D_i up to isomorphism.

Finally, we prove (2) and (4). For any j, pick an index i such that $\mathcal{A}_j \subseteq \overline{\mathcal{A}_i}$ and $\mathcal{A}_i$ contains a big divisor, and let E be the sum of all f-exceptional prime divisors. Since $\operatorname{Mob}(f^*(dD_i) + E) = M_i$ and $\operatorname{Fix}|f^*(dD_i) + E| = F_i + E$, the divisors F_i and E are ψ_i-exceptional by Lemma 4.1, and in particular, φ_i is a contraction.

Let D be any divisor in $\mathcal{A}_j$; without loss of generality, we may assume that $D = D_j$. Since all functions o_Γ are linear on $\overline{\mathcal{A}_i}$, we have $\operatorname{Supp} F_j \subseteq \operatorname{Supp} F_i$, hence F_j is ψ_i-exceptional by the argument above. As $M_j = \psi_j^* \mathcal{O}_{X_j}(1)$, by (3) we have

$$f^*(dD_j) = \psi_i^*(\varphi_{ij}^* \mathcal{O}_{X_j}(1)) + F_j,$$

and the divisor $(\varphi_i)_*(dD_j) = (\psi_i)_* M_j = \varphi_{ij}^* \mathcal{O}_{X_j}(1)$ is basepoint free. We conclude that φ_i is a semiample model for D_j, and φ_j is the ample model for D_j. Q.E.D.

An immediate corollary is the following result from [8]; we prove the converse statement in the next section.

Corollary 4.4. *Let X be a $\mathbb{Q}$-factorial projective variety. If X is a Mori Dream Space, then its Cox ring is finitely generated.*

Proof. We first show that the divisorial ring $R(X, \overline{\operatorname{Mov}}(X))$ is finitely generated. Indeed, with notation from Definition 2.2, we have that

$$\overline{\operatorname{Mov}}(X) = \bigcup \mathcal{C}_j, \quad \text{where} \quad \mathcal{C}_j = f_j^* \operatorname{Nef}(X_j),$$

and hence it is enough to show that each ring $R(X, \mathcal{C}_j) \simeq R(X_j, \operatorname{Nef}(X_j))$ is finitely generated. But this is clear because each $\operatorname{Nef}(X_j)$ is spanned by finitely many semiample divisors.

Let $\mathcal{F}_i$ be all the faces of all $\mathcal{C}_j$ with the property that $\mathcal{F}_i \subseteq \partial\overline{\operatorname{Mov}}(X)$ and $\mathcal{F}_i \cap \operatorname{Big}(X) \neq \emptyset$. Let $\varphi_i \colon X \dashrightarrow X_i$ be the ample models associated to interiors of $\mathcal{F}_i$, cf. Theorem 4.2, and let E_{ik} be the exceptional divisors of φ_i. Denote $\mathcal{D}_i = \mathcal{F}_i + \sum_k \mathbb{R}_+ E_{ik}$, and note that each $\mathcal{D}_i$ is a rational polyhedral cone.

We claim that

$$\overline{\operatorname{Eff}}(X) = \overline{\operatorname{Mov}}(X) \cup \bigcup\nolimits_i \mathcal{D}_i.$$

To see this, let $D \in \operatorname{Big}(X) \backslash \overline{\operatorname{Mov}}(X)$ be a $\mathbb{Q}$-divisor. Then $P_\sigma(D)$ is a big $\mathbb{Q}$-divisor which belongs to $\partial\overline{\operatorname{Mov}}(X)$ by Lemma 2.10, and hence the ring $R(X, D)$ is finitely generated by the above. There is a face $\mathcal{F}_{i_0}$ which contains $P_\sigma(D)$ in its relative interior, and φ_{i_0} is the ample model of $P_\sigma(D)$ by Theorem 4.2. The divisor $N_\sigma(D)$ is contracted by φ_{i_0} by Lemma 4.1, and thus $D \in \mathcal{D}_{i_0}$. Therefore, we have

$$\mathrm{Big}(X) \subseteq \overline{\mathrm{Mov}}(X) \cup \bigcup\nolimits_i \mathcal{D}_i,$$

and by taking closures we obtain $\overline{\mathrm{Eff}}(X) \subseteq \overline{\mathrm{Mov}}(X) \cup \bigcup_i \mathcal{D}_i$. The converse inclusion is obvious.

In particular, the cone $\overline{\mathrm{Eff}}(X)$ is rational polyhedral, and the ring $R(X, \overline{\mathrm{Eff}}(X))$ is a Cox ring of X. Fix an index i and pick generators $G_1, \ldots, G_p$ of $\mathcal{D}_i$. It is enough to show that the ring $R(X; G_1, \ldots, G_p)$ is finitely generated. The map φ_i is a semiample model for each G_ℓ by Theorem 4.2(4), and thus $G_\ell = \varphi_i^* M_\ell + F_\ell$, where M_ℓ is a semiample $\mathbb{Q}$-divisor on X_i and F_ℓ is φ_i-exceptional. But then

$$R(X; G_1, \ldots, G_p) \simeq R(X_i; M_1, \ldots, M_p),$$

and the finite generation follows. Q.E.D.

The next theorem shows that in the classical setting of adjoint divisors, some of the ample models X_i from Theorem 4.2 are $\mathbb{Q}$-factorial. This is a known consequence of the classical Minimal Model Program [9, Theorem 3.3], however here we obtain the result directly.

Theorem 4.5. *Let X be a projective $\mathbb{Q}$-factorial variety, and let $\Delta_1, \ldots, \Delta_r$ be big $\mathbb{Q}$-divisors such that all pairs (X, Δ_i) are klt. Let*

$$\mathfrak{R} = R(X; K_X + \Delta_1, \ldots, K_X + \Delta_r),$$

and note that $\mathfrak{R}$ is finitely generated by Theorem 3.5. Assume that $\operatorname{Supp}\mathfrak{R}$ *contains a big divisor. Then there exist a finite decomposition* $\operatorname{Supp}\mathfrak{R} = \coprod \mathcal{A}_i$ *and maps $\varphi_i \colon X \dashrightarrow X_i$ as in Theorem 4.2, such that:*

(i) *if φ_i is birational, then X_i has rational singularities,*

(ii) *if the numerical classes of the elements of $\overline{\mathcal{A}_i}$ span $N^1(X)_\mathbb{R}$, then X_i is $\mathbb{Q}$-factorial.*

Proof. We assume the notation from the proof of Theorem 4.2. For (i), pick a big $\mathbb{Q}$-divisor Δ such that (X, Δ) is klt and $K_X + \Delta \in \mathcal{A}_i$. Then $(X_i, (\varphi_i)_*\Delta)$ is also klt because φ_i is $(K+\Delta)$-nonpositive, hence X_i has rational singularities.

We now show (ii). Let B be a Weil divisor on X_i, and let $\widetilde{B}$ be its proper transform on $\widetilde{X}$. As $\widetilde{X}$ is smooth, $\widetilde{B}$ is $\mathbb{Q}$-Cartier. Let $E_1, \ldots, E_k$ be all the f-exceptional prime divisors on $\widetilde{X}$. Since f is a resolution, we have

$$N^1(\widetilde{X})_\mathbb{R} = f^* N^1(X)_\mathbb{R} \oplus \bigoplus\nolimits_{j=1}^k \mathbb{R}[E_j]. \tag{6}$$

Let $B_1, \dots, B_r$ be integral divisors in $\mathcal{A}_i$ whose numerical classes generate $N^1(X)_\mathbb{R}$. Then, by (6) there are rational numbers p_j, r_j such that

$$\widetilde{B} \equiv \sum p_j f^*(dB_j) + \sum r_j E_j.$$

Denote

$$M = \sum p_j \operatorname{Mob} f^*(dB_j) \quad \text{and} \quad F = \sum p_j \operatorname{Fix}|f^*(dB_j)| + \sum r_j E_j.$$

By Theorem 4.2(4), there exist ample $\mathbb{Q}$-divisors A_j on X_i such that $\operatorname{Mob} f^*(dB_j) = \psi_i^* A_j$, hence $M \equiv_{X_i} 0$. Therefore

$$\widetilde{B} - F \equiv_{X_i} 0.$$

Observe that $\operatorname{Supp} F \subseteq \operatorname{Supp}(F_i + \sum E_j)$, and that the divisor $F_i + \sum E_j$ is ψ_i-exceptional by Lemma 4.1. By (i) and by [13, Proposition 12.1.4], there is a divisor $T \in \operatorname{Div}_\mathbb{Q}(X_i)$ such that $\widetilde{B} - F \sim_\mathbb{Q} \psi_i^* T$, and thus the divisor $B = (\psi_i)_* \widetilde{B} \sim_\mathbb{Q} T$ is $\mathbb{Q}$-Cartier. Q.E.D.

It is natural to ask whether the conclusion on $\mathbb{Q}$-factoriality from Theorem 4.5 can be extended to the general situation of Theorem 4.2. We argue below that such a statement is, in general, not true, and we pin down precisely the obstacle to $\mathbb{Q}$-factoriality. The astonishing conclusion is that, in some sense, $\mathbb{Q}$-factoriality of ample models is essentially a condition on the numerical equivalence classes of the divisors in $\operatorname{Supp} \mathfrak{R}$.

With the notation from Theorem 4.2, what we are aiming for is the following statement. We would like to have a (possibly finer) decomposition $\operatorname{Supp} \mathfrak{R} = \coprod \mathcal{N}_i$ together with birational maps $\varphi_i \colon X \dashrightarrow X_i$ such that φ_i is an optimal model for every $D \in \mathcal{N}_i$, and in particular, every X_i is $\mathbb{Q}$-factorial. It is immediate that, if the numerical classes of the elements of $\mathcal{N}_i$ span $N^1(X)_\mathbb{R}$, then φ_i is also the ample model for every $D \in \mathcal{N}_i$.

The following easy result gives us a necessary condition for the ample model of a big divisor to be $\mathbb{Q}$-factorial.

Lemma 4.6. *Let X be a $\mathbb{Q}$-factorial projective variety, and let D be a big $\mathbb{Q}$-divisor such that the ring $R(X, D)$ is finitely generated, and the map $\varphi \colon X \dashrightarrow \operatorname{Proj} R(X, D)$ is D-nonpositive. Let D' be a $\mathbb{Q}$-divisor such that $D \equiv D'$.*

Then the ring $R(X, D')$ is finitely generated if and only if the $\mathbb{Q}$-divisor $\varphi_* D'$ is $\mathbb{Q}$-Cartier.

Proof. If $R(X, D')$ is finitely generated, then by Lemma 3.11, φ is equal to the map $X \dashrightarrow \operatorname{Proj} R(X, D')$ up to isomorphism. Therefore $\varphi_* D'$ is ample, and in particular $\mathbb{Q}$-Cartier.

For the converse implication, denote $Y = \operatorname{Proj} R(X, D)$ and let $(p, q) \colon W \to X \times Y$ be a resolution of φ. By Lemma 2.6, we have

$$p^*(D - D') = q^*\varphi_*(D - D'),$$

hence $\varphi_* D \equiv \varphi_* D'$. Since $\varphi_* D$ is ample, so is $\varphi_* D'$, hence the ring $R(Y, \varphi_* D')$ is finitely generated. By Lemma 4.1, the divisor $E = p^* D - q^*\varphi_* D$ is effective and q-exceptional, and since $E = p^* D' - q^*\varphi_* D'$, we have $R(X, D') \simeq R(Y, \varphi_* D')$. Q.E.D.

Therefore, in the notation of Lemma 4.6, if the ample model of D is $\mathbb{Q}$-factorial, then the ring $R(X, D')$ is finitely generated for every $\mathbb{Q}$-divisor D' in the numerical class of D. This motivates the following key definition.

Definition 4.7. Let X be a $\mathbb{Q}$-factorial projective variety. We say that a divisor $D \in \operatorname{Div}_{\mathbb{Q}}(X)$ is *gen* if for all $\mathbb{Q}$-Cartier $\mathbb{Q}$-divisors $D' \equiv D$, the section ring $R(X, D')$ is finitely generated.

There are three main examples of gen divisors of interest to us:

(i) ample $\mathbb{Q}$-divisors are gen,
(ii) every adjoint divisor $K_X + \Delta + A$ is gen, where A is an ample $\mathbb{Q}$-divisor on X, and the pair (X, Δ) is klt; indeed, this follows from Theorem 3.5,
(iii) if $\operatorname{Pic}(X)_{\mathbb{Q}} = N^1(X)_{\mathbb{Q}}$, then every divisor with a finitely generated section ring is gen.

As we show in Section 5, having lots of gen divisors is essentially equivalent to being able to run a Minimal Model Program. We have seen above that this is a necessary condition for the models to be optimal, and in particular $\mathbb{Q}$-factorial. We show in Theorem 5.4 that, remarkably, this is also a sufficient condition. This, together with (ii) and (iii), explains *precisely* why we are able to run the MMP for adjoint divisors and on Mori Dream Spaces, and the details are worked out in Corollaries 5.6 and 5.7.

We conclude this section with an example where all the conditions of Theorem 4.2 are satisfied, but the absence of gen divisors implies that there is no decomposition of $\operatorname{Supp} \mathfrak{R}$ into regions of divisors that share an optimal model. In particular, we cannot run the MMP as explained in Section 5, and therefore the conditions from Theorems 5.2 are 5.4 are not only sufficient, but they are optimal. The example shows that the finite generation of a divisorial ring in itself is not sufficient to perform the Minimal Model Program.

Example 4.8. Let X, L_1 and L_2 be as in Example 3.10, and note that X is a smooth surface with $\dim N^1(X)_{\mathbb{R}} = 2$. We show that there exist a big divisor D and an ample divisor A on X such that the ring $\mathfrak{R} = R(X; D, A)$ is finitely generated, the divisor L_1 belongs to the interior of the cone $\operatorname{Supp}\mathfrak{R} = \mathbb{R}_+D + \mathbb{R}_+A$, and *none* of the divisors in the cone $\mathbb{R}_+D + \mathbb{R}_+L_1 \subseteq \operatorname{Supp}\mathfrak{R}$ is gen. In particular, we cannot perform the MMP for D.

We first claim that there exists an irreducible curve C on X such that

$$L_1 \cdot C = 0 \quad \text{and} \quad C^2 < 0. \tag{7}$$

Indeed, since L_1 is semiample but not ample, there exists an irreducible curve $C \subseteq X$ such that $L_1 \cdot C = 0$. Since L_1 is big and nef, we have $L_1^2 > 0$, so the Hodge index theorem then implies $C^2 < 0$.

Now, set $D = L_1 + C$. Since $\dim N^1(X)_{\mathbb{R}} = 2$ and D is not nef, it is immediate that there exists an ample divisor A on X such that $L_1 \in \mathbb{R}_+D + \mathbb{R}_+A$. In order to show that $\mathfrak{R}$ is finitely generated, it is enough to show that the rings $R(X; D, L_1)$ and $R(X; L_1, A)$ are finitely generated, and this latter ring is finitely generated since both L_1 and A are semiample.

For $k_1, k_2 \in \mathbb{N}$, consider the divisor

$$D_{k_1,k_2} = k_1D + k_2L_1 = (k_1 + k_2)L_1 + k_1C.$$

Then (7) implies that $P_\sigma(D_{k_1,k_2}) = (k_1 + k_2)L_1$, hence $H^0(X, D_{k_1,k_2}) \simeq H^0(X, (k_1 + k_2)L_1)$. Therefore the ring

$$R(X; D, L_1) \simeq R(X; L_1, L_1)$$

is finitely generated.

Finally, note that $D_{k_1,k_2} \equiv (k_1 + k_2)L_2 + k_1C$, and that $P_\sigma((k_1 + k_2)L_2 + k_1C) = (k_1 + k_2)L_2$. Therefore the ring

$$R(X, (k_1 + k_2)L_2 + k_1C) \simeq R(X, (k_1 + k_2)L_2)$$

is not finitely generated, thus the divisor D_{k_1,k_2} is not gen.

Remark 4.9. The notion of genness is a very subtle one. For instance, every $\mathbb{Q}$-divisor D with $\kappa_\sigma(D) = 0$ is gen (for the definition and properties of κ_σ see [15]). Indeed, for every $\mathbb{Q}$-divisor $D' \equiv D$ we have $\kappa(D') \leq \kappa_\sigma(D') = 0$, hence the ring $R(X, D)$ is isomorphic to either $\mathbb{C}$ or to the polynomial ring $\mathbb{C}[T]$.

§5. Running the D-MMP

Let X be a projective $\mathbb{Q}$-factorial variety, and let $\mathcal{C} \subseteq \mathrm{Div}_{\mathbb{R}}(X)$ be a rational polyhedral cone such that the divisorial ring $\mathfrak{R} = R(X, \mathcal{C})$ is finitely generated. Then by Theorem 4.2 we know that $\operatorname{Supp}\mathfrak{R}$ has a decomposition into finitely many rational polyhedral cones giving the geography of ample models associated to $\mathfrak{R}$.

In this section we explain how, when all divisors in the interior of $\operatorname{Supp}\mathfrak{R}$ are gen, the aforementioned decomposition can be refined to give a geography of optimal models. As indicated in the previous sections, the main technical obstacle is to prove $\mathbb{Q}$-factoriality of models, and this is the point where the gen condition on divisors plays a crucial role.

We assume that $\operatorname{Supp}\mathfrak{R}$ contains an ample divisor, and fix a divisor $D \in \operatorname{Supp}\mathfrak{R}$. Then we can run the Minimal Model Program for D as follows.

We define a certain finite rational polyhedral decomposition $\mathcal{C} = \bigcup \mathcal{N}_i$ in Theorem 5.4. If D is not nef, we show in Theorem 5.2 that there is a D-negative birational map $\varphi\colon X \dashrightarrow X^+$ such that X^+ is $\mathbb{Q}$-factorial, and φ is elementary – this corresponds to contractions of extremal rays in the classical MMP. We also show that there is a rational polyhedral subcone $D \in \mathcal{C}' \subseteq \mathcal{C}$ which is a union of *some, but not all* of the cones $\mathcal{N}_i$, such that $R(X, \mathcal{C}') \simeq R(X^+, \varphi_*\mathcal{C}')$ and the cone $\varphi_*\mathcal{C}' \subseteq \mathrm{Div}_{\mathbb{R}}(X^+)$ contains an ample divisor. Now we replace X by X^+, D by φ_*D, and $\mathcal{C}$ by $\varphi_*\mathcal{C}'$, and we repeat the procedure. Since there are only finitely many cones $\mathcal{N}_i$, this process must terminate with a variety X_D on which the proper transform of D is nef, and this is the optimal model for D. It is then automatic that X_D is also an optimal model for all divisors in the cone $\mathcal{N}_{i_0} \ni D$. The details are given in Theorem 5.4.

In the context of adjoint divisors and the classical MMP, we can additionally *direct* the MMP by an ample $\mathbb{Q}$-divisor A on X, as in [4]. The proofs of Theorems 5.2 and 5.4 can be easily modified to obtain the *D-MMP with scaling of A*, however we do not pursue this here.

First we define elementary contractions.

Definition 5.1. A birational contraction $\varphi\colon X \dashrightarrow Y$ between normal projective varieties is *elementary* if it not an isomorphism, and it is either an isomorphism in codimension 1, or a morphism whose exceptional locus is a prime divisor on X.

The following theorem is the key result: it shows that in our situation elementary contractions exist.

Theorem 5.2. *Let X be a projective $\mathbb{Q}$-factorial variety and let $\mathcal{C} \subseteq \mathrm{Div}_{\mathbb{R}}(X)$ be a rational polyhedral cone. Denote by* $\pi\colon \mathrm{Div}_{\mathbb{R}}(X) \to$

$N^1(X)_{\mathbb{R}}$ the natural projection. Assume that the ring $\mathfrak{R} = R(X, \mathcal{C})$ is finitely generated, that $\operatorname{Supp}\mathfrak{R}$ *contains an ample divisor, that* $\pi(\operatorname{Supp}\mathfrak{R})$ *spans $N^1(X)_{\mathbb{R}}$, and that every divisor in the interior of* $\operatorname{Supp}\mathfrak{R}$ *is gen. Let* $\operatorname{Supp}\mathfrak{R} = \bigcup \mathcal{C}_i$ *be a decomposition as in Theorem 3.2. Let $D \in$* $\operatorname{Supp}\mathfrak{R}$ *be a $\mathbb{Q}$-divisor which is not nef. Then:*

(1) the cone $\operatorname{Supp}\mathfrak{R} \cap \pi^{-1}\big(\operatorname{Nef}(X)\big)$ *is rational polyhedral, and every element of this cone is semiample,*

(2) there exists a rational hyperplane $\mathcal{H} \subseteq N^1(X)_{\mathbb{R}}$ which intersects the interior of $\pi(\operatorname{Supp}\mathfrak{R})$ *and contains a codimension* 1 *face of* $\pi(\operatorname{Supp}\mathfrak{R}) \cap \operatorname{Nef}(X)$*, such that $\pi(D)$ and* $\operatorname{Nef}(X)$ *are on the opposite sides of $\mathcal{H}$,*

(3) let $\mathcal{W} \subseteq N^1(X)_{\mathbb{R}}$ be the half-space bounded by $\mathcal{H}$ which does not contain $\operatorname{Nef}(X)$*, and let* $\mathcal{C}' = \operatorname{Supp}\mathfrak{R} \cap \pi^{-1}(\mathcal{W})$*. Then there exists a $\mathbb{Q}$-factorial projective variety X^+ together with an elementary contraction $\varphi\colon X \dashrightarrow X^+$, such that φ is W-nonpositive for every $W \in \mathcal{C}'$, and it is W-negative for every $W \in \mathcal{C}' \backslash \pi^{-1}(\mathcal{H})$,*

(4) we have $R(X, \mathcal{C}') \simeq R(X^+, \mathcal{C}^+)$, where $\mathcal{C}^+ = \varphi_*\mathcal{C}' \subseteq \operatorname{Div}_{\mathbb{R}}(X^+)$, and $\mathcal{C}^+$ contains an ample divisor,

(5) for every cone $\mathcal{C}_i$ and for every geometric valuation Γ over X, the function o_Γ is linear on the cone $\varphi_*(\mathcal{C}' \cap \mathcal{C}_i) \subseteq \mathcal{C}^+$.

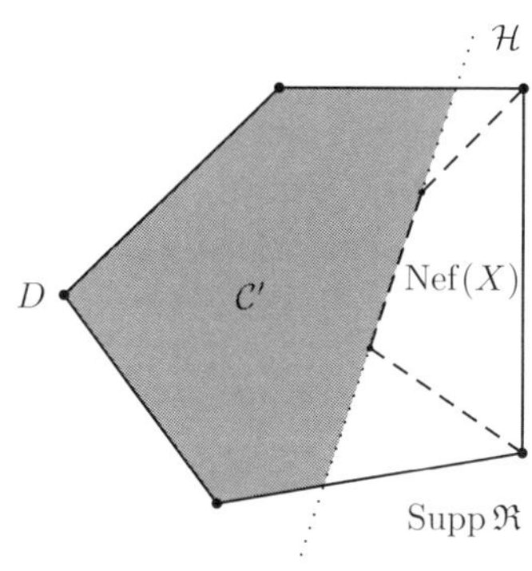

Proof. *Step 1.* The statement (1) follows immediately from Corollary 3.4, statement (4) follows from (3) by Remark 2.4(i) and from the construction below, while (5) follows from (3) by Lemma 2.11. To show (2), let α be any ample class in the interior of $\pi(\operatorname{Supp}\mathfrak{R}) \subseteq N^1(X)$, and let β be the intersection of the segment $[\pi(D), \alpha]$ with $\partial \operatorname{Nef}(X)$. Then β lies in the interior of $\pi(\operatorname{Supp}\mathfrak{R})$, and by (1) there is a rational codimension 1 face of $\pi(\operatorname{Supp}\mathfrak{R}) \cap \operatorname{Nef}(X)$ containing β. We define $\mathcal{H}$

to be the rational hyperplane containing that face.

Step 2. It remains to show (3). By Corollary 3.4, there are cones $\mathcal{C}_j \not\subseteq \pi^{-1}\big(\mathrm{Nef}(X)\big)$ and $\mathcal{C}_k \subseteq \pi^{-1}\big(\mathrm{Nef}(X)\big)$ such that $\dim \pi(\mathcal{C}_j) = \dim \pi(\mathcal{C}_k) = \dim N^1(X)_{\mathbb{R}}$ and $\pi(\mathcal{C}_j) \cap \pi(\mathcal{C}_k) \subseteq \mathcal{H}$; denote $\mathcal{C}_{jk} = \mathcal{C}_j \cap \mathcal{C}_k$. Let $\varphi\colon X \dashrightarrow X^+$ and $\theta\colon X \dashrightarrow Y$ be the ample models associated to relative interiors of $\mathcal{C}_j$ and $\mathcal{C}_{jk}$ as in the proof of Theorem 4.2, and note that θ is a morphism by (1) since

$$\mathcal{C}_{jk} \subseteq \mathrm{Supp}\,\mathfrak{R} \cap \pi^{-1}\big(\mathrm{Nef}(X)\big).$$

Then, by Theorem 4.2(3), there is a morphism $\theta^+\colon X^+ \to Y$ such that the diagram

$$\begin{array}{ccc} X & \overset{\varphi}{\dashrightarrow} & X^+ \\ & {\scriptstyle\theta}\searrow \quad \swarrow{\scriptstyle\theta^+} & \\ & Y & \end{array}$$

is commutative. The following is the key claim:

Claim 5.3. *Let F be an $\mathbb{R}$-divisor on X such that $\pi(F) \in \mathcal{H}$. Then $F \sim_{\mathbb{R}} \theta^* F_Y$ for some $F_Y \in \mathrm{Div}_{\mathbb{R}}(Y)$. If additionally $\pi(F) \in \pi(\mathcal{C}_{jk})$, then F_Y is ample. In particular, a curve C is contracted by θ if and only if $C \cdot \delta = 0$ for every $\delta \in \mathcal{H}$.*

Pick $\mathbb{Q}$-divisors $B_1, \dots, B_r$ in $\mathcal{C}_{jk}$ and nonzero real numbers λ_i such that $\pi(B_i)$ span $\mathcal{H}$ and $\pi(F) = \sum \lambda_i \pi(B_i)$. We may assume that $\lambda_i \geq 0$ for all i when $\pi(F) \in \pi(\mathcal{C}_{jk})$. Hence, there is a $\mathbb{Q}$-divisor $B_1' \equiv B_1$ such that

$$F = \lambda_1 B_1' + \sum_{i \geq 2} \lambda_i B_i.$$

Note that, by the definition of θ, there are ample divisors A_i on Y such that $B_i \sim_{\mathbb{Q}} \theta^* A_i$ for all $i \geq 2$.

Since B_1 is gen, the ring $R(X, B_1')$ is finitely generated, and therefore B_1' is semiample by Lemma 3.3(2) as it is nef and big. Denote by $\theta'\colon X \to Y'$ the semiample fibration associated to B_1'. By Lemma 3.11, there is an isomorphism $\eta\colon Y \to Y'$ such that $\theta' = \eta \circ \theta$. Since $B_1' \sim_{\mathbb{Q}} (\theta')^* A_1'$ for an ample divisor A_1' on Y', we have $B_1' \sim_{\mathbb{Q}} \theta^* A_1$, where $A_1 = \eta^* A_1'$. Therefore $F \sim_{\mathbb{R}} \theta^*(\sum \lambda_i A_i)$, which proves the claim.

Step 3. We next show that X^+ is $\mathbb{Q}$-factorial.

Consider a Weil divisor P^+ on X^+, and let P be its proper transform on X. Since X is $\mathbb{Q}$-factorial, the divisor P is $\mathbb{Q}$-Cartier. Since $\dim \pi(\mathcal{C}_j) = \dim N^1(X)_{\mathbb{R}}$, there exist a $\mathbb{Q}$-divisor $G \in \mathcal{C}_j$ and $\alpha \in \mathbb{Q}$ such that $\pi(P+\alpha G)\in\mathcal{H}$. By Claim 5.3, there exists $M\in\mathrm{Div}_{\mathbb{Q}}(Y)$ such that $P+\alpha G\sim_{\mathbb{Q}}\theta^*M$. Let $(p,q)\colon\widetilde{X}\to X\times X^+$ be a resolution of φ. By the definition of φ and by Theorem 4.2, there is an ample $\mathbb{Q}$-divisor A on X^+ and a q-exceptional $\mathbb{Q}$-divisor E on $\widetilde{X}$ such that $p^*G=q^*A+E$. It follows that

$$p^*P\sim_{\mathbb{Q}}(\theta\circ p)^*M-\alpha(q^*A+E)=(\theta^+\circ q)^*M-\alpha q^*A-\alpha E.$$

Since φ is a contraction, we have $P^+=q_*p^*P$, and therefore the divisor

$$P^+\sim_{\mathbb{Q}}(\theta^+)^*M-\alpha A$$

is $\mathbb{Q}$-Cartier.

Step 4. In this step we show that φ is an elementary map.

If θ is an isomorphism in codimension 1, then so are φ and θ^+ as φ is a contraction.

Hence, we may assume that there exists a θ-exceptional prime divisor E. Let C be a curve contracted by θ, and let R be a ray in $N_1(X)_{\mathbb{R}}$ orthogonal to the hyperplane $\mathcal{H}$. Then the class of C belongs to R by Claim 5.3, and so $E\cdot R<0$ by Lemma 2.5. In particular, we have $E\cdot C<0$, thus $C\subseteq E$, and the exceptional locus of θ equals E. Therefore, θ is an elementary contraction.

Observe that $\pi(E)$ and $\mathrm{Nef}(X)$ lie on opposite sides of $\mathcal{H}$. This implies that there is a $\mathbb{Q}$-divisor G_E in the relative interior of $\mathcal{C}_j$ such that $\pi(G_E-E)$ belongs to the relative interior of $\mathcal{C}_{jk}$. Then, by Claim 5.3, there exists an ample divisor $M_E\in\mathrm{Div}_{\mathbb{Q}}(Y)$ such that $G_E-E\sim_{\mathbb{Q}}\theta^*M_E$, and thus

$$H^0(X,mG_E)\simeq H^0(X,m\theta^*M_E) \tag{8}$$

for every positive integer m. Since φ is the map $X\dashrightarrow\mathrm{Proj}\,R(X,G_E)$ by definition, we may assume that $X^+=Y$ and $\varphi=\theta$ by (8), which shows that φ is an elementary contraction.

Step 5. The only thing left to prove is the last statement in (3). For $W\in\mathcal{C}'$, there exists an $\mathbb{R}$-divisor $G_W\in\mathcal{C}_j$ such that $\pi(W-G_W)\in\mathcal{H}$. Thus $W\equiv_Y G_W$ by Claim 5.3. Since φ is G_W-nonpositive by Theorem 4.2(4), this implies that φ is W-nonpositive by Corollary 2.6. If φ is an isomorphism in codimension 1, it is automatic that it is then also W-negative.

If $W\in\mathcal{C}'\backslash\pi^{-1}(\mathcal{H})$ and φ contracts a divisor E, there exists a positive rational number λ such that $\pi(W-\lambda E)\in\mathcal{H}$. Again by Claim 5.3, and since $X^+=Y$ and $\varphi=\theta$, there is a divisor $M_W\in\mathrm{Div}_{\mathbb{R}}(X^+)$ such that $W-\lambda E\sim_{\mathbb{R}}\varphi^*M_W$. But then it is clear that φ is W-negative. Q.E.D.

The following is the main result of this paper – the geography of optimal models.

Theorem 5.4. *Let X be a projective $\mathbb{Q}$-factorial variety, and let $\mathcal{C} \subseteq \mathrm{Div}_{\mathbb{R}}(X)$ be a rational polyhedral cone. Denote by $\pi\colon \mathrm{Div}_{\mathbb{R}}(X) \to N^1(X)_{\mathbb{R}}$ the natural projection. Assume that the ring $\mathfrak{R} = R(X, \mathcal{C})$ is finitely generated, that $\operatorname{Supp}\mathfrak{R}$ contains an ample divisor, that $\pi(\operatorname{Supp}\mathfrak{R})$ spans $N^1(X)_{\mathbb{R}}$, and that every divisor in the interior of $\operatorname{Supp}\mathfrak{R}$ is gen.*

Then for any $\mathbb{Q}$-divisor $D \in \mathcal{C}$, we can run a D-MMP which terminates.

Furthermore, there is a finite decomposition

$$\operatorname{Supp}\mathfrak{R} = \coprod \mathcal{N}_i$$

into cones having the following properties:

(1) *each $\overline{\mathcal{N}_i}$ is a rational polyhedral cone,*

(2) *for each i, there exists a $\mathbb{Q}$-factorial projective variety X_i and a birational contraction $\varphi_i\colon X \dashrightarrow X_i$ such that φ_i is an optimal model for every divisor in $\mathcal{N}_i$,*

(3) *every element of the cone $(\varphi_i)_*\mathcal{N}_i$ is semiample.*

Proof. Denote by $V \subseteq \mathrm{Div}_{\mathbb{R}}(X)$ the minimal vector space containing $\mathcal{C}$, and define $\mathcal{C}^1 = \operatorname{Supp}\mathfrak{R}$. Let $\mathcal{C}^1 = \bigcup_{i\in I_1}\mathcal{C}^1_i$ be the rational polyhedral decomposition as in Theorem 3.2. By subdividing $\mathcal{C}^1$ further, we may assume that the following property is satisfied:

($\natural$) let $\mathcal{G} \subseteq V$ be any hyperplane which contains a codimension 1 face of some $\mathcal{C}^1_{i_0}$. Then every $\mathcal{C}^1_i$ is contained in one of the two half-spaces of V bounded by $\mathcal{G}$.

For each $i \in I_1$, let $\mathcal{N}_i$ be the relative interior of $\mathcal{C}_i$. We claim that $\mathcal{C}^1 = \coprod_{i\in I_1}\mathcal{N}_i$ is the desired decomposition.

Let D be a point in some $\mathcal{N}_{i_0}$. If D is nef, then every divisor in $\mathcal{N}_{i_0}$ is semiample by Corollary 3.4, so the theorem follows.

Therefore, we may assume that D is not nef. Denote $Y_1 = X$ and $D_1 = D$. We show that there exists a D_1-MMP which terminates.

By Theorem 5.2, the cone $\mathcal{C}^1 \cap \pi^{-1}\big(\mathrm{Nef}(Y_1)\big)$ is rational polyhedral. Let $\mathcal{H} \subseteq N^1(Y_1)_{\mathbb{R}}$ be a rational hyperplane as in Theorem 5.2, and let $\mathcal{C}^1_\ell$, for $\ell \in I_2 \subsetneq I_1$, be those cones for which $\pi(\mathcal{C}^1_\ell)$ and $\pi(D)$ are on the same side of $\mathcal{H}$, cf. ($\natural$). Let $f_1\colon Y_1 \dashrightarrow Y_2$ be an elementary map as in Theorem 5.2(3), and denote $D_2 = (f_1)_*D_1$. Define rational polyhedral cones $\mathcal{C}^2_\ell = (f_1)_*\mathcal{C}^1_\ell \subseteq \mathrm{Div}_{\mathbb{R}}(Y_2)$, and set

$$\mathcal{C}^2 = \textstyle\bigcup_{\ell\in I_2}\mathcal{C}^2_\ell. \tag{9}$$

Then the ring $\mathfrak{R}^2 = R(Y_2, \mathcal{C}^2)$ is finitely generated by Theorem 5.2(4). By Theorem 5.2(5), the relation (9) gives a decomposition of $\mathcal{C}^2$ as in Theorem 3.2. Also note that $(f_1)_*(\mathcal{N}_{i_0}) \subseteq \mathcal{C}^2$.

In this way we construct a sequence of divisors D_p on $\mathbb{Q}$-factorial varieties Y_p. Since the size of the index sets I_p drops with each step, this process must terminate with a model X_{p_0} on which the divisor D_{p_0} is nef. Similarly as above, X_{p_0} is an optimal model for all divisors in $\mathcal{N}_{i_0}$, and the proper transform on Y_{p_0} of every element of $\mathcal{N}_{i_0}$ is semiample. Q.E.D.

Corollary 5.5. *Let X be a projective $\mathbb{Q}$-factorial variety, let $S_1, \dots, S_p$ be distinct prime divisors on X, denote $V = \sum_{i=1}^p \mathbb{R}S_i \subseteq \mathrm{Div}_{\mathbb{R}}(X)$, and let A be an ample $\mathbb{Q}$-divisor on X. Let $\mathcal{C} \subseteq \mathcal{L}(V)$ be a rational polytope such that for every $\Delta \in \mathcal{C}$, the pair (X, Δ) is klt.*

Then there exists a positive integer M such that for every $\Delta \in \mathcal{C} \cap \mathcal{E}_A(V)$, there is a $(K_X + \Delta)$-MMP consisting of at most M steps.

Proof. By enlarging V and $\mathcal{C}$, we may assume that the numerical classes of the elements of $\mathcal{C} \cap \mathcal{E}_A(V)$ span $N^1(X)_{\mathbb{R}}$. The set $\mathcal{C} \cap \mathcal{E}_A(V)$ is a rational polytope by Corollary 3.6, and let $B_1, \dots, B_r$ be its vertices. Choose a positive integer $\lambda \gg 0$ such that all $K_X + A + B_i + \lambda A$ are ample. Denote

$$\mathcal{D} = \sum \mathbb{R}_+(K_X + A + B_i) + \sum \mathbb{R}_+(K_X + A + B_i + \lambda A).$$

Then the ring $\mathfrak{R} = R(X, \mathcal{D})$ is finitely generated by Theorem 3.5, and we have $\mathbb{R}_+(K_X + A + \mathcal{C} \cap \mathcal{E}_A(V)) \subseteq \mathrm{Supp}\,\mathfrak{R}$. Let $\mathrm{Supp}\,\mathfrak{R} = \coprod_{i=1}^N \mathcal{N}_i$ be the decomposition as in Theorem 5.4. Then it is immediate from the proof of Theorem 5.4 that we can set $M = N$. Q.E.D.

The following corollary is finiteness of models, cf. [1, Lemma 7.1].

Corollary 5.6. *Let X be a projective $\mathbb{Q}$-factorial variety, let $S_1, \dots, S_p$ be distinct prime divisors on X, denote $V = \sum_{i=1}^p \mathbb{R}S_i \subseteq \mathrm{Div}_{\mathbb{R}}(X)$, and let A be an ample $\mathbb{Q}$-divisor on X. Let $\mathcal{C} \subseteq \mathcal{L}(V)$ be a rational polytope such that for every $\Delta \in \mathcal{C}$, the pair (X, Δ) is klt.*

Then there are finitely many rational maps $\varphi_i \colon X \dashrightarrow Y_i$, with the property that if $\Delta \in \mathcal{C} \cap \mathcal{E}_A(V)$, then there is an index i such that φ_i is a log terminal model of $K_X + \Delta$.

Proof. By enlarging V and $\mathcal{C}$, we may assume that the numerical classes of the elements of $\mathcal{C} \cap \mathcal{E}_A(V)$ span $N^1(X)_{\mathbb{R}}$, and that there exists a divisor $B \in \mathcal{C} \cap \mathcal{E}_A(V)$ such that $K_X + A + B$ is ample. The ring $R(X, \mathbb{R}_+(K_X + A + \mathcal{C} \cap \mathcal{E}_A(V)))$ is finitely generated by Corollary 3.6, so the result follows immediately from Theorem 5.4. Q.E.D.

Finally, we recover one of the main results of [8].

Corollary 5.7. *Let X be a $\mathbb{Q}$-factorial projective variety such that $\mathrm{Pic}(X)_{\mathbb{Q}} = N^1(X)_{\mathbb{Q}}$. Then X is a Mori Dream Space if and only if its Cox ring is finitely generated.*

In particular, if (X, Δ) is a klt log Fano pair, then X is a Mori Dream Space.

Proof. Let $D_1, \ldots, D_r$ be a basis of $\mathrm{Pic}(X)_{\mathbb{Q}}$ such that $\overline{\mathrm{Eff}}(X) \subseteq \sum \mathbb{R}_+ D_i$. The associated divisorial ring $\mathfrak{R} = R(X; D_1, \ldots, D_r)$ is a Cox ring of X. Corollary 4.4 shows that if X is a Mori Dream Space, then $\mathfrak{R}$ is finitely generated. We now prove the converse statement.

Assume that $\mathfrak{R}$ is finitely generated, and let $\mathrm{Supp}\,\mathfrak{R} = \coprod_{i=1}^N \mathcal{N}_i$ be the decomposition from Theorem 5.4. Then $\mathrm{Nef}(X)$ is the span of finitely many semiample divisors by Corollary 3.4, and by the definition of the sets $\mathcal{N}_i$ and by Corollary 3.4, there is a set $I \subseteq \{1, \ldots, N\}$ such that

$$\overline{\mathrm{Mov}}(X) = \bigcup\nolimits_{i \in I} \overline{\mathcal{N}_i}.$$

By taking a smaller index set I, we may assume that the dimension of $\overline{\mathcal{N}_i}$ equals $\dim N^1(X)_{\mathbb{R}}$ for all $i \in I$. For $i \in I$, let $\varphi_i \colon X \dashrightarrow X_i$ be the maps as in Theorem 5.4. Then $\overline{\mathcal{N}_i} \subseteq \varphi_i^*\big(\mathrm{Nef}(X_i)\big)$, and hence

$$\overline{\mathrm{Mov}}(X) \subseteq \bigcup\nolimits_{i \in I} \varphi_i^*\big(\mathrm{Nef}(X_i)\big).$$

Each φ_i is an optimal model for every divisor in $\mathcal{N}_i$, thus each φ_i is an isomorphism in codimension 1. Therefore, $R(X_i; (\varphi_i)_* D_1, \ldots, (\varphi_i)_* D_r)$ is a Cox ring of X_i, and it is finitely generated since it is isomorphic to $\mathfrak{R}$. In particular, every $\mathrm{Nef}(X_i)$ is spanned by finitely many semiample divisors by above, and hence

$$\overline{\mathrm{Mov}}(X) \supseteq \bigcup\nolimits_{i \in I} \varphi_i^*\big(\mathrm{Nef}(X_i)\big).$$

This shows that X is a Mori Dream Space.

Now, if (X, Δ) is a klt log Fano pair, then $H^i(X, \mathcal{O}_X) = 0$ for all $i > 0$ by Kawamata-Viehweg vanishing. The long exact sequence in cohomology associated to the exponential sequence

$$0 \to \mathbb{Z} \to \mathcal{O}_X \to \mathcal{O}_X^* \to 0$$

shows that $\mathrm{Pic}(X)_{\mathbb{Q}} = N^1(X)_{\mathbb{Q}}$. Let $D_1, \ldots, D_r$ be a basis of $\mathrm{Pic}(X)_{\mathbb{Q}}$ such that $\overline{\mathrm{Eff}}(X) \subseteq \sum \mathbb{R}_+ D_i$, and pick a rational number $0 < \varepsilon \ll 1$ such that $A_i = \varepsilon D_i - (K_X + \Delta)$ is ample for every i. Then the ring

$$R(X; \varepsilon D_1, \ldots, \varepsilon D_r) = R(X; K_X + \Delta + A_1, \ldots, K_X + \Delta + A_r)$$

is finitely generated by Theorem 3.5, hence the ring $R(X; D_1, \dots, D_r)$ is finitely generated by Lemma 3.1. Q.E.D.

References

[1] C. Birkar, P. Cascini, C. D. Hacon, and J. McKernan, *Existence of minimal models for varieties of log general type*, J. Amer. Math. Soc. **23** (2010), no. 2, 405–468.

[2] Th. Bauer, A. Küronya, and T. Szemberg, *Zariski decompositions, volumes, and stable base loci*, J. für die reine und angew. Math. **576** (2004), 209–233.

[3] P. Cascini and V. Lazić, *New outlook on the Minimal Model Program, I*, Duke Math. J. **161** (2012), no. 12, 2415–2467.

[4] A. Corti and V. Lazić, *New outlook on the Minimal Model Program, II*, Math. Ann. **356** (2013), no. 2, 617–633.

[5] O. Debarre, *Higher-dimensional algebraic geometry*, Universitext, Springer-Verlag, New York, 2001.

[6] L. Ein, R. Lazarsfeld, M. Mustaţă, M. Nakamaye, and M. Popa, *Asymptotic invariants of base loci*, Ann. Inst. Fourier (Grenoble) **56** (2006), no. 6, 1701–1734.

[7] R. Hartshorne, *Algebraic geometry*, Graduate Texts in Mathematics, vol. 52, Springer-Verlag, New York, 1977.

[8] Y. Hu and S. Keel, *Mori dream spaces and GIT*, Michigan Math. J. **48** (2000), 331–348.

[9] C. D. Hacon and J. McKernan, *The Sarkisov program*, J. Algebraic Geom. **22** (2013), no. 2, 389–405.

[10] J. Kollár et al., *Flips and abundance for algebraic threefolds*, Astérisque 211, Soc. Math. France, Paris, 1992.

[11] Y. Kawamata, *Crepant blowing-up of 3-dimensional canonical singularities and its application to degenerations of surfaces*, Ann. of Math. (2) **127** (1988), no. 1, 93–163.

[12] ———, *On the cone of divisors of Calabi-Yau fiber spaces*, Internat. J. Math. **8** (1997), 665–687.

[13] J. Kollár and S. Mori, *Classification of three-dimensional flips*, J. Amer. Math. Soc. **5** (1992), no. 3, 533–703.

[14] R. Lazarsfeld, *Positivity in algebraic geometry. I, II*, Ergebnisse der Mathematik und ihrer Grenzgebiete, vol. 48, 49, Springer-Verlag, Berlin, 2004.

[15] N. Nakayama, *Zariski-decomposition and abundance*, MSJ Memoirs, vol. 14, Mathematical Society of Japan, Tokyo, 2004.

[16] M. Reid, *Canonical 3-folds*, Journées de Géométrie Algébrique d'Angers (A. Beauville, ed.), Sijthoof and Nordhoof, Alphen aan den Rijn, 1980, pp. 273–310.

[17] V. V. Shokurov, *3-fold log models*, J. Math. Sci. **81** (1996), no. 3, 2667–2699.

Department of Mathematics, Imperial College London, 180 Queen's Gate, London SW7 2AZ, UK
E-mail address: a.kaloghiros@imperial.ac.uk

Budapest University of Technology and Economics, Department of Algebra, P.O. Box 91, 1521 Budapest, Hungary

Albert-Ludwigs-Universität Freiburg, Mathematisches Institut, Eckerstraße 1, 79104 Freiburg, Germany
E-mail address: alex.kueronya@math.uni-freiburg.de

Mathematisches Institut, Universität Bonn, Endenicher Allee 60, 53115 Bonn, Germany
E-mail address: lazic@math.uni-bonn.de

Advanced Studies in Pure Mathematics 70, 2016
Minimal Models and Extremal Rays (Kyoto, 2011)
pp. 247–275

Local structure of principally polarized stable Lagrangian fibrations

Jun-Muk Hwang and Keiji Oguiso

Dedicated to Professor Shigefumi Mori

Abstract.

A holomorphic Lagrangian fibration is stable if the characteristic cycles of the singular fibers are of type $I_m, 1 \leq m < \infty$, or A_∞. We will give a complete description of the local structure of a stable Lagrangian fibration when it is principally polarized. In particular, we give an explicit form of the period map of such a fibration and conversely, for a period map of the described type, we construct a principally polarized stable Lagrangian fibration with the given period map. This enables us to give a number of examples exhibiting interesting behavior of the characteristic cycles.

§1. Introduction

For a holomorphic symplectic manifold (M, ω), i.e., a $2n$-dimensional complex manifold with a holomorphic symplectic form $\omega \in H^0(M, \Omega^2_M)$, a proper flat morphism $f : M \to B$ over an n-dimensional complex manifold B is called a (holomorphic) Lagrangian fibration if all smooth fibers are Lagrangian submanifolds of M. The discriminant $D \subset B$, i.e., the set of critical values of f, is a hypersurface if it is non-empty. In [7], the structure of the singular fiber of f at a general point $b \in D$ was studied. By introducing the notion of characteristic cycles, [7] shows that the structure of such a singular fiber can be described in a manner completely parallel to Kodaira's classification ([12], see also V. 7 in [1])

Received December 26, 2011.
Revised September 5, 2012.
2010 *Mathematics Subject Classification.* 14D05,32G20.
Key words and phrases. holomorphic Lagrangian fibration.
Jun-Muk Hwang is supported by National Researcher Program 2010-0020413 of NRF, and Keiji Oguiso is supported by by JSPS Gran-in-Aid (B) No 22340009, JSPS Grant-in-Aid (S), No 22224001, and by KIAS Scholar Program.

of singular fibers of elliptic fibrations. Furthermore, to study the multiplicity of the singular fibers, [8] generalized the stable reduction theory of elliptic fibrations (cf. V.10 in [1]), explicitly describing how arbitrary singular fiber over a general point of D can be transformed to a stable singular fiber, a singular fiber of particularly simple type. These results exhibit that the theory of general singular fibers of a holomorphic Lagrangian fibration gives a very natural generalization of Kodaira's theory of elliptic fibrations.

The current work is yet another manifestation of this principle. An important part of Kodaira's theory is the study of the asymptotic behavior of the elliptic modular function of a given elliptic fibration near a singular fiber. As a generalization of this we will study the asymptotic behavior of the periods of the abelian fibers near a general singular fiber of a holomorphic Lagrangian fibration. Here we need to make two additional assumptions on the Lagrangian fibration.

First, we will assume that the singular fibers are of stable type, i.e., its characteristic cycles are of type $I_k, 0 \leq k \leq \infty$ (I_∞ meaning A_∞).

The second assumption we will make is that the Lagrangian fibration is *principally polarized*, in the sense explained in Definition 2. This condition is satisfied if there exists an f-ample line bundle on $M \backslash f^{-1}(D)$ whose restriction on smooth fibers gives principal polarizations on the abelian varieties.

We believe that these assumptions are natural and understanding the structure of Lagrangian fibration under these assumptions is essential for the study of general cases.

The main results of the paper are the following two theorems.

Theorem 1. *Let $f : M \to B$ be a principally polarized stable Lagrangian fibration (cf. Definition 1 and Definition 2). Then at a general point $b \in D$ of the discriminant, there exist a coordinate system $(z_1, \ldots, z_n)$ of B where D is defined by $z_n = 0$, a positive integer ℓ and a holomorphic function Ψ with the following properties: for a suitable choice of an integral frame of the local system $R^1 f_* \mathbb{Z}$ on $B \setminus D$, the period matrices $(\theta^i_j) = (\theta^i_j(z_1, \ldots, z_n))$ of the fibers over $B \setminus D$ have the form*

$$\theta^i_j = \frac{\partial^2 \Psi}{\partial z_i \partial z_j} \text{ for } (i,j) \neq (n,n) \text{ and } \theta^n_n = \frac{\partial^2 \Psi}{\partial z_n \partial z_n} + \frac{\ell}{2\pi\sqrt{-1}} \log z_n.$$

Theorem 2. *On the polydisc $\Delta^n = \{(z_1, \ldots, z_n), |z_i| < 1\}$, let ℓ be a positive integer and $\Psi(z_1, \ldots, z_n)$ be a holomorphic function such that* *the matrix function* $(\theta^i_j(z_1, \ldots, z_n))$ *defined by*

$$\theta^i_j = \frac{\partial^2 \Psi}{\partial z_i \partial z_j} \text{ for } (i,j) \neq (n,n) \text{ and } \theta^n_n = \frac{\partial^2 \Psi}{\partial z_n \partial z_n} + \frac{\ell}{2\pi\sqrt{-1}} \log z_n$$

has positive definite imaginary part on the complement of $z_n = 0$. *Then there exists a principally polarized stable Lagrangian fibration* $f^{\mathrm{model}} : M^{\mathrm{model}} \to \Delta^n$ *whose period matrices in a suitable frame of* $R^1 f_* \mathbb{Z}$ *are* $(\theta^i_j(z_1, \ldots, z_n))$. *Moreover,* ℓ *is the number of irreducible components of a singular fiber of* f^{model}.

Theorem 1 and Theorem 2 are a sort of converse to each other. More precisely, we can prove the following.

Proposition 1. *In the setting of Theorem 1, choose a polydisc neighborhood* $U \subset B$ *of* b *such that the coordinates* $(z_1, \ldots, z_n)$ *and the function* Ψ *are defined on* U. *Applying Theorem 2, we get a stable Lagrangian fibration* $f^{\mathrm{model}} : M^{\mathrm{model}} \to U$. *Then the two fibrations,* $f^U : f^{-1}(U) \to U$ *and* $f^{\mathrm{model}} : M^{\mathrm{model}} \to U$ *are biholomorphic as fiber spaces in a neighborhood of a general point of* $D \cap U$. *More precisely, there exist a proper analytic subset* $G \subset D \cap U$ *and a biholomorphic morphism*

$$g : f^{-1}(U \setminus G) \cong (f^{\mathrm{model}})^{-1}(U \setminus G)$$

such that g *preserves the symplectic structures and* $f = f^{\mathrm{model}} \circ g$.

See also Proposition 19 and Corollary 1 for slightly stronger results.

These results say that, in a neighborhood of a general singular fiber, a principally polarized stable Lagrangian fibration determines/is determined by a *single* holomorphic function $\Psi(z_1, \ldots, z_n)$ modulo terms of degree ≤ 1 and a positive integer ℓ with the specified properties.

That the period matrix of a *smooth* Lagrangian fibration is the Hessian of a potential function is a consequence of the action-angle variables, one of the most important properties of Lagrangian fibrations (cf. [2]). The logarithmic behavior of the multi-valued function θ^n_n reflects the stability assumption on the singular fiber. The novelty in Theorem 1 lies in the choice of the variable z_n through which these two aspects are intertwined. In this sense, it can be interpreted as a generalization of action-angle variables to *singular (stable)* Lagrangian fibrations. The existence of z_n follows from the fact proved in Proposition 15 that the characteristic foliation accounts for the degenerate part of the polarization restricted to the fixed part of the monodromy.

Theorem 2 is shown by explicitly constructing a principally polarized stable Lagrangian fibration from a given potential function $\Psi(z)$ with positive definite $\operatorname{Im}(\theta_i^j)$. This part is a generalization of Nakamura's construction [13] to principally polarized stable Lagrangian fibration in arbitrary dimension.

Using our construction, we shall give a concrete 4-dimensional example of a principally polarized stable Lagrangian fibration in which the types of characteristic cycles of singular fibers change fiber by fiber, too. To our knowledge, such an example has not been noticed previously.

Acknowledgement. We would like to express our thanks to Professor Yujiro Kawamata for a nice comment (Proposition 20). This paper is dedicated to Professor Shigefumi Mori on the occasion of his 60th birthday.

§2. Stable Lagrangian fibrations

Definition 1. *A* Lagrangian fibration *is a proper flat morphism $f : M \to B$ from a holomorphic symplectic manifold (M, ω) of dimension $2n$ to a complex manifold B of dimension n such that the smooth locus of each fiber is a Lagrangian submanifold of M. The* discriminant $D \subset B$ *is the set of the critical values of f, which is a hypersurface in B if it is non-empty.* Throughout this paper, we assume that D is non-empty. *We say that f is a* stable Lagrangian fibration *if $D \subset B$ is a submanifold and each singular fiber $f^{-1}(b), b \in D$, is stable, i.e., it is reduced and its normalization is a disjoint union of a finite number of compact complex manifolds $Y^1, \ldots, Y^\ell$ for some positive integer $\ell = \ell(b)$ (which might depend on $b \in D$) such that*

(i) *each Y^i is a $\mathbb{P}^1$-bundle over an $(n-1)$-dimensional complex torus A^i whose fibers are sent to characteristic leaves of $f^{-1}(b)$ in the sense of [7], i.e., for a defining function $h \in \mathcal{O}(B)$ of the divisor D, the Hamiltonian vector field $\iota_\omega(f^*dh)$, where $\iota_\omega : \Omega^1_M \to T(M)$ is the vector bundle isomorphism induced by ω, is tangent to the image of the fibers in M;*

(ii) *there exist submanifolds $S^i_1, S^i_2 \subset Y^i$, with $S^i_1 \neq S^i_2$ except possibly when $\ell = 1, 2$, such that $S^i_1 \cup S^i_2$ is a 2-to-1 unramified cover of A^i under the $\mathbb{P}^1$-bundle projection;*

(iii) *the normalization $\nu : \bigcup Y^i \to f^{-1}(b)$ is obtained by the identification via a collection of biholomorphic morphisms $g_i : S^i_2 \to S^{i+1}_1$ for $1 \leq i \leq \ell - 1$ and $g_\ell : S^\ell_2 \to S^1_1$ with the additional requirement $g_1 = g_2^{-1}$ if $S^1_1 = S^1_2$ and $S^2_1 = S^2_2$ for $\ell = 2$.*

A maximal connected union of the $\mathbb{P}^1$-fibers in (i) under the identification in (iii) is called a characteristic cycle. *A characteristic cycle can be either of finite type (I_m-type, $1 \leq m < \infty$) or of infinite type A_∞, which we also denote by I_∞.*

Remark 1. *There is a notion of characteristic cycle, as explained in p.983 of [7], for a general singular fiber of a Lagrangian fibration whose fibers are of Fujiki class. As proved in Theorem 1.4 of [7], a characteristic cycle is isomorphic to one of the Kodaira's list of singular fibers of elliptic fibrations of surfaces or of I_∞ type. Furthermore, it was proved in [7] that the fibration is stable in the sense of Definition 1 if and only if the characteristic cycle is of type I_m, $1 \leq m \leq \infty$. Thus it is possible to define a stable Lagrangian fibration in a simpler way, in terms of its characteristic cycle. However, Definition 1 is more convenient for the proof of Theorem 1.1.*

Recall (cf. [8] Section 4) that in a neighborhood of a general singular fiber, any Lagrangian fibration whose fibers are of Fujiki class can be transformed to a stable Lagrangian fibration by certain explicitly given bimeromorphic modifications and branched covering. We will be interested in the local property of the fibration at a point of D. Thus we will make the following

(*Assumption*) the pair $(D \subset B)$ is a germ of the pair satisfying the condition that D is a a smooth hypersurface in an n-dimensional complex manifold such that $B \setminus D$ has a cyclic fundamental group.

The following is immediate from Proposition 2.2 of [7].

Proposition 2. *Given a stable Lagrangian fibration, we can assume that there exists an action of the complex Lie group $\mathbb{C}^{n-1}$ on M preserving the fibers and the symplectic form such that S_1^i, S_2^i are orbits of this action for all $1 \leq i \leq \ell$. This action of $\mathbb{C}^{n-1}$ on Y^i descends to the translation action on A^i. The patching biholomorphisms g_i in Definition 1 (iii) as well as the $\mathbb{P}^1$-bundle structure in (i) are equivariant under this action. In particular, if $S_1^1 = S_2^2$ (resp. $S_1^2 = S_2^2$), the Galois action of the double cover $S_1^1 \to A^1$ (resp. $S_1^2 \to A^2$) is given by a translation on the torus S_1^1 (resp. S_1^2).*

Proposition 3. *In the notation of Definition 1, let $f^{-1}(b)$ be a singular fiber of a stable Lagrangian fibration. Then all the complex tori A^i's are biholomorphic to each other by biholomorphic maps induced by g_i's. Thus we may denote them by a single letter A. When $S_1^1 = S_2^1$ and $S_1^2 = S_2^2$ with $\ell = 2$, there exists a morphism $\mu : f^{-1}(b) \to A$ whose fibers are isomorphic to a characteristic cycle of I_2-type.*

Proof. If $S_1^i \neq S_2^i$ for some (and hence all) i, then by the $\mathbb{P}^1$-bundle projection, $S_1^i \cong S_2^i \cong A^i$. Thus the biholomorphisms g_i give biholomorphisms between A^i's.

It remains to handle the case $S_1^1 = S_2^1$ and $S_1^2 = S_2^2$ with $\ell = 2$. By Proposition 2, the Galois action on S_1^1 (resp. S_1^2) of the double cover

over A^1 (resp. A^2) is given by a translation, say, by $\gamma_1 \in \mathbb{C}^{n-1}$ (resp. $\gamma_2 \in \mathbb{C}^{n-1}$). By the $\mathbb{C}^{n-1}$-equivariance of $g_1 = g_2^{-1}$, for each $\alpha \in S_1^1$, we have $g_1(\gamma_1 \cdot \alpha) = \gamma_2 \cdot g_1(\alpha)$. Thus by the normalization morphism $\nu : Y^1 \cup Y^2 \to f^{-1}(b)$, a point $\alpha \in S_1^1$ is identified with $g_1(\alpha) \in S_1^2$, and the point $\gamma_1 \cdot \alpha \in S_1^1$, which lies in the $\mathbb{P}^1$-fiber through α, is identified with $\gamma_2 \cdot g_1(\alpha)$, which lies in the $\mathbb{P}^1$-fiber through $g_1(\alpha)$. Thus $A^1 \cong A^2$ and we get a morphism $\mu : f^{-1}(b) \to A^1$ whose fiber is a union of two $\mathbb{P}^1$'s identified at two points. Q.E.D.

In general, the $\mathbb{P}^1$-bundles $Y^i \to A$ do not patch together to define a morphism $\mu : f^{-1}(b) \to A$, except the case described in Proposition 3. However, we have the following topological substitute.

Proposition 4. *For the singular fiber $f^{-1}(b)$ in Definition 1, denote by Y^i and $\mu^i : Y^i \to A$ the $\mathbb{P}^1$-bundle structure on the components of the normalization of $f^{-1}(b)$ and set $Y_o := Y^1 \setminus (S_1^1 \cup S_2^1)$. There exists an analytic variety $f^{-1}(b)^1$ with the following properties.*

(1) *The normalization of $f^{-1}(b)^1$ is biholomorphic to the normalization of $f^{-1}(b)$ and there exists a morphism $\mu : f^{-1}(b)^1 \to A$ such that the induced morphism on each component Y^i of the normalization agrees with μ^i.*

(2) *There exists a homeomorphism $h : f^{-1}(b) \to f^{-1}(b)^1$ such that, denoting by $j : Y_0 \to f^{-1}(b)$ and $j^1 : Y_0 \to f^{-1}(b)^1$ the natural inclusions, the continuous maps $h \circ j$ and j^1 are homotopic.*

Proof. Let us use the notation introduced in Definition 1 (iii) for the description of the normalization morphism $\nu : \bigcup Y^i \to f^{-1}(b)$.

First, we consider the case $S_1^i \neq S_2^i$ for all $1 \leq i \leq \ell$. Define $\widetilde{f^{-1}(b)}$ as the variety obtained from $\bigcup Y^i$ with all the patching identification $g_1, \ldots, g_{\ell-1}$ such that the normalization factors through

$$\nu : \bigcup Y^i \to \widetilde{f^{-1}(b)} \to f^{-1}(b)$$

with the second arrow given by the identification via g_ℓ. When $\ell = 1$, $\widetilde{f^{-1}(b)} = Y^1$. It is immediate that the morphisms μ^i patch together to define a morphism $\widetilde{f^{-1}(b)} \to A$, which determines a biholomorphism $\zeta : S_2^\ell \to S_1^1$. Fix a point $\alpha \in S_2^\ell$ and let $\beta = \zeta(\alpha) \in S_1^1$. For $t \in [0, 1] \subset \mathbb{R}$, let $\tau_t : S_1^1 \to S_1^1$ be the translation by $t(\beta - g_\ell(\alpha))$.

Define a new family of biholomorphic morphisms $g_\ell^t : S_2^\ell \to S_1^1$ by $g_\ell^t = \tau_t \circ g_\ell$. Clearly, $g_\ell^0 = g_\ell$. We claim that $g_\ell^1 = \zeta$. In fact, $\zeta^{-1} \circ g_\ell^1$ is an automorphism of S_2^ℓ which fixes the point α. But both g_ℓ^1 and ζ must be equivariant under the $\mathbb{C}^{n-1}$-action of Proposition 2. Thus $\zeta^{-1} \circ g_\ell^1$ must be the identity map of S_2^ℓ, proving the claim.

Let $f^{-1}(b)^t$ be the variety obtained from $\widetilde{f^{-1}(b)}$ by identifying S_1^1 and S_2^ℓ via g_ℓ^t. Then

$$f^{-1}(b)^0 = f^{-1}(b)$$

and $f^{-1}(b)^1$ is homeomorphic to $f^{-1}(b)$. The $\mathbb{C}^{n-1}$-action descends to $f^{-1}(b)^t$ for each t as τ_t commutes with the $\mathbb{C}^{n-1}$-action. By our choice of α and β, the morphisms μ^i patch up to give a morphism $\mu : f^{-1}(b)^1 \to A$.

For each t, we have a homeomorphism $h_t : f^{-1}(b) \to f^{-1}(b)^t$ such that h_0 is the identity morphism. The homeomorphism $h := h_1$ certainly satisfies the required properties.

Now consider the case when $\ell = 1$ and $S_1^1 = S_2^1$. Set $\widetilde{f^{-1}(b)} = Y^1$ and define $\zeta : S_2^1 \to S_1^1$ as the Galois action of the double covering $S_1^1 \to A^1$ in Definition 1 (ii). Then the same argument as in the previous case applies.

Finally, when $\ell = 2$, $S_1^1 = S_2^1$ and $S_1^2 = S_2^2$, we may put $f^{-1}(b)^1 = f^{-1}(b)$ and use Proposition 3. Q.E.D.

We have a generalization of the classical action-angle correspondence as follows.

Proposition 5. *Let $f : M \to B$ be a stable Lagrangian fibration that admits a Lagrangian section $\Sigma \subset M$ of f. Then we have a natural surjective unramified morphism $\Phi : T^*B \to M \setminus (E \cup \operatorname{Sing}(f))$ where E is the union of the irreducible components of the fibers of f disjoint from Σ such that*

(1) *$f \circ \Phi$ agrees with the natural projection $g : T^*B \to B$,*
(2) *Φ sends the zero section of T^*B to Σ and*
(3) *$\Phi^*\omega$ coincides with the standard symplectic form on T^*B.*

In particular, $\Gamma := \Phi^{-1}(\Sigma)$ is a Lagrangian submanifold (with infinitely many connected components) in T^*B. For each $b \in B$, $\Phi_b := \Phi|_{T_b^*(B)} : T_b^*(B) \to f^{-1}(b) \setminus E$ is the universal covering and $\Gamma_b := \Gamma \cap T_b^*(B)$ is naturally isomorphic to $H_1(f^{-1}(b) \setminus E, \mathbb{Z})$.

Proof. Over $B \setminus D$, this is just a holomorphic version (cf. Proposition 3.5 in [6]) of the classical action-angle correspondence as described in Section 44 of [5]. The statement over D follows by the same argument as for the smooth fibers. In fact, for each $b \in B$, the vector group $T_b^*(B)$ acts on the fiber $f^{-1}(b)$ with n-dimensional orbits on the smooth locus of $f^{-1}(b)$ (cf. Proposition 3.3 in [6]). The morphism Φ_b is defined by taking the orbit map of the point $\Sigma \cap f^{-1}(b)$ under this action, which is a universal covering map for the smooth locus of the component of

$f^{-1}(b)$ containing $\Sigma \cap f^{-1}(b)$. This shows (1) and (2). The proof of (3) is the same as that of Theorem 44.2 of [5]. Q.E.D.

Proposition 6. *Given a stable Lagrangian fibration $f : M \to B$ and a fixed point $b \in D$, since we are assuming that $(D \subset B)$ is a germ as explained in (Assumption), we have a Lagrangian section $\Sigma \subset M$ from Darboux theorem. From Proposition 5, we know that the connected component of Γ containing a point of $\Gamma \cap T_b^*(B)$ is a Lagrangian section of $T^*(B) \to B$, i.e., a closed 1-form on B. Let $\Gamma' \subset \Gamma$ be the union of all connected components of Γ that intersect the fiber $T_b^*(B)$. Then for each $s \in B \setminus D$, $\Gamma'_s := \Gamma' \cap T_s^*(B)$ is a sublattice of Γ_s satisfying $\Gamma_s/\Gamma'_s \cong \mathbb{Z}$.*

Proof. Since $f^{-1}(b) \setminus E$ is a $\mathbb{C}^*$-bundle over an $(n-1)$-dimensional torus, we see that $\Gamma \cap T_b^*(B)$ has rank $2n-1$. Thus Γ'_s has rank $2n-1$. It remains to show that Γ_s/Γ'_s is torsion-free. Suppose it has k-torsion, $0 < k \in \mathbb{Z}$, i.e., there exists a point $\alpha \in \Gamma_s \setminus \Gamma'_s$ such that $k\alpha \in \Gamma'_s$. Let $\widetilde{k\alpha}$ be a closed 1-form given by the component of Γ' containing $k\alpha$. Then the closed 1-form $\tilde{\alpha} = \frac{1}{k}\widetilde{k\alpha}$ is also a component of Γ' containing α, which implies $\alpha \in \Gamma'_s$, a contradiction. Q.E.D.

Proposition 7. *In the notation of Proposition 5, let Y_o be the fiber of $M \setminus E$ at a point $b \in D$. Let $\Phi_b : T_b^*(B) \to Y_o$ be the universal covering map and $\varrho : Y_o \to A$ be the $\mathbb{C}^*$-bundle over an $(n-1)$-dimensional torus. Let $\Upsilon \subset \Gamma_b = \Gamma'_b$ be the rank-1 sublattice corresponding to the kernel of*

$$\varrho_* : H_1(Y_o, \mathbb{Z}) \to H_1(A, \mathbb{Z}).$$

Then for any $v \in \Gamma'_b \setminus \Upsilon$, there exists $\varpi \in H^1(f^{-1}(b), \mathbb{Z})$ such that $\langle \varpi, j_* v\rangle \neq 0$ where $j_* : H_1(Y_o, \mathbb{Z}) \to H_1(f^{-1}(b), \mathbb{Z})$ is induced by the inclusion $j : Y_o \subset f^{-1}(b)$.

Proof. Since $\varrho_*(v) \in H_1(A, \mathbb{Z})$ is non-zero, there exists an element $\varphi \in H^1(A, \mathbb{Z})$ such that $\langle \varphi, \varrho_*(v)\rangle \neq 0$. Let $\varpi = (\mu \circ h)^*\varphi$ where the continuous map $\mu \circ h : f^{-1}(b) \to A$ is as defined in Proposition 4 such that $\mu \circ h \circ j$ is homotopic to ϱ. Then

$$\langle \varpi, j_*(v)\rangle = \langle (\mu \circ h)^*\varphi, j_*(v)\rangle =$$

$$\langle (\mu \circ h \circ j)^*\varphi, v\rangle = \langle \varrho^*\varphi, v\rangle = \langle \varphi, \varrho_*(v)\rangle \neq 0.$$

Q.E.D.

Proposition 8. *For a stable Lagrangian fibration $f : M \to B$ and $s \in B \setminus D$, set $\Lambda_s := H_1(f^{-1}(s), \mathbb{Z})$ for $s \in B \setminus D$. Fix a generator of the cyclic fundamental group of $\pi_1(B \setminus D, s)$ and denote by $\tau_s : \Lambda_s \to \Lambda_s$ the monodromy operator of the generator. Then the fixed part $\Lambda'_s \subset \Lambda_s$ of τ_s at $s \in B \setminus D$ has corank 1.*

Proof. For any $s \in B\backslash D$, we can identify each fiber $\Lambda_s = H_1(M_s, \mathbb{Z})$ with the fiber Γ_s of Proposition 5. Thus the result follows from Proposition 6. Q.E.D.

§3. Principally polarized stable Lagrangian fibration

Definition 2. *For a stable Lagrangian fibration $f : M \to B$, denote by Λ the local system on $B \setminus D$ defined by the lattice Λ_s in Proposition 8. A* principal polarization *on $M \setminus f^{-1}(D)$ is a unimodular anti-symmetric form $Q : \wedge^2\Lambda \to \mathbb{Z}_{B\setminus D}$ where $\mathbb{Z}_{B\setminus D}$ denotes the constant sheaf of integers on $B \setminus D$, which induces a principal polarization on each smooth fiber of f. A stable Lagrangian fibration with a choice of principal polarization is called a* principally polarized stable Lagrangian fibration.

Remark 2. *In Definition 2, the polarization on $M \setminus f^{-1}(D)$ may not extend to an f-ample class of the whole M. In fact, f need not be projective. This definition is useful because there are many situations where the polarization exists* a priori *only on the smooth fibers, e.g., in Kodaira's study of elliptic fibrations and also in our construction in Section 5.*

Proposition 9. *Let $f : M \to B$ be a principally polarized stable Lagrangian fibration. Then the monodromy operator in Proposition 8 satisfies $\tau_s \neq \mathrm{Id}$ and $\tau_s \circ \tau_s \neq \mathrm{Id}$.*

Proof. If $\tau_s = \mathrm{Id}$, then we see that f is a smooth fibration, as in the proof of Proposition 3.2 in [6]. In fact, since there is no monodromy and f is polarized over $B \setminus D$, we can extend the period map of the abelian family on $B \setminus D$ to the whole B ([3], Theorem 9.5). Thus, we obtain a smooth abelian fibration $f' : M' \to B$ such that f and f' are bimeromorphic outside D. Since M' contains no rational curves and both M and M' have trivial canonical bundles, this implies M and M' are biholomorphic, a contradiction to the non-emptiness of the discriminant D of f.

If $\tau_s \circ \tau_s = \mathrm{Id}$, take a double cover $g : B' \to B$ branched along D and let $D' = g^{-1}(D)$. Denote by $\hat{f} : \hat{M} \to B'$ the fiber product of f and g, which has no monodromy on $B' \setminus D'$. By the $\mathbb{C}^{n-1}$-action of Proposition 2 which lifts to $\hat{M}$, the following property of $\hat{M}$ can be seen from the corresponding properties in the case of $n = 1$ (cf. Proof of Proposition 9.2 in [1]): $\hat{M}$ is normal, Gorenstein with singularities of type $A_1 \times$ (germ of $2(n-1)$-dimensional manifold) and has trivial canonical bundle. Thus we have a crepant resolution $f' : M' \to B'$, which is a family with trivial canonical bundle and no monodromy. Then we get a contradiction as in the previous case. Q.E.D.

Lemma 1. *Let $\tau : \Lambda \to \Lambda$ be an automorphism of a lattice such that $\Lambda' := \{v \in \Lambda, \tau(v) = v\}$ is a sublattice of corank 1, i.e., $\Lambda/\Lambda' \cong \mathbb{Z}$. If $\tau \circ \tau \neq \mathrm{Id}$, then $\eta := \tau - \mathrm{Id}$ satisfies $\eta \circ \eta = 0$.*

Proof. Note that $\Lambda' \subset \mathrm{Ker}(\eta)$. The induced automorphism $\bar{\tau} : \Lambda/\Lambda' \to \Lambda/\Lambda'$ is either Id or $-$Id. If $\bar{\tau} = \mathrm{Id}$, then for a non-zero $v \in \Lambda \setminus \Lambda'$, we have $\tau(v) = v + \lambda$ for some $\lambda \in \Lambda'$. Then $\eta(v) = \lambda \in \Lambda' \subset \mathrm{Ker}(\eta)$. This proves that $\eta \circ \eta = 0$. If $\bar{\tau} = -\mathrm{Id}$, then for a non-zero $v \in \Lambda \setminus \Lambda'$, we have $\tau(v) = -v + \lambda$ for some $\lambda \in \Lambda'$. Then

$$\tau \circ \tau(v) = -\tau(v) + \tau(\lambda) = -(-v + \lambda) + \lambda = v.$$

Thus $\tau \circ \tau = \mathrm{Id}$, a contradiction. Q.E.D.

Proposition 10. *In the setting of Proposition 9, let $\eta := \tau_s - \mathrm{Id}$. Then for any $\beta \in \mathrm{Im}(\eta)$ and an element $\varphi \in H^1(M, \mathbb{Z})$,*

$$\langle i^*\varphi, \beta \rangle = \langle \varphi, i_*\beta \rangle = 0$$

where $i^* : H^1(M, \mathbb{Z}) \to H^1(M_s, \mathbb{Z})$ and $i_* : H_1(M_s, \mathbb{Z}) \to H_1(M, \mathbb{Z})$ are the homomorphisms induced by the inclusion $i : M_s := f^{-1}(s) \subset M$.

Proof. Let $\mathcal{H}$ be the local system on $B \setminus D$ given by $H^1(M_s, \mathbb{Z})$, $s \in B \setminus D$. Denote by $\tau^* : \mathcal{H}_s \to \mathcal{H}_s$ the transformation dual to τ, i.e., for any $\varpi \in H^1(M_s, \mathbb{Z})$ and $u \in H_1(M_s, \mathbb{Z})$,

$$\langle \tau^*(\varpi), u \rangle = \langle \varpi, \tau(u) \rangle.$$

By Proposition 8, Proposition 9 and Lemma 1, we have $\eta \neq 0$ and $\eta \circ \eta = 0$, i.e.,

$$0 \neq \mathrm{Im}(\eta) \subset \mathrm{Ker}(\eta) = \Lambda'_s.$$

Similarly, $\eta^* := \tau^* - \mathrm{Id}$ is an endomorphism of $\mathcal{H}_s$ with $\eta^* \neq 0$ and $\eta^* \circ \eta^* = 0$. Since $i^*\varphi \in \mathrm{Ker}(\eta^*)$ by (the easy half of) the global invariant cycles theorem (cf. Theorem 4.24 of [14]), for any $\psi \in \mathrm{Ker}(\eta^*)$ and $u \in \Lambda_s$,

$$\langle \psi, \eta(u) \rangle = \langle \eta^*(\psi), u \rangle = 0.$$

It follows that $\langle i^*\varphi, \mathrm{Im}(\eta) \rangle = 0$. Q.E.D.

Remark 3. *If the family $f : M \to B$ is projective, we could have used the Monodromy Theorem (cf. Theorem 3.15 in [14]) in place of Proposition 3.3 and Lemma 1 in the above proof. We have used the above approach because we do not want to assume that f is projective.*

Proposition 11. *For a principally polarized stable Lagrangian fibration $f : M \to B$ and $s \in B \setminus D$, let $\tau_s : \Lambda_s \to \Lambda_s$ be the monodromy operator of Proposition 8. Set $\eta := \tau_s - \mathrm{Id}$ as in Proposition 10 and $\Lambda'_s = \mathrm{Ker}(\eta)$. Then τ_s preserves the polarization $Q_s : \wedge^2 \Lambda_s \to \mathbb{Z}$ and $\mathrm{Im}(\eta) \subset \Lambda_s$ is contained in*

$$\Xi_s := \{v \in \Lambda'_s \mid Q(v, w) = 0 \text{ for all } w \in \Lambda'_s\}.$$

Proof. It is clear that τ_s preserves the polarization Q_s. Since $\eta \circ \eta = 0$ by Lemma 1,

$$Q_s(\eta(v), u) + Q_s(v, \eta(u)) = 0 \text{ for all } v, u \in \Lambda_s.$$

Thus for any $v \in \Lambda_s$ and $u \in \mathrm{Ker}(\eta) = \Lambda'_s$, we have $Q_s(\eta(v), u) = -Q_s(v, \eta(u)) = 0$, which means $\eta(v) \in \Xi_s$. Q.E.D.

Definition 3. *Let Λ be a free abelian group of rank $2n$. Given a unimodular non-degenerate anti-symmetric form $Q : \wedge^2 \Lambda \to \mathbb{Z}$, a basis $\{p_1, \ldots, p_n, q_1, \ldots, q_n\}$ of Λ is called a* symplectic basis *of Λ with respect to Q if, in terms of the dual basis $\{p^1, \ldots, p^n, q^1, \ldots, q^n\}$ of* $\mathrm{Hom}(\Lambda, \mathbb{Z})$,

$$Q = p^1 \wedge q^1 + p^2 \wedge q^2 + \cdots + p^n \wedge q^n.$$

Lemma 2. *In the setting of Definition 3, let $\tau : \Lambda \to \Lambda$ be a group automorphism preserving Q. Assume that the subgroup $\Lambda' \subset \Lambda$ of elements fixed under τ has corank 1. Then there exists a symplectic basis $\{p_1, \ldots, p_n, q_1, \ldots, q_n\}$ of Λ such that*

$$\{p_1, \ldots, p_n, q_1, \ldots, q_{n-1}\} \subset \Lambda'.$$

Proof. Fix a symplectic basis $\{a_1, \ldots, a_n, b_1, \ldots, b_n\}$ such that

$$Q = a^1 \wedge b^1 + \cdots + a^n \wedge b^n.$$

The anti-symmetric form $Q|_{\Lambda'}$ must have a kernel of rank 1, i.e.,

$$\Xi := \{v \in \Lambda',\ Q(v, u) = 0 \text{ for all } u \in \Lambda'\}$$

has rank 1. Pick a generator p_n of Ξ. Since Ξ is primitive, i.e., Λ/Ξ has no torsion, we can write

$$p_n = \alpha_1 a_1 + \cdots + \alpha_n a_n + \beta_1 b_1 + \cdots + \beta_n b_n$$

with some integers α_i, β_i satisfying $gcd(\alpha_1, \ldots, \alpha_n, \beta_1, \ldots, \beta_n) = 1$. Thus there exists integers $\alpha'_1, \ldots, \alpha'_n, \beta'_1, \ldots, \beta'_n$ such that

$$\alpha'_1 \cdot \alpha_1 + \cdots + \alpha'_n \cdot \alpha_n + \beta'_1 \cdot \beta_1 + \cdots + \beta'_n \cdot \beta_n = 1.$$

Let

$$q_n := -\beta_1' a_1 - \cdots - \beta_n' a_n + \alpha_1' b_1 + \cdots + \alpha_n' b_n.$$

Then $Q(p_n, q_n) = 1$. Define

$$\Lambda'' := \{v \in \Lambda, Q(p_n, v) = 0 = Q(q_n, v)\}.$$

Then $\Lambda'' \subset \Lambda'$ is a lattice of rank $2n-2$ such that $Q|_{\Lambda''}$ is unimodular and non-degenerate (cf. [4], the proof of Lemma in p.304). Let $\{p_1, \ldots, p_{n-1}, q_1, \ldots, q_{n-1}\}$ be a symplectic basis of Λ''. Then

$$\{p_1, \ldots, p_n, q_1, \ldots, q_n\}$$

is a symplectic basis of Λ with the required property. Q.E.D.

Proposition 12. *In the setting of Proposition 6, identify* $\Lambda_s = H_1(M_s, \mathbb{Z})$ *with* Γ_s *for* $s \in B \setminus D$ *as in the proof of Proposition 8. Assume that we have a principal polarization* Q. *Then we can find a collection of connected components* $\{p_1, \ldots, p_n, q_1, \ldots, q_{n-1}\}$ *of* Γ' *such that for each* $x \in B \setminus D$, *there exists* $q_{n,x} \in \Gamma_x$ *such that*

$$\{p_{1,x}, \ldots, p_{n,x}, q_{1,x}, \ldots, q_{n-1,x}, q_{n,x}\}$$

is a symplectic basis of $\Lambda_x = \Gamma_x$ *with respect to* Q_x.

Proof. Fix a point $s \in B \setminus D$. The monodromy $\tau_s : \Lambda_s \to \Lambda_s$ preserves the polarization Q_s on Λ_s and fixes $\Lambda_s' = \Gamma_s'$. Applying Lemma 2, we have a symplectic basis $\{p_{1,s}, \ldots, p_{n,s}, q_{1,s}, \ldots, q_{n,s}\}$ with $p_{1,s}, \ldots, p_{n,s}, q_{1,s}, \ldots, q_{n-1,s} \in \Lambda_s'$. Since Γ' consists of sections of $g : T^*B \to B$, the vectors $p_{1,s}, \ldots, p_{n,s}, q_{1,s}, \ldots, q_{n-1,s}$ uniquely determine components $p_1, \ldots, p_n, q_1, \ldots, q_{n-1}$ of Λ'. To check the existence of $q_{n,x}$ for any $x \in B \setminus D$, just pick as $q_{n,x}$ any vector in Λ_x contained in the component of Λ containing $q_{n,s}$. Q.E.D.

Proposition 13. *In the notation of Proposition 12, when* $b \in D$, *the vector* $p_{n,b} \in \Gamma_b'$ *regarded as an element of* $H_1(Y_o, \mathbb{Z})$ *in the notation of Proposition 7, lies in the lattice* Υ *of Proposition 7.*

Proof. Suppose not. By our (Assumption) after Definition 1, we may assume that M is topologically retractable to $f^{-1}(b)$ and identify $H^1(f^{-1}(b), \mathbb{Z})$ with $H^1(M, \mathbb{Z})$. Then by Proposition 7, there exists $\varpi \in H^1(f^{-1}(b), \mathbb{Z}) = H^1(M, \mathbb{Z})$ such that $\langle \varpi, j_* p_{n,b} \rangle \neq 0$. For a point $s \in B \setminus D$, the choice in Proposition 12 implies that $p_{n,s} \in \Xi_s$ of Proposition 11. Denote by ϖ_s the element in $H^1(M_s, \mathbb{Z})$ induced by $\varpi \in H^1(M, \mathbb{Z})$ under the identification $H^1(f^{-1}(b), \mathbb{Z}) = H^1(M, \mathbb{Z})$. Since $j_* p_{n,b} \in H_1(f^{-1}(b), \mathbb{Z}) = H_1(M, \mathbb{Z})$ and the image of $p_{n,s} \in H_1(M_s, \mathbb{Z})$

in $H_1(M, \mathbb{Z})$ belongs to the same class, Proposition 10 and Proposition 11 say that

$$\langle \varpi, j_* p_{n,b} \rangle = \langle \varpi_s, p_{n,s} \rangle = 0.$$

This is a contradiction. Q.E.D.

Proposition 14. *In Proposition 13, the $\mathbb{C}$-linear span of Υ in $T_b^*(B)$ is exactly $\mathbb{C} \cdot dh$ where $h \in \mathcal{O}(B)$ is a defining equation of the divisor D.*

Proof. From the definition of Υ in Proposition 7, the linear span of Υ is sent to a fiber of the $\mathbb{C}^*$-bundle. By Definition 1 (i), this fiber is a leaf of the characteristic foliation, which is given by the Hamiltonian vector field $\iota_\omega(f^* dh)$ on M. Under the symplecto-morphism Φ in Proposition 5, this corresponds to $\mathbb{C} \cdot dh$. Q.E.D.

Proposition 15. *Let $\{p_1, \ldots, p_n, q_1, \ldots, q_{n-1}\}$ be as in Proposition 12. Then there exists a holomorphic coordinate system $\{z_1, \ldots, z_n\}$ on B such that, regarded as sections of $T^*(B)$,*

$$p_1 = dz_1, \; \ldots, \; p_n = dz_n$$

and D is given by $z_n = 0$.

Proof. Since $p_1, \ldots, p_n$ are closed 1-forms which are point-wise linearly independent at every point of B, we can find coordinates $z_1, \ldots, z_n$ with $p_i = dz_i$. By Proposition 14, we may choose z_n to be a defining equation of D. Q.E.D.

Let us recall the classical Riemann condition (e.g. [4], p.306).

Proposition 16. *Let V be a complex vector space of dimension n and let $\Lambda \subset V$ be a lattice of rank $2n$ such that V/Λ is an abelian variety with a principal polarization. For a symplectic basis*

$$\{p_1, \ldots, p_n, q_1, \ldots, q_n\}$$

of Λ with respect to the principal polarization $Q : \wedge^2 \Lambda \to \mathbb{Z}$, $\{p_1, \ldots, p_n\}$ becomes a $\mathbb{C}$-basis of V and the period matrix (θ_i^j) defined by

$$q_i = \sum_{j=1}^{n} \theta_i^j p_j \quad \text{in } V$$

is symmetric in (i, j) and $\operatorname{Im}(\theta_i^j) > 0$.

Theorem 1 now follows from the next Theorem:

Theorem 3. *Given a principally polarized stable Lagrangian fibration $f : M \to B$ with a Lagrangian section $\Sigma \subset M$, there exists a holomorphic coordinate system $(z_1, \ldots, z_n)$ on B such that*

(i) *$z_n = 0$ is a local defining equation of D;*

(ii) *on $B \setminus D$, $dz_1, \ldots, dz_{n-1}, dz_n$ belong to Γ' in the notation of Proposition 6;*

(iii) *there exists a symplectic basis $\{p_{1,s}, \ldots, p_{n,s}, q_{1,s}, \ldots, q_{1,n}\}$ on each $\Lambda_s = \Gamma_s, s \in B \setminus D$ satisfying*

$$p_{1,s} = (dz_1)_s, \ldots, p_{n,s} = (dz_n)_s$$

and the associated period matrix in the sense of Proposition 16 is given by

$$\theta_i^j = \frac{\partial^2 \Psi}{\partial z_i \partial z_j} + \delta_{in}\delta_{jn}\frac{\ell}{2\pi\sqrt{-1}} \log z_n$$

for some holomorphic function Ψ on B, which we call a potential function *of the Lagrangian fibration, and some integer ℓ.*

Proof. Let $\{p_1, \ldots, p_n, q_1, \ldots, q_{n-1}\}$ be as in Proposition 12 and Proposition 15. At a point $s \in B \setminus D$, we add $q_{n,s}$ to get a symplectic basis of Λ_s. By analytic continuation, we get a multi-valued 1-form q_n over $B \setminus D$ such that any choice of a value $q_{n,t}$ of q_n at a point $t \in B \setminus D$, together with $p_{1,t}, \ldots, p_{n,t}, q_{1,t}, \ldots, q_{n-1,t}$, gives a symplectic basis of Λ_t. Using the coordinate system in Proposition 15, we can write

$$q_i = \sum_{j=1}^{n} \theta_i^j dz_j,$$

where θ_i^j is a (univalent) holomorphic function on B for each $1 \le i \le n-1$ and $1 \le j \le n$, while θ_n^j is a multi-valued holomorphic function on $B \setminus D$ for each $1 \le j \le n$. By Proposition 16, $\theta_j^i = \theta_i^j$ for each $1 \le i, j \le n$. It follows that θ_n^j is univalent holomorphic function on B for each $1 \le j \le n-1$. By the choice of $p_{n,s} \in \Xi_s$ and Proposition 11, the monodromy operator $\tau_s : \Lambda_s \to \Lambda_s$ is of the form

$$\tau_s(q_{n,s}) = q_{n,s} + \ell p_{n,s}$$

for some integer ℓ. Thus

$$\tilde{\theta}_n^n := \theta_n^n - \frac{\ell}{2\pi\sqrt{-1}} \log z_n$$

is univalent. $\tilde{\theta}^n_n$ is also holomorphic on B by the positive definiteness of $\mathrm{Im}\,(\theta^j_i)$. Set $\tilde{\theta}^j_i := \theta^j_i$ if $(i,j) \neq (n,n)$. Then $\tilde{\theta}^j_i$ is a univalent holomorphic function on B for all values of $1 \leq i, j \leq n$ and

$$q_i = \sum_{j=1}^{n} \tilde{\theta}^j_i dz_j \text{ for } 1 \leq i \leq n-1$$

$$q_n = \sum_{j=1}^{n} \tilde{\theta}^j_n dz_j + \frac{\ell}{2\pi\sqrt{-1}} \log z_n \, dz_n.$$

Since q_i's are closed 1-forms on B, we have

$$\frac{\partial \tilde{\theta}^j_i}{\partial z_k} = \frac{\partial \tilde{\theta}^k_i}{\partial z_j} = \frac{\partial \tilde{\theta}^i_j}{\partial z_k}$$

for any $1 \leq i, j, k \leq n$. By Poincaré's lemma, there exists a holomorphic function Ψ such that

$$\tilde{\theta}^j_i = \frac{\partial^2 \Psi}{\partial z_i \partial z_j}.$$

Q.E.D.

§4. Construction of principally polarized stable Lagrangian fibrations with given potential functions

In this section, we shall prove Theorem 2 and Proposition 1. For an n-dimensional polydisc B with a given holomorphic function $\Psi(z)$ with the property specified in Theorem 2, we shall construct a principally polarized stable Lagrangian fibration $f : (M, \omega_M) \to B$ having $\Psi(z)$ as its potential function. This will turn out to be $f^{\mathrm{model}} : M^{\mathrm{model}} \to \Delta^n$ in Theorem 2. Our construction closely follows Nakamura's toroidal construction [Na]. However, main differences are the following:

(i) the base space B is of dimension n (rather than 1).

(ii) the total space should be not only smooth but also symplectic.

(I) Construction of a non-proper Lagrangian fibration $\tilde{M} \to B$.

For each integer $k \in \mathbb{Z}$, let E_k be a copy of $\mathbb{C} \times \mathbb{C}$ equipped with linear coordinates (x_k, y_k). We define a complex manifold E by identifying points in $\cup_{k \in \mathbb{Z}} E_k$ by the following rule: a point (x_k, y_k) of E_k with $x_k \neq 0$ is identified with a point (x_{k+1}, y_{k+1}) of E_{k+1} with $y_{k+1} \neq 0$, if and only if

$$x_{k+1} = x_k^2 y_k, \quad \text{and} \quad y_{k+1} = \frac{1}{x_k}.$$

On E, $z_n := x_k y_k$ is a well-defined holomorphic function independent of k and

$$w_n := x_k^{-k+1} y_k^{-k}$$

is a meromorphic function independent of k, with zeros and poles supported on

$$\cup_{k\in\mathbb{Z}}(x_k y_k = 0).$$

Moreover, the 2-forms $dy_k \wedge dx_k$ glue together yielding a holomorphic symplectic form ω_E on E, satisfying

$$\omega_E = dz_n \wedge \frac{dw_n}{w_n}.$$

Fix coordinates

$$(z_1, \dots, z_{n-1}, w_1, \dots, w_{n-1})$$

on $\mathbb{C}^{n-1} \times \mathbb{C}^{n-1}$ and regard them as functions on the open subset $\mathbb{C}^{n-1} \times (\mathbb{C}^\times)^{n-1}$ defined by

$$w_1 \neq 0, \ \dots, \ w_{n-1} \neq 0.$$

Define

$$\tilde{X} := \mathbb{C}^{n-1} \times (\mathbb{C}^\times)^{n-1} \times E.$$

On $\tilde{X}$, we have the holomorphic functions $z_1, \dots, z_n, w_1, \dots, w_{n-1}$ and the meromorphic function w_n.

Define a morphism $\tilde{p} : \tilde{X} \to \mathbb{C}^n$ by $(z^1, \dots, z^n)$. The fiber of $\tilde{p}$ over b with $z_n(b) \neq 0$ is isomorphic to

$$(\mathbb{C}^\times)^{n-1} \times \mathbb{C}^\times$$

with coordinates $(w_1, \dots, w_{n-1}, w_n)$ and the fiber over b with $z_n(b) = 0$ is isomorphic to

$$(\mathbb{C}^\times)^{n-1} \times \cup_{k\in\mathbb{Z}}\mathbb{P}^1_k$$

where $\mathbb{P}^1_k$ is a copy of the projective line $\mathbb{P}^1$ with affine coordinate x_k.

We have a holomorphic symplectic 2-form

$$\omega_{\tilde{X}} := \sum_{i=1}^{n-1} dz_i \wedge \frac{dw_i}{w_i} + \omega_E = \sum_{i=1}^{n} dz_i \wedge \frac{dw_i}{w_i}$$

on $\tilde{X}$. From now, we regard $\tilde{X}$ as a symplectic manifold by this symplectic form. From the coordinate expression of $\omega_{\tilde{X}}$ and $\tilde{p}$, it is immediate that $\tilde{p}$ is a *non-proper* Lagrangian fibration.

We set

$$\tilde{M} := \tilde{X} \times_{\mathbb{C}^n} B$$

where

$$B = \{(z_1, \ldots, z_{n-1}, z_n) \mid |z_i| < 1 (\forall i)\}.$$

We denote the natural projection $\tilde{M} \to B$ induced from $\tilde{p}$ by

$$\tilde{f} : \tilde{M} \to B\ .$$

Note that the restriction $\omega_{\tilde{M}}$ of $\omega_{\tilde{X}}$ is a symplectic 2-form on $\tilde{M}$ and $\tilde{f}$ is a non-proper Lagrangian fibration.

(II) Group action of $\Gamma = \mathbb{Z}^n$ *on* $\tilde{M}$.

Let $\Psi(z_1, z_2, \cdots, z_n)$ be a holomorphic function on B and ℓ be a positive integer such that the imaginary part $\operatorname{Im}\theta(z)$ of the period matrix

$$\theta(z) = \tilde{\theta}(z) + \frac{\log z_n}{2\pi\sqrt{-1}} \begin{pmatrix} O_{n-1} & 0 \\ 0 & \ell \end{pmatrix}$$

is positive definite. Here

$$\tilde{\theta}(z) = (\frac{\partial^2 \Psi}{\partial z_i \partial z_j})$$

is the Hessian matrix of $\Psi(z_1, z_2, \cdots, z_n)$. We will write

$$\tilde{\theta}(z) = \begin{pmatrix} \tilde{\Theta}_1(z) & \tilde{\Theta}_2(z) \\ \tilde{\Theta}_2^t(z) & \tilde{\theta}_n^n(z) \end{pmatrix},$$

where $\tilde{\Theta}_1(z)$ is an $(n-1) \times (n-1)$ matrix, $\tilde{\Theta}_2(z)$ is an $(n-1) \times 1$ matrix, $\tilde{\Theta}_2^t(z)$ is the transpose of $\tilde{\Theta}_2(z)$ and $\tilde{\theta}_n^n(z)$ is a 1×1 matrix.

Set $\Gamma = \mathbb{Z}^{n-1} \oplus \mathbb{Z}$. We define a group action of Γ on $\tilde{M}$ as follows. Let $\gamma = (j, m) \in \Gamma$. Then the action $T_\gamma : \tilde{M} \to \tilde{M}$ is defined in terms of the coordinate functions on $\mathbb{C}^{n-1} \times (\mathbb{C}^\times)^{n-1} \times E_k \subset \tilde{M}$ by

$$T_\gamma^* z_i = z_i \text{ for } i = 1, \ldots, n-1$$

$$T_\gamma^*(\Pi_{i=1}^{n-1} w_i^{b_i}) = \exp(2\pi\sqrt{-1}(j\tilde{\Theta}_1(z)b + m\tilde{\Theta}_2^t(z)b)\Pi_{i=1}^{n-1} w_i^{b_i}$$

where $b = (b_i)_{i=1}^{n-1}$ is an $(n-1) \times 1$ matrix, and

$$T_\gamma^* x_k = \exp(2\pi\sqrt{-1}(j\tilde{\Theta}_2(z) + m\tilde{\theta}_n^n(z))x_{k+m\ell}$$

$$T_\gamma^* y_k = (\exp(2\pi\sqrt{-1}(j\tilde{\Theta}_2(z) + m\tilde{\theta}_n^n(z)))^{-1} y_{k+m\ell}\ .$$

It is immediate that $T_\gamma^* T_{\gamma'}^* = T_{\gamma+\gamma'}^*$. Then $T_\gamma \in \mathrm{Aut}\,(\tilde{M}/B)$ and $\gamma \mapsto T_\gamma$ defines an injective group homomorphism from Γ to $\mathrm{Aut}\,(\tilde{M}/B)$. Here $\mathrm{Aut}\,(\tilde{M}/B)$ is the group of automorphisms of $\tilde{M}$ over B, i.e., the group of automorphisms g of $\tilde{M}$ such that $\tilde{f} \circ g = \tilde{f}$.

Proposition 17. *The action Γ on $\tilde{M}$ is properly discontinuous, free and symplectic, in the sense that $T_\gamma^* \omega_{\tilde{M}} = \omega_{\tilde{M}}$ for each $\gamma \in \Gamma$.*

Proof. Freeness of the action is clear from the description of the action. The proof of proper discontinuity is essentially the same as the proof of [13], Theorem 2.6. This can be also seen from the concrete description of fibers below in (III).

Let us show that the action is symplectic, i.e., $\omega_{\tilde{M}} = T_\gamma^* \omega_{\tilde{M}}$ for each $\gamma = (j, m)$. This is a new part not considered by [13]. We have $T_\gamma^* z_i = z_i$, $T_\gamma^* w_i = \exp(2\pi\sqrt{-1} f_i(z)) w_i$ for all i, where, in terms of the standard basis $\langle e_i \rangle_{i=1}^{n-1}$ of $\mathbb{C}^{n-1}$,

$$f_i(z) = j\tilde{\Theta}_1(z)e_i + m\tilde{\Theta}_2^t(z)e_i$$

for $1 \le i \le n-1$ and

$$f_n(z) = j\tilde{\Theta}_2(z) + m(\tilde{\theta}_n^n(z) + \frac{\ell \log z_n}{2\pi\sqrt{-1}}) \ .$$

Thus, for i with $1 \le i \le n$, we have

$$T_\gamma^* dz_i = dz_i \ ,$$

$$T_\gamma^* \frac{dw_i}{w_i} = T_\gamma^*(d\log w_i)$$

$$= d\log(T_\gamma^* w_i) = d(2\pi\sqrt{-1} f_i(z)) + d(\log w_i)$$

$$= \frac{dw_i}{w_i} + 2\pi\sqrt{-1}\sum_{k=1}^n \frac{\partial f_i}{\partial z_k} dz_k \ .$$

Using these identities, we can compute

$$T_\gamma^* \omega_{\tilde{M}} = \sum_{i=1}^n T_\gamma^*(dz_i) \wedge T_\gamma^*(\frac{dw_i}{w_i})$$

$$= \omega_{\tilde{M}} - 2\pi\sqrt{-1}\sum_{i=1}^n \sum_{k=1}^n \frac{\partial f_i}{\partial z_k} dz_k \wedge dz_i$$

$$= \omega_{\tilde{M}} - 2\pi\sqrt{-1} \sum_{1 \le i < k \le n} (\frac{\partial f_i}{\partial z_k} - \frac{\partial f_k}{\partial z_i}) dz_k \wedge dz_i \ .$$

On the other hand, by definition of $f_i(z)$ $(1 \le i \le n)$ and definition of $\tilde{\theta}(z)$ from the potential function $\Psi(z)$, we know that

$$f_i(z) = \sum_{\alpha=1}^{n-1} j_\alpha \tilde{\theta}^i_\alpha(z) + m\tilde{\theta}^i_n$$

$$= \sum_{\alpha=1}^{n-1} j_\alpha \frac{\partial^2 \Psi}{\partial z_\alpha \partial z_i} + m\frac{\partial^2 \Psi}{\partial z_n \partial z_i} + \delta_{i,n}\frac{m\ell \log z_n}{2\pi\sqrt{-1}} \ .$$

Since $\Psi(z)$ is holomorphic, it follows that

$$\frac{\partial}{\partial z_k}\left(\frac{\partial^2 \Psi}{\partial z_\alpha \partial z_i}\right) = \frac{\partial^3 \Psi}{\partial z_k \partial z_\alpha \partial z_i} = \frac{\partial}{\partial z_i}\left(\frac{\partial^2 \Psi}{\partial z_\alpha \partial z_k}\right) \ .$$

Substituting this into the formula above, we obtain that $T_\gamma^* \omega_{\tilde{M}} = \omega_{\tilde{M}}$.

Q.E.D.

(III) Group quotient of $\tilde{M}$ by $\Gamma = \mathbb{Z}^n$.

Let $M = \tilde{M}/\Gamma$. By Proposition 17, M is a smooth symplectic manifold with symplectic form ω_M induced by $\omega_{\tilde{M}}$ and M admits a fibration $f : M \to B$ induced by $\tilde{f}$. We denote the (scheme theoretic) fiber $f^{-1}(b)$ over $b \in B$ by M_b. Let us describe the fibers M_b.

(III-1) Smooth fibers M_b

First consider the case where $z_n(b) \neq 0$, i.e., the case where M_b is smooth. We have

$$\tilde{M}_b = (\mathbb{C}^\times)^{n-1} \times \{(x_k, y_k) | x_k y_k = b_n\} \simeq (\mathbb{C}^\times)^n_{(w_1,\dots,w_{n-1},w_n)} \ .$$

and $w_n = z_n(b)^{-(k+1)} x_k$. Let $\langle e_i \rangle_{i=1}^n$ be the ordered standard basis of Γ. From the description in (II), the action of Γ is given by:

$$T_{e_i}^* w_j = \exp(2\pi\sqrt{-1}\theta_i^j(b)) w_j$$

$$T_{e_i}^* w_n = \exp(2\pi\sqrt{-1}\theta_i^n(b)) w_n$$

for $1 \le i \le n-1$ with $\theta_i^j = \tilde{\theta}_i^j$ and

$$T_{e_n}^* w_j = \exp(2\pi\sqrt{-1}\theta_n^j(b)) w_j$$

$$T_{e_n}^* w_n = \exp(2\pi\sqrt{-1}\tilde{\theta}_n^n(b)) z_n(b)^\ell w_n = \exp(2\pi\sqrt{-1}\theta_n^n(b)) w_n \ .$$

Let us consider the universal covering map of fibers over $B \setminus D$:

$$\pi_b : \mathbb{C}^n \to (\mathbb{C}^\times)^n = \tilde{M}_b \ ; \ (t_i)_{i=1}^n \mapsto (\exp(2\pi\sqrt{-1}t_i))_{i=1}^n \ .$$

Then

$$M_b = \tilde{M}_b/\Gamma = \mathbb{C}^n/\Lambda_b$$

where $\Lambda_b \subset \mathbb{C}^n$ is the sublattice generated by the following $2n$ vectors:

$$e_i := (\delta_{ij})_{j=1}^n \ , \ \theta_i(b) := (\theta_{ij}(b))_{j=1}^n \ 1 \le i \le 2n \ .$$

Hence M_b is an n-dimensional principally polarized abelian variety of period $\theta(b)$, as desired. By the description of ω_M, the fibers M_b, $z_n(b) \neq 0$, are also Lagrangian submanifolds.

Proposition 18. *For $b \in B, z_n(b) \neq 0$, choose the basis*

$$p_{1,b}, \ldots, p_{n,b}, q_{1,b}, \ldots, q_{n,b}$$

of $H_1(M_b, \mathbb{Z})$ such that $\tilde{M}_b = \mathbb{C}^n/\langle p_{j,b}\rangle_{j=1}^n$ and

$$q_{i,b} = \sum_{j=1}^n \theta_i^j(b) p_{j,b}$$

for each i $(1 \le i \le n)$. Let

$$p_b^1, \ldots, p_b^n, q_b^1, \ldots, q_b^n$$

be the dual basis of $H^1(M_b, \mathbb{Z})$. Then the integral 2-form

$$L_b := \sum_{i=1}^n p_b^i \wedge q_b^i$$

gives a monodromy invariant principal polarization of M over $B \setminus D$ where $D = (z_n = 0)$.

Proof. When $z_n(b) \neq 0$, the fiber $\tilde{M}_b$ of $\tilde{p} : \tilde{M} \to B$ is $(\mathbb{C}^\times)^n$ and this family has no monodromy over $B \setminus (z_n = 0)$. Thus we can fix a basis $p_{1,b}, \ldots, p_{n,b}$ of $H_1(\tilde{M}_b, \mathbb{Z})$ uniformly in $b, z_n(b) \neq 0$.

To get a basis of $H_1(M_b, \mathbb{Z})$, we choose additional elements

$$q_{1,b}, \ldots, q_{n,b} \in H_1(M_b, \mathbb{Z})$$

determined by the deck-transformation of $\tilde{M}_b$ induced by the action $T_{e_1}, \ldots, T_{e_n}$. From the description of $T_{e_i}^*$ on w_j, they satisfy the relation

$$q_{i,b} = \sum_{j=1}^n \theta_i^j(b) p_{j,b}.$$

We see that $q_{1,b}, \dots, q_{n-1,b}$ are invariant under the monodromy, while $q_{n,b} \mapsto q_{n,b} + \ell p_{n,b}$ under the monodromy of the generator γ of $\pi_1(B \setminus D)$, i.e., the circle around discriminant divisor $z_n = 0$. The 2-form L_b is a principal polarization on M_b. It remains to show that L_b is invariant under the monodromy. By definition of $\theta(b)$, we compute that

$$\gamma^*(L_b) = \gamma^*(\sum_{i=1}^{n-1} p_b^i \wedge q_b^i) + \gamma^*(p_b^n \wedge q_b^n)$$

$$= \sum_{i=1}^{n-1} p_b^i \wedge q_b^i + p_b^n \wedge (q_b^n - \ell p_b^n) = L_b.$$

This implies the invariance. Q.E.D.

Remark 4. *As in* [13], *one can also describe $\tilde{f} : \tilde{M} \to B$ in terms of toric geometry. Following an argument similar to* [13], *Section 4, it seems possible to give a relatively principally polarized divisor (the relative theta divisor) which is defined globally over $B \setminus D$. However, its closure is not necessarily f-ample even if total space is of dimension* 4 *(cases of stable principally polarized Lagrangian 4-folds). In fact, a failure of f-ampleness of the closure already happens when the fiber dimension is* 2 *and the base dimension is* 1 *as explicitly described in* [13] *Section 4, Page 219. See also Remark 3.6.*

(III-2) Singular fibers M_b

Next consider the singular fibers of f. They are M_b with $z_n(b) = 0$. Recall from (I) that $\tilde{M}_b$ is the product of $(\mathbb{C}^\times)^{n-1}$ with coordinate $(w_i)_{i=1}^{n-1}$ and the infinite tree $\cup_{k \in \mathbb{Z}} \mathbb{P}_k^1$ of projective lines $\mathbb{P}_k^1$ with affine coordinate x_k, and $M_b = \tilde{M}_b / \Gamma$. Let us denote by $(0)_k, (\infty)_k \in \mathbb{P}_k^1$ the two points on the projective line $\mathbb{P}_k^1$ such that $(0)_k$ is identified with $(\infty)_{k-1}$ in the tree.

From (II), the action of Γ is given by:

$$T_{e_i}^* w_j = \exp(2\pi\sqrt{-1}\tilde{\theta}_i^j(b)) w_j$$

$$T_{e_i}^* x_k = \exp(2\pi\sqrt{-1}\tilde{\theta}_i^n(b)) x_k$$

for $1 \le i \le n-1$ and

$$T_{e_n}^* w_j = \exp(2\pi\sqrt{-1}\tilde{\theta}_n^j(b)) w_j$$

$$T_{e_n}^* x_k = \exp(2\pi\sqrt{-1}\tilde{\theta}_n^n(b)) x_{k+\ell} ,$$

where $\langle e_i \rangle_{i=1}^n$ is the ordered standard basis of Γ. Here we note that the last equality shows that the monodromy operation corresponds to the shift of the components of the infinite tree $\cup_{k\in\mathbb{Z}}\mathbb{P}^1_k$. Thus $\tilde{M}/\langle e_n \rangle$ can be described as the variety obtained from

$$(\mathbb{C}^\times)^{n-1} \times \cup_{k=0}^{\ell-1}\mathbb{P}^1_k$$

by identifying the point

$$(w^1, \dots, w^{n-1}) \times (0)_0 \in (\mathbb{C}^\times)^{n-1} \times (0)_0$$

with the point

$$(\exp(2\pi\sqrt{-1}\tilde{\theta}_n^1)w^1, \dots, \exp(2\pi\sqrt{-1}\tilde{\theta}_n^{n-1})w^{n-1}) \in (\mathbb{C}^\times)^{n-1} \times (\infty)_{\ell-1}.$$

From this description, M_b consists of ℓ irreducible components, each of whose normalization is isomorphic to a $\mathbb{P}^1$-bundle over a complex torus of dimension $(n-1)$ isogenous to $(\mathbb{C}^\times)^{n-1}/\langle e_i \rangle_{i=1}^{n-1}$, where the action of $\langle e_i \rangle_{i=1}^{n-1}$ is given by the coordinate action $T^*_{e_i}$ $(1 \le i \le n-1)$ on w_j $(1 \le j \le n-1)$ described above. Note that the quotient $(\mathbb{C}^\times)^{n-1}/\langle e_i \rangle_{i=1}^{n-1}$ is compact because the imaginary part of $\tilde{\Theta}_1(z)$ is positive definite from the assumption that the imaginary part of $\theta(z)$ is positive definite. We also note that the characteristic cycles are of type I_m for some $1 \le m \le \infty$.

Let us denote the fibration $f : M \to B$ constructed above by $f^{\mathrm{model}} : M^{\mathrm{model}} \to B$. From the construction and the description of singlar fibers in (III-2), we obtain Theorem 2.

To prove Proposition 1, we need the following proposition, which is essentially proved in [8] Proposition 5.1:

Proposition 19. *Let $B = \{(z_1, \dots, z_n) | |z_i| < 1\}$ be the polydisc. Let $f : M \to B$ and $f' : M' \to B$ be two Lagrangian fibrations with the same discriminant $D \subset B$, having Lagrangian sections $\Sigma \subset M$ and $\Sigma' \subset M'$. Suppose there exists a biholomorphic morphism $\Phi : M \setminus f^{-1}(D) \to M' \setminus f'^{-1}(D)$ over B such that $\Phi(\Sigma) = \Sigma'$ and Φ is symplectomorphic, i.e., $\Phi^*\omega_{M'} = \omega_M$. Then*

(1) Φ extends to a bimeromorphic map $M \cdots \to M'$ over B and the extended bimeromorphic map, denoted by the same latter Φ, is isomorphic in codimension 1.

(2) Let $b \in D$ and $L \subset B$ be the intersection of B with a general line passing through b, in terms of the coordinates $(z_1, \dots, z_n)$. Let $M_L = f^{-1}(L)$, $M'_L = (f')^{-1}(L)$ and $\Phi_L : M_L \cdots \to M'_L$, the bimeromprphic map over L induced from Φ. Then Φ_L extends to a biholomorphic map over L. In particular, $M_b \simeq M'_b$ for all $b \in D$.

Proof. The statement (2) is exactly the same as [8] Proposition 5.1. Since both K_M and $K_{M'}$ are trivial, the same argument of [8] Proposition 5.1 concludes that Φ and Φ^{-1} extend to bimeromorphic maps such that the indeterminacies are of codimension at least 2. (However, the argument there does not automatically exclude the possible indeterminacies of Φ and Φ^{-1}.) Q.E.D.

Proof of Proposition 1. We can apply Proposition 19 (1) for $f : M \to B$ and $f^{\mathrm{model}} : M^{\mathrm{model}} \to B$ with the same potential function $\Psi(z)$ and ℓ to obtain Proposition 1. Q.E.D.

Applying Proposition 19 (2) for $f : M \to B$ and $f^{\mathrm{model}} : M^{\mathrm{model}} \to B$, we also obtain

Corollary 1. *For a general principally polarized stable Lagrangian fibration $f : M \to B$, the singular fibers are isomorphic to the singular fibers of its model $f^{\mathrm{model}} : M^{\mathrm{model}} \to B$. In particular, the positive integer $|\ell|$ in Theorem 3 is the number of components of the singular fiber and $S_1^i \neq S_2^i$ for each i in the notation of Definition 1.*

It is natural to ask whether Φ in Proposition 19 can be extended to a biholomorphic map from M to M'. We do not know the answer in the general case. But this is indeed the case if f and f' are projective morphisms, as explained by the next proposition which we learned from Kawamata.

Proposition 20. *Let $f : M \to B$ a principally polarized stable Lagrangian fibration and $f^{\mathrm{model}} : M^{\mathrm{model}} \to B$ be the model with the same potential function $\Psi(z)$ and ℓ as f. Assume that the morphisms f and f^{model} are projective. Then M and M^{model} are isomorphic over B.*

Proof. By assumption, M and M^{model} are minimal, birational and projective over B. Let $\Phi : M \cdots \to M^{\mathrm{model}}$ be a birational map over B. If Φ is not an isomorphism, then Φ is decomposed into a finite sequence of flops over B by Theorem 1 of [10]. In particular, M^{model} admits at least one flopping contraction. The exceptional locus of a flopping contraction is covered by rational curves by Theorem 1 of [9]. Moreover, all rational curves numerically equivalent to any of these rational curves are contracted under this flopping contraction by Contraction Theorem (Theorem 3.2.1 (i) of [11]). On the other hand, by the construction of M^{model}, the total space of deformation of any rational curve C in M^{model} is a codimension 1 subset of M^{model}. This contradicts the fact that the exceptional locus of a flopping contraction has codimension ≥ 2 (cf. p. 420 Proof in [10]). It follows that M^{model} has no flopping contraction and Φ is an isomorphism. Q.E.D.

§5. Periods and the characteristic cycles

In this section, we will examine the relation between types of the characteristic cycles and the periods. For simplicity, we will restrict our discussion to the case of $\ell = 1$. The generalization to arbitrary ℓ is straightforward. Explicit constructions (Constructions I -III) will be given when $n = 2$, i.e., constructions of 4-dimensional principally polarized stable Lagrangian fibrations. Construction I gives an explicit example in which the types of characteristic cycles change fiber by fiber. Construction II gives an explicit example in which the types of characteristic cycles are constant type I_m ($m < \infty$) and Construction III gives an explicit example in which the types of characteristic cycles are constant type A_∞.

Proposition 21. *Let $f : M \to B$ be a $2n$-dimensional principally polarized stable Lagrangian fibration with potential function $\Psi(z)$ and $\ell = 1$. We denote the (univalent) period matrix by $\tilde{\theta}(z) = (\tilde{\theta}_i^j(z))_{i,j=1}^n$ and the multi-valued period matrix $\theta(z)$ of f as*

$$\theta(z) = \tilde{\theta}(z) + \frac{\log z_n}{2\pi\sqrt{-1}} \begin{pmatrix} O_{n-1} & 0 \\ 0 & 1 \end{pmatrix}.$$

For $b \in D$ for which M_b is singular, define $n(b)$ ($1 \le n(b) \le \infty$) to be the order of

$$(\exp(2\pi\sqrt{-1}\tilde{\theta}_n^j(b)))_{j=1}^{n-1} \bmod \langle(\exp(2\pi\sqrt{-1}\tilde{\theta}_i^j(b)))_{j=1}^{n-1} \,|\, 1 \le i \le n-1\rangle$$

in the multiplicative group

$$(\mathbb{C}^\times)^{n-1}/\langle(\exp(2\pi\sqrt{-1}\tilde{\theta}_i^j(b)))_{j=1}^{n-1} \,|\, 1 \le i \le n-1\rangle .$$

Then the characteristic cycle of M_b is of type $I_{n(b)}$.

Remark 5. *When $n = 2$, we can state the above result more explicitly as follows. We shall use this description in Constructions (I)-(III) below. We write the (univalent) period matrix $\tilde{\theta}(z)$ and the multi-valued period matrix $\theta(z)$ of f as*

$$\tilde{\theta}(z) = \begin{pmatrix} \tilde{\theta}_1(z) & \tilde{\theta}_2(z) \\ \tilde{\theta}_2(z) & \tilde{\theta}_3(z) \end{pmatrix}, \theta(z) = \tilde{\theta}(z) + \frac{\log z_2}{2\pi\sqrt{-1}} \begin{pmatrix} 0 & 0 \\ 0 & 1 \end{pmatrix}.$$

For $b = (b_1, 0) \in B$ for which M_b is singular, the characteristic cycle of M_b is then of type $I_{n(b)}$, where $n(b)$ ($1 \le n(b) \le \infty$) is exactly the order of

$$\exp(2\pi\sqrt{-1}\tilde{\theta}_2(b)) \bmod \langle\exp(2\pi\sqrt{-1}\tilde{\theta}_1(b))\rangle$$

in the multiplicative group $\mathbb{C}^\times/\langle\exp(2\pi\sqrt{-1}\tilde{\theta}_1(b))\rangle$.

Proof. In the description (III-2), M_b is the quotient of

$$\tilde{M}_b = \cup_{k\in\mathbb{Z}}(\mathbb{C}^\times)^{n-1} \times \mathbb{P}^1_k$$

with coordinates $((w_j)_{j=1}^{n-1}, x_k)$ $(k \in \mathbb{Z})$ by the action of $\Gamma = \mathbb{Z}^n$ with ordered standard basis

$$\langle e_1, \ldots, e_{n-1}, e_n\rangle \ .$$

In terms of the standard basis, the action is given by:

$$T^*_{e_i} : (w_j)_{j=1}^{n-1} \mapsto (\exp(2\pi\sqrt{-1}\tilde{\theta}^j_i(b))w_j)_{j=1}^{n-1}$$

$$T^*_{e_i} : x_k \mapsto \exp(2\pi\sqrt{-1}\tilde{\theta}^n_i(b))x_k$$

for e_i $(1 \le i \le n-1)$, and for e_n

$$T^*_{e_n} : (w_j)_{j=1}^{n-1} \mapsto (\exp(2\pi\sqrt{-1}\tilde{\theta}^j_n(b))w_j)_{j=1}^{n-1}$$

$$T^*_{e_n} : x_k \mapsto \exp(2\pi\sqrt{-1}\tilde{\theta}^n_n(b))x_{k+1} \ .$$

Thus $\tilde{M}_b/\langle e_n\rangle$ is $(\mathbb{C}^\times)^{n-1} \times \mathbb{P}^1_0$ in which $((w_j)_{j=1}^{n-1}, 0)$ and $((w'_j)_{j=1}^{n-1}, \infty)$ are identified exactly when the two points

$$(w_j)_{j=1}^{n-1} \text{ and } (w'_j)_{j=1}^{n-1} \text{ of } (\mathbb{C}^\times)^{n-1}$$

are in the same orbit under the action of the cyclic subgroup

$$G(b) := \langle(\exp(2\pi\sqrt{-1}\tilde{\theta}^j_n(b)))_{j=1}^{n-1}\rangle$$

of $(\mathbf{C}^\times)^{n-1}$. On the other hand, $((\mathbb{C}^\times)^{n-1} \times \mathbb{P}^1_0)/\langle e_i\rangle_{i=1}^{n-1}$ is the normalization of M_b. Thus, M_b is obtained from $((\mathbb{C}^\times)^{n-1} \times \mathbb{P}^1_0)/\langle e_i\rangle_{i=1}^{n-1}$ by identifying the two $(n-1)-$dimensional complex tori $((\mathbb{C}^\times)^{n-1} \times \{\infty\})/\langle e_i\rangle_{i=1}^{n-1}$ and $((\mathbb{C}^\times)^{n-1} \times \{0\})/\langle e_i\rangle_{i=1}^{n-1}$,by the action of $G(b)$ above. Here, as a subgroup of $(\mathbb{C}^\times)^{n-1}$, the group $\langle e_i\rangle_{i=1}^{n-1}$ is the multiplicative subgroup generated by the $n-1$ elements

$$(\exp(2\pi\sqrt{-1}\tilde{\theta}^j_i(b)))_{j=1}^{n-1} \ , \ 1 \le i \le n-1 \ .$$

This implies the result. Q.E.D.

Construction I.

Under the notation of Proposition 21, we set $n = 2, \ell = 1$ and

$$\Psi(z_1, z_2) := \frac{(z_1 + 5\sqrt{-1})^3 + (z_2 + 5\sqrt{-1})^3 + 3z_1^2 z_2 + 3z_1 z_2^2}{6} ,$$

Then

$$\tilde{\theta}(z) = \begin{pmatrix} z_1 + z_2 + 5\sqrt{-1} & z_1 + z_2 \\ z_1 + z_2 & z_1 + z_2 + 5\sqrt{-1} \end{pmatrix},$$

$$\theta(z) = \tilde{\theta}(z) + \frac{\log z_2}{2\pi\sqrt{-1}} \begin{pmatrix} 0 & 0 \\ 0 & 1 \end{pmatrix}$$

and

$$\operatorname{Im}\theta(z) = \begin{pmatrix} y_1 + y_2 + 5 & y_1 + y_2 \\ y_1 + y_2 & y_1 + y_2 + 5 - \log|z_2|/2\pi \end{pmatrix}.$$

Here and hereafter x_i and y_i are the real and imaginary part of z_i respectively. Since $t + 5 > 0$ and $(t+5)^2 - t^2 > 0$ when $-2 < t < 2$ and $|z_2| < 1$, it follows that $\operatorname{Im}\theta(z)$ is positive definite on the polydisc

$$\{(z_1, z_2) \,|\, |z_i| < 1\}.$$

Taking a smaller 2-dimensional polydisc B with multi-radius ϵ, we then obtain a 4-dimensional Lagrangian fibration $f : M \to B$, associated with the potential function $\Psi(z)$ and $\ell = 1$. The discriminant set is $z_2 = 0$. Define $N = N(z)$ to be the order of

$$e^{2\pi\sqrt{-1}z_1} \bmod \langle e^{2\pi\sqrt{-1}(z_1+5\sqrt{-1})}\rangle$$

in the multiplicative group $\mathbb{C}^\times/\langle e^{2\pi\sqrt{-1}(z_1+5\sqrt{-1})}\rangle$. By abuse of language, we include $N = \infty$ when the order is not finite. Then, the characteristic cycle of $M_{(z_1,0)}$ is of type I_N.

Proposition 22. *In Construction 1, the characteristic cycle on $M_{(z_1,0)}$ is of Type I_k with $k < \infty$ if and only if*

$$z_1 \in \mathbb{Q}(\sqrt{-1}).$$

So, the singular fibers of finite characteristic cycle I_k $(k < \infty)$ and the singular fibers of infinite characteristic cycle I_∞ are both dense over the discriminant set. Moreover, the characteristic cycle of $M_{(z_1,0)}$ is precisely of type I_k $(k < \infty)$ for $z_1 = 1/k$. So, the singular fibers with characteristic cycles of type I_k with any sufficiently large k appear in this family.

Proof. By the definition of $N = N(z)$, it follows that $N < \infty$ for $M_{(z_1,0)}$ if and only if there are integers $k > 0$ and m such that

$$(e^{2\pi\sqrt{-1}z_1})^k = (e^{2\pi\sqrt{-1}(z_1+5\sqrt{-1})})^m.$$

The last condition is equivalent to

$$kz_1 - m(z_1 + 5\sqrt{-1}) \in \mathbb{Z}$$

which is also equivalent to

$$(k-m)x_1 \in \mathbb{Z} \text{ and } (k-m)y_1 - 5m = 0 .$$

Note that $k - m \neq 0$ in the last equivalent condition, as otherwise $k = m = 0$. It is immediate to see that two integers $k > 0$ and m satisfying last equivalent condition exist if and only if $z_1 \in \mathbb{Q}(\sqrt{-1})$. Since $|e^{2\pi\sqrt{-1}(z_1+5\sqrt{-1})}| > 1$ for $|z_1| < 1$, whereas $|e^{2\pi\sqrt{-1}/k}| = 1$ for $k \in \mathbb{Z}$, it follows that the order $N(z)$ for $z_1 = 1/k$ is precisely the order of $e^{2\pi\sqrt{-1}/k}$ in the multiplicative group $\mathbb{C}^\times$. This implies the last statement. Q.E.D.

Construction II.

Under the notation of Proposition 21, we set $n = 2, \ell = 1$ and

$$\Psi(z_1, z_2) := \frac{\sqrt{-1}(z_1^2 + z_2^2)}{2} + \frac{z_1 z_2}{k} ,$$

where n is a positive integer. Then

$$\tilde{\theta}(z) = \begin{pmatrix} \sqrt{-1} & 1/k \\ 1/k & \sqrt{-1} \end{pmatrix}, \theta(z) = \tilde{\theta}(z) + \frac{\log z_2}{2\pi\sqrt{-1}} \begin{pmatrix} 0 & 0 \\ 0 & 1 \end{pmatrix}$$

and

$$\operatorname{Im}\theta(z) = \begin{pmatrix} 1 & 0 \\ 0 & 1 - \log|z_2|/2\pi \end{pmatrix} .$$

The matrix $\operatorname{Im}\theta(z)$ is positive definite. So, taking a smaller polydisc B of dimension 2, we obtain a 4-dimensional Lagrangian fibration $f : M \to B$, associated with the potential function $\Psi(z)$ and $\ell = 1$ above. The discriminant set is $z_2 = 0$. The order of

$$e^{2\pi\sqrt{-1}/k} \bmod \langle e^{-2\pi} \rangle$$

is exactly k in the multiplicative group $\mathbb{C}^\times / \langle e^{-2\pi} \rangle$, where

$$-2\pi = 2\pi\sqrt{-1} \cdot \sqrt{-1}.$$

Then, the characteristic cycle of $M_{(z_1,0)}$ is of type I_k, and in particular, the type is constant.

Construction III.

Under the notation of Proposition 21, we set $n = 2, \ell = 1$ and

$$\Psi(z_1, z_2) := \frac{\sqrt{-1}(z_1^2 + z_2^2)}{2} + \alpha z_1 z_2 \,,$$

where α is any irrational, real number, say $\sqrt{2}$. Then

$$\tilde{\theta}(z) = \begin{pmatrix} \sqrt{-1} & \alpha \\ \alpha & \sqrt{-1} \end{pmatrix}, \theta(z) = \tilde{\theta}(z) + \frac{\log z_2}{2\pi\sqrt{-1}} \begin{pmatrix} 0 & 0 \\ 0 & 1 \end{pmatrix}$$

and

$$\operatorname{Im}\theta(z) = \begin{pmatrix} 1 & 0 \\ 0 & 1 - \log|z_2|/2\pi \end{pmatrix}.$$

The matrix $\operatorname{Im}\theta(z)$ is positive definite. So, taking a smaller polydisc B of dimension 2, we obtain a 4-dimensional Lagrangian fibration $f : M \to B$, associated with the potential function $\Psi(z)$ and $\ell = 1$ above. The discriminant set is $z_2 = 0$. Since α is an irrational real number, the element

$$e^{2\pi\sqrt{-1}\cdot\alpha} \bmod \langle e^{-2\pi}\rangle$$

is of infinite order in the multiplicative group $\mathbb{C}^\times/\langle e^{-2\pi}\rangle$, where $-2\pi = 2\pi\sqrt{-1}\cdot\sqrt{-1}$. Then, the characteristic cycle of $M_{(z_1,0)}$ is of type A_∞, and in particular the type is constant.

References

[1] W. Barth, K. Hulek, C. Peters, A. Van de Ven, Compact complex surfaces, Second enlarged edition, Springer Verlag, Berlin-Heidelberg, 2004.

[2] R. Donagi, E. Markman, Cubics, integrable systems, and Calabi-Yau threefolds, In: Proceedings of the Hirzebruch 65 Conference on Algebraic Geometry (Ramat Gan, 1993), Israel Math. Conf. Proc. **9** Bar-Ilan Univ., Ramat Gan, 1996, pp 199-221.

[3] P. A. Griffiths, Periods of integrals on algebraic manifolds. III. Some global differential-geometric properties of the period mapping, Inst. Hautes Études Sci. Publ. Math., **38** (1970), 125–180.

[4] P. A. Griffiths, J. Harris, Principles of algebraic geometry, Wiley Interscience, 1978.

[5] V. Guillemin, S. Sternberg, Symplectic techniques in physics, Second edition, Cambridge University Press, Cambridge, 1990.

[6] J.-M. Hwang, Base manifolds for fibrations of projective irreducible symplectic manifolds, Invent. math., **174** (2008), 625–644.

[7] J.-M. Hwang, K. Oguiso, Characteristic foliation on the discriminant hypersurface of a holomorphic Lagrangian fibration, Amer. J. Math., **131** (2009), 981–1007.

[8] J.-M. Hwang, K. Oguiso, Multiple fibers of holomorphic Lagrangian fibrations, Communications in Contemporary Math., **13** (2011), 309–329.
[9] Y. Kawamata, On the length of an extremal rational curves, Invent. Math., **105** (1991), 609–611.
[10] Y. Kawamata, Flops connect minimal models, Publ. Res. Inst. Math. Sci., **44** (2008), 419–423.
[11] Y. Kawamata, K. Matsuda, K. Matsuki, Introduction to the minimal model problem, In: Algebraic geometry (Sendai, 1985), Adv. Stud. Pure Math., **10**, Math. Soc. Japan, 1987, pp. 283–360.
[12] K. Kodaira, On compact complex analytic surfaces II, III, Ann. of Math., **77** (1963), 563–626, ibid **78** (1963), 1–40.
[13] I. Nakamura, Relative compactification of the Néron model and its application, In: Complex analysis and algebraic geometry, Iwanami Shoten, Tokyo, 1977, pp. 207–225.
[14] C. Voisin, Hodge theory and complex algebraic geometry II, Cambridge University Press, Cambridge, 2003.

Jun-Muk Hwang, Korea Institute for Advanced Study, Hoegiro 87, Seoul, 130-722, Korea
E-mail address: jmhwang@kias.re.kr

Keiji Oguiso, Department of Mathematics, Osaka University
Toyonaka 560-0043 Osaka, Japan and
Korea Institute for Advanced Study, Hoegiro 87, Seoul, 130-722, Korea
E-mail address: oguiso@math.sci.osaka-u.ac.jp

Advanced Studies in Pure Mathematics 70, 2016
Minimal Models and Extremal Rays (Kyoto, 2011)
pp. 277–298

On symplectic hypersurfaces

Manfred Lehn, Yoshinori Namikawa, Christoph Sorger and Duco van Straten

§1. Introduction

A symplectic variety is a normal complex variety X with a holomorphic symplectic form ω on the regular part X_{reg} and with rational Gorenstein singularities. Affine symplectic varieties arise in many different ways such as closures of nilpotent orbits of a complex simple Lie algebra, as Slodowy slices to such nilpotent orbits or as symplectic reductions of holomorphic symplectic manifolds with Hamiltonian actions. Many examples of affine symplectic varieties tend to require large embedding codimensions compared to their dimensions.

In this article we treat the rarest case, namely affine symplectic *hypersurfaces*. For technical reasons we also impose the condition that X admit a good $\mathbb{C}^*$-action, i.e. that its affine coordinate ring $A = \mathbb{C}[X]$ be positively graded, $A = \oplus_{i \geq 0} A_i$ with $A_0 = \mathbb{C}$, and that ω is also homogeneous of positive weight s. This condition is satisfied in all examples we know. Finally, such a homogeneous symplectic hypersurface X is called *indecomposable* if the unique fixed point of the $\mathbb{C}^*$-action is a Poisson subscheme of X. As the term indecomposable indicates, such singularities are essential factors of more general hypersurfaces in the sense that every homogeneous hypersurface (X, ω) equivariantly decomposes into a product $W_1 \times ... \times W_k \times X'$, where X' is an indecomposable homogeneous hypersurface and each W_i is isomorphic to $\mathbb{C}^2$ with a standard symplectic form of the same weight s as ω (Lemma 2.5).

Indecomposable homogeneous hypersurfaces $X = \{f = 0\} \subset \mathbb{C}^{2n+1}$ have the remarkable property that the Poisson structure $\{-,-\} : A \times A \to A$ defined on the coordinate ring A by the symplectic structure

Received December 20, 2011.
Revised November 19, 2012.

extends to the ambient space (Lemma 2.7). Consequently, the deformation $X_t = \{f = t\}$ is a Poisson deformation, from which it follows that X admits a crepant resolution (Theorem 2.8).

Since homogeneous symplectic hypersurfaces have no local moduli (cf. [9], Proposition (3.5)), they arise in a discrete way. As is well known, an indecomposable homogeneous hypersurface of dimension 2 is a Kleinian singularity of type A, D or E. In higher dimensions, the classification is an open problem. At this moment we know of a series X_n, $n \geq 2$, of 4-dimensional examples and of a single 6-dimensional example $\hat{X}$. We found them originally as the transversal slices to certain nilpotent orbits in complex simple Lie algebras [6]. In this article, we give several different descriptions of the same hypersurfaces.

Given that these constructions all lead to the same examples it might be natural to ask: *Is every indecomposable homogeneous symplectic hypersurface isomorphic to an ADE surface singularity, one of the 4-dimensional hypersurfaces X_n, or the 6-dimensional hypersurface $\hat{X}$?*

In the final section we look at X_n from the view point of contact geometry. Let $Y \subset \mathbb{P}(2n-1, 2n-1, 2, 2, 2)$ be the 3-dimensional projective variety defined by the same equation as X_n. The symplectic structure on X_n induces a contact structure on the regular part Y^0 of Y with the contact line bundle $\mathcal{O}(2) := \mathcal{O}_{\mathbb{P}}(2)|_{Y^0}$. We construct an explicit birational map between Y and the projectivised cotangent bundle $\mathbb{P}(T^*_{\mathbb{P}^1 \times \mathbb{P}^1})$ so that this contact structure is transformed to the canonical contact structure on $\mathbb{P}(T^*_{\mathbb{P}^1 \times \mathbb{P}^1})$. More exactly, we take a resolution $\mu : \tilde{Y} \to Y$ by blowing up the singular locus of Y and construct a birational contraction map $\nu : \tilde{Y} \to \mathbb{P}(T^*_{\mathbb{P}^1 \times \mathbb{P}^1})$. The pull-back of both contact structures by μ and ν then determine the same contact structure on $\tilde{Y}$ outside some divisor F with $F \subset \mathrm{Exc}(\mu) \cap \mathrm{Exc}(\nu)$. Now this construction tells us that if we start from $\mathbb{P}(T^*_{\mathbb{P}^1 \times \mathbb{P}^1})$, then after a suitable birational modification we can reach a singular contact Fano 3-fold Y. It would be interesting to know if such phenomena occur more generally.

§2. The Poisson matrix

A symplectic variety in the sense of Beauville [1] is a normal complex variety X with a symplectic form ω on the regular part X_{reg} and the property that for some proper resolution of singularities $\pi : X' \to X$ the form $\pi^*\omega$ extends to a regular form on X'. One can show that the same property then holds for any proper resolution. Equivalently, it is sufficient to require that X have rational Gorenstein singularities [7].

A $\mathbb{C}^*$-action on an affine variety $X = \mathrm{Spec}(A)$ is called good if the homogeneous components of the corresponding grading of the coordinate ring $A = \mathbb{C}[X]$ satisfy $A_0 = \mathbb{C}$ and $A_d = 0$ for $d < 0$. In this case we write $\mathfrak{m} := \bigoplus_{d>0} A_d$ for the maximal ideal corresponding to the unique fixed point $O \in X$. Then $\mathfrak{m}/\mathfrak{m}^2$ is a finite dimensional $\mathbb{C}^*$-representation. We may choose homogeneous elements $\bar{x}_1, \ldots, \bar{x}_m \in A$ whose residue classes form a basis of eigenvectors for the action and which therefore generate the ring A. This yields an equivariant embedding $X \to \mathbb{C}^m$ of minimal codimension, with $\mathbb{C}^*$ acting linearly and contracting on $\mathbb{C}^m$.

Definition 2.1. — *A 2n-dimensional* homogeneous symplectic hypersurface *is a symplectic variety (X, ω) with a good $\mathbb{C}^*$-action $\lambda : \mathbb{C}^* \times X \to X$ such that*

(1) *ω is homogeneous of degree s, i.e. $\lambda(t)^*\omega = t^s\omega$, and*
(2) *$\dim T_O X = 2n + 1$, where $O \in X$ is the unique fixed point of X.*

Lemma 2.2. — *Let (X, ω) be homogeneous symplectic hypersurface. Then the degree s of ω is positive.*

Proof. Let $\pi : X' \to X$ be a $\mathbb{C}^*$-equivariant resolution of the singularities of X. The fixed point locus for the induced $\mathbb{C}^*$-action on X' consists of a finite number of smooth projective varieties F_i lying above the origin $0 \in X$. We prove that there is a fixed point q such that the action of $\mathbb{C}^*$ on the cotangent space $T_q^* X'$ has only non-negative weights.

For each fixed point q, we define $T_q^*(X')^{\geq 0}$ to be the subspace of $T_q^* X'$ spanned by eigenvectors with non-negative weights. By Theorem 4.1 of [2], for each F_i there exists a locally closed, smooth and $\mathbb{C}^*$-invariant subvariety X_i' of X' containing F_i such that $T_q^*(X_i') = T_q^*(X')^{\geq 0}$ for all $q \in F_i$. Let $p \in X'$ be a point such that $\pi(p) \neq 0$. Then the closure of the $\mathbb{C}^*$-orbit passing through p is contained in some X_i' by Theorem 4.2 of [2]. This means that there is a locally closed $\mathbb{C}^*$-invariant decomposition of X', $X' = \cup X_i'$. In particular, $\dim X_{i_0}' = \dim X'$ for some i_0. Then $T_q^*(X')^{\geq 0} = T_q^* X'$ for $q \in F_{i_0}$.

Let us take such a fixed point q. Then at least one weight must be positive, as the action on X' is non-trivial. By assumption ω^n extends to a regular $2n$-form ψ of degree ns on X'. At q it can be expressed in terms of local coordinates as $\psi = g(z_1, \ldots, z_{2n}) dz_1 \wedge \ldots \wedge dz_{2n}$, so that $ns = \deg(\psi) \geq \sum_i \deg(z_i) > 0$. Q.E.D.

Every symplectic variety (X, ω) carries a canonical Poisson structure: On the open regular part X_{reg} there is an isomorphism $\omega^{-1} : \Omega_X \to T_X$, and the Poisson bracket is defined by $\{f, g\} := df(\omega^{-1}(dg))$ for $f, g \in \mathcal{O}_X(U)$, $U \subset X_{\text{reg}}$. As X is normal, this bracket can be uniquely extended for any two regular functions on X. If X is affine with coordinate ring A, the Poisson bracket is completely determined by its values on a set $\bar{x}_1, \ldots, \bar{x}_m$ of generators of A. The matrix $\bar{\Theta} \in A^{m\times m}$ with entries

$$\bar{\Theta}_{ij} := \{\bar{x}_i, \bar{x}_j\} \tag{2.1}$$

is skew-symmetric and satisfies the Jacobi identity

$$\sum_m \left(\bar{\Theta}_{im} \frac{\partial \bar{\Theta}_{jk}}{\partial \bar{x}_m} + \bar{\Theta}_{jm} \frac{\partial \bar{\Theta}_{ki}}{\partial \bar{x}_m} + \bar{\Theta}_{km} \frac{\partial \bar{\Theta}_{ij}}{\partial \bar{x}_m} \right) = 0. \tag{2.2}$$

In the following, we will refer to $\bar{\Theta}$ as the Poisson matrix of X. Assume now that (X, ω) is a homogeneous symplectic hypersurface of dimension $2n$ with an equivariant embedding $X \to \mathbb{C}^{2n+1}$ such that the coordinates $x_1, \ldots, x_{2n+1}$ of the ambient space have degree $d_i := \deg(x_i) > 0$ and X is defined by a homogeneous polynomial $f \in \mathbb{C}[x_1, \ldots, x_{2n+1}]$ of degree $d := \deg(f) > 0$. As ω is homogeneous of degree $s > 0$, the Poisson structure is homogeneous of degree $-s$ and

$$\deg(\bar{\Theta}_{ij}) = \deg(x_i) + \deg(x_j) - s = d_i + d_j - s. \tag{2.3}$$

There exists a direct explicit relation between the Poisson matrix of X and its defining equation f, which we will explain next.

Recall that the pfaffian of a skew-symmetric $2n \times 2n$-matrix B is a homogeneous polynomial $\mathrm{pf}(B)$ of the entries of B of degree n such that $\mathrm{pf}(B)^2 = \det(B)$. Explicitly,

$$\mathrm{pf}(B) = \sum_\pi \mathrm{sgn}(\pi) B_{\pi(1)\pi(2)} \cdots B_{\pi(2n-1)\pi(2n)}, \tag{2.4}$$

where π runs through a subset of permutations in S_{2n} such that the collections $T_\pi := \{\{\pi(1), \pi(2)\}, \ldots, \{\pi(2n-1), \pi(2n)\}\}$ represent every decomposition of $\{1, \ldots, 2n\}$ into n unordered pairs exactly once.

If B is a skew-symmetric $(2n+1)\times(2n+1)$-matrix, let $\mathrm{pf}(B)$ denote the vector whose i-th entry is given as $\mathrm{pf}(B)_i = (-1)^{i-1}\,\mathrm{pf}(B^i)$, where B^i is obtained from B by deleting the i-th row and column. It is well-known that $B\,\mathrm{pf}(B) = 0$.

Lemma 2.3. — *Let $X \subset \mathbb{C}^{2n+1}$ be a homogeneous symplectic hypersurface defined by a homogeneous polynomial $f \in \mathbb{C}[x_1, \ldots, x_{2n+1}]$ and let $\bar{\Theta}$ denote its Poisson matrix. Then there is a non-zero constant c such that*

$$\mathrm{pf}(\bar{\Theta}) = c\,\mathrm{grad}(f) \tag{2.5}$$

as vectors with values in $\mathbb{C}[X]$.

Proof. As $f = 0$ in $\mathbb{C}[X]$, it follows that

$$0 = \{x_i, f\} = \sum_j \bar{\Theta}_{ij} \frac{\partial f}{\partial x_j} \in \mathbb{C}[X], \tag{2.6}$$

or briefly: $\bar{\Theta}\,\mathrm{grad}(f) = 0$. On the other hand, $\bar{\Theta}\,\mathrm{pf}(\bar{\Theta}) = 0$. Now over the regular part of X, the derivative $\mathrm{grad}(f)$ vanishes nowhere according to the Jacobian criterion for smoothness. Moreover, $\bar{\Theta}$ has rank $2n$ since X is symplectic so that the kernel of $\bar{\Theta}$ is one-dimensional and at least one of the pfaffians $\mathrm{pf}(\bar{\Theta}^i)$ is non-zero. So $\mathrm{pf}(\bar{\Theta})$ also vanishes nowhere on X_{reg}. As both $\mathrm{grad}(f)$ and $\mathrm{pf}(\bar{\Theta})$ span the kernel of $\bar{\Theta}$, there is an invertible regular function c on X_{reg} such that $\mathrm{pf}(\bar{\Theta}) = c\,\mathrm{grad}(f)$ on X_{reg}. Since X is normal, the function c extends to an invertible regular function on X, and $\mathrm{pf}(\bar{\Theta}) = c\,\mathrm{grad}(f)$ holds everywhere on X. As c is homogeneous of some weight, it must be constant. Q.E.D.

Replacing f by some scalar multiple, we can and will assume from now on that for every homogeneous symplectic hypersurface the following fundamental relation between the defining equation and the Poisson matrix holds:

$$\boxed{\mathrm{pf}(\bar{\Theta}) = \mathrm{grad}(f)} \tag{2.7}$$

Definition 2.4. — *A homogeneous symplectic hypersurface X is* indecomposable *if its unique fixed point is a Poisson subscheme of X.*

Using the previous notations this is equivalent to saying that the homogeneous maximal ideal $\mathfrak{m}$ satisfies $\{\mathfrak{m}, A\} \subset \mathfrak{m}$ which in turn is equivalent to the condition that $\bar{\Theta}_{ij} \in \mathfrak{m}$ for all i, j.

The following decomposition lemma is due to Weinstein [10] if the underlying variety is smooth. For singular Poisson varieties an analogous statement in the formal category has been proved by Kaledin. In the weighted homogeneous situation the argument of Weinstein extends easily. In fact, the proof is easier than both in Weinstein's and Kaledin's situation as the choice of new coordinates can be carried out in finitely many steps.

Lemma 2.5. — *Let (X, ω) be a homogeneous symplectic hypersurface. Then there is an equivariant symplectic isomorphism $X \cong W_1 \times \ldots \times W_k \times X'$, where X' is an indecomposable homogeneous symplectic hypersurface and each W_i is isomorphic to $\mathbb{C}^2$ with symplectic form $dz_1 \wedge dz_2$ and homogeneous coordinates with* $\deg(z_1) + \deg(z_2) = s$.

Proof. Let $X \to \mathbb{C}^{2n+1}$ be a homogeneous embedding with linear coordinates x_i of degree $d_i > 0$, and let $\bar{\Theta}$ denote the corresponding

Poisson matrix. If $\bar{\Theta}_{ij} \in \mathfrak{m}$ for all index pairs, X is indecomposable, and we are done. Otherwise there are indices i, j such that $\bar{\Theta}_{ij}$ is a non-zero constant, and after an appropriate linear coordinate change, we may assume that $\bar{\Theta}_{12} = 1$.

For every $i > 1$ we may expand $\bar{\Theta}_{1i} = \sum_m x_2^m u_m$ as a polynomial in x_2 and put $\tilde{x}_i := x_i - \sum_m x_2^m a_m$ with $a_0 = 0$. Here the coefficients a_m are polynomials in the coordinates $x_1, x_3, \ldots, x_{2n+1}$ that have to be chosen in such a way so as to give

$$(2.8) \qquad 0 \overset{!}{=} \{x_1, \tilde{x}_i\} = \sum_m x_2^m (u_m - \{x_1, a_m\} - (m+1)a_{m+1}).$$

Thus we may set recursively $a_{m+1} = \frac{1}{m+1}(u_m - \{x_1, a_m\})$. As $\deg(a_m)$ is strictly decreasing for $m = 1, 2, \ldots$, all sums are in fact finite.

Hence, after renaming our variables we may assume that $\{x_1, x_i\} = 0$ for all $i \neq 2$. In a similar way, we may now consider the expansion $\{x_2, x_i\} = \sum_m v_m x_1^m$ and new coordinates $\tilde{x}_i = x_i - \sum_m a_m x_1^m$ with recursively defined polynomials a_m. In order that the new coordinate change should not destroy the just achieved orthogonality property $\{x_1, x_i\} = 0$ for $i > 2$, it is important to note that the coefficients v_m do not contain positive powers of x_2. Indeed this is a consequence of the Jacobi identity:

(2.9)

$$\{x_1, \{x_2, x_i\}\} = \{\{x_1, x_2\}, x_i\} + \{x_2, \{x_1, x_i\}\} = \{1, x_i\} + \{x_2, 0\} = 0,$$

so that

$$(2.10) \qquad 0 = \{x_1, \sum_m v_m x_1^m\} = \sum_m \{x_1, v_m\} x_1^m = \sum_m \frac{\partial v_m}{\partial x_2} x_1^m.$$

Hence repeating the argument of the first step and renaming the variables we arrive at a set of coordinates satisfying $\{x_1, x_2\} = 1$ and $\{x_i, x_j\} = 0$ for $i \leq 2 < j$.

Let $\bar{\Theta}_{ij}$ be the Poisson matrix with respect to this new set of homogeneous generators so that $\bar{\Theta}_{ij} = 0$ if $i \leq 1$ and $j \geq 2$ and $\bar{\Theta}_{12} = 1$. It follows from the Jacobi identity that

$$(2.11) \qquad \frac{\partial \bar{\Theta}_{ij}}{\partial \bar{x}_1} = -\{x_2, \bar{\Theta}_{ij}\} = 0$$

and analogously that $\frac{\partial \bar{\Theta}_{ij}}{\partial \bar{x}_2} = 0$. This implies that $\bar{\Theta}_{ij} \in \mathbb{C}[\bar{x}_3, \ldots, \bar{x}_{2n+1}]$. Similarly, $\bar{f} = 0$ implies $\frac{\partial f}{\partial x_1} = -\{x_2, f\} = 0$ and so on, so that $f \in \mathbb{C}[x_3, \ldots, x_{2n+1}]$. This shows that there is a graded Poisson isomorphism

$A \cong \mathbb{C}[x_1, x_2] \otimes A'$ with $A' = \mathbb{C}[x_3, \ldots, x_{2n+1}]/(f)$ where the symplectic form on the first factor is $dx_1 \wedge dx_2$ and where $\deg(x_1) + \deg(x_2) = s$.

The assertion follows by induction on the dimension of X. Q.E.D.

Lemma 2.6. — *Let $X \subset \mathbb{C}^{2n+1}$ be an indecomposable homogeneous symplectic hypersurface defined by a polynomial $f \in \mathbb{C}[x_1, \ldots, x_{2n+1}]$.*

(1) *$f \in \mathfrak{n}^{n+1}$ where $\mathfrak{n} = (x_1, \ldots, x_{2n+1})$.*

(2) *All partial derivatives $\partial f/\partial x_i$ are non-zero polynomials.*

Proof. 1. As X is indecomposable, all entries of the Poisson matrix are contained in the maximal ideal $\mathfrak{m} = (x_1, \ldots, x_{2n+1}) \subset \mathbb{C}[X]$. Hence all coefficients of its pfaffian are contained in $\mathfrak{n}^m$ as each summand of $\mathrm{pf}(\bar{\Theta})_i$ is the product of n entries of the Poisson matrix. The assertion now follows from identity (2.7).

2. Consider the stratification $X = X_0 \supset X_1 \supset X_2 \supset \ldots$, where X_{m+1} is the singular part of X_m with its reduced subscheme structure. Kaledin has shown that each X_m is a Poisson subscheme of X, and that the canonically induced Poisson structure on its normalisation $\tilde{X}_m \to X_m$ turns $\tilde{X}_m$ into a symplectic variety. In particular, all X_m are even-dimensional (possibly reducible) varieties. Let X_k denote the last non-empty piece of the stratification. It is a smooth symplectic variety and contains the origin as a Poisson subscheme. According to Kaledin [4], Lemma 1.4 and Theorem 2.5, this is impossible unless $X_k = \{O\}$. Now if $\partial f/\partial x_i$ were identically zero for some index i, i.e. if f were independent of x_i, every stratum X_m, including X_k would contain the line given by $x_j = 0$ for all $j \neq i$, a contradiction. Q.E.D.

Lemma 2.7. — *Let $X \subset \mathbb{C}^{2n+1}$ be an indecomposable homogeneous symplectic hypersurface. Then the Poisson structure on X can be uniquely extended to a homogeneous Poisson structure on the ambient space $\mathbb{C}^{2n+1}$. In particular, if Θ denotes the matrix $\Theta_{ij} = \{x_i, x_j\}$, where $x_1, \ldots, x_{2n+1}$ are linear homogeneous coordinates on $\mathbb{C}^{2n+1}$, then, possibly after rescaling it, the defining equation f of X satisfies* $\mathrm{grad}(f) = \mathrm{pf}(\Theta)$.

Proof. The natural epimorphism $\mathbb{C}[x_1, \ldots, x_{2n+1}] \to \mathbb{C}[X]$ is an isomorphism in all degrees less than $d = \deg(f)$. Thus the Poisson matrix $\bar{\Theta}$ of X can be uniquely lifted to a skew-symmetric matrix Θ with values in the polynomial ring if the degree condition $\deg(\bar{\Theta}_{ij}) = d_i + d_j - s < d$ is satisfied. And the bracket defined by $\{g, h\} := \sum_{ij} \Theta_{ij} \frac{\partial g}{\partial x_i} \frac{\partial h}{\partial x_j}$ will automatically satisfy the Jacobi-identity provided that all summands in equation (2.2) have degree $< d$. Hence it suffices to show that

$$d_i + d_j - s < d \qquad \text{and} \qquad d_i + d_j + d_k - 2s < d \tag{2.12}$$

for all pairwise distinct indices i, j, k.

For any finite subset $I \subset K := \{1, \ldots, 2n+1\}$ with an odd number of elements, let $\bar{\Theta}^I$ denote the skew-symmetric matrix obtained from $\bar{\Theta}$ by elimination of the i-th row and column for all $i \in I$. Every monomial that appears in the pfaffian $\mathrm{pf}(\bar{\Theta}^I)$ is of the form $\pm\bar{\Theta}_{i_1 i_2}\bar{\Theta}_{i_3 i_4}\cdots\bar{\Theta}_{i_{\ell-1} i_\ell}$ where $\{i_1, \ldots, i_\ell\} = K \setminus I$. Thus if $\mathrm{pf}(\bar{\Theta}^I) \neq 0$, then

$$\deg(\mathrm{pf}(\bar{\Theta}^I)) = \sum_{i \notin I} d_i - \frac{1}{2}(2n+1-|I|)s. \tag{2.13}$$

We apply this observation to submatrices of the form $\bar{\Theta}^i$, $\bar{\Theta}^{ijk}$ and $\bar{\Theta}^{ijkpq}$. For brevity, let $\delta = \sum_i d_i$. From the connection between the derivatives of f and $\bar{\Theta}$ we conclude that

$$d - d_i = \deg\left(\frac{\partial f}{\partial x_i}\right) = \deg(\mathrm{pf}(\bar{\Theta}^i)) = \delta - d_i - ns. \tag{2.14}$$

Hence $\delta = d + ns$.

If $n = 1$, one has $d = d_1 + d_2 + d_3 - s$, and hence $d_i + d_j - s = d - d_k < d$ and $d_i + d_j + d_k - 2s = d - s < d$ for $\{i, j, k\} = \{1, 2, 3\}$. Assume $n \geq 2$ for the rest of the proof.

Let i, j be distinct indices and assume that $\mathrm{pf}(\bar{\Theta}^{ijk}) \neq 0$ for some $k \in K \setminus \{i, j\}$. Then

$$0 \leq \deg(\mathrm{pf}(\bar{\Theta}^{ijk})) = \delta - d_i - d_j - d_k - (n-1)s = (d - d_k) - (d_i + d_j - s),$$

so that $d_i + d_j - s \leq d - d_k < d$. If on the other hand we had $\mathrm{pf}(\bar{\Theta}^{ijk}) = 0$ for all k, then $\bar{\Theta}^{ij}$ would have rank $\leq 2n-4$, and hence $\mathrm{rk}(\bar{\Theta}^i) \leq 2n-2$, so that $\mathrm{pf}(\bar{\Theta}^i) = 0$ contradicting the fact that $\partial f/\partial x_i \neq 0$ by Lemma 2.6.

Let i, j, k be distinct indices. If $n = 2$ and $\{1, 2, 3, 4, 5\} \setminus \{i, j, k\} = \{p, q\}$, then $d_i + d_j + d_k - 2s = d - d_p - d_q < d$. Hence assume that $n \geq 3$. Suppose that $\mathrm{pf}(\bar{\Theta}^{ijkpq}) \neq 0$ for some pair of indices $p, q \in K \setminus \{i, j, k\}$. Then

$$\begin{aligned} 0 \leq \deg(\mathrm{pf}(\bar{\Theta}^{ijkpq})) &= \delta - d_i - d_j - d_k - d_p - d_q - (n-2)s \\ &= (d - d_p - d_q) - (d_i + d_j + d_k - 2s), \end{aligned}$$

so that $d_i + d_j + d_k - 2s \leq d - d_p - d_q < d$. If on the other hand we had $\mathrm{pf}(\bar{\Theta}^{ijkpq}) = 0$ for all p, q, then $\bar{\Theta}^{ijk}$ would have rank $\leq 2n-6$, and hence $\mathrm{rk}\,\Theta^i \leq 2n-2$, leading to the same contradiction as before. Q.E.D.

Theorem 2.8. — *Let $X \subset \mathbb{C}^{2n+1}$ be an indecomposable homogeneous symplectic hypersurface. Then X admits a crepant resolution.*

Proof. The equation of X defines a flat deformation $f : \mathbb{C}^{2n+1} \to \mathbb{C}$. By Lemma 2.7, the Poisson structure on X uniquely extends to a homogeneous Poisson structure on the polynomial ring, and since $\{x_i, f\} = \sum_j \Theta_{ij}\partial f/\partial x_j = 0$, the deformation is in fact a Poisson deformation. For any $t \neq 0$, the fibre $f^{-1}(t)$ is smooth. Hence it follows from Corollary 5.6 in [8] that X admits a crepant resolution. Q.E.D.

§3. Examples

The following indecomposable symplectic hypersurfaces are known to us:

(1) ADE-surface singularities. These come in two series A_n and D_n and three exceptional examples E_6, E_7 and E_8.
(2) A series of four-dimensional hypersurfaces X_n, $n \geq 2$, with equations $f_n = a^2x + 2aby + b^2z + (xz - y^2)^n \in \mathbb{C}[a, b, x, y, z]$.
(3) A single six-dimensional example $\hat{X}$.

If we search for higher-dimensional symplectic hypersurfaces, relation (2.7) suggests to start from a skew-symmetric $(2n+1) \times (2n+1)$-matrix Θ with values in the polynomial ring $\mathbb{C}[x_1, \ldots, x_{2n+1}]$. It is then easy to reconstruct the polynomial f from the pfaffian minors of Θ. Of course, this puts quite strong differential conditions on Θ: It must satisfy the Jacobi identity (2.2), and its pfaffian minors must satisfy the Schwarz integrability conditions

$$(-1)^{i-1}\frac{\partial\, \mathrm{pf}(\Theta^i)}{\partial x_j} = (-1)^{j-1}\frac{\partial\, \mathrm{pf}(\Theta^j)}{\partial x_i}. \tag{3.1}$$

And finally one has to check that $X = \{f = 0\}$ is indeed symplectic.

Conversely, if $f \in A = \mathbb{C}[x_1, \ldots, x_{2n+1}]$ defines a symplectic hypersurface $X = \{f = 0\} \subset \mathbb{C}^{2n+1}$, the Poisson matrix is determined as the middle part of a skew-symmetric minimal resolution of the Jacobian ideal J:

$$0 \longrightarrow A \xrightarrow{df} A^{\oplus 2n+1} \xrightarrow{\Theta} A^{\oplus 2n+1} \xrightarrow{df} J \tag{3.2}$$

3.1. Two-dimensional examples

Two-dimensional symplectic surface singularities are classical and well studied mathematical objects ever since Klein discussed the invariants of finite subgroups $G \subset \mathrm{SL}_2(\mathbb{C})$ and computed the equation of the embedding $\mathbb{C}^2/G \subset \mathbb{C}^3$. For a two-dimensional symplectic hypersurface $X = \{f = 0\} \subset \mathbb{C}^3$, relation (2.7) is equivalent to saying that

$$\{x_i, x_j\} = \sum_k \varepsilon_{ijk} \frac{\partial f}{\partial x_k}. \tag{3.3}$$

Here ε_{ijk} denotes the totally skew-symmetric tensor that equals the sign of the permutation $(1, 2, 3) \mapsto (i, j, k)$ if i, j and k are pairwise distinct and 0 else. The corresponding symplectic form is obtained as the residue

$$\omega = \operatorname{res}_f \frac{dx_1 \wedge dx_2 \wedge dx_3}{f}. \tag{3.4}$$

Note that any choice of a homogeneous polynomial f defines a Poisson structure. But X will be symplectic if and only if it is isomorphic to one of the quotient singularities $\mathbb{C}^2/G$ in the following list:

group G	type	equation f
cyclic C_n	A_{n-1}	$x^2 + y^2 + z^n$
binary dihedral D_n^*	D_{n+2}	$x^2 + y^2 z + z^{n+1}$
binary tetrahedral $\mathbb{T}^*$	E_6	$x^2 + y^3 + z^4$
binary octahedral $\mathbb{O}^*$	E_7	$x^2 + y^3 + yz^3$
binary icosahedral $\mathbb{I}^*$	E_8	$x^2 + y^3 + z^5$

3.2. Four-dimensional examples

We know three constructions to obtain the hypersurfaces X_n:

The first construction establishes X_n as the transversal slice to the orbit of certain nilpotent elements x in a simple Lie algebra $\mathfrak{g}$. We only sketch the construction and refer to [6] for details. By the theorem of Jacobson-Morosov, one may choose elements h, y such that the map $\mathfrak{sl}_2 \to \mathfrak{g}$, $\left(\begin{smallmatrix} 0 & 1 \\ 0 & 0 \end{smallmatrix}\right) \mapsto x$, $\left(\begin{smallmatrix} 1 & 0 \\ 0 & -1 \end{smallmatrix}\right) \mapsto h$, $\left(\begin{smallmatrix} 0 & 0 \\ 1 & 0 \end{smallmatrix}\right) \mapsto y$, defines a Lie algebra homomorphism. The so-called Slodowy slice $S := x + \ker(\operatorname{ad} y)$ intersects the orbit of x for the adjoint action transversely. Let $N \subset \mathfrak{g}$ denote the cone of nilpotent elements. Then $S_0 := S \cap N$ is a symplectic variety. If $\mathfrak{g} = \mathfrak{sp}_{2n}$ is the Lie algebra of type C_n and x is a nilpotent element of Jordan type $[2n-2, 1, 1]$, then S_0 is isomorphic to the hypersurface X_n defined by the vanishing of $f_n := a^2 x + 2aby + b^2 z + (xz - y^2)^n$.

The second construction is based on the following ansatz: Let V denote an even-dimensional representation of the Lie algebra $\mathfrak{sl}_2$. A Poisson bracket on the symmetric algebra $A = S^*(\mathfrak{sl}_2 \oplus V)$ is determined by its value on pairs of vectors in $\mathfrak{sl}_2 \oplus V$: it then extends uniquely to A by its biderivative properties. We put $\{x, x'\} := [x, x']$ and $\{x, v\} := x.v$ for $x, x' \in \mathfrak{sl}_2$ and $v \in V$ using the Lie bracket on $\mathfrak{sl}_2$ and the action of $\mathfrak{sl}_2$ on V. It remains to choose a skew-symmetric map $\varphi := \{-, -\}|_{\Lambda^2 V}$:

$\Lambda^2 V \to A$ which we assume to take values in the subring $S^*(\mathfrak{sl}_2)$. The Jacobi relation can be thought of as a homomorphism $J : \Lambda^3(\mathfrak{sl}_2 \oplus V) \to A$. Its restriction to $\Lambda^3(\mathfrak{sl}_2) \oplus \Lambda^2(\mathfrak{sl}_2) \otimes V$ vanishes since $[-,-]$ is a Lie bracket and V is a representation. The vanishing of $J|_{\mathfrak{sl}_2 \otimes \Lambda^2 V}$ forces φ to be equivariant. So it remains to verify that $J|_{\Lambda^3 V}$ vanishes.

Assume now that $V = \mathbb{C}^2$ is the two-dimensional standard representation. As $\Lambda^3 V = 0$, the Jacobi condition is automatically satisfied for any equivariant homomorphism $\varphi : \Lambda^2 V = \mathbb{C} \to S^*(\mathfrak{sl}_2)$. So φ has to be a homogeneous element in the invariant subring $S^*(\mathfrak{sl}_2)^{\mathfrak{sl}_2}$. As is well-known, this subring is freely generated by the Casismir element Δ. Explicitly, one obtains in terms of a standard basis x, h, y of $\mathfrak{sl}_2$ and a basis e_0, e_1 of $\mathbb{C}^2$ the following Poisson matrices

$$\Theta_n = \begin{pmatrix} 0 & -2x & h & 0 & e_0 \\ 2x & 0 & -2y & e_0 & -e_1 \\ -h & 2y & 0 & e_1 & 0 \\ 0 & -e_0 & -e_1 & 0 & 2n\Delta^{n-1} \\ -e_0 & e_1 & 0 & -2n\Delta^{n-1} & 0 \end{pmatrix}, \tag{3.5}$$

where $\Delta = h^2 + 4xy$. Integrating the pfaffian vector $df_n = c_n \operatorname{pf}(\Theta_n)$ yields the expression

$$f_n = -ye_0^2 + he_0e_1 + xe_1^2 + \Delta^n. \tag{3.6}$$

Up to a rescaling of the coordinates this is the same equation as in the first construction. The weights of the coordinates in this case are $\deg(x) = \deg(h) = \deg(y) = 2$ and $\deg(e_0) = \deg(e_1) = 2n - 1$.

The third construction is due to Hanany and Mekareeya [3]. Let Γ denote a unitrivalent graph. This means that Γ is an undirected graph, possibly with loops, such that each vertex is the end point of exactly one or three edges. Here loops are counted twice. Attaching to each edge e a two-dimensional symplectic vector space V_e and to each inner vertex i the 8-dimensional $W_i = \bigotimes_{e \to i} V_e$, where the tensor product is taken over the three edges that end in i, we may form the symplectic vector space $W(\Gamma) := \bigoplus_i W_i$, where i runs through the set of inner vertices. The group $G(\Gamma) := \prod_e \mathrm{SL}(V_e)$, where e runs through the set of inner edges, acts on $W(\Gamma)$ preserving the symplectic form. Let $X(\Gamma) := W(\Gamma) /\!\!/\!\!/ G(\Gamma)$ denote the symplectic reduction. Based on physical considerations Hanany and Mekareeya argue that $X(\Gamma)$ is a symplectic variety that up to symplectic isomorphism depends only on the number $e(\Gamma)$ of exterior edges of Γ and its first Betti number $g(\Gamma)$. If Γ is read as the dual graph of a stable curve, $g(\Gamma)$ is the genus of that curve. Hanany and Mekareeya give the formula $\dim(X(\Gamma)) = 2(1+e(\Gamma))$, deduce from a calculation of the equivariant Hilbert series that $X(\Gamma)$ is a four-dimensional hypersurface if $e(\Gamma) = 1$, and state its defining equation.

For completeness sake and in order to see that graphs with $e(\Gamma) = 1$ and $g(\Gamma) = n$ lead to our hypersurfaces X_n, we carry out the necessary invariant theoretic calculations in detail for the following graphs:

(3.7) 

For each loop of the form

(3.8) 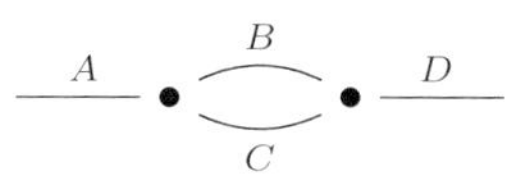

we need to consider the vector space $ABC \oplus BCD$, where we have dropped the tensor sign. We may consider $BC = \mathbb{C}^2 \otimes \mathbb{C}^2 = \mathbb{C}^4$ as the fundamental representation of $\mathrm{SL}(B) \times \mathrm{SL}(C)/(-I, -I) = \mathrm{SO}_4$. This allows to simplify the diagram above to 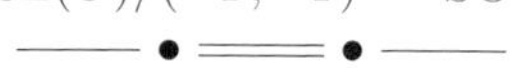 where the double line indicates the fundamental representation of SO_4. Similarly, a loop

(3.9) 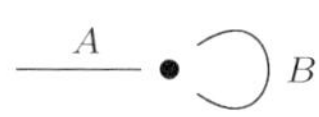

leads to the vector space ABB. Again we consider $BB = \mathbb{C}^2 \otimes \mathbb{C}^2 \cong \mathbb{C} \oplus \mathbb{C}^3$ as the sum of the trivial and the fundamental representation for the group $\mathrm{SL}(B)/(-I) = \mathrm{SO}_3$. We indicate this by a wriggled line . Thus we may replace the graph (3.7) by

(3.10) 

It follows from this reasoning that $W(\Gamma)$ is the space of representations for the following quiver:

$$\bullet \xrightarrow{x_1} \circ \xrightarrow{y_1} \bullet \xrightarrow{x_2} \circ \xrightarrow{y_2} \cdots \xrightarrow{y_{n-1}} \bullet \xrightarrow{x_n} \diamond\,, \tag{3.11}$$

where $\bullet$ correspond to copies of the fundamental representation $\mathbb{C}^2$ of SL_2, $\circ$ correspond to copies of the fundamental representation $\mathbb{C}^4$ of SO_4, and $\diamond$ corresponds to the representation $\mathbb{C} \oplus \mathbb{C}^3$ of SO_3. Using the symplectic and orthogonal forms on these representations we reinterpret tensor products as Hom spaces associated to the arrows. Then $W(\Gamma) = \{(x_1, y_1, \ldots, y_{n-1}, x_n)\}$, where $x_i \in \mathrm{Hom}(\mathbb{C}^2, \mathbb{C}^4)$ for $i = 1, \ldots, n$ and $y_i \in \mathrm{Hom}(\mathbb{C}^4, \mathbb{C}^2)$ for $i = 1, \ldots, n-1$. Let $x_i^* = J^{-1} x_i^t Q \in \mathrm{Hom}(\mathbb{C}^4, \mathbb{C}^2)$ and $y_i^* = Q^{-1} y_i^t J \in \mathrm{Hom}(\mathbb{C}^2, \mathbb{C}^4)$ denote the associated adjoint homomorphisms. Here J and Q denote the symplectic and quadratic form on

$\mathbb{C}^2$ and $\mathbb{C}^4$, respectively. Then the ideal I defined by momentum maps for the action of $G(\Gamma)$ is generated by the components of the elements $x_i x_i^* + y_i^* y_i$ and $y_i y_i^* + x_{i+1}^* x_{i+1}$ for $i = 1, \ldots, n-1$ and $\pi(x_n x_n^*)$, where $\pi : \mathfrak{so}_4 \to \mathfrak{so}_3$ denotes the projection dual to the inclusion $\mathfrak{so}_3 \to \mathfrak{so}_4$ associated to the representation of SO_3 on the vector space $\mathbb{C} \oplus \mathbb{C}^3$ at the end of the quiver. We calculate the symplectic reduction in three steps, taking invariants for the the groups SO_4 first, then for the groups SL_2, and finally for SO_3.

The invariant ring for the groups SO_4 is generated by x_n and the elements

$$a_i := x_i^* x_i,\ b_i := y_i x_i,\ c_i := y_i y_i^*,\ d_i := \det(x_i | y_i^*),\ \text{for } i = 1, \ldots, n-1.$$

The intersection with the momentum ideal is generated by

$$\pi(x_n x_n^*),\ c_i + a_{i+1},\ a_i^2 + b_i^* b_i,\ b_i a_i + c_i b_i,\ b_i b_i^* + c_i^2, d_i,\ \text{for } i = 1, \ldots, n-1,$$

where we have put $a_n := x_n^* x_n$ to simplify notations. This allows us to ignore the invariants d_i and c_i. We are left with the following set of generators

$$a_i \in \mathfrak{sl}_2,\ b_i \in \mathrm{Hom}(\mathbb{C}^2, \mathbb{C}^2),\ \text{for } i = 1, \ldots, n-1,\ \text{and } x_n \in \mathrm{Hom}(\mathbb{C}^2, \mathbb{C}^4),$$

with relations

$$\pi(x_n x_n^*),\ \ b_i a_i - a_{i+1} b_i,\ \ a_i^2 + b_i^* b_i,\ \ b_i b_i^* + a_{i+1}^2.$$

Since a_i^2, $b_i b_i^*$, $b_i^* b_i$ are multiples of the identity, the corresponding relations can be rephrased as $a_1^2 = a_i^2 = -b_i b_i^* = -b_i^* b_i$. The generators a_i, b_i, x_n define a representation of the shortened quiver

(3.12) 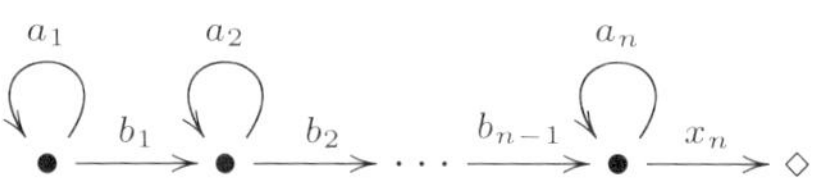

The invariant ring for the action of the groups SL_2 is generated by the components of all maps that are compositions of arrows forming a path from one end of the quiver to another or traces of compositions of arrows forming a closed loop. We can use the relations to move aside any of the loops a_i^2, $b_i b_i^*$ or $b_i^* b_i$. This reduces the number of generators to a_1, $u := x_n b_{n-1} \cdots b_1$ and $v := x_n x_n^*$,

(3.13) 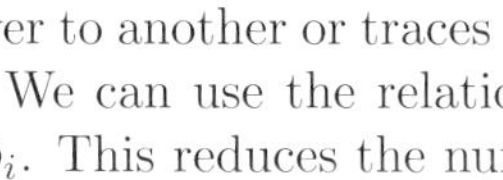

satisfying the relations

$$u a_1 - v u,\ u u^* - \det(a_1)^{n-1} v,\ u^* u - \det(a_1)^{n-1} a_1,\ \mathrm{tr}(a_1^2) - \mathrm{tr}(v^2),\ \pi(v).$$

It remains to take invariants for the action of SO_3. The decomposition $\mathbb{C}^4 = \mathbb{C} \oplus \mathbb{C}^3$ yields corresponding decompositions

$$(3.14) \qquad u = \begin{pmatrix} u_1 \\ u_2 \end{pmatrix}, \quad v = \begin{pmatrix} 0 & -w^* \\ w & z \end{pmatrix}.$$

As $z = \pi(v)$ is a relation, we may ignore z and continue to calculate with the SO_3-invariant generators a_1 and u_1 and the vector valued generators $u_2 \in \mathrm{Hom}(\mathbb{C}^2, \mathbb{C}^3)$ and $w \in \mathrm{Hom}(\mathbb{C}, \mathbb{C}^3)$. The remaining relations translate into

$$(3.15) \qquad u_1 a_1 + w^* u_2, \quad u_2 a_1 - w u_1, \quad u_2 u_2^*, \quad \det(a_1) - w^* w.$$

$$(3.16) \qquad u_2 u_1^* - \det(a_1)^{n-1} w, \quad u_1^* u_1^* + u_2^* u_2 - \det(a_1)^{n-1} a_1.$$

The invariants for the SO_3 action are generated by a_1, u_1: the further invariants

$$(3.17) \qquad w^* w, \quad w^* u_2, \quad u_2^* u_2, \quad \det(w|u_2),$$

can be expressed in terms of a_1 and u_1 due to the given relations. So we end up with five generators x, y, z, a and b that are the components of

$$(3.18) \qquad a_1 = \begin{pmatrix} y & x \\ -z & -y \end{pmatrix} \in \mathfrak{sl}_2 \quad \text{and} \quad u_1 = \begin{pmatrix} a \\ b \end{pmatrix} \in \mathrm{Hom}(\mathbb{C}^2, \mathbb{C})$$

and satisfy the single equation

$$(3.19) \qquad 0 = \det(a_1)^n + u_1 a_1 u_1^* = a^2 x - 2aby + b^2 z + (xz - y^2)^n.$$

3.3. The six-dimensional example

At present we know only one six-dimensional indecomposable hypersurface, denoted $\hat{X}$. It looks rather special, and the following discussion might indicate that it is an exceptional example that is not contained in a series. We first encountered $\hat{X}$ as the slice to the six-dimensional nilpotent orbit in the nilpotent cone of the simple Lie algebra $\mathfrak{g}_2$ [6]. Its defining equation $\hat{f}$ is rather complicated, and it is easier to obtain it indirectly. $\hat{X}$ has the interesting property that it is completely determined by its singular locus $\Sigma \subset \mathbb{C}^7$, and we will explain how to recover $\hat{X}$ starting from Σ.

Let V denote the irreducible two-dimensional representation of the symmetric group S_3, realised as the kernel of the linear form $x_1 + x_2 + x_3$ in $\mathbb{C}^3$. The invariant ring $\mathbb{C}[V \oplus V^*]^{S_3}$ is generated by 3 polynomials a_1, a_2, a_3 of degree 2 that are obtained by the process of polarisation from the second elementary symmetric polynomial $x_1x_2 + x_2x_3 + x_3x_1$ and by 4 polynomials $b_1, \dots, b_4$ of degree 3 that are obtained by similarly polarising the third elementary symmetric polynomial $x_1x_2x_3$. This is a classical result treated for example by Weyl in [11, p. 36 ff.]. These invariants are explicitly given as follows: $a_1 = \sum_{i<j} x_ix_j$, $a_2 = \sum_{i,j} x_iy_j$, $a_3 = \sum_{i<j} y_iy_j$ and $b_1 = x_1x_2x_3$, $b_2 = x_1x_2y_3 + x_1y_2x_3 + y_1x_2x_3$, $b_3 = x_1y_2y_3 + y_1x_2y_3 + y_1y_2x_3$, $b_4 = y_1y_2y_3$ for the dual coordinates y_1, y_2, y_3 on V^*.

These polynomials define an embedding $\Sigma := (V \oplus V^*)/S_3 \to \mathbb{C}^7$. The relations among the invariants are generated by the following 2 polynomials of weighted degree 5 and 3 polynomials of degree 6:

$$\begin{aligned} t_1 &= a_3b_2 - a_2b_3 + 3a_1b_4,\\ t_2 &= 3a_3b_1 - a_2b_2 + a_1b_3,\\ t_3 &= a_3(a_2^2 - 4a_1a_3) - 3b_3^2 + 9b_2b_4,\\ t_4 &= a_2(a_2^2 - 4a_1a_3) - 3b_2b_3 + 27b_1b_4,\\ t_5 &= a_1(a_2^2 - 4a_1a_3) - 3b_2^2 + 9b_1b_3; \end{aligned} \tag{3.20}$$

We note in passing that the same quotient variety Σ can also be obtained as symplectic reduction for the action of SL_2 on $S^4\mathbb{C}^2 \oplus (S^4\mathbb{C}^2)^*$ or as the symplectic reduction for the action of SL_3 on $\mathfrak{sl}_3 \oplus \mathfrak{sl}_3^*$.

As a subring of $\mathbb{C}[V \oplus V^*]$ the graded coordinate ring $\mathbb{C}[\Sigma]$ inherits a canonical Poisson structure. The Poisson brackets $\{a_i, a_j\}$, $\{a_i, b_j\}$ and $\{b_i, b_j\}$ will have degrees 2, 3 and 4 respectively. But since the smallest degree of a relation among the a_i's and b_i's has degree 5, it follows that the Poisson structure uniquely extends to a homogeneous Poisson structure on the ambient space $\mathbb{C}^7$. Calculation gives the following Poisson matrix:

$$\begin{pmatrix} 0 & -2a_1 & -a_2 & 0 & -3b_1 & -2b_2 & -b_3\\ 2a_1 & 0 & -2a_3 & 3b_1 & b_2 & -b_3 & -3b_4\\ a_2 & 2a_3 & 0 & b_2 & 2b_3 & 3b_4 & 0\\ 0 & -3b_1 & -b_2 & 0 & \frac{2}{3}a_1^2 & \frac{2}{3}a_1a_2 & \frac{1}{3}a_2^2 - \frac{2}{3}a_1a_3\\ 3b_1 & -b_2 & -2b_3 & -\frac{2}{3}a_1^2 & 0 & \frac{10}{3}a_1a_3 - \frac{1}{3}a_2^2 & \frac{2}{3}a_2a_3\\ 2b_2 & b_3 & -3b_4 & -\frac{2}{3}a_1a_2 & \frac{1}{3}a_2^2 - \frac{10}{3}a_1a_3 & 0 & \frac{2}{3}a_3^2\\ b_3 & 3b_4 & 0 & \frac{2}{3}a_1a_3 - \frac{1}{3}a_2^2 & -\frac{2}{3}a_2a_3 & -\frac{2}{3}a_3^2 & 0 \end{pmatrix}$$

If we denote this matrix by $\hat{\Theta}$, its pfaffian $\mathrm{pf}(\hat{\Theta})$ allows to determine the hypersurface equation $\hat{f}$ via $d\hat{f} = c\,\mathrm{pf}(\hat{\Theta})$, up to some normalising constant c. The equation is rather complicated. One can express it using the relations between the invariants, i.e. the equations of Σ, as follows:

$$\hat{f} = a_1t_1^2 - a_2t_1t_2 + a_3t_2^2 + \frac{1}{12}(t_4^2 - 4t_3t_5). \tag{3.21}$$

Since $\hat{f} \in (t_1, \dots, t_5)^2$, the singular locus of $\hat{X} = \{\hat{f} = 0\}$ contains Σ, and an explicit calculation shows that Σ actually equals the *reduced* singular locus of $\hat{X}$.

One can also describe the Poisson matrix $\hat{\Theta}$ by the Poisson algebra approach described in the four-dimensional case: consider the four-dimensional irreducible representation $V = S^3\mathbb{C}^2$ of $\mathfrak{sl}_2$. The choice of an equivariant map $\varphi : \Lambda^2 V \to S^*(\mathfrak{sl}_2)$ gives rise to a Poisson structure on $A = S^*(\mathfrak{sl}_2 \oplus V)$ if certain conditions imposed by the Jacobi identity are satisfied: As there are equivariant decompositions $\Lambda^2 V = \mathbb{C} \oplus S^4\mathbb{C}^2$ and $S^*(\mathfrak{sl}_2) = \mathbb{C}[\Delta] \otimes \bigoplus_{m\geq 0} S^{2m}(\mathbb{C}^2)$, the space of homogeneous equivariant maps $\varphi : \Lambda^2 V \to S^*(\mathfrak{sl}_2)_N$ is two-dimensional for each even $N \geq 2$, generated by maps $\mathbb{C} \to \mathbb{C} \cdot \Delta^{N/2}$ and $S^4\mathbb{C}^2 \to S^4\mathbb{C}^2 \cdot \Delta^{N/2-1}$. However, and in contrast to the four-dimensional case, only for the degree $N = 2$ there is a map φ leading to a non-degenerate hypersurface: the one described above.

§4. Contact Fano 3-folds

Consider the 3-dimensional projective varieties $Y \subset \mathbb{P} := \mathbb{P}(2n-1, 2n-1, 2, 2, 2)$ defined by the weighted homogeneous polynomial $a^2x + 2aby + b^2z + (xz - y^2)^n = 0$ for each $n \geq 2$. Here the coordinates are given the degrees $|a| = |b| = 2n - 1$ and $|x| = |y| = |z| = 2$. As before, let X_n denote the symplectic hypersurface in $\mathbb{C}^5$ defined by the same equation. In this section we introduce a contact structure on Y and relate it with the projectivised cotangent bundle $\mathbb{P}(T^*_{\mathbb{P}^1\times\mathbb{P}^1})$ by explicit birational maps.

The singular locus of Y has two components: Y has quotient singularities of type $\frac{1}{2n-1}(1,1)$ along the smooth rational curve $C = \{x = y = z = 0\}$ and Du Val singularities of type D_{2n} along the smooth rational curve $D = \{a = b = 0,\ xz - y^2 = 0\}$. The projection map $p : \mathbb{C}^5 \setminus \{0\} \to \mathbb{P}(2n-1, 2n-1, 1^3)$ is a $\mathbb{C}^*$-bundle outside C and D. Define $Y^0 := Y \setminus (C \cup D)$ and $X_n^0 := p^{-1}(Y^0)$.

Recall that a contact structure on a complex manifold M of dimension $2d + 1$ is an exact sequence of vector bundles

$$0 \longrightarrow D \longrightarrow T_M \xrightarrow{\theta} L \longrightarrow 0,$$

with $\mathrm{rk}(D) = 2d$ and $\mathrm{rk}(L) = 1$ so that $d\theta|_D$ induces a non-degenerate pairing on D. By using the formula for exterior derivation

$$d\theta(x, y) = x(\theta(y)) - y(\theta(x)) - \theta([x, y])$$

one can check that this is equivalent to saying that $[-,-] : D \times D \to L = T_M/D$ is non-degenerate. We call L the contact line bundle.

We shall introduce a contact structure on Y^0 with the contact line bundle $\mathcal{O}(2) := \mathcal{O}_{\mathbb{P}}(2)|_{Y^0}$. Let ω be a symplectic 2-form on X_n^0 of weight

2. By construction, the projection $p : X_n^0 \to Y^0$ is a $\mathbb{C}^*$-bundle, and X_n^0 is in fact isomorphic to the complement of the zero section of the line bundle $\mathcal{O}(-1)$ on Y^0. There is a canonical trivialisation $p^*\mathcal{O}(1) \cong \mathcal{O}_{X_n^0}$, and hence a trivialisation $p^*\mathcal{O}(i) \cong \mathcal{O}_{X_n^0}$ for any $i \in \mathbb{Z}$. Let ζ be the vector field which generates the $\mathbb{C}^*$-action. Since ω has weight 2, one can write $\omega(\zeta, \cdot) = p^*\theta$ for some appropriate element $\theta \in H^0(Y^0, \Omega^1_{Y^0} \otimes \mathcal{O}(2))$. This θ gives a contact structure on Y^0 with the contact line bundle $\mathcal{O}(2)$.

The rational map

$$\mathbb{P}((2n-1)^2, 2^3) \dashrightarrow \mathbb{P}^2 = \mathbb{P}(2^3), (a : b : x : y : z) \to (x : y : z)$$

induces a rational map $Y \dashrightarrow \mathbb{P}^2$. To eliminate the indeterminancy of the rational map, we take the blow-up Y_1 of Y along C. Let $F_1 \subset Y_1$ be the exceptional divisor of the blowing-up. Notice that F_1 is a $\mathbb{P}^1$-bundle over C. Then the rational map actually becomes a morphism $f_1 : Y_1 \to \mathbb{P}^2$. Let us consider the fibres of f_1. For $(1 : \mu : \lambda) \in \mathbb{P}^2$, the fibre $f_1^{-1}(1 : \mu : \lambda)$ is isomorphic to the quasi-homogeneous hypersurface of $\mathbb{P}(2n-1, 2n-1, 2)$ defined by

$$a^2 + 2ab\mu + b^2\lambda + (\lambda - \mu^2)^n x^{2n-1} = 0.$$

If $(1 : \mu : \lambda) \in \{xz - y^2 = 0\}$, then it is a multiple fibre with multiplicity 2. If $(1 : \mu : \lambda) \notin \{xz - y^2 = 0\}$, then the fibre is a smooth rational curve. In other words, f_1 is a conic bundle whose discriminant locus D' is $\{xz - y^2 = 0\}$ and all singular fibres are non-reduced.

Set $S_1 := f_1^{-1}(D')_{red}$. Then S_1 is a $\mathbb{P}^1$-bundle over D'. Since the blowing-up $Y_1 \to Y$ does not change an open neighborhood of $D \subset Y$, its inverse image D_1 by the blowing-up is isomorphic to D. Moreover, D_1 is a section of the $\mathbb{P}^1$-bundle $S_1 \to D'$. The singular locus of Y_1 coincides with D_1. As Y_1 has Du Val singularities of type D_{2n} along D_1, one can take its minimal resolution $\tilde{Y} \to Y_1$. The exceptional locus of the minimal resolution consists of $2n$ divisors $E^{(1)}, \ldots, E^{(2n)}$ intersecting each other according to the following D_{2n}-configuration:

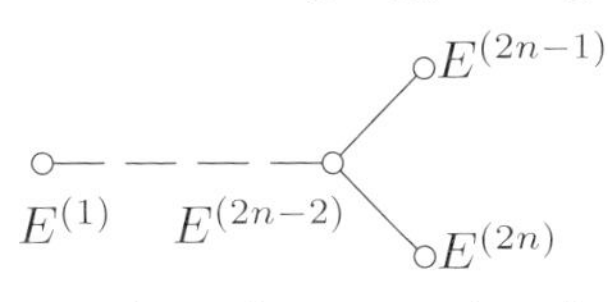

Here the vertices correspond to the exceptional surfaces and the edges correspond to the intersection curves. Each surface $E^{(i)}$ is a $\mathbb{P}^1$-bundle over D and each intersection curve is a section of the $\mathbb{P}^1$-bundle map.

Let S and F be respectively the proper transforms of S_1 and F_1 by the map $\tilde{Y} \to Y_1$. Then S intersects with F along a section of the

$\mathbb{P}^1$-bundle structure. There are no intersetion of F with $E^{(i)}$'s. On the other hand, S intersects with only $E^{(1)}$. Notice that $E^{(1)} \cap S$ is a section of the ruled surface $E^{(1)}$, which is disjoint from $E^{(1)} \cap E^{(2)}$:

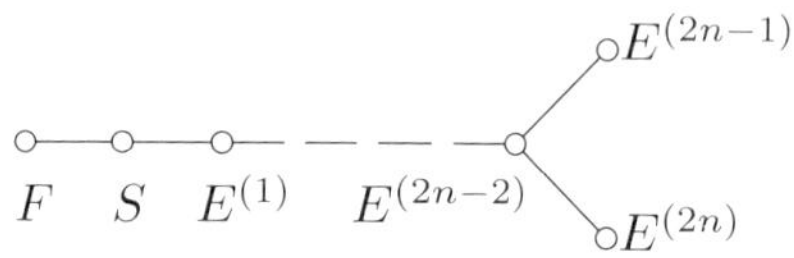

One can blow down successively these divisors along their rulings in the following order: S, $E^{(1)}$, ..., $E^{(2n-3)}$, and finally F. We call the resulting variety Z. The existence of such birational contraction maps are justified in the following way. Let us consider Y_1 and S_1. Let ℓ_1 be a fibre of $S_1 \to D'$. We prove that $(K_{Y_1}, \ell_1) = -1$. Let $\ell \subset Y$ be the image of ℓ_1 by the map $\pi_1 : Y_1 \to Y$. By an explicit calculation we see that $(\mathcal{O}(1), \ell) = \frac{1}{2} \cdot \frac{1}{2n-1}$. Since $K_Y = \mathcal{O}_Y(-4)$, one has $(K_Y, \ell) = -\frac{2}{2n-1}$. Since $K_{X_1} = (\pi_1)^* K_Y - \frac{2n-3}{2n-1} \cdot F_1$ and $(F_1, \ell_1) = 1$, we see that $(K_{Y_1}, \ell_1) = -1$. Denote by π_2 the minimal resolution $\tilde{Y} \to Y_1$. The proper transform S of S_1 by π_2 is isomorphic to S_1; hence there is a $\mathbb{P}^1$-bundle map $S \to D'$. Let $\tilde{\ell}$ be a fibre of this map. Then, since $K_{\tilde{Y}} = (\pi_2)^* K_{Y_1}$, we see that

$$(K_{\tilde{Y}}, \tilde{\ell}) = -1.$$

Let m_i be a fibre of the $\mathbb{P}^1$-bundle structure of $E^{(i)}$. Then we have

$$(K_{\tilde{Y}}, m_i) = 0.$$

By Nakano-Fujiki criterion one has a bimeromorphic map $\nu_1 : \tilde{Y} \to Z_1$ to a Moishezon manifold Z_1, where ν_1 contracts all rulings of S to points. As S intersects with $E^{(1)}$ along a section, we have

$$(K_{Z_1}, \nu_1(m_1)) = (K_{\tilde{Y}}, m_1) - 1 = -1.$$

Then we get a bimeromorphic map $\nu_2 : Z_1 \to Z_2$, where ν_2 contracts all rulings of $\nu_1(E^{(1)})$ to points. We can further continue the same procedures in the order of $E^{(2)}$,..., $E^{(2n-3)}$ and finally F. As a consequence we have a sequence of birational contraction maps

$$\tilde{Y} \to Z_1 \to Z_2 \to \ldots \to Z.$$

In the remainder we denote by ν the map $\tilde{Y} \to Z$ and by μ the map $\tilde{Y} \to Y$.

Lemma 4.1. — *Z has a contact structure.*

Proof. The birational map $\pi : \tilde{Y} \to Y$ is a crepant resolution of Y around D. As remarked above, Y^0 has a contact form $\eta \in \Gamma(Y^0, \Omega^1_{Y^0} \otimes \mathcal{O}(2))$ with the contact line bundle $\mathcal{O}(2)$. Take a point $x \in D$. Since $\mathcal{O}(2)$ is a line bundle around D, one can trivialise $\mathcal{O}(2)$ on an open neighbourhood $x \in U \subset Y$. Then η is regarded as a 1-form on U_{reg} such that $\eta \wedge d\eta$ is a nowhere-vanishing 3-form on U_{reg}. This 3-form extends to a generator of the invertible dualising sheaf ω_U. Set $\tilde{U} := \pi^{-1}(U)$ and $\pi_U := \pi|_{\tilde{U}}$. Then $(\pi_U)^*(\eta \wedge d\eta)$ is a nowhere-vanishing 3-form on $\tilde{U}$ because π_U gives a crepant resolution of U. This shows that $(\pi_U)^*\eta$ is a contact 1-form on $\tilde{U}$ with the contact line bundle $(\pi_U)^*(\mathcal{O}(2)|_U)$. As a consequence, $\tilde{Y}$ has a contact structure outside $F = \pi^{-1}(C)$.

Let $\beta \subset Z$ be the image of F by the birational morphism $\nu : \tilde{Y} \to Z$. Note that $\dim \beta = 1$. Let us consider the birational morphism

$$\tilde{Y} - \nu^{-1}(\beta) \to Z - \beta.$$

There is an open subset Z^0 of $Z - \beta$ such that $\nu^{-1}(Z^0) \cong Z^0$ and such that the complement of Z^0 in $Z - \beta$ has at least codimension 2. The restriction of the contact structure on $\tilde{Y} - F$ to $\nu^{-1}(Z^0)$ gives a contact structure of Z^0. Since the complement of Z^0 in Z has at least codimension 2, the contact structure uniquely extends to a contact structure on Z. Q.E.D.

Lemma 4.2. — *Z is isomorphic to $\mathbb{P}(T^*_{\mathbb{P}^1 \times \mathbb{P}^1})$.*

Proof. We cover D by three orbifold charts $W_x \to Y$, $W_y \to Y$ and $W_z \to Y$, where $W_x := \{x = 1\} \subset \mathbb{C}^5$, $W_y := \{y = 1\}$ and $W_z := \{z = 1\}$. Note that each map is a $\mathbb{Z}_2$-cover onto its image. Let V be the union of these images. The blowing up of each chart along the singular locus is $\mathbb{Z}_2$-equivariant, and three pieces $\tilde{W}_x/\mathbb{Z}_2$, $\tilde{W}_y/\mathbb{Z}_2$ and $\tilde{W}_z/\mathbb{Z}_2$ are glued together to give a partial resolution $V' \to V$. Since it does not change anything outside D, it gives a partial resolution $Y' \to Y$. The exceptional locus E' of the partial resolution is a $\mathbb{P}^1$-bundle over D and Y' has A_{2n-1}-singularities along a section of this $\mathbb{P}^1$-bundle. Note that the partial resolution $Y' \to Y$ eliminate the indeterminancy of the rational map

$$Y - - \to \mathbb{P}^1, (a : b : x : y : z) \to (a : b)$$

and gives a morphism $Y' \to \mathbb{P}^1$. Each fibre of E' is isomorphically mapped onto $\mathbb{P}^1$ by the morphism. This, in particular, shows that E' has two $\mathbb{P}^1$-bundle structures. Hence we see that $E' \cong \mathbb{P}^1 \times \mathbb{P}^1$. By the definition, $\tilde{Y} \to Y$ factors through $Y' : \tilde{Y} \to Y' \to Y$. The proper transform of E' by the birational map $\tilde{Y} \to Y$ is nothing but $E^{(2n-1)}$. By the argument above, $E^{(2n-1)} \cong \mathbb{P}^1 \times \mathbb{P}^1$.

Next look at $E^{(2n-2)}$. It has a $\mathbb{P}^1$-bundle structure whose fibres correspond to exceptional curves of the map $\tilde{Y} \to Y$. Since it has three disjoint sections (corresponding to the intersections with $E^{(2n-3)}$, $E^{(2n-1)}$ and $E^{(2n)}$), we also see that $E^{(2n-2)} \cong \mathbb{P}^1 \times \mathbb{P}^1$. Define $\sigma := E^{(2n-2)} \cap E^{(2n-3)}$.

We write $E^{(2n-2)}$, $E^{(2n-1)}$ and $E^{(2n)}$ for their images in Z by the map $\tilde{Y} \to Z$.

Pick a fibre α of $E^{(2n-2)} \subset Z$. Since $(E^{(2n-2)}, \alpha)_Z = 0$, the curve α can move aside $E^{(2n-2)}$ in a parameter space of dimension 2. We prove that there is a morphism $Z \to \mathbb{P}^1 \times \mathbb{P}^1$ whose fibres are all deformation equivalent to α. The linear system which gives the morphism is $|\mathcal{O}_Z(E^{(2n-2)})|$. To prove that this linear system is free from base points, it suffices to show that $|\mathcal{O}_{E^{(2n-2)}}(E^{(2n-2)})|$ is free from base points by the exact sequence

$$0 \to H^0(Z, \mathcal{O}_Z) \to H^0(Z, \mathcal{O}_Z(E^{(2n-2)})) \to H^0(E^{(2n-2)}, \mathcal{O}_{E^{(2n-2)}}(E^{(2n-2)})) \to 0.$$

Note that

$$K_{\tilde{Y}} = \mu^* K_Y - \frac{2n-3}{2n-1} F.$$

We can also write $K_{\tilde{Y}}$ by a linear combination of $\nu^* K_Z$, S, F and $E^{(1)}$, $\ldots$, $E^{(2n-3)}$. By using the two expression of $K_{\tilde{Y}}$, one can write

$$\nu^* K_Z = \mu^* K_Y - 2E^{(2n-3)} + \text{other terms}.$$

Restricting this to $E^{(2n-2)}$ we get $K_Z|_{E^{(2n-2)}} = -4\alpha - 2\sigma$ since $(K_Y, D) = -4$, which easily follows from the fact that $K_Y = \mathcal{O}(-4)$ and $(\mathcal{O}(1), D) = 1$.

Now, by the adjunction formula $K_{E^{(2n-2)}} = K_Z + E^{(2n-2)}|_{E^{(2n-2)}}$ we see that $E^{(2n-2)}|_{E^{(2n-2)}} \sim 2\alpha$. The corresponding linear system is free from base points.

Since $h^0(Z, \mathcal{O}_Z(E^{(2n-2)})) = 4$, we have a morphism $Z \to \mathbb{P}^3$. Since $(E^{(2n-2)})^3 = 0$ and $(E^{(2n-2)})^2 \sim 2\alpha$, the image has 2 dimension. Moreover, since $(E^{(2n-1)}, \alpha) = 1$ and $E^{(2n-1)} \cong \mathbb{P}^1 \times \mathbb{P}^1$, the morphism is a $\mathbb{P}^1$-bundle over $\mathbb{P}^1 \times \mathbb{P}^1$ with a section $E^{(2n-1)}$. As we have seen in Lemma 4.1, Z has a contact structure. Moreover the morphism defined here is a Legendre $\mathbb{P}^1$-bundle. By [5], it then follows that Z is a projectivised cotangent bundle $\mathbb{P}(T^*_{\mathbb{P}^1 \times \mathbb{P}^1})$. Q.E.D.

Remark 4.3. — *Let X'_n be a Slodowy slice to a niloptent orbit $\mathcal{O}_{[4n-3,1^3]}$ of $\mathfrak{so}_{4n}$ with $n \geq 2$. Then one can check that X'_n is isomorphic to the complete intersection of $\mathbb{C}^6(\alpha, \beta, \gamma, x, y, z)$ defined by two equations $f = g = 0$ with*

$$f = \alpha x + 2\beta y + \gamma z = 0 \qquad \textit{and} \qquad g = \alpha\gamma - \beta^2 + 1/4(xz - y^2)^{2n-1} = 0.$$

With the new coordinates $A = \alpha - \frac{1}{2}z(xz - y^2)^{n-1}$, $B = \beta + \frac{1}{2}y(xz - y^2)^{n-1}$ and $C = \gamma - \frac{1}{2}x(xz - y^2)^{n-1}$ the equations $f = 0$ and $g = 0$ respectively become

$$Ax + 2By + Cz + (xz - y^2)^n = 0 \qquad \textit{and} \qquad AC - B^2 = 0.$$

It follows from this description that $\tau(a, b, x, y, z) := (a^2, ab, b^2, x, y, z)$ defines a double covering $\tau : X_n \to X'_n$. Note that τ is ramified precisely over the singular locus of X'_n. Moreover, X'_n is equipped with the Kostant-Kirillov 2-form ω' on the regular locus. Then $\tau^*\omega'$ is equivalent to the Kostant-Kirillov 2-form ω on X_n by Theorem (3.1) in [9].

Let Y' be the 3-dimensional projective variety in $\mathbb{P}(2n-1, 2n-1, 2n-1, 1, 1, 1)$ defined by $f = g = 0$. The degrees of the coordinates are $|\alpha| = |\beta| = |\gamma| = 2n - 1$ and $|x| = |y| = |z| = 1$. Then Y' admits a contact structure on its regular part. Moreover, by the observation above, we immediately see that $Y \cong Y'$ as contact varieties.

References

[1] A. Beauville: Symplectic Singularities. Inv. Math. 139 (2000), 541-549.

[2] A. Bialynicki-Biura: Some theorems on actions of algebraic groups. Ann. of Math. 98 (1973), 480-497.

[3] A. Hanany, N. Mekareeya: Tri-vertices and $SU(2)$'s. arXiv:1012.2119v1.

[4] D. Kaledin: Symplectic varieties from the Poisson point of view. J. Reine Angew. Math. 600 (2006), 135 - 160.

[5] S. Kebekus, T. Peternell, A. Sommese, J. Wisniewski: Projective contact manifolds, Invent. Math. 142 (2000), 1 - 15.

[6] M. Lehn, Y. Namikawa, C. Sorger: Slodowy slices and universal Poisson deformations. Compositio Math. 148 (2012), 121 - 144.

[7] Y. Namikawa: Extension of 2-forms and symplectic varieties. J. Reine Angew. Math. 539 (2001), 123 - 147.

[8] Y. Namikawa: Poisson deformations of affine symplectic varieties. Duke Math. J. 156 (2011), 51 - 85.

[9] Y. Namikawa: Equivalence of symplectic singularities. Kyoto J. Math. 53(2) (2013), 483 - 514.

[10] A. Weinstein: The local structure of Poisson manifolds. J. Differential Geometry 18 (1983), 523 - 557.

[11] H. Weyl: The Classical Groups. 2nd ed. Princeton University Press 1946.

Manfred Lehn, Fachbereich Physik, Mathematik u. Informatik, Johannes Gutenberg-Universität Mainz, D-55099 Mainz, Germany
E-mail address: lehn@mathematik.uni-mainz.de

Yoshinori Namikawa, Department of Mathematics, Faculty of Science, Kyoto University, Kitashirakawa-Oiwakecho, Kyoto, 606-8502, Japan
E-mail address: namikawa@math.kyoto-u.ac.jp

Christoph Sorger, Laboratoire de Mathématiques Jean Leray (UMR 6629 du CNRS) Université de Nantes, 2, Rue de la Houssinière, BP 92208, F-44322 Nantes Cedex 03, France
E-mail address: christoph.sorger@univ-nantes.fr

Duco van Straten, Fachbereich Physik, Mathematik u. Informatik Johannes Gutenberg-Universität Mainz, D-55099 Mainz, Germany
E-mail address: straten@mathematik.uni-mainz.de

Advanced Studies in Pure Mathematics 70, 2016
Minimal Models and Extremal Rays (Kyoto, 2011)
pp. 299–321

Vanishing theorems

Osamu Fujino

Dedicated to Professor Shigefumi Mori

Abstract.

We prove some injectivity, torsion-free, and vanishing theorems for simple normal crossing pairs. Our results heavily depend on the theory of mixed Hodge structures on cohomology groups with compact support. We also treat several basic properties of semi divisorial log terminal pairs.

CONTENTS

§1. Introduction

In this paper, we prove some vanishing theorems for simple normal crossing pairs, which will play important roles in the study of higher dimensional algebraic varieties. We note that the notion of *simple normal crossing pairs* includes here the case when the ambient variety itself has several irreducible components with simple normal crossings. Theorem 1.1 is a generalization of the works of several authors: Kawamata, Viehweg, Kollár, Esnault–Viehweg, Ambro, Fujino, and others (cf. [Ko1], [KMM], [EV], [Ko2], [A], [F2], [F4], [F7], [F8], [F11], and so on).

Received January 5, 2012.
Revised January 24, 2013.
2010 *Mathematics Subject Classification.* Primary 14F17; Secondary 14E30.
Key words and phrases. vanishing theorems, torsion-free theorem, semi divisorial log terminal, Hodge theory, semipositivity.

Theorem 1.1 (see Theorem 3.7). *Let (Y, Δ) be a simple normal crossing pair such that Δ is a boundary $\mathbb{R}$-divisor on Y. Let $f : Y \to X$ be a proper morphism to an algebraic variety X and let L be a Cartier divisor on Y such that $L - (K_Y + \Delta)$ is f-semi-ample.*

(i) *every associated prime of $R^q f_* \mathcal{O}_Y(L)$ is the generic point of the f-image of some stratum of (Y, Δ) for every q.*

(ii) *let $\pi : X \to V$ be a projective morphism to an algebraic variety V such that*

$$L - (K_Y + \Delta) \sim_{\mathbb{R}} f^* H$$

for some π-ample $\mathbb{R}$-divisor H on X. Then $R^q f_* \mathcal{O}_Y(L)$ is π_*-acyclic, that is,

$$R^p \pi_* R^q f_* \mathcal{O}_Y(L) = 0$$

for every $p > 0$ and $q \geq 0$.

When X is a divisor on a smooth variety M, Theorem 1.1 is contained in [A] and plays crucial roles in the theory of quasi-log varieties. For the details, see [F8, Chapter 3] and [F9]. When X is quasi-projective, it is proved in [FF, Section 6]. Here, we need no extra assumptions on X. Therefore, Theorem 1.1 is new. The theory of resolution of singularities for *reducible* varieties has recently been developing (cf. [BM] and [BP]). It refines several vanishing theorems in [F8]. It is one of the main themes of this paper. We will give a proof of Theorem 1.1 in Section 3. Note that we do not treat *normal crossing varieties*. We only discuss *simple* normal crossing varieties because the theory of resolution of singularities for reducible varieties works well only for *simple* normal crossing varieties. We note that the fundamental theorems for the log minimal model program for log canonical pairs can be proved without using the theory of quasi-log varieties (cf. [F10] and [F11]). The case when Y is smooth in Theorem 1.1 is sufficient for [F10] and [F11]. For that case, see [F7] and [F11, Sections 5 and 6]. Our proof of Theorem 1.1 heavily depends on the theory of mixed Hodge structures on cohomology groups with compact support.

1.2 (Hodge theoretic viewpoint). Let X be a projective simple normal crossing variety with $\dim X = n$. We are mainly interested in $H^\bullet(X, \omega_X)$ or $H^\bullet(X, \omega_X \otimes L)$ for some line bundle L on X. By the theory of mixed Hodge structures,

$$\mathrm{Gr}_F^n H^\bullet(X, \mathbb{C}) \simeq H^{\bullet - n}(X, \nu_* \omega_{X^\nu}),$$

where $\nu : X^\nu \to X$ is the normalization, and

$$\mathrm{Gr}_F^0 H^\bullet(X, \mathbb{C}) \simeq H^\bullet(X, \mathcal{O}_X).$$

Note that F is the Hodge filtration on the natural mixed Hodge structure on $H^\bullet(X, \mathbb{Q})$. Let D be a simple normal crossing divisor on X. Then we obtain

$$\mathrm{Gr}_F^n H^\bullet(X \setminus D, \mathbb{C}) \simeq H^{\bullet-n}(X, \nu_*\omega_{X^\nu} \otimes \mathcal{O}_X(D))$$

and

$$\mathrm{Gr}_F^0 H_c^\bullet(X \setminus D, \mathbb{C}) \simeq H^\bullet(X, \mathcal{O}_X(-D)).$$

Note that

$$\mathrm{Gr}_F^0 H^\bullet(X \setminus D, \mathbb{C}) \simeq H^\bullet(X, \mathcal{O}_X)$$

and that

$$H^{\bullet-n}(X, \nu_*\omega_{X^\nu} \otimes \mathcal{O}_X(D)) \not\simeq H^{\bullet-n}(X, \omega_X \otimes \mathcal{O}_X(D)).$$

We also note that $H_c^\bullet(X \setminus D, \mathbb{Q})$ need not be the dual vector space of $H^{2n-\bullet}(X \setminus D, \mathbb{Q})$ when X is not smooth. In this setting, we are interested in $H^\bullet(X, \omega_X(D))$ or $H^\bullet(X, \omega_X(D) \otimes L)$. Therefore, we consider the natural mixed Hodge structure on $H_c^\bullet(X \setminus D, \mathbb{C})$ and take the dual vector space of

$$\mathrm{Gr}_F^0 H_c^\bullet(X \setminus D, \mathbb{C}) \simeq H^\bullet(X, \mathcal{O}_X(-D))$$

by Serre duality. Then we obtain $H^{n-\bullet}(X, \omega_X(D))$. We note that if L is semi-ample then we can reduce the problem to the case when L is trivial by the usual covering trick. The above observation is crucial for our treatment of the vanishing theorems and the semipositivity theorems in [F8] and [FF]. In this paper, we do not discuss the Hodge theoretic part of vanishing and semipositivity theorems. We prove Theorem 1.1 by assuming the Hodge theoretic injectivity theorem: Theorem 3.1. For the details of the Hodge theoretic part, see [F8], [FF], and [F15].

The author learned the following example from Kento Fujita.

Example 1.3. Let $X_1 = \mathbb{P}^2$ and let C_1 be a line on X_1. Let $X_2 = \mathbb{P}^2$ and let C_2 be a smooth conic on X_2. We fix an isomorphism $\tau : C_1 \to C_2$. By gluing X_1 and X_2 along $\tau : C_1 \to C_2$, we obtain a simple normal crossing surface X such that $-K_X$ is ample (cf. [Ft]). We can check that X can not be embedded into any smooth varieties as a simple normal crossing divisor.

Example 1.3 shows that Theorem 1.1 is not covered by the results in [A], [F4], and [F8].

Remark 1.4 (cf. [F8, Proposition 3.65]). We can construct a proper simple normal crossing variety X with the following property. Let $f : Y \to X$ be a proper morphism from a simple normal crossing variety Y such that f induces an isomorphism $f|_V : V \simeq U$ where V (resp. U) is a dense Zariski open subset of Y (resp. X) which contains the generic point of any stratum of Y (resp. X). Then Y is non-projective. Therefore, we can not directly use Chow's lemma to reduce our main theorem (cf. Theorem 1.1) to the quasi-projective case (cf. [FF, Section 6]).

There exists another standard approach to various Kodaira type vanishing theorems. It is an analytic method (see, for example, [F5] and [F6]). At the present time, the relationship between our Hodge theoretic approach and the analytic method is not clear.

We summarize the contents of this paper. In Section 2, we collect some basic definitions and results for the study of simple normal crossing varieties and divisors on them. Section 3 is the main part of this paper. It is devoted to the study of injectivity, torsion-free, and vanishing theorems for simple normal crossing pairs. We note that we do not prove the Hodge theoretic injectivity theorem: Theorem 3.1. We just quote it from [F8] (see also [F15]). Section 4 is an easy application of the vanishing theorem in Section 3. We prove the basic properties of semi divisorial log terminal pairs in the sense of Kollár. In Section 5, we explain our new semipositivity theorem, which is a generalization of the Fujita–Kawamata semipositivity theorem, without proof. It depends on the theory of variations of mixed Hodge structures on cohomology groups with compact support and is related to the results obtained in Section 3. Anyway, the vanishing theorem and the semipositivity theorem discussed in this paper follow from the theory of mixed Hodge structures on cohomology groups with compact support.

For various applications of Theorem 1.1 and related topics, see [F8], [FF], [F13], [F14], [F15], and so on.

Acknowledgments. The author was partially supported by the Grant-in-Aid for Young Scientists (A) ♯20684001 from JSPS. He would like to thank Professors Takeshi Abe, Taro Fujisawa, and Shunsuke Takagi for discussions and useful comments. He would also like to thank Professor János Kollár for giving him a preliminary version of his book [Ko3]. The main part of this paper is a revised version of Section 5 of the author's unpublished preprint [F4]. Although he tried to publish [F4], its importance could have been misunderstood by the referee in 2007, when almost all the minimal modelers were mainly interested in the big development by Birkar–Cascini–Hacon–M^c^Kernan. When he wrote [F4] in Nagoya, he was partially supported by the Grant-in-Aid for Young Scientists (A) ♯17684001 from JSPS and by the Inamori Foundation.

We will work over $\mathbb{C}$, the complex number field, throughout this paper. But we note that, by using the Lefschetz principle, all the results in this paper hold over an algebraically closed field k of characteristic zero.

§2. Preliminaries

First, we quickly recall basic definitions of divisors. We note that we have to deal with reducible algebraic schemes in this paper. For details, see, for example, [H, Section 2] and [L, Section 7.1].

2.1. Let X be a noetherian scheme with structure sheaf $\mathcal{O}_X$ and let $\mathcal{K}_X$ be the sheaf of total quotient rings of $\mathcal{O}_X$. Let $\mathcal{K}_X^*$ denote the (multiplicative) sheaf of invertible elements in $\mathcal{K}_X$, and $\mathcal{O}_X^*$ the sheaf of invertible elements in $\mathcal{O}_X$. We note that $\mathcal{O}_X \subset \mathcal{K}_X$ and $\mathcal{O}_X^* \subset \mathcal{K}_X^*$.

2.2 (Cartier, $\mathbb{Q}$-Cartier, and $\mathbb{R}$-Cartier divisors)**.** A *Cartier divisor* D on X is a global section of $\mathcal{K}_X^*/\mathcal{O}_X^*$, that is, D is an element of $H^0(X, \mathcal{K}_X^*/\mathcal{O}_X^*)$. A $\mathbb{Q}$-*Cartier divisor* (resp. $\mathbb{R}$-*Cartier divisor*) is an element of $H^0(X, \mathcal{K}_X^*/\mathcal{O}_X^*) \otimes_{\mathbb{Z}} \mathbb{Q}$ (resp. $H^0(X, \mathcal{K}_X^*/\mathcal{O}_X^*) \otimes_{\mathbb{Z}} \mathbb{R}$).

2.3 (Linear, $\mathbb{Q}$-linear, and $\mathbb{R}$-linear equivalence)**.** Let D_1 and D_2 be two $\mathbb{R}$-Cartier divisors on X. Then D_1 is *linearly* (resp. $\mathbb{Q}$-*linearly*, or $\mathbb{R}$-*linearly*) *equivalent* to D_2, denoted by $D_1 \sim D_2$ (resp. $D_1 \sim_{\mathbb{Q}} D_2$, or $D_1 \sim_{\mathbb{R}} D_2$) if

$$D_1 = D_2 + \sum_{i=1}^{k} r_i (f_i)$$

such that $f_i \in \Gamma(X, \mathcal{K}_X^*)$ and $r_i \in \mathbb{Z}$ (resp. $r_i \in \mathbb{Q}$, or $r_i \in \mathbb{R}$) for every i. We note that (f_i) is a *principal Cartier divisor* associated to f_i, that is, the image of f_i by $\Gamma(X, \mathcal{K}_X^*) \to \Gamma(X, \mathcal{K}_X^*/\mathcal{O}_X^*)$. Let $f : X \to Y$ be a morphism. If there is an $\mathbb{R}$-Cartier divisor B on Y such that $D_1 \sim_{\mathbb{R}} D_2 + f^*B$, then D_1 is said to be *relatively* $\mathbb{R}$-*linearly equivalent* to D_2. It is denoted by $D_1 \sim_{\mathbb{R}, f} D_2$.

2.4 (Supports)**.** Let D be a Cartier divisor on X. The *support* of D, denoted by Supp$D$, is the subset of X consisting of points x such that a local equation for D is not in $\mathcal{O}_{X,x}^*$. The support of D is a closed subset of X.

2.5 (Weil divisors, $\mathbb{Q}$-divisors, and $\mathbb{R}$-divisors)**.** Let X be an equidimensional reduced separated algebraic scheme. We note that X is not necessarily regular in codimension one. A (*Weil*) *divisor* D on X is a finite formal sum

$$\sum_{i=1}^{n} d_i D_i$$

where D_i is an irreducible reduced closed subscheme of X of pure codimension one and d_i is an integer for every i such that $D_i \neq D_j$ for $i \neq j$.

If $d_i \in \mathbb{Q}$ (resp. $d_i \in \mathbb{R}$) for every i, then D is called a $\mathbb{Q}$-*divisor* (resp. $\mathbb{R}$-*divisor*). We define the *round-up* $\lceil D \rceil = \sum_{i=1}^{r} \lceil d_i \rceil D_i$ (resp. the *round-down* $\lfloor D \rfloor = \sum_{i=1}^{r} \lfloor d_i \rfloor D_i$), where for every real number x, $\lceil x \rceil$ (resp. $\lfloor x \rfloor$) is the integer defined by $x \leq \lceil x \rceil < x+1$ (resp. $x-1 < \lfloor x \rfloor \leq x$). The *fractional part* $\{D\}$ of D denotes $D - \lfloor D \rfloor$. We define $D^{<1} = \sum_{d_i<1} d_i D_i$ and so on. We call D a *boundary* $\mathbb{R}$-divisor if $0 \leq d_i \leq 1$ for every i.

Next, we recall the definition of *simple normal crossing pairs.*

Definition 2.6 (Simple normal crossing pairs)**.** We say that the pair (X, D) is *simple normal crossing* at a point $a \in X$ if X has a Zariski open neighborhood U of a that can be embedded in a smooth variety Y, where Y has regular system of parameters $(x_1, \cdots, x_p, y_1, \cdots, y_r)$ at $a = 0$ in which U is defined by a monomial equation

$$x_1 \cdots x_p = 0$$

and

$$D = \sum_{i=1}^{r} \alpha_i (y_i = 0)|_U, \quad \alpha_i \in \mathbb{R}.$$

We say that (X, D) is a *simple normal crossing pair* if it is simple normal crossing at every point of X. If $(X, 0)$ is a simple normal crossing pair, then X is called a *simple normal crossing variety.* If X is a simple normal crossing variety, then X has only Gorenstein singularities. Thus, it has an invertible dualizing sheaf ω_X. Therefore, we can define the *canonical divisor* K_X such that $\omega_X \simeq \mathcal{O}_X(K_X)$ (cf. [L, Section 7.1 Corollary 1.19]). It is a Cartier divisor on X and is well-defined up to linear equivalence.

We note that a simple normal crossing pair is called a *semi-snc pair* in [Ko3, Definition 1.10].

Definition 2.7 (Strata and permissibility)**.** Let X be a simple normal crossing variety and let $X = \bigcup_{i \in I} X_i$ be the irreducible decomposition of X. A *stratum* of X is an irreducible component of $X_{i_1} \cap \cdots \cap X_{i_k}$ for some $\{i_1, \cdots, i_k\} \subset I$. A Cartier divisor D on X is *permissible* if D contains no strata of X in its support. A finite $\mathbb{Q}$-linear (resp. $\mathbb{R}$-linear) combination of permissible Cartier divisors is called a *permissible* $\mathbb{Q}$-*Cartier divisor* (resp. $\mathbb{R}$-*Cartier divisor*) on X.

2.8. Let X be a simple normal crossing variety. Let $\mathrm{PerDiv}(X)$ be the abelian group generated by permissible Cartier divisors on X and let $\mathrm{Weil}(X)$ be the abelian group generated by Weil divisors on X. Then we can define natural injective homomorphisms of abelian groups

$$\psi : \mathrm{PerDiv}(X) \otimes_{\mathbb{Z}} \mathbb{K} \to \mathrm{Weil}(X) \otimes_{\mathbb{Z}} \mathbb{K}$$

for $\mathbb{K} = \mathbb{Z}$, $\mathbb{Q}$, and $\mathbb{R}$. Let $\nu : \widetilde{X} \to X$ be the normalization. Then we have the following commutative diagram.

$$\begin{array}{ccc}
\mathrm{Div}(\widetilde{X}) \otimes_{\mathbb{Z}} \mathbb{K} & \xrightarrow[\widetilde{\psi}]{\sim} & \mathrm{Weil}(\widetilde{X}) \otimes_{\mathbb{Z}} \mathbb{K} \\
{\scriptstyle \nu^*} \uparrow & & \downarrow {\scriptstyle \nu_*} \\
\mathrm{PerDiv}(X) \otimes_{\mathbb{Z}} \mathbb{K} & \xrightarrow[\psi]{} & \mathrm{Weil}(X) \otimes_{\mathbb{Z}} \mathbb{K}
\end{array}$$

Note that $\mathrm{Div}(\widetilde{X})$ is the abelian group generated by Cartier divisors on $\widetilde{X}$ and that $\widetilde{\psi}$ is an isomorphism since $\widetilde{X}$ is smooth.

By ψ, every permissible Cartier (resp. $\mathbb{Q}$-Cartier or $\mathbb{R}$-Cartier) divisor can be considered as a Weil divisor (resp. $\mathbb{Q}$-divisor or $\mathbb{R}$-divisor). Therefore, various operations, for example, $\lfloor D \rfloor$, $D^{<1}$, and so on, make sense for a permissible $\mathbb{R}$-Cartier $\mathbb{R}$-divisor D on X.

We note the following easy example.

Example 2.9. Let X be a simple normal crossing variety in $\mathbb{C}^3 = \mathrm{Spec}\mathbb{C}[x, y, z]$ defined by $xy = 0$. We set $D_1 = (x + z = 0) \cap X$ and $D_2 = (x - z = 0) \cap X$. Then $D = \frac{1}{2}D_1 + \frac{1}{2}D_2$ is a permissible $\mathbb{Q}$-Cartier $\mathbb{Q}$-divisor on X. In this case, $\lfloor D \rfloor = (x = z = 0)$ on X. Therefore, $\lfloor D \rfloor$ is not a Cartier divisor on X.

Definition 2.10 (Simple normal crossing divisors)**.** Let X be a simple normal crossing variety and let D be a Cartier divisor on X. If (X, D) is a simple normal crossing pair and D is reduced, then D is called a *simple normal crossing divisor* on X.

Remark 2.11. Let X be a simple normal crossing variety and let D be a $\mathbb{K}$-divisor on X where $\mathbb{K} = \mathbb{Q}$ or $\mathbb{R}$. If $\mathrm{Supp}D$ is a simple normal crossing divisor on X and D is $\mathbb{K}$-Cartier, then $\lfloor D \rfloor$ and $\lceil D \rceil$ (resp. $\{D\}$, $D^{<1}$, and so on) are Cartier (resp. $\mathbb{K}$-Cartier) divisors on X (cf. [BP, Section 8]).

The following lemma is easy but important.

Lemma 2.12. *Let X be a simple normal crossing variety and let B be a permissible $\mathbb{R}$-Cartier $\mathbb{R}$-divisor on X such that $\lfloor B \rfloor = 0$. Let A be a Cartier divisor on X. Assume that $A \sim_{\mathbb{R}} B$. Then there exists a permissible $\mathbb{Q}$-Cartier $\mathbb{Q}$-divisor C on X such that $A \sim_{\mathbb{Q}} C$, $\lfloor C \rfloor = 0$, and* $\mathrm{Supp} C = \mathrm{Supp} B$.

Proof. We can write $B = A + \sum_{i=1}^{k} r_i(f_i)$, where $f_i \in \Gamma(X, \mathcal{K}_X^*)$ and $r_i \in \mathbb{R}$ for every i. Here, $\mathcal{K}_X$ is the sheaf of total quotient rings of $\mathcal{O}_X$ (see 2.1). Let $P \in X$ be a scheme theoretic point corresponding to some stratum of X. We consider the following affine map

$$\mathbb{K}^k \to H^0(X_P, \mathcal{K}_{X_P}^*/\mathcal{O}_{X_P}^*) \otimes_{\mathbb{Z}} \mathbb{K}$$

given by $(a_1, \cdots, a_k) \mapsto A + \sum_{i=1}^{k} a_i(f_i)$, where $X_P = \mathrm{Spec} \mathcal{O}_{X,P}$ and $\mathbb{K} = \mathbb{Q}$ or $\mathbb{R}$. Then we can check that

$$\mathcal{P} = \{(a_1, \cdots, a_k) \in \mathbb{R}^k \,|\, A + \sum_i a_i(f_i) \text{ is permissible}\} \subset \mathbb{R}^k$$

is an affine subspace of $\mathbb{R}^k$ defined over $\mathbb{Q}$. Therefore, we see that

$$\mathcal{S} = \{(a_1, \cdots, a_k) \in \mathcal{P} \,|\, \mathrm{Supp}(A + \sum_i a_i(f_i)) \subset \mathrm{Supp} B\} \subset \mathcal{P}$$

is an affine subspace of $\mathbb{R}^k$ defined over $\mathbb{Q}$. Since $(r_1, \cdots, r_k) \in \mathcal{S}$, we know that $\mathcal{S} \neq \emptyset$. We take a point $(s_1, \cdots, s_k) \in \mathcal{S} \cap \mathbb{Q}^k$ which is general in $\mathcal{S}$ and sufficiently close to $(r_1, \cdots, r_k)$ and set $C = A + \sum_{i=1}^{k} s_i(f_i)$. By construction, C is a permissible $\mathbb{Q}$-Cartier $\mathbb{Q}$-divisor such that $C \sim_{\mathbb{Q}} A$, $\lfloor C \rfloor = 0$, and $\mathrm{Supp} C = \mathrm{Supp} B$. Q.E.D.

We need the following important definition in Section 3.

Definition 2.13 (Strata and permissibility for pairs). Let (X, D) be a simple normal crossing pair such that D is a boundary $\mathbb{R}$-divisor on X. Let $\nu : X^\nu \to X$ be the normalization. We define Θ by the formula

$$K_{X^\nu} + \Theta = \nu^*(K_X + D).$$

Then a *stratum* of (X, D) is an irreducible component of X or the ν-image of a log canonical center of (X^ν, Θ). We note that (X^ν, Θ) is log canonical. When $D = 0$, this definition is compatible with Definition 2.7. A Cartier divisor B on X is *permissible with respect to* (X, D) if B contains no strata of (X, D) in its support. A finite $\mathbb{R}$-linear (resp. $\mathbb{Q}$-linear) combination of permissible Cartier divisors with respect to (X, D) is called a *permissible* $\mathbb{R}$*-divisor* (resp. $\mathbb{Q}$*-divisor*) *with respect to* (X, D).

For the reader's convenience, we recall Grothendieck's Quot scheme. For the details, see, for example, [N, Theorem 5.14] and [AK, Section 2]. We will use it in the proof of the main theorem: Theorem 3.7.

Theorem 2.14 (Grothendieck). *Let S be a noetherian scheme, let $\pi : X \to S$ be a projective morphism, and let L be a relatively very ample line bundle on X. Then for any coherent $\mathcal{O}_X$-module E and any polynomial $\Phi \in \mathbb{Q}[\lambda]$, the functor $\mathfrak{Quot}^{\Phi,L}_{E/X/S}$ is representable by a projective S-scheme $\mathrm{Quot}^{\Phi,L}_{E/X/S}$.*

§3. Vanishing theorems

Let us start with the following injectivity theorem (cf. [F4, Proposition 3.2] and [F8, Proposition 2.23]). The proof of Theorem 3.1 in [F8] is purely Hodge theoretic. It uses the theory of mixed Hodge structures on cohomology groups with compact support (cf. 1.2). For the details, see [F8, Chapter 2] and [F15].

Theorem 3.1 (Hodge theoretic injectivity theorem). *Let $(X, S+B)$ be a simple normal crossing pair such that X is proper, $S+B$ is a boundary $\mathbb{R}$-divisor, S is reduced, and $\lfloor B \rfloor = 0$. Let L be a Cartier divisor on X and let D be an effective Cartier divisor whose support is contained in* $\mathrm{Supp} B$. *Assume that $L \sim_{\mathbb{R}} K_X + S + B$. Then the natural homomorphisms*

$$H^q(X, \mathcal{O}_X(L)) \to H^q(X, \mathcal{O}_X(L+D)),$$

which are induced by the inclusion $\mathcal{O}_X \to \mathcal{O}_X(D)$, are injective for all q.

Remark 3.2. In [F15], we prove a slight generalization of Theorem 3.1. However, Theorem 3.1 is sufficient for the proof of Theorem 3.4 below.

The next lemma is an easy generalization of the vanishing theorem of Reid–Fukuda type for simple normal crossing pairs, which is a very special case of Theorem 3.7 (i). However, we need Lemma 3.3 for our proof of Theorem 3.7.

Lemma 3.3 (Relative vanishing lemma). *Let $f : Y \to X$ be a proper morphism from a simple normal crossing pair (Y, Δ) to an algebraic variety X such that Δ is a boundary $\mathbb{R}$-divisor on Y. We assume that f is an isomorphism at the generic point of any stratum of the pair (Y, Δ). Let L be a Cartier divisor on Y such that $L \sim_{\mathbb{R},f} K_Y + \Delta$. Then $R^q f_* \mathcal{O}_Y(L) = 0$ for every $q > 0$.*

Proof. By shrinking X, we may assume that $L \sim_{\mathbb{R}} K_Y + \Delta$. By applying Lemma 2.12 to $\{\Delta\}$, we may further assume that Δ is a $\mathbb{Q}$-divisor and $L \sim_{\mathbb{Q}} K_Y + \Delta$.

Step 1. We assume that Y is irreducible. In this case, $L-(K_Y+\Delta)$ is nef and log big over X with respect to the pair (Y, Δ), that is, $L - (K_Y + \Delta)$ is nef and big over X and $(L-(K_Y+\Delta))|_W$ is big over $f(W)$ for every stratum W of the pair (Y, Δ). Therefore, $R^q f_* \mathcal{O}_Y(L) = 0$ for every $q > 0$ by the vanishing theorem of Reid–Fukuda type (see, for example, [F8, Lemma 4.10]).

Step 2. Let Y_1 be an irreducible component of Y and let Y_2 be the union of the other irreducible components of Y. Then we have a short exact sequence

$$0 \to \mathcal{O}_{Y_1}(-Y_2|_{Y_1}) \to \mathcal{O}_Y \to \mathcal{O}_{Y_2} \to 0.$$

We set $L' = L|_{Y_1} - Y_2|_{Y_1}$. Then we have a short exact sequence

$$0 \to \mathcal{O}_{Y_1}(L') \to \mathcal{O}_Y(L) \to \mathcal{O}_{Y_2}(L|_{Y_2}) \to 0$$

and $L' \sim_{\mathbb{Q}} K_{Y_1} + \Delta|_{Y_1}$. On the other hand, we can check that

$$L|_{Y_2} \sim_{\mathbb{Q}} K_{Y_2} + Y_1|_{Y_2} + \Delta|_{Y_2}.$$

We have already known that $R^q f_* \mathcal{O}_{Y_1}(L') = 0$ for every $q > 0$ by Step 1. By induction on the number of the irreducible components of Y, we have $R^q f_* \mathcal{O}_{Y_2}(L|_{Y_2}) = 0$ for every $q > 0$. Therefore, $R^q f_* \mathcal{O}_Y(L) = 0$ for every $q > 0$ by the exact sequence:

$$\cdots \to R^q f_* \mathcal{O}_{Y_1}(L') \to R^q f_* \mathcal{O}_Y(L) \to R^q f_* \mathcal{O}_{Y_2}(L|_{Y_2}) \to \cdots.$$

So, we finish the proof of Lemma 3.3. Q.E.D.

Although Lemma 3.3 is a very easy generalization of the relative Kawamata–Viehweg vanishing theorem, it is sufficiently powerful for the study of reducible varieties once we combine it with the recent results in [BM] and [BP]. In Section 4, we will see an application of Lemma 3.3 for the study of semi divisorial log terminal pairs.

It is the time to state the main injectivity theorem for simple normal crossing pairs. It is a direct application of Theorem 3.1. Our formulation of Theorem 3.4 is indispensable for the proof of our main theorem: Theorem 3.7.

Theorem 3.4 (Injectivity theorem for simple normal crossing pairs)**.** *Let (X, Δ) be a simple normal crossing pair such that X is proper and* *that Δ is a boundary $\mathbb{R}$-divisor on X. Let L be a Cartier divisor on X and let D be an effective Cartier divisor that is permissible with respect to (X, Δ). Assume the following conditions.*

(i) $L \sim_{\mathbb{R}} K_X + \Delta + H$,

(ii) *H is a semi-ample $\mathbb{R}$-divisor, and*

(iii) *$tH \sim_{\mathbb{R}} D + D'$ for some positive real number t, where D' is an effective $\mathbb{R}$-Cartier $\mathbb{R}$-divisor that is permissible with respect to (X, Δ).*

Then the homomorphisms

$$H^q(X, \mathcal{O}_X(L)) \to H^q(X, \mathcal{O}_X(L + D)),$$

which are induced by the natural inclusion $\mathcal{O}_X \to \mathcal{O}_X(D)$, are injective for all q.

Remark 3.5. For the definition and the basic properties of semi-ample $\mathbb{R}$-divisors, see [F11, Definition 4.11, Lemma 4.13, and Lemma 4.14].

Proof of Theorem 3.4. We set $S = \lfloor \Delta \rfloor$ and $B = \{\Delta\}$ throughout this proof. We obtain a projective birational morphism $f : Y \to X$ from a simple normal crossing variety Y such that f is an isomorphism over $X \setminus \operatorname{Supp}(D + D' + B)$, and that the union of the support of $f^*(S + B + D + D')$ and the exceptional locus of f has a simple normal crossing support on Y (cf. [BP, Theorem 1.4]). Let B' be the strict transform of B on Y. We may assume that $\operatorname{Supp} B'$ is disjoint from any strata of Y that are not irreducible components of Y by taking blowing-ups. We write

$$K_Y + S' + B' = f^*(K_X + S + B) + E,$$

where S' is the strict transform of S and E is f-exceptional. By the construction of $f : Y \to X$, S' is Cartier and B' is $\mathbb{R}$-Cartier. Therefore, E is also $\mathbb{R}$-Cartier. It is easy to see that $E_+ = \lceil E \rceil \geq 0$. We set $L' = f^*L + E_+$ and $E_- = E_+ - E \geq 0$. We note that E_+ is Cartier and E_- is $\mathbb{R}$-Cartier because $\operatorname{Supp} E$ is simple normal crossing on Y (cf. Remark 2.11). Since f^*H is an $\mathbb{R}_{>0}$-linear combination of semi-ample Cartier divisors, we can write $f^*H \sim_{\mathbb{R}} \sum_i a_i H_i$, where $0 < a_i < 1$ and H_i is a general Cartier divisor on Y for every i. We set

$$B'' = B' + E_- + \frac{\varepsilon}{t} f^*(D + D') + (1 - \varepsilon) \sum_i a_i H_i$$

for some $0 < \varepsilon \ll 1$. Then $L' \sim_{\mathbb{R}} K_Y + S' + B''$. By construction, $\lfloor B'' \rfloor = 0$, the support of $S' + B''$ is simple normal crossing on Y, and

$\mathrm{Supp}B'' \supset \mathrm{Supp}f^*D$. So, Theorem 3.1 implies that the homomorphisms

$$H^q(Y, \mathcal{O}_Y(L')) \to H^q(Y, \mathcal{O}_Y(L' + f^*D))$$

are injective for all q. By Lemma 3.3, $R^q f_* \mathcal{O}_Y(L') = 0$ for every $q > 0$ and it is easy to see that $f_*\mathcal{O}_Y(L') \simeq \mathcal{O}_X(L)$. By the Leray spectral sequence, the homomorphisms

$$H^q(X, \mathcal{O}_X(L)) \to H^q(X, \mathcal{O}_X(L + D))$$

are injective for all q. Q.E.D.

Lemma 3.6. *Let $f : Z \to X$ be a proper morphism from a simple normal crossing pair (Z, B) to an algebraic variety X. Let $\overline{X}$ be a projective variety such that $\overline{X}$ contains X as a Zariski dense open subset. Then there exist a proper simple normal crossing pair $(\overline{Z}, \overline{B})$ that is a compactification of (Z, B) and a proper morphism $\overline{f} : \overline{Z} \to \overline{X}$ that compactifies $f : Z \to X$. Moreover, $\overline{Z} \setminus Z$ is a divisor on $\overline{Z}$,* $\mathrm{Supp}\overline{B} \cup \mathrm{Supp}(\overline{Z} \setminus Z)$ *is a simple normal crossing divisor on $\overline{Z}$, and $\overline{Z} \setminus Z$ has no common irreducible components with $\overline{B}$. We note that we can make $\overline{B}$ a $\mathbb{K}$-Cartier $\mathbb{K}$-divisor on $\overline{Z}$ when so is B on Z, where $\mathbb{K}$ is $\mathbb{Z}$, $\mathbb{Q}$, or $\mathbb{R}$. When f is projective, we can make $\overline{Z}$ projective.*

Proof. Let $\overline{B} \subset \overline{Z}$ be any compactification of $B \subset Z$. By blowing up $\overline{Z}$ inside $\overline{Z} \setminus Z$, we may assume that $f : Z \to X$ extends to $\overline{f} : \overline{Z} \to \overline{X}$, $\overline{Z}$ is a simple normal crossing variety, and $\overline{Z} \setminus Z$ is of pure codimension one (see [BM, Theorem 1.5]). By [BP, Theorem 1.4], we can construct a desired compactification. Note that we can make $\overline{B}$ a $\mathbb{K}$-Cartier $\mathbb{K}$-divisor by the argument in [BP, Section 8]. Q.E.D.

Theorem 3.7 below is our main theorem of this paper, which is a generalization of Kollár's torsion-free and vanishing theorem (see [Ko1, Theorem 2.1]). The reader find various applications of Theorem 3.7 in [F8], [FF], and [F13]. We note that Theorem 3.7 for *embedded normal crossing pairs* was first formulated by Florin Ambro for his theory of *quasi-log varieties* (cf. [A]). For the details of the theory of quasi-log varieties, see [F8, Chapter 3] and [F9].

Theorem 3.7. *Let (Y, Δ) be a simple normal crossing pair such that Δ is a boundary $\mathbb{R}$-divisor on Y. Let $f : Y \to X$ be a proper morphism to an algebraic variety X and let L be a Cartier divisor on Y such that $L - (K_Y + \Delta)$ is f-semi-ample.*

(i) *every associated prime of $R^q f_* \mathcal{O}_Y(L)$ is the generic point of the f-image of some stratum of (Y, Δ) for every q.*

(ii) *let $\pi : X \to V$ be a projective morphism to an algebraic variety V such that*

$$L - (K_Y + \Delta) \sim_{\mathbb{R}} f^*H$$

for some π-ample $\mathbb{R}$-divisor H on X. Then $R^q f_* \mathcal{O}_Y(L)$ is π_*-acyclic, that is,

$$R^p \pi_* R^q f_* \mathcal{O}_Y(L) = 0$$

for every $p > 0$ and $q \geq 0$.

Proof. We set $S = \lfloor \Delta \rfloor$, $B = \{\Delta\}$, and $H' \sim_{\mathbb{R}} L - (K_Y + \Delta)$ throughout this proof. Let us start with the proof of (i).

Step 1. First, we assume that X is projective. We may assume that H' is semi-ample by replacing L (resp. H') with $L + f^*A'$ (resp. $H' + f^*A'$), where A' is a very ample Cartier divisor on X. Suppose that $R^q f_* \mathcal{O}_Y(L)$ has a local section whose support does not contain the f-images of any strata of $(Y, S+B)$. More precisely, let U be a non-empty Zariski open set and let $s \in \Gamma(U, R^q f_* \mathcal{O}_Y(L))$ be a non-zero section of $R^q f_* \mathcal{O}_Y(L)$ on U whose support $V \subset U$ does not contain the f-images of any strata of $(Y, S+B)$. Let $\overline{V}$ be the closure of V in X. We note that $\overline{V} \setminus V$ may contain the f-image of some stratum of $(Y, S+B)$. Let Y_2 be the union of the irreducible components of Y that are mapped into $\overline{V} \setminus V$ and let Y_1 be the union of the other irreducible components of Y. We set

$$K_{Y_1} + S_1 + B_1 = (K_Y + S + B)|_{Y_1}$$

such that S_1 is reduced and that $\lfloor B_1 \rfloor = 0$. By replacing Y, S, B, L, and H' with Y_1, S_1, B_1, $L|_{Y_1}$, and $H'|_{Y_1}$, we may assume that no irreducible components of Y are mapped into $\overline{V} \setminus V$. Let C be a stratum of $(Y, S+B)$ that is mapped into $\overline{V} \setminus V$. Let $\sigma : Y' \to Y$ be the blowing-up along C. We set $E = \sigma^{-1}(C)$. We can write

$$K_{Y'} + S' + B' = \sigma^*(K_Y + S + B)$$

such that S' is reduced and $\lfloor B' \rfloor = 0$. Thus,

$$\sigma^* H' \sim_{\mathbb{R}} \sigma^* L - (K_{Y'} + S' + B')$$

and

$$\sigma^* H' \sim_{\mathbb{R}} \sigma^* L - E - (K_{Y'} + (S' - E) + B').$$

We note that $S' - E$ is effective. We replace Y, H', L, S, and B with Y', $\sigma^* H'$, $\sigma^* L - E$, $S' - E$, and B'. By repeating this process finitely many times, we may assume that $\overline{V}$ does not contain the f-images of any strata of $(Y, S+B)$. Then we can find a very ample Cartier divisor A on X with the following properties.

(a) f^*A is permissible with respect to $(Y, S+B)$, and

(b) $R^q f_*\mathcal{O}_Y(L) \to R^q f_*\mathcal{O}_Y(L) \otimes \mathcal{O}_X(A)$ is not injective.

We may assume that $H' - f^*A$ is semi-ample by replacing L (resp. H') with $L + f^*A$ (resp. $H' + f^*A$). If necessary, we replace L (resp. H') with $L + f^*A''$ (resp. $H' + f^*A''$), where A'' is a very ample Cartier divisor. Then, we have

$$H^0(X, R^q f_*\mathcal{O}_Y(L)) \simeq H^q(Y, \mathcal{O}_Y(L))$$

and

$$H^0(X, R^q f_*\mathcal{O}_Y(L) \otimes \mathcal{O}_X(A)) \simeq H^q(Y, \mathcal{O}_Y(L + f^*A)).$$

We obtain that

$$H^0(X, R^q f_*\mathcal{O}_Y(L)) \to H^0(X, R^q f_*\mathcal{O}_Y(L) \otimes \mathcal{O}_X(A))$$

is not injective by (b) if A'' is sufficiently ample. So,

$$H^q(Y, \mathcal{O}_Y(L)) \to H^q(Y, \mathcal{O}_Y(L + f^*A))$$

is not injective. It contradicts Theorem 3.4. Therefore, the support of every non-zero local section of $R^q f_*\mathcal{O}_Y(L)$ contains the f-image of some stratum of (Y, Δ), equivalently, the support of every non-zero local section of $R^q f_*\mathcal{O}_Y(L)$ is equal to the union of the f-images of some strata of (Y, Δ). This means that every associated prime of $R^q f_*\mathcal{O}_Y(L)$ is the generic point of the f-image of some stratum of (Y, Δ). We finish the proof when X is projective.

Step 2. Next, we assume that X is not projective. Note that the problem is local. So, we shrink X and may assume that X is affine. We can write $H' \sim_{\mathbb{R}} H'_1 + H'_2$, where H'_1 (resp. H'_2) is a semi-ample $\mathbb{Q}$-divisor (resp. a semi-ample $\mathbb{R}$-divisor) on Y. We can write $H'_2 \sim_{\mathbb{R}} \sum_i a_i A_i$, where $0 < a_i < 1$ and A_i is a general effective Cartier divisor on Y for every i. Replacing B (resp. H') with $B + \sum_i a_i A_i$ (resp. H'_1), we may assume that H' is a semi-ample $\mathbb{Q}$-divisor. Without loss of generality, we may further assume that $(Y, B + S + H')$ is a simple normal crossing pair. We compactify X and apply Lemma 3.6. Then we obtain a compactification $\overline{f} : \overline{Y} \to \overline{X}$ of $f : Y \to X$. Let $\overline{H'}$ be the closure of H' on $\overline{Y}$. If $\overline{H'}$ is not a semi-ample $\mathbb{Q}$-divisor, then we take blowing-ups of $\overline{Y}$ inside $\overline{Y} \setminus Y$ and obtain a semi-ample $\mathbb{Q}$-divisor $\widetilde{H'}$ on $\overline{Y}$ such that $\widetilde{H'}|_Y = H'$. Let $\overline{B}$ (resp. $\overline{S}$) be the closure of B (resp. S)

on $\overline{Y}$. We may assume that $\overline{S}$ is Cartier and $\overline{B}$ is $\mathbb{R}$-Cartier (cf. Lemma 3.6). We construct a coherent sheaf $\mathcal{F}$ on $\overline{Y}$ which is an extension of $\mathcal{O}_Y(L)$. We consider Grothendieck's Quot scheme $\mathrm{Quot}^{1,\mathcal{O}_{\overline{Y}}}_{\mathcal{F}/\overline{Y}/\overline{Y}}$ (see Theorem 2.14). Note that the restriction of $\mathrm{Quot}^{1,\mathcal{O}_{\overline{Y}}}_{\mathcal{F}/\overline{Y}/\overline{Y}}$ to Y is nothing but Y because $\mathcal{F}|_Y = \mathcal{O}_Y(L)$ is a line bundle on Y. Therefore, the structure morphism from $\mathrm{Quot}^{1,\mathcal{O}_{\overline{Y}}}_{\mathcal{F}/\overline{Y}/\overline{Y}}$ to $\overline{Y}$ has a section s over Y. By taking the closure of $s(Y)$ in $\mathrm{Quot}^{1,\mathcal{O}_{\overline{Y}}}_{\mathcal{F}/\overline{Y}/\overline{Y}}$, we have a compactification $Y^\dagger$ of Y and a line bundle $\mathcal{L}$ on $Y^\dagger$ with $\mathcal{L}|_Y = \mathcal{O}_Y(L)$. If necessary, we take more blowing-ups of $Y^\dagger$ outside Y (cf. [BP, Theorem 1.4]). Then we obtain a new compactification $\overline{Y}$ and a Cartier divisor $\overline{L}$ on $\overline{Y}$ with $\overline{L}|_Y = L$ (cf. Lemma 3.6). In this situation, $\widetilde{H'} \sim_{\mathbb{R}} \overline{L} - (K_{\overline{Y}} + \overline{S} + \overline{B})$ does not necessarily hold. We can write

$$H' + \sum_i b_i(f_i) = L - (K_Y + S + B),$$

where b_i is a real number and $f_i \in \Gamma(Y, \mathcal{K}^*_Y)$ for every i. We set

$$E = \widetilde{H'} + \sum_i b_i(f_i) - (\overline{L} - (K_{\overline{Y}} + \overline{S} + \overline{B})).$$

We note that we can see $f_i \in \Gamma(\overline{Y}, \mathcal{K}^*_{\overline{Y}})$ for every i (cf. [L, Section 7.1 Proposition 1.15]). We replace $\overline{L}$ (resp. $\overline{B}$) with $\overline{L} + \lceil E \rceil$ (resp. $\overline{B} + \{-E\}$). Then we obtain the desired property of $R^q \overline{f}_* \mathcal{O}_{\overline{Y}}(\overline{L})$ since $\overline{X}$ is projective. We note that $\lceil E \rceil$ is Cartier because $\mathrm{Supp}E$ is in $\overline{Y} \setminus Y$ and E is $\mathbb{R}$-Cartier (cf. Remark 2.11). So, we finish the whole proof of (i).

From now on, we prove (ii).

Step 1. We assume that $\dim V = 0$. In this case, we can write $H \sim_{\mathbb{R}} H_1 + H_2$, where H_1 (resp. H_2) is an ample $\mathbb{Q}$-divisor (resp. an ample $\mathbb{R}$-divisor) on X. So, we can write $H_2 \sim_{\mathbb{R}} \sum_i a_i A_i$, where $0 < a_i < 1$ and A_i is a general very ample Cartier divisor on X for every i. Replacing B (resp. H) with $B + \sum_i a_i f^* A_i$ (resp. H_1), we may assume that H is an ample $\mathbb{Q}$-divisor. We take a general member $A \in |mH|$, where m is a sufficiently large and divisible integer, such that $A' = f^*A$ and $R^q f_* \mathcal{O}_Y(L + A')$ is π_*-acyclic, that is, Γ-acyclic, for all q. By (i), we have the following short exact sequences,

$$0 \to R^q f_* \mathcal{O}_Y(L) \to R^q f_* \mathcal{O}_Y(L + A') \to R^q f_* \mathcal{O}_{A'}(L + A') \to 0.$$

for every q. Note that $R^q f_* \mathcal{O}_{A'}(L + A')$ is π_*-acyclic by induction on $\dim X$ and $R^q f_* \mathcal{O}_Y(L + A')$ is also π_*-acyclic by the above assumption.

Thus, $E_2^{pq} = 0$ for $p \geq 2$ in the following commutative diagram of spectral sequences.

$$\begin{array}{ccc} E_2^{pq} = H^p(X, R^q f_* \mathcal{O}_Y(L)) & \Longrightarrow & H^{p+q}(Y, \mathcal{O}_Y(L)) \\ \Big\downarrow{\scriptstyle \varphi^{pq}} & & \Big\downarrow{\scriptstyle \varphi^{p+q}} \\ \overline{E}_2^{pq} = H^p(X, R^q f_* \mathcal{O}_Y(L+A')) & \Longrightarrow & H^{p+q}(Y, \mathcal{O}_Y(L+A')) \end{array}$$

We note that φ^{1+q} is injective by Theorem 3.4. We have that

$$E_2^{1q} \xrightarrow{\alpha} H^{1+q}(Y, \mathcal{O}_Y(L))$$

is injective by the fact that $E_2^{pq} = 0$ for $p \geq 2$. We also have that $\overline{E}_2^{1q} = 0$ by the above assumption. Therefore, we obtain $E_2^{1q} = 0$ since the injection

$$E_2^{1q} \xrightarrow{\alpha} H^{1+q}(Y, \mathcal{O}_Y(L)) \xrightarrow{\varphi^{1+q}} H^{1+q}(Y, \mathcal{O}_Y(L+A'))$$

factors through $\overline{E}_2^{1q} = 0$. This implies that $H^p(X, R^q f_* \mathcal{O}_Y(L)) = 0$ for every $p > 0$.

Step 2. We assume that V is projective. By replacing H (resp. L) with $H + \pi^* G$ (resp. $L + (\pi \circ f)^* G$), where G is a very ample Cartier divisor on V, we may assume that H is an ample $\mathbb{R}$-divisor. If G is a sufficiently ample Cartier divisor on V, then we have

$$H^k(V, R^p \pi_* R^q f_* \mathcal{O}_Y(L) \otimes G) = 0$$

for every $k \geq 1$,

$$\begin{aligned} H^0(V, R^p \pi_* R^q f_* \mathcal{O}_Y(L) \otimes \mathcal{O}_V(G)) &\simeq H^p(X, R^q f_* \mathcal{O}_Y(L) \otimes \mathcal{O}_X(\pi^* G)) \\ &\simeq H^p(X, R^q f_* \mathcal{O}_Y(L + f^* \pi^* G)) \end{aligned}$$

for every p and q, and $R^p \pi_* R^q f_* \mathcal{O}_Y(L) \otimes \mathcal{O}_V(G)$ is generated by its global sections for every p and q. Since

$$L + f^* \pi^* G - (K_Y + \Delta) \sim_{\mathbb{R}} f^*(H + \pi^* G),$$

and $H + \pi^* G$ is ample, we can apply Step 1 and obtain

$$H^p(X, R^q f_* \mathcal{O}_Y(L + f^* \pi^* G)) = 0$$

for every $p > 0$. Thus, $R^p \pi_* R^q f_* \mathcal{O}_Y(L) = 0$ for every $p > 0$ by the above arguments.

Step 3. When V is not projective, we shrink V and may assume that V is affine. By the same argument as in Step 1, we may assume that H is $\mathbb{Q}$-Cartier. Let $\overline{\pi} : \overline{X} \to \overline{V}$ be a compactification of $\pi : X \to V$ such that $\overline{X}$ and $\overline{V}$ are projective. We may assume that there exists a $\overline{\pi}$-ample $\mathbb{Q}$-divisor $\overline{H}$ on $\overline{X}$ such that $\overline{H}|_X = H$. By Lemma 3.6, we can compactify $f : (Y, S + B) \to X$ and obtain $\overline{f} : (\overline{Y}, \overline{S} + \overline{B}) \to \overline{X}$. We note that $\overline{f}^* \overline{H} \sim_{\mathbb{R}} \overline{L} - (K_{\overline{Y}} + \overline{S} + \overline{B})$ does not necessarily hold, where $\overline{L}$ is a Cartier divisor on $\overline{Y}$ constructed as in Step 2 in the proof of (i). By the same argument as in Step 2 in the proof of (i), we obtain that $R^p\pi_* R^q f_* \mathcal{O}_Y(L) = 0$ for every $p > 0$ and $q \geq 0$.

We finish the whole proof of (ii). Q.E.D.

§4. Semi divisorial log terminal pairs

Let us start with the definition of *semi divisorial log terminal pairs* in the sense of Kollár. For details of singularities which appear in the minimal model program, see [F3] and [Ko3].

Definition 4.1 (Semi divisorial log terminal pairs)**.** Let X be an equidimensional reduced separated S_2 algebraic scheme which is simple normal crossing in codimension one. Let $\Delta = \sum_i a_i \Delta_i$ be an $\mathbb{R}$-Weil divisor on X such that $0 < a_i \leq 1$ for every i and that Δ_i is not contained in the singular locus of X, where Δ_i is a prime divisor on X for every i and $\Delta_i \neq \Delta_j$ for $i \neq j$. Assume that $K_X + \Delta$ is $\mathbb{R}$-Cartier. The pair (X, Δ) is *semi divisorial log terminal* (*sdlt*, for short) if $a(E, X, \Delta) > -1$ for every exceptional divisor E over X such that (X, Δ) is not a simple normal crossing pair at the generic point of $c_X(E)$, where $c_X(E)$ is the center of E on X.

We note that if (X, Δ) is sdlt and X is irreducible then (X, Δ) is a divisorial log terminal pair (dlt, for short). The following theorem is a direct generalization of [F8, Theorem 4.14] (cf. [F12, Proposition 2.4]). It is an easy application of Lemma 3.3.

Theorem 4.2 (cf. [F12, Theorem 5.2])**.** *Let (X, D) be a semi divisorial log terminal pair. Let $X = \bigcup_{i \in I} X_i$ be the irreducible decomposition. We set*

$$Y = \bigcup_{i \in J} X_i \subset X$$

for $J \subset I$. Then Y is Cohen–Macaulay, seminormal, and has only Du Bois singularities. In particular, each irreducible component of X is normal and X itself is Cohen–Macaulay.

We note that an irreducible component of a seminormal scheme need not be seminormal (see [Ko3, Example 10.12]). We also note that an irreducible component of a Cohen–Macaulay scheme need not be Cohen–Macaulay. The author learned the following example from Shunsuke Takagi.

Example 4.3. We set

$$R = \mathbb{C}[x, y, z, w]/(yz - xw, xz^2 - y^2w).$$

Then $X = \mathrm{Spec} R$ is a reduced reducible two-dimensional Cohen–Macaulay scheme. An irreducible component

$$Y = \mathrm{Spec} R/(y^3 - x^2z, z^3 - yw^2)$$

of X is not Cohen–Macaulay. It is because

$$R/(y^3 - x^2z, z^3 - yw^2) \simeq \mathbb{C}[s^4, s^3t, st^3, t^4].$$

The Cohen–Macaulayness of X is very important for various duality theorems. We use it in the proof of Theorem 7.1 in [FF].

Let us start the proof of Theorem 4.2.

Proof of Theorem 4.2. By [BP, Theorem 1.4], there is a morphism $f : Z \to X$ given by a composite of blowing-ups with smooth centers such that $(Z, f_*^{-1}D + \mathrm{Exc}(f))$ is a simple normal crossing pair and that f is an isomorphism over U, where U is the largest Zariski open subset of X such that $(U, D|_U)$ is a simple normal crossing pair. Then we can write

$$K_Z + D' = f^*(K_X + D) + E,$$

where D' and E are effective and have no common irreducible components. By construction, E is f-exceptional and $\mathrm{Supp}(E + D')$ is a simple normal crossing divisor on Z. Since X is S_2 and simple normal crossing in codimension one, X is seminormal. Then we obtain $f_*\mathcal{O}_Z \simeq \mathcal{O}_X$. Let $Z = \bigcup_{i \in I} Z_i$ be the irreducible decomposition. We consider the short exact sequence

$$0 \to \mathcal{O}_V(-W|_V) \to \mathcal{O}_Z \to \mathcal{O}_W \to 0,$$

where $W = \bigcup_{i \in J} Z_i$ and $V = \bigcup_{i \in I \setminus J} Z_i$. Therefore,

$$0 \to \mathcal{O}_V(\lceil E \rceil - W|_V) \to \mathcal{O}_Z(\lceil E \rceil) \to \mathcal{O}_W(\lceil E \rceil) \to 0$$

is exact. We note that $\lceil E \rceil$ is Cartier (cf. Remark 2.11). By Lemma 3.3, $R^i f_* \mathcal{O}_Z(\lceil E \rceil) = 0$ for every $i > 0$. We note that

$$\lceil E \rceil \sim_{\mathbb{R}, f} K_Z + D' + \{-E\}.$$

Since

$$(\lceil E\rceil - W)|_V \sim_{\mathbb{R},f} K_V + (D' + \{-E\})|_V,$$

$R^if_*\mathcal{O}_V(\lceil E\rceil - W|_V) = 0$ for every $i > 0$ by Lemma 3.3 again. Therefore, we obtain that

$$0 \to f_*\mathcal{O}_V(\lceil E\rceil - W|_V) \to f_*\mathcal{O}_Z(\lceil E\rceil) \simeq \mathcal{O}_X \to f_*\mathcal{O}_W(\lceil E\rceil) \to 0$$

is exact and that $R^if_*\mathcal{O}_W(\lceil E\rceil) = 0$ for every $i > 0$. Since $\lceil E\rceil|_W$ is effective and $\mathcal{O}_X \to f_*\mathcal{O}_W(\lceil E\rceil) \to 0$ factors through $\mathcal{O}_Y$, we have $\mathcal{O}_Y \simeq f_*\mathcal{O}_W \simeq f_*\mathcal{O}_W(\lceil E\rceil)$. Therefore, Y is seminormal because so is W. In the derived category of coherent sheaves on Y, the composition

$$\mathcal{O}_Y \to Rf_*\mathcal{O}_W \to Rf_*\mathcal{O}_W(\lceil E\rceil) \simeq \mathcal{O}_Y \tag{1}$$

is a quasi-isomorphism. Therefore, Y has only Du Bois singularities because W is a simple normal crossing variety. On the other hand, $R^if_*\omega_W = 0$ for every $i > 0$ by Lemma 3.3. By applying Grothendieck duality to (1):

$$\mathcal{O}_Y \to Rf_*\mathcal{O}_W \to \mathcal{O}_Y,$$

we obtain

$$\omega^\bullet_Y \overset{a}{\to} Rf_*\omega^\bullet_W \overset{b}{\to} \omega^\bullet_Y, \tag{2}$$

where $\omega^\bullet_Y$ (resp. $\omega^\bullet_W$) is the dualizing complex of Y (resp. W). Note that $b \circ a$ is a quasi-isomorphism. Thus we have

$$h^i(\omega^\bullet_Y) \subset R^if_*\omega^\bullet_W = R^{i+d}f_*\omega_W$$

where $d = \dim Y = \dim W$. This implies that $h^i(\omega^\bullet_Y) = 0$ for every $i > -\dim Y$. Thus, Y is Cohen–Macaulay and $\omega^\bullet_Y \simeq \omega_Y[d]$. Q.E.D.

As a byproduct of the proof of Theorem 4.2, we obtain the following useful vanishing theorem. Roughly speaking, Proposition 4.4 says that Y has only *semi-rational* singularities.

Proposition 4.4. *In the notation of the proof of* Theorem 4.2, $f_*\mathcal{O}_W \simeq \mathcal{O}_Y$ *and* $R^if_*\mathcal{O}_W = 0$ *for every* $i > 0$.

Proof. By (2) in the proof of Theorem 4.2, we obtain

$$\omega_Y \overset{\alpha}{\to} f_*\omega_W \overset{\beta}{\to} \omega_Y$$

where $\beta \circ \alpha$ is an isomorphism. Since ω_W is locally free and f is an isomorphism over U, $f_*\omega_W$ is a pure sheaf of dimension d. Thus $f_*\omega_W \simeq \omega_Y$ because they are isomorphic over U. Then we obtain $Rf_*\omega^\bullet_W \simeq \omega^\bullet_Y$

in the derived category of coherent sheaves on Y. By Grothendieck duality, $Rf_*\mathcal{O}_W \simeq R\mathcal{H}om(Rf_*\omega^\bullet_W, \omega^\bullet_Y) \simeq \mathcal{O}_Y$ in the derived category of coherent sheaves on Y. Therefore, $f_*\mathcal{O}_W \simeq \mathcal{O}_Y$ and $R^if_*\mathcal{O}_W = 0$ for every $i > 0$. Q.E.D.

As an easy application of Theorem 4.2, we have an adjunction formula for sdlt pairs.

Corollary 4.5 (Adjunction for sdlt pairs). *In the notation of* Theorem 4.2, *we define* D_Y *by*

$$(K_X + D)|_Y = K_Y + D_Y.$$

Then the pair (Y, D_Y) *is semi divisorial log terminal.*

Proof. By Theorem 4.2, Y is Cohen–Macaulay. In particular, Y satisfies Serre's S_2 condition. Then it is easy to see that the pair (Y, D_Y) is semi divisorial log terminal. Q.E.D.

We close this section with an important remark.

Remark 4.6. Let (X, D) be a semi divisorial log terminal pair in the sense of Kollár (see Definition 4.1). Then it is a semi divisorial log terminal pair in the sense of [F1, Definition 1.1]. A key point is that any irreducible component of X is normal (see Theorem 4.2). When the author defined semi divisorial log terminal pairs in [F1, Definition 1.1], the theory of resolution of singularities for *reducible* varieties (cf. [BM] and [BP]) was not available.

§5. Semipositivity theorem

In [F8, Chapter 2], we discuss mixed Hodge structures on cohomology groups with compact support for the proof of Theorem 3.1 (see also [F15]). In [FF], we investigate variations of mixed Hodge structures on cohomology groups with compact support. By the Hodge theoretic semipositivity theorem obtained in [FF, Section 7], we can prove the following theorem as an application of Theorem 3.7.

Theorem 5.1 (Semipositivity theorem). *Let* (X, D) *be a simple normal crossing pair such that* D *is reduced and let* $f : X \to Y$ *be a projective surjective morphism onto a smooth complete algebraic variety* Y. *Assume that every stratum of* (X, D) *is dominant onto* Y. *Let* Σ *be a simple normal crossing divisor on* Y *such that every stratum of* (X, D) *is smooth over* $Y_0 = Y \setminus \Sigma$. *Then* $R^if_*\omega_{X/Y}(D)$ *is locally free for every* i. *We set* $X_0 = f^{-1}(Y_0)$, $D_0 = D|_{X_0}$, *and* $d = \dim X - \dim Y$. *We further* *assume that all the local monodromies on* $R^{d-i}(f|_{X_0\setminus D_0})_!\mathbb{Q}_{X_0\setminus D_0}$ *around* Σ *are unipotent. Then we obtain that* $R^if_*\omega_{X/Y}(D)$ *is a semipositive (in the sense of Fujita–Kawamata) locally free sheaf on* Y*, that is, a* nef *locally free sheaf on* Y*.*

We note that Theorem 5.1 is a generalization of the Fujita–Kawamata semipositivity theorem (cf. [Ka]). We also note that Theorem 5.1 contains the main theorem of [F2]. In [F2], we use variations of mixed Hodge structures on cohomology groups of smooth quasi-projective varieties. However, our formulation in [FF] based on mixed Hodge structures on cohomology groups with compact support is more suitable for reducible varieties than the formulation in [F2] (cf. 1.2). Theorem 3.7 and Theorem 5.1 show that the theory of mixed Hodge structures on cohomology groups with compact support is useful for the study of higher dimensional algebraic varieties. For details, see [F8, Chapter 2], [FF], and [F15].

Finally, we note that in [F14] we prove the projectivity of the moduli spaces of stable varieties as an application of Theorem 5.1 by Kollár's projectivity criterion.

References

[AK] A. Altman, S. Kleiman, Compactifying the Picard scheme, Adv. in Math. **35** (1980), no. 1, 50–112.

[A] F. Ambro, Quasi-log varieties, Tr. Mat. Inst. Steklova **240** (2003), Birats. Geom. Linein. Sist. Konechno Porozhdennye Algebry, 220–239; translation in Proc. Steklov Inst. Math. 2003, no. 1 (240), 214–233

[BM] E. Bierstone, P. D. Milman, Resolution except for minimal singularities I, Adv. Math. **231** (2012), no. 5, 3022–3053.

[BP] E. Bierstone, F. Vera Pacheco, Resolution of singularities of pairs preserving semi-simple normal crossings, Rev. R. Acad. Cienc. Exactas Fís. Nat. Ser. A Math. RACSAM **107** (2013), no. 1, 159–188.

[EV] H. Esnault, E. Viehweg, *Lectures on vanishing theorems*, DMV Seminar, **20**. Birkhäuser Verlag, Basel, 1992.

[F1] O. Fujino, Abundance theorem for semi log canonical threefolds, Duke Math. J. **102** (2000), no. 3, 513–532.

[F2] O. Fujino, Higher direct images of log canonical divisors, J. Differential Geom. **66** (2004), no. 3, 453–479.

[F3] O. Fujino, What is log terminal?, *Flips for* 3*-folds and* 4*-folds*, 49–62, Oxford Lecture Ser. Math. Appl., **35**, Oxford Univ. Press, Oxford, 2007,

[F4] O. Fujino, Vanishing and injectivity theorems for LMMP, preprint (2007).

[F5] O. Fujino, A transcendental approach to Kollár's injectivity theorem, Osaka J. Math. **49** (2012), no. 3, 833–852.

[F6] O. Fujino, A transcendental approach to Kollár's injectivity theorem II, J. Reine Angew. Math. **681** (2013), 149–174.

[F7] O. Fujino, On injectivity, vanishing and torsion-free theorems for algebraic varieties, Proc. Japan Acad. Ser. A Math. Sci. **85** (2009), no. 8, 95–100.

[F8] O. Fujino, Introduction to the log minimal model program for log canonical pairs, preprint (2009).

[F9] O. Fujino, Introduction to the theory of quasi-log varieties, *Classification of algebraic varieties*, 289–303, EMS Ser. Congr. Rep., Eur. Math. Soc., Zürich, 2011.

[F10] O. Fujino, Non-vanishing theorem for log canonical pairs, J. Algebraic Geom. **20** (2011), no. 4, 771–783.

[F11] O. Fujino, Fundamental theorems for the log minimal model program, Publ. Res. Inst. Math. Sci. **47** (2011), no. 3, 727–789.

[F12] O. Fujino, On isolated log canonical singularities with index one, J. Math. Sci. Univ. Tokyo **18** (2011), 299–323.

[F13] O. Fujino, Fundamental theorems for semi log canonical pairs, Algebraic Geometry **2** (2014), 194–228.

[F14] O. Fujino, Semipositivity theorems for moduli problems, preprint (2012).

[F15] O. Fujino, Injectivity theorems, to appear in Adv. Stud. Pure Math.

[FF] O. Fujino, T. Fujisawa, Variations of mixed Hodge structure and semipositivity theorems, Publ. Res. Inst. Math. Sci. **50** (2014), no. 4, 589–661.

[Ft] K. Fujita, Simple normal crossing Fano varieties and log Fano manifolds, Nagoya Math. J. **214** (2014), 95–123.

[H] R. Hartshorne, Generalized divisors on Gorenstein schemes, *Proceedings of Conference on Algebraic Geometry and Ring Theory in honor of Michael Artin, Part III (Antwerp, 1992)*, K-Theory **8** (1994), no. 3, 287–339.

[Ka] Y. Kawamata, Characterization of abelian varieties, Compositio Math. **43** (1981), no. 2, 253–276.

[KMM] Y. Kawamata, K. Matsuda, K. Matsuki, Introduction to the minimal model problem, *Algebraic geometry, Sendai, 1985*, 283–360, Adv. Stud. Pure Math., **10**, North-Holland, Amsterdam, 1987.

[Ko1] J. Kollár, Higher direct images of dualizing sheaves, I. Ann. of Math. **123** (1986), 11–42.

[Ko2] J. Kollár, *Shafarevich maps and automorphic forms*, M. B. Porter Lectures. Princeton University Press, Princeton, NJ, 1995.

[Ko3] J. Kollár, *Singularities of the minimal model program*, With a collaboration of Sándor Kovács, Cambridge Tracts in Mathematics, **200**. Cambridge University Press, Cambridge, 2013.

[L] Q. Liu, *Algebraic geometry and arithmetic curves*, Translated from the French by Reinie Erné, Oxford Graduate Texts in Mathematics, **6**. Oxford Science Publications. Oxford University Press, Oxford, 2002.

[N] N. Nitsure, Construction of Hilbert and Quot schemes, *Fundamental algebraic geometry*, 105–137, Math. Surveys Monogr., **123**, Amer. Math. Soc., Providence, RI, 2005.

Department of Mathematics, Graduate School of Science, Osaka University, Toyonaka, Osaka 560–0043, Japan
E-mail address: `fujino@math.sci.osaka-u.ac.jp`

Advanced Studies in Pure Mathematics 70, 2016
Minimal Models and Extremal Rays (Kyoto, 2011)
pp. 323–360

Explicit resolution of three dimensional terminal singularities

Jungkai Alfred Chen

Dedicated to Prof. Shigefumi Mori on his 60th birthday

Abstract.

We prove that any three dimensional terminal singularity $P \in X$ can be resolved by a sequence of divisorial extractions with minimal discrepancies which are weighted blowups over points.

§1. Introduction

Terminal singularities are the smallest category that minimal model program could work in higher dimension. In fact, the development of minimal model program in dimension three was built on the understanding of three dimensional terminal singularities: Reid set up some fundamental results on canonical and terminal singularities (cf. [20, 21, 22]), Mori classified three dimensional terminal singularities explicitly (cf. [18]) and then Kollár and Mori proved the existence of flips by classifying "extremal neighborhood" (cf. [19, 13]), which is essentially the classification of singularities on a rational curve representing extremal ray. Together with the termination of flips of Shokurov (cf. [23]), one has the minimal model program in dimension three.

It is interesting, and perhaps of fundamental importance, to know those birational maps explicitly in minimal model program. For example, if X is a non-singular threefold and $X \to W$ is a divisorial contraction to a point then W could have simple singularities like $(x^2 + y^2 + z^2 + u^2 = 0)$, $(x^2 + y^2 + z^2 + u^3 = 0)$, or a quotient singularity $\frac{1}{2}(1,1,1)$ (cf. [17]). It is expected that the singularities get worse by further contractions.

Received February 21, 2012.
Revised January 25, 2013.
2010 *Mathematics Subject Classification.* 14E30, 14J17.
The author was partially supported by National Science Council and National Center for Theoretical Sciences, Taiwan.

On the other way around, given a germ of three-dimensional terminal singularity $P \in X$, it is expected that one can have a resolution by successive divisorial extractions. For example, given a terminal quotient singularity $P \in X$, one has the "economical resolution" by Kawamata blowups successively. In [5], Hayakawa shows the following

Theorem 1. *For a terminal singularity $P \in X$ of index $r > 1$, there exists a partial resolution*

$$X_n \to \ldots \to X_1 \to X_0 = X \ni P$$

such that X_n is Gorenstein and each $f_i : X_{i+1} \to X_i$ is a divisorial contraction to a point $P_i \in X_i$ of index $r_i > 1$ with minimal discrepancy $1/r_i$. All these maps f_i are weighted blowups.

It is natural to ask whether one can resolve terminal singularities of index 1 in a similar manner, after Markushevich's result that there exists a divisorial contraction with discrepancy 1 over any cDV point which is a weighted blowup (cf. [16]).

Definition 1.1. *Given a three-dimensional terminal singularity $P \in X$. We say that there exists a feasible resolution for $P \in X$ if there is a sequence*

$$X_n \to X_{n-1} \to \ldots \to X_1 \to X_0 = X \ni P,$$

such that X_n is non-singular and each $X_{i+1} \to X_i$ is a divisorial contraction to a point with minimal discrepancy, i.e. a contraction to a point $P_i \in X_i$ of index $r_i \geq 1$ with discrepancy $1/r_i$.

The purpose of this note is prove that a feasible resolution exists for three dimensional terminal singularities.

Theorem 2 (Main Theorem). *Given a three-dimensional terminal singularity $P \in X$. There exists a feasible resolution for $P \in X$.*

One might expect to construct such resolution by finding a divisorial contraction with discrepancy 1 over a terminal singularity of index 1 and combining with Theorem 1. However, there are some technical difficulties.

First of all, given a divisorial contraction $Y \to X \ni P$ with discrepancy 1 over a terminal singularity $P \in X$ of index 1, then Y usually have singularities of higher indexes. Resolving these higher index points by Hayakawa's result, one might pickup some extra singularities of index 1 in the process. However, the studies of singularities of index 1 was not there in Hayakaya's work.

Another difficulty is that singular Riemann-Roch is not sensitive to Gorenstein points. Therefore, the powerful technique introduced by

Kawakita (cf. [8, 9, 10]) to study singularities and divisorial contraction by using singular Riemann-Roch formula is not valid.

What we did in this note is basically pick convenient weighted blowups, keep good track of terminal singularities, and put them into a right hierarchy. The hierarchy is as following: 1. terminal quotient singularities; 2. cA points; 3. cA/r points; 4. cD and $cAx/2$ poitns; 5. $cAx/4$, $cD/2$, and $cD/3$ points ; 6. cE_6 points; 7. $cE/2$ points; 8. cE_7 points; 9. cE_8 points.

These involves careful elaborative case-by-case studies. The reader might find that the structure is very similar to part of Kollár work in [14]. Indeed, a lots of materials can be found in the preprints of Hayakawa [6, 7], in which he tried to classified all divisorial contractions with discrepancy 1 over a cD or cE point. Many of our choices of resolutions are inspired by his works. This work can not be done without his work in [6, 7]. For reader's convenience and for the sake of self-contained, we choose to reproduce the proofs that we needed here. The existence of divisorial contractions and explicit descriptions are already given by Hayakawa in his series of works. What is really new in this article is that we choose those convenient weighted blowups and work out the inductive process.

Acknowledgement. This work was done during the author's visit to RIMS as Visiting Professor. The author would like thank the hospitality of RIMS. We would like to thank Prof. Shigefumi Mori for the invitation and encouraging discussion during the preparation of the work. We are indebted to Hayakawa, Kawakita, Kawamata, Kollár, Matsuki for helpful discussions. The author would also like to thank Kollár for showing his work [14].

§2. Preliminaries

2.1. weighted blowups

We will need weighted blowup which are divisorial extraction with minimal discrepancy. For this purpose, we first fix some notations.

Let $N = \mathbb{Z}^n$ and $v_0 = \frac{1}{r}(a_1, ..., a_n) \in \frac{1}{r}\mathbb{Z}^n$. We write $\overline{N} := N + \mathbb{Z}v_0$. Let σ be the cone of first quadrant, i.e. the cone generated by the standard basis $e_1, ..., e_n$ and Σ be the fan consists of σ and all the subcones of σ. We have that $\mathcal{X}_0 := \mathrm{Spec}\mathbb{C}[\sigma^\vee \cap \overline{M}]$ is a quotient variety of $\mathbb{C}^n$ by the cyclic group $\mathbb{Z}/r\mathbb{Z}$, which we denote it as $\mathbb{C}^n/v_0$ or $\mathbb{C}^n/\frac{1}{r}(a_1, ..., a_n)$.

For any primitive vector $v = \frac{1}{r}(b_1, b_2, ..., b_n) \in \overline{N}$ with $b_i > 0$, we can consider the weighted blowup $\mathcal{X}_1 \to \mathcal{X}_0 := \mathbb{C}^n/v_0$ with weight v, which is the toric variety obtained by subdivision along v. More concretely, let

σ_i be the cone generated by $\{e_1, ..., e_{i-1}, v_1, e_{i+1}, ..., e_n\}$, then

$$\mathcal{X}_1 := \cup_{i=1}^n \mathcal{U}_i,$$

where $\mathcal{U}_i = \mathrm{Spec}\mathbb{C}[\sigma_i^\vee \cap \overline{M}]$. We always denote the origin of $\mathcal{U}_i$ by Q_i and the exceptional divisor $\mathcal{E} \cong \mathbb{P}((b_1, b_2, ..., b_n))$ by $\mathbb{P}(v)$.

For any semi-invariant $\varphi = \sum \alpha_{i_1,...,i_n} x_1^{i_1} ... x_n^{i_n}$, and for any vector $v = \frac{1}{r}(b_1, b_2, ..., b_n) \in \overline{N}$ we define

$$wt_v(\varphi) := \min\{\sum_{j=1}^n \frac{b_j i_j}{r} | \alpha_{i_1,...,i_n} \neq 0\}.$$

Let $X \in \mathcal{X}_0$ be a complete intersection defined by semi-invariants $(\varphi_1 = ... = \varphi_c = 0)$. Let Y be its proper transform in $\mathcal{X}_1$. By abuse the notation, we also call the induced map $f : Y \to X$ the weighted blowups of X with weight v, or denote it as $\mathrm{wBl}_v : Y \to X$. Notice that $Y \cap U_i$ is defined by $\tilde{\varphi}_1 = ... = \tilde{\varphi}_c = 0$ with

$$\tilde{\varphi}_j := \varphi(x_1 x_i^{\frac{b_1}{r}}, \ldots, x_{i-1} x_i^{\frac{b_{i-1}}{r}}, x_i^{\frac{b_i}{r}}, x_{i+1} x_i^{\frac{b_{i+1}}{r}}, \ldots, x_n x_i^{\frac{b_n}{r}}) x_i^{-wt_v(\varphi)},$$

for each j. Let $E := \mathcal{E} \cap Y \subset \mathbb{P}(v)$ denote the exceptional divisor and $U_i := \mathcal{U}_i \cap Y$.

Let $X = (\varphi_1 = \varphi_2 = ... = \varphi_c = 0) \subset \mathbb{C}^n / v_0$ be an irreducible variety such that $o = P \in X$ is the only singularities. Let $Y \to X \ni P$ be a weighted blowup with weight v and exceptional divisor $E = \mathcal{E} \cap Y \subset \mathbb{P}(v)$. We are interesting in $\mathrm{Sing}(Y)$. We may decompose it into

$$\mathrm{Sing}(Y) = \mathrm{Sing}(Y)_{\mathrm{ind}=1} \cup \mathrm{Sing}(Y)_{\mathrm{ind}>1},$$

where $\mathrm{Sing}(Y)_{\mathrm{ind}=1}$ (resp. $\mathrm{Sing}(Y)_{\mathrm{ind}>1}$) denotes the locus of singularities of index $= 1$ (resp. > 1). Clearly, the locus of points of index > 1 in $\mathcal{X}_1$ coincide with $\mathrm{Sing}(\mathbb{P}(v))$. Hence we have

$$\mathrm{Sing}(Y)_{\mathrm{ind}>1} = Y \cap \mathrm{Sing}(\mathbb{P}(v)) = E \cap \mathrm{Sing}(\mathbb{P}(v)).$$

We will need the following Lemma to determine singularities on Y of index 1.

Lemma 3. *Keep the notation as above. Consider* $\mathrm{wBl}_v : Y \to X$.

(1) *If* $\mathcal{U}_i \cong \mathbb{C}^n$, *then* $\mathrm{Sing}(Y) \cap U_i \subset \mathrm{Sing}(E) \cap U_i$.

(2) *If* Y *is a terminal threefold, then* $\mathrm{Sing}(Y)_{\mathrm{ind}=1} \subset \mathrm{Sing}(E)$

Proof. For each i, we may write $\varphi_i = \varphi_{i,h} + \varphi_{i,r}$, where $\varphi_{i,h}$ (resp. $\varphi_{i,r}$) denotes the homogeneous part (remaining part) of φ_i with weight equals to $wt_v(\varphi_i)$.

In order to prove (2), it suffices to check that $\mathrm{Sing}(Y)_{\mathrm{ind}=1} \cap U_j \subset \mathrm{Sing}(E) \cap U_j$ for each j. Without loss of generality, we work on U_n. For simplicity on notations, we also assume that X is a hypersurface.

Let $\rho_n : \mathbb{C}^n \to U_n \cong \mathbb{C}^n/\mu_r$ be the canonical projection. On $\mathbb{C}^n$, $\rho_n^{-1}(Y)$ is defined by semi-invariant

$$\tilde{\varphi} = \widetilde{\varphi}_h + \widetilde{\varphi}_r = \varphi_h(x_1, ..., x_{n-1}, 1) + \widetilde{\varphi}_r,$$

with $x_n | \widetilde{\varphi}_r$ and $\rho_i^{-1}(E) \subset \rho_n^{-1}(U_n)$ is defined by $\widetilde{\varphi}_h = x_n = 0$.

Note that $\mathrm{Sing}(Y) \subset E$ for $f : Y \to X$ is isomorphic away from $p \in X$. Also note that a quotient of a three dimensional smooth point can not be terminal singularity of index 1. Therefore,

$$\begin{aligned}
\rho_n^{-1}(\mathrm{Sing}(Y)_{\mathrm{ind}=1} \cap U_n) \quad &= \rho_n^{-1}(\mathrm{Sing}(Y)_{\mathrm{ind}=1} \cap U_n \cap E) \\
&= \rho_n^{-1}(\mathrm{Sing}(Y \cap U_n)_{\mathrm{ind}=1}) \cap \rho_n^{-1}(E) \\
&\subset \mathrm{Sing}(\rho_n^{-1}(Y \cap U_n)) \cap \rho_n^{-1}(E) \\
&= \tilde{\varphi} = \tilde{\varphi}_{x_1} \cdots = \tilde{\varphi}_{x_n} = x_n = 0 \\
&\subset \tilde{\varphi} = \tilde{\varphi}_{x_1} = \ldots = \tilde{\varphi}_{x_{n-1}} = x_n = 0 \\
&= \tilde{\varphi}_h = \tilde{\varphi}_{h,x_1} = \ldots = \tilde{\varphi}_{h,x_{n-1}} = 0 \\
&= \mathrm{Sing}(\rho_n^{-1}(E \cap U_n)),
\end{aligned}$$

where $\tilde{\varphi}_{x_i}$ denotes $\frac{\partial \tilde{\varphi}}{\partial x_i}$.

Since ρ_n is étale, therefore, $\rho_n(\mathrm{Sing}(\rho_n^{-1}(E \cap U_n))) \subset \mathrm{Sing}(E \cap U_n)$. The statement follows for hypersurface. The same argument works for higher codimension as well.

The proof for (1) also follows from the similar argument. Q.E.D.

Corollary 4. *Let Y be a terminal threefold obtained by* $\mathrm{wBl}_v : Y \to X$ *with weight v. Suppose that E is a quasi-smooth weighted complete intersection in $\mathbb{P}(v)$, then* $\mathrm{Sing}(Y)_{\mathrm{ind}=1} = \emptyset$.

Proof. If E is a quasi-smooth weighted complete intersection in $\mathbb{P}(v)$, then $\mathrm{Sing}(E) = E \cap \mathrm{Sing}(\mathbb{P}(v))$. Therefore,

$$\mathrm{Sing}(Y)_{\mathrm{ind}=1} \subset \mathrm{Sing}(E) \subset \mathrm{Sing}(\mathbb{P}(v)).$$

However, $\mathrm{Sing}(\mathbb{P}(v))$ consists of quotient singularities of index > 1. This implies in particular that $\mathrm{Sing}(Y)_{\mathrm{ind}=1} = \emptyset$. Q.E.D.

2.2. weighted blowup of threefolds

Given a threefold terminal singularity $P \in X = (\varphi = 0) \subset \mathbb{C}^4/v_0$ of index r, we usually consider weighted blowup $\mathrm{wBl}_v : Y \to X$ with weight $v = \frac{1}{r}(a_1, a_2, a_3, a_4)$ and $a_i \in \mathbb{Z}_{>0}$ [1]. It worths to determine when a weighted blowup is a divisorial contraction.

Theorem 5. *Let $P \in X = (\varphi = 0) \subset \mathbb{C}^4$ be a germ of three dimensional terminal singularity and $f = \mathrm{wBl}_v : Y \to X$ with weight $v = \frac{1}{r}(a_1, a_2, a_3, 1)$ with exceptional divisor $E \subset \mathbb{P}(v)$. Suppose that*

- *E is irreducible;*
- *$\frac{1}{r}\sum a_i - wt_v(\varphi) - 1 = \frac{1}{r}$;*
- *either $Y \cap U_4$ is terminal or E has Du Val singularities on U_4.*

Then $Y \to X$ is a divisorial contraction.

Moreover, $\mathrm{Sing}(Y)_{\mathrm{ind}=1} \cap U_4 \subset \mathrm{Sing}(E) \cap U_4$. *For any $R \in \mathrm{Sing}(Y)_{ind=1} \cap U_4$, R is at worst of type cA (resp. cD, cE_6, cE_7, cE_8) if $R \in \mathrm{Sing}(E)$ is of type A (resp. D, E_6, E_7, E_8).*

Proof. Suppose that E is irreducible, then $K_Y = f^*K_X + a(E,X)E$ with $a(E,X) = \frac{1}{r}\sum a_i - wt_v(\varphi) - 1$. Let $D = (x_4 = 0) \subset Div(X)$ and D_Y be its proper transform in Y. One has $f^*D = D_Y + \frac{1}{r}E$. Hence $\frac{1}{r} = \frac{1}{r}\sum a_i - 1 - wt_v(\varphi)$ implies that

$$f^*(K_X + D) = K_Y + D_Y$$

and hence $D_Y \sim_X -K_Y$.

Let $g : Z \to Y$ be a resolution of Y. For any exceptional divisor F in Z such that $g(Z) \subset D_Y$. One has $g^*D_Y = D_Z + mF + \ldots$ for some $m > 0$. It follows that $a(F, Y) = m > 0$. Hence Y is terminal if $Y - D_Y = Y \cap U_4$ is terminal.

In fact, by Lemma 3.(1), one sees that $\mathrm{Sing}(Y) \cap U_4 \subset \mathrm{Sing}(E) \cap U_4$. If $E \cap U_4$ is DuVal, then $\mathrm{Sing}(E) \cap U_4$ is isolated hence so is $\mathrm{Sing}(Y) \cap U_4$. More precisely, for $R \in \mathrm{Sing}(E) \cap U_4$ with local equation

$$\psi := \varphi_h(x_1, ..., x_4)|_{x_4=1}$$

of type A (resp. D, E), the local equation for $R \in Y \cap U_4$ is of the form

$$\psi + x_4 g(x_1, ..., x_4),$$

which is a compound DuVal equation. Therefore, if R is singular in Y, then R is a at worst isolated cDV of type cA (resp. cD, cE) [2]. By results of Reid [20], Kollár and Shephard-Barron [15], an isolated cDV

[1]Divisorial contractions to a point of index $r > 1$ have been studied extensively by Hayakawa and are known to be weighted blowups

[2]If there are lower degree terms appearing in g, then the singularity R could be simpler or even non-singular

singularity is terminal. Hence Y is terminal and therefore $f : Y \to X$ is a divisorial contraction. Q.E.D.

In the sequel, all weighted blowups with discrepancy $1/r$ over a terminal singularity of index r can easily checked to be divisorial contractions with minimal discrepancy $1/r$ by the above Theorem 5 or by direct computation.

We will need some further easy but handy Lemmas.

Lemma 6. *Let* $\mathrm{wBl}_v : Y \to X$ *be a divisorial contraction. If* $wt_v(x^2) = wt_v(\varphi)$*, then* $\mathrm{Sing}(Y) \cap U_1 = \emptyset$.

Proof. Now E is defined by $(\Phi : \mathbf{x}^2 + f(\mathbf{y}, \mathbf{z}, \mathbf{u}) = 0) \subset \mathbb{P}(v)$. It is clear that $\mathrm{Sing}(Y)_{\mathrm{ind}=1} \cap U_1 \subset \mathrm{Sing}(E) \cap U_1 = \emptyset$. Notice also that Q_1 is the only point in U_1 with index > 1 and $Q_1 \notin Y$. Hence $\mathrm{Sing}(Y) \cap U_1 = \emptyset$. Q.E.D.

Lemma 7. *Let* $\mathrm{wBl}_v : Y \to X$ *be a divisorial contraction. If* $x_i^m x_j \in \varphi$ *with* $wt_v(x_i^m x_j) = wt_v(\varphi)$ *or* $x_i^m \in \varphi$ *with* $wt_v(x_i^m) = wt_v(\varphi) + 1$*, then* $Y \cap U_i$ *is non-singular away from* Q_i *and* Q_i *is either non-singular or a terminal quotient singularity of index* $wt_v(x_i)$*. In particular,* $\mathrm{Sing}(Y)_{\mathrm{ind}=1} \cap U_i = \emptyset$.

Proof. On U_i, $Y \cap U_i$ is given by $(\tilde{\varphi} = 0) \subset \mathbb{C}^4/\mathbb{Z}_{a_i}$ with

$$\tilde{\varphi} = \begin{cases} x_j + \text{others}, & \text{if } wt_v(x_i^m x_j) = wt_v(\varphi); \\ x_i + \text{others}, & \text{if } wt_v(x_i^m) = wt_v(\varphi) + 1. \end{cases}$$

Hence $Y \cap U_i \cong \mathbb{C}^3/\mathbb{Z}_{a_i}$ and the statement follws. Q.E.D.

Lemma 8. *Consider* $X = (\varphi = 0) \subset \mathbb{C}^4$*. Suppose that* $R \in X$ *is an isolated singularity and* $xy \in \varphi$ *or* $x^2 + y^2 \in \varphi$*. Then* R *is of* cA*-type.*

Proof. Up to a unit, we may assume that $\varphi = xy + xg(x, z, u) + yh(y, z, u) + \bar{\varphi}(z.u)$. Since $\frac{\partial^2 \varphi}{\partial x \partial y}(R) = 1 \neq 0$, the local expansion near R is of the form $\bar{x}\bar{y} + \bar{f}(\bar{z}, \bar{u})$ where $\bar{x} = x - x(R)$ respectively. Also $mult_o \bar{f} \geq 2$. Hence it is a cA point. Q.E.D.

Corollary 9. *Consider* $X = (\varphi = 0) \subset \mathbb{C}^4/\mathbb{Z}_r$*. Suppose that* $R \in \mathrm{Sing}(X)_{\mathrm{ind}=1}$ *is an isolated singularity and either* $xy \in \varphi$ *or* $x^2 + y^2 \in \varphi$*. Then* R *is of* cA*-type.*

Proof. Let $\pi : \mathbb{C}^4 \to \mathbb{C}^4/\mathbb{Z}_r$ is the quotient map. Since R is a index 1 point, $\pi^{-1}(R)$ does not intersects the fixed locus of the $\mathbb{Z}_r$ action. This implies in particular that $\mathbb{Z}_r$ acts on $\pi^{-1}(R)$ freely and each point of $Q \in \pi^{-1}(R)$ is singular in $\pi^{-1}(X)$. By Lemma 8, Q is of type cA and hence so is R. Q.E.D.

By the similar argument, one can also see the following

Lemma 10. *Consider* $X = (\varphi = 0) \subset \mathbb{C}^4/\mathbb{Z}_r$ *with* $r = 1, 2, 3$. *Suppose that* $\varphi = x^2 + f(y, z, u)$ *with* f_3, *the 3-jet of* f, *is nonzero and not a cube. Let* $R \in \mathrm{Sing}(X)$ *be an isolated singularity. Then* R *could only be of type* $cA, cA/r, cD, cD/r$ *or a terminal quotient singularity.*

§3. resolution of cA and cA/r points

Lemma 11. *Let* $f : Y \to X$ *be the economic resolution of a terminal quotient singularity* $P \in X$. *Then this is a feasible resolution for* $P \in X$.

Proof. Given a terminal quotient singularity $P \in X$ of type $\frac{1}{r}(s, r-s, 1)$ with $s < r$ and $(s, r) = 1$, we start by considering the Kawamata blowup $Y \to X$ (cf. [11]), i.e. weighted blowup with weight $v = \frac{1}{r}(s, r-s, 1)$ over P. It is clearly a divisorial contraction with minimal discrepancy $\frac{1}{r}$.

Note that $\mathrm{Sing}(Y)$ consists of at most two points Q_1, Q_2 of index $s, r-s$ respectively. By induction on r, we get a resolution $Y = Y_{r-1} \to \ldots \to Y \to X \ni P$. It is easy to see that this is the economic resolution.
Q.E.D.

Theorem 12. *There is a feasible resolution for any* cA *points.*

Proof. For any cA point $p \in X$, there is an embedding $j : X \subset \mathbb{C}^4$ such that $P \in X$ is defined by $(\varphi : xy + f(z, u) = 0) \subset \mathbb{C}^4$. We fix this embedding once and for all and define

$$\tau(P) := \min\{i + j | z^i u^j \in f(z, u)\}.$$

We may and do assume that $z^\tau \in f$. We write $f = f_\tau + f_{>\tau}$, where f_τ denote the homogeneous part of weight τ.

We need to introduce

$$\tau^\sharp(P) := \min\{i + j | z^i u^j \in f(z, u), i \leq 1\}.$$

Since $P \in X$ is isolated, one has that f contains a term of the type zu^{p-1} or u^p for some p. Hence $\tau^\sharp(P)$ is well-defined. Notice also that $\tau(P) \leq \tau^\sharp(P)$.

We shall prove by induction on τ and $\tau^\sharp$.

Case 1. $\tau = 2$.

By easy change of coordinates, we may and do assume that $f(z, u) = z^2 + u^b$. We take $Y \to X$ to be the weighted blowup with weights $(1, 1, 1, 1)$ (or the usual blowup over P). It is clear that $\mathrm{Sing}(Y) = \{Q_4\}$, which is defined by

$$(\tilde{\varphi} : xy + z^2 + u^{b-2} = 0) \subset \mathbb{C}^4.$$

By induction on b, we are done.

Case 2. $\tau > 2$.

We may write $f_\tau = \prod(z - \alpha_t u)^{l_t}$ since $z^\tau \in f$. We take $\mathrm{wBl}_v : Y_1 \to X$ with weights $v = (1, \tau - 1, 1, 1)$. It is clear that $\mathrm{Sing}(Y)_{\mathrm{ind}>1} = \{Q_2\}$, which is a terminal quotient singularity of index $\tau - 1$. Hence it remains to consider index 1 points.

We have that

$$E = (\mathbf{xy} + \prod(\mathbf{z} - \alpha_t\mathbf{u})^{l_t} = 0) \subset \mathbb{P}(1, \tau - 1, 1, 1).$$

Now $\mathrm{Sing}(E) = \{R_t = (0, 0, \alpha_t, 1)\}_{l_t \geq 2}$. In fact, for any R_t with $l_t \geq 2$, one sees that $R_t \subset E$ is a singularity of A-type, it follows that if R_t is singular in Y, then it is of type cA with $\tau(R_t) \leq l_t$.

Subcase 2-1. f_τ factored into more than one factors.

Then $\tau(R_t) \leq l_t < \tau$, then we are done by induction on τ.

Subcase 2-2. If f_τ factored into only one factor.

We may assume $f_\tau = z^\tau$ by changing coordinates. It is easy to see that $\tau(Q_4) \leq \tau^\sharp(Q_4) < \tau^\sharp(P)$ hence

$$\tau(Q_4) + \tau^\sharp(Q_4) < \tau(P) + \tau^\sharp(P).$$

Then we are done by induction on $\tau + \tau^\sharp$. Q.E.D.

Corollary 13. *There is a feasible resolution for any cA/r point.*

Proof. Given $P \in X$ defined by

$$(\varphi : xy + f(z^r, u) = xy + \sum a_{ij} z^{ir} u^j = 0) \subset \mathbb{C}^4 / \frac{1}{r}(s, r - s, 1, r).$$

Let

$$\begin{cases} \kappa^\sharp(\varphi) := \min\{k | u^k \in f\}, \\ \kappa(\varphi) := \min\{i + j | z^{ir} u^j \in f\}. \end{cases}$$

We shall prove by induction on $\kappa^\sharp + \kappa$. Note that there is some $u^k \in f$ otherwise P is not isolated. Thus $\kappa^\sharp + \kappa$ is finite and $\kappa \leq \kappa^\sharp$.

1. $\kappa^\sharp = 1, \kappa = 1$.

Then $P \in X$ is a terminal quotient singularity. We are done.

2. $\kappa^\sharp + \kappa > 2$.

We always consider $Y \to X$ the weighted blowup with weights $\frac{1}{r}(s, \kappa r - s, 1, r)$, which is a divisorial contraction by [4]. Computation on each charts similarly, one sees the following:

(1) $Y \cap U_1$ is singular only at Q_1, which is a terminal quotient singularity of index s (non-singular on U_1 if $s = 1$).
(2) $Y \cap U_2$ is singular only at Q_2, which is a terminal quotient singularity of index $\kappa r - s$ (non-singular on U_2 if $\kappa r - s = 1$).

(3) $Y \cap U_3$ is defined by $xy + f(z, uz)z^{-\kappa} = 0 \subset \mathbb{C}^4$. Hence $\mathrm{Sing}(Y) \cap U_3$ must be of type cA by Lemma 8. There exists a feasible resolution over these points.
(4) it remains to consider Q_4, which is locally defined by

$$(\tilde{\varphi} : xy + \sum a_{ij} z^{ir} u^{i+j-\kappa} = 0) \subset \mathbb{C}^4 / \frac{1}{r}(s, r-s, 1, r).$$

In fact, one sees that $\kappa^\sharp(Q_4) = \kappa^\sharp(P) - \kappa(P)$ and $\kappa(Q_4) \leq \kappa(P)$.

By induction on $\kappa^\sharp + \kappa$, we have a feasible resolution over Q_4. Together with feasible resolution over other singularities on Y, we have a feasible resolution over Y and hence over X. Q.E.D.

§4. resolution of cD and $cAx/2$ points

Given a cD_n point $P \in X$ which is defined by $(\varphi : x^2 + y^2 z + z^{n-1} + ug(x, y, z, u) = 0) \subset \mathbb{C}^4$ for some $n \geq 4$. We start by considering the *normal form* of cD singularities.

Definition 4.1. *We say that a cD point $P \in X$ admits a normal form if there is an embedding*

$$(\varphi : x^2 + y^2 z + \lambda y u^l + f(z, u) = 0) \subset \mathbb{C}^4$$

with the following properties:

(1) $l \geq 2$. *(We adapt the convention that $l = \infty$ if $\lambda = 0$.)*
(2) $zu^{p-1} \in f$ *or* $u^p \in f$ *for some* $p > 0$ *if* $\lambda = 0$.
(3) $z^{q-1}u \in f$ *or* $z^q \in f$ *for some* $q > 0$.

An isolated singularity $P \in X$ given by this form (with $l \geq 0$ and possibly not of cD type) is called a cD-like singularity, which is terminal.

For a cD-like singularity $P \in X$, we define

$$\begin{cases} \mu^\sharp(P \in X) := \min\{2i + j | z^i u^j \in \varphi, i = 0 \text{ or } 1\}; \\ \mu(P \in X) := \min\{2i + j | z^i u^j \in \varphi\}; \\ \mu^\flat(P \in X) := \min\{\mu(P \in X), 2l - 2\}; \\ \tau^\sharp(P \in X) := \min\{i + j | z^i u^j \in \varphi, i = 0 \text{ or } 1\}. \end{cases}$$

Clearly, one has $\mu^\flat \leq \mu \leq \mu^\sharp \leq \infty$. Also $\mu^\sharp, \tau^\sharp < \infty$ if $\lambda = 0$.

Lemma 14. *Given a cD-like point $P \in X$ defined by*

$$(\varphi \colon x^2 + y^2 z + \lambda y u^l + f(z, u) = 0) \subset \mathbb{C}^4,$$

with $\mu^\flat \leq 3$. Then there exists a feasible resolution for $P \in X$.

Proof. If there is a linear or quadratic term in f, then P is non-singular or of cA-type by Lemma 8. In particular, feasible resolution exists. We thus assume that $l \geq 2$ and we may write

$$f(z,u) = f_3(z,u) + f_{\geq 4}(z,u),$$

where $f_3(z,u)$ (resp. $f_{\geq 4}(z,u)$) is the 3-jet (resp. 4 and higher jets) of $f(z,u)$.

Case 1. $l \geq 3$.

If $\mu \leq 2$, then P is at worst of type cA. Thus we may and do assume that $\mu = 3$ and hence $u^3 \in f_3 \neq 0$. Clearly, $\varphi_3 = y^2z + f_3$ is irreducible.

Subcase 1-1. f_3 is factored into more than one factors.

We consider $\mathrm{wBl}_v : Y \to X$ with weight $v = (2,1,1,1)$. One can verify that $E = (\mathbf{y}^2\mathbf{z} + f_3(\mathbf{z},\mathbf{u}) = 0) \subset \mathbb{P}(2,1,1,1)$ is irreducible. By Lemma 7, one has that $\mathrm{Sing}(Y) \cap U_i$ is non-singular away from Q_i for $i = 1, 2$. In fact, Y is non-singular at Q_2. Therefore, $Y \to X$ is a divisorial contraction by Theorem 5.

Since $Q_4 \notin Y$, it remains to consider $Y \cap U_3$, which is defined by

$$\begin{aligned} \tilde{\varphi}: \quad & x^2z + y^2 + \lambda yz^{l-2}u^l + \tilde{f}_3(z,u) + \tilde{f}_{\geq 4} \\ & = x^2z + y^2 + \lambda yz^{l-2}u^l + \prod(u-\alpha_t)^{l_t} + \tilde{f}_{\geq 4}, \end{aligned}$$

where $\tilde{f}_3(z,u)$ (resp. $\tilde{f}_{\geq 4}(z,u)$) denotes the proper transform of $f_3(z,u)$ (resp. $f_{\geq 4}(z,u)$). More explicitly, $\tilde{f}_3(z,u) = f_3(z,zu)z^{-wt_v(\varphi)}$.

Let R be a singular point in $\mathrm{Sing}(Y) \cap U_3$. If f_3 is factored into more than one factors, then $l_t \leq 2$ for all t. It follows that R is at worst of cA type by Lemma 8. Notice also that $\mathrm{Sing}(Y)_{\mathrm{ind}>1} = \{Q_1\}$ which is a quotient singularity of index 2. Thus feasible resolution exists.

Subcase 1-2. f_3 is factored into one factor.

We thus assume that $f_3 = (u - \alpha z)^3$. Change coordinate by

$$\begin{cases} \bar{u} = u - \alpha z; \\ \bar{y} = y + \frac{\lambda}{2}\sum_{j \geq 1} C^l_j \alpha^j z^{j-1} \bar{u}^{l-j}; \end{cases}$$

then we have

$$\varphi = x^2 + \bar{y}^2 z + \lambda \bar{y}\bar{u}^l + \bar{u}^3 + \bar{f}_{\geq 4}(z,\bar{u}).$$

Therefore, we may and do assume that $f_3 = u^3$ in the normal form.

We consider again $\mathrm{wBl}_v : Y \to X$ with weight $v = (2,1,1,1)$. Since

$$E = (\mathbf{y}^2\mathbf{z} + \mathbf{u}^3 = 0) \subset \mathbb{P}(2,1,1,1),$$

therefore $Y \to X$ is a divisorial contraction by Theorem 5 (where U_4 is replaced by U_2). Moreover, $\mathrm{Sing}(E) \subset (\mathbf{y} = \mathbf{u} = 0)$, hence $\mathrm{Sing}(E) \subset$

$U_1 \cup U_3$. Together with Lemma 7, it remains to consider Q_3, which is a cD point given by

$$\tilde{\varphi} : x^2 z + y^2 + \lambda y u^l z^{l-2} + u^3 + \tilde{f}_{\geq 4}.$$

Change coordinate by $\bar{y} := x, \bar{x} := y + \frac{1}{2}\lambda u^l = z^{l-2}$, one sees that Q_3 is cD-like given by

$$\tilde{\varphi} : \bar{x}^2 + \bar{y}^2 z + u^3 + \tilde{f}_{\geq 4} - \frac{1}{4}\lambda^2 u^{2l} z^{2l-4}.$$

Clearly, Q_3 is still in Subcase 1-2 and $\tau(Q_3) \leq \tau(P) - 2$. By induction on τ, we conclude that feasible resolution exists for this case.

Case 2. $l = 2$.
Subcase 2-1. $f_3 = 0$.
We consider $\mathrm{wBl}_v : Y \to X$ with weight $v = (2, 2, 1, 1)$. Now

$$E = (\mathbf{x}^2 + \lambda \mathbf{y}\mathbf{u}^2 + f_4(\mathbf{z}, \mathbf{u}) = 0) \subset \mathbb{P}(2, 2, 1, 1)$$

is clearly irreducible. By considering $Y \cap U_4$, which is nonsingular, one has that $Y \to X$ is a divisorial contraction by Lemma 7 and Theorem 5.

One sees that $\mathrm{Sing}(E) \subset (\mathbf{x} = \mathbf{u} = 0)$, hence $\mathrm{Sing}(E) \subset U_2 \cup U_3$. Notice also that $Q_1 \notin Y$ and $Q_2 \in Y$ is a $cA/2$ point. Together with Lemma 8, it remains to consider Q_3, which is a cD-like point and still in Subcase 2-1. Clearly, $\tau(Q_3) \leq \tau(P) - 2$. By induction on τ, we conclude that feasible resolution exists for this case.
Subcase 2-2. $f_3 \neq 0$ and $\varphi_3 = y^2 z + \lambda y u^2 + f_3$ is irreducible.
We consider $\mathrm{wBl}_v : Y \to X$ with weights $v = (2, 1, 1, 1)$. Now

$$E = (\mathbf{y}^2 \mathbf{z} + \lambda \mathbf{y}\mathbf{u}^2 + f_3(\mathbf{z}, \mathbf{u}) = 0) \subset \mathbb{P}(2, 1, 1, 1).$$

One has that $Y \to X$ is a divisorial contraction by the same reason. By Lemma 7, it's clear that $\mathrm{Sing}(Y) \cap (U_1 \cup U_2 \cup U_4) = \{Q_1\}$, which is a quotient singularity of index 2.

It remains to consider Q_3. If f_3 is factored into more than one factors then the same argument in Subcase 1-1 works. We thus assume that $f_3 = (\beta u + \alpha z)^3$. In fact, one sees that Q_3 is singular only when $f_3 = u^3$. Argue as in Subcase 1-2. We have a feasible resolution for $P \in X$.
Subcase 2-3. $f_3 \neq 0$ and $\varphi_3 = y^2 z + \lambda y u^2 + f_3$ is reducible.
In this situation, $y^2 z + \lambda y u^2 + f_3 = (y + l(z, u))(zy + \lambda u^2 - l(z, u)z)$ for some linear form $l(z, u) \neq 0$. Let $\bar{y} = y + l(z, u)$, then we have

$$\varphi_3 = \bar{y}^2 z + \lambda \bar{y} u^2 - 2\bar{y} z l(z, u).$$

We consider weighted blowup $Y \to X$ with weights $(2,2,1,1)$. Now

$$E = (\mathbf{x}^2 + \lambda\mathbf{y}\mathbf{u}^2 - 2\mathbf{y}\mathbf{z}l(\mathbf{z},\mathbf{u}) + f_4(\mathbf{z},\mathbf{u}) = 0) \subset \mathbb{P}(2,2,1,1),$$

is clearly irreducible. By considering $Y \cap U_4$, which is nonsingular by Lemma 7, one has that $Y \to X$ is a divisorial contraction by Theorem 5. Since $l(z,u) \neq 0$, one sees that $Y \cap U_2$ has at worst $cA/2$ singularities and $Y \cap U_3$ has at worst cA singularities. Therefore feasible resolution exists. Q.E.D.

By [16, Proposition 1.3], we have that

(1) if $P \in X$ is cD_4, then $\varphi = x^2 + \varphi_3(y,z,u) + \varphi_{\geq 4}(y,z,u)$ with $\varphi_3(y,z,u)$ is not divisible by a square of a linear form;

(2) if $P \in X$ is cD_n with $n \geq 5$, then $\varphi = x^2 + y^2z + \varphi_{\geq 4}(y,z,u)$.

Therefore, the plan is as following: for cD_4 points, the parallel argument as in Lemma 14 works. For $cD_n \geq 5$ points, which always admits normal forms, we prove by induction on $\mu^\flat$. We will need to consider $cAx/2$ points simultaneously in the induction.

Proposition 15. *There is a feasible resolution for any cD_4 singularity.*

Proof. We have $y^2z, z^3 \in \varphi_3$. Replacing z by $z+u$ and completing square, we may and do assume that $\varphi_3 = y^2z + \lambda yu^2 + f_3(z,u)$, with $z^3 \in f_3$ and

$$\varphi = x^2 + y^2z + \lambda yu^2 + f_3(z,u) + \varphi_{\geq 4}(y,z,u).$$

Case 1. $\lambda = 0$ and φ_3 is irreducible.
We can work as in Subcase 1-1 and 1-2 of Lemma 14.
Case 2. $\lambda = 0$ and φ_3 is reducible.
In this situation, $y^2z + f_3 = z(y^2 + q(z,u))$ for some quadratic form $q(z,u) \neq 0$. We consider weighted blowup $Y \to X$ with weights $v = (2,1,2,1)$. Now

$$E = (\mathbf{x}^2 + \mathbf{y}^2\mathbf{z} + \text{possibly others} = 0) \subset \mathbb{P}(2,2,1,1),$$

is clearly irreducible. By considering $Y \cap U_2$, which is nonsingular by Lemma 7, one has that $Y \to X$ is a divisorial contraction by Theorem 5.

Since $z^3 \in \varphi_3$, one sees that $Y \cap U_3$ define by

$$(\tilde{\varphi} : x^2 + y^2 + z^2 + \text{others} = 0) \subset \mathbb{C}^4/\frac{1}{2}(2,1,1,1),$$

which has at worst $cA/2$ singularities.

It remains to consider Q_4. Since $P \in X$ is isolated, one sees that there exists yu^p, zu^p or $u^p \in \varphi$ for some p. It follows that there exists yu^{p-1}, zu^{p-2} or $u^{p-3} \in \tilde{\varphi}$ in $Y \cap U_4$. Hence feasible resolution exists by induction on p.
Case 3. $\lambda \neq 0$. We can work as in Subcase 2-2 and 2-3 of Lemma 14. Note that $z^3 \in f_3$ hence Subcase 2-1 can not happen. Q.E.D.

Lemma 16. *A singularity $P \in X$ of type $cD_{n\geq 5}$ admits a normal form with $l \geq 3$.*

Proof. It is straightforward to solve for formal power series $\bar{y} = y + y_2 + y_3 + \ldots$ and $\bar{z} = z + z_2 + z_3 + \ldots$ satisfying

$$x^2 + \bar{y}^2\bar{z} + \lambda\bar{y}u^l + f(\bar{z}, u) = x^2 + y^2z + g_{\geq 4}(y, z, u),$$

where $y_k = y_k(z, u)$ and $z_k = z_k(y, z, u)$ are the k-th jets and $l = \min\{k|yu^k \in g_{\geq 4}(y, z, u)\}$. By Artin's Approximation Theorem [1], this gives an embedding as desired.

Observe that $\varphi = 0$ is singular along the line $(x = y = z = 0)$ if $z^2|f(z, u)$. Similarly, $\varphi = 0$ is singular along the line $(x = y = u = 0)$ if $u^2|f(z, u)$. Since $P \in X$ is isolated, it follows that $z^{q-1}u \in \varphi$ or $z^q \in \varphi$ for some $q > 0$ and $zu^{p-1} \in \varphi$ or $u^p \in \varphi$ for some $p > 0$ if $\lambda = 0$. Q.E.D.

In order to obtain a feasible resolution for $cD_{n\geq 5}$ points in general, we will need to consider $cAx/2$ point as well. Given a $cAx/2$ point $P \in X$, with an embedding

$$(\varphi\colon x^2 + y^2 + f(z, u) = 0) \subset \mathbb{C}^4/\frac{1}{2}(1, 0, 1, 1),$$

we define

$$\tau(P \in X) := \min\{i + j|z^iu^j \in f\}.$$

Note that $f(z, u)$ is $\mathbb{Z}_2$-invariant and hence consists of even degree terms only. We set $\tau' := \tau/2 \in \mathbb{Z}$.

For inductive purpose, we start by considering points with τ small.

Lemma 17. *Given a $cAx/2$-like point P defined by*

$$(\varphi\colon x^2 + y^2 + f(z, u) = 0) \subset \mathbb{C}^4/\frac{1}{2}(1, 0, 1, 1),$$

with $\tau \leq 2$. Suppose that P is terminal. Then P is non-singular or $cA/2$. In any case, feasible resolutions exist for such points.

Proof. If $\tau = 0$, then P is clearly non-singular. If $\tau = 2$, then we may assume that $z^2 \in f(z,u)$. Hence it is a $cA/2$ point. By Corollary 13, feasible resolution exists. Q.E.D.

We are now ready to handle $cAx/2$ and cD points.

Proposition 18. *Given a $cAx/2$ point $P \in X$ with $\tau(P \in X) = \tau_0 \geq 4$. Suppose that feasible resolutions exist for cD-like point with $\mu^\flat < \tau_0$ and feasible resolutions exist for $cAx/2$-like point with $\tau < \tau_0$. Then there is a feasible resolution for $P \in X$.*

Proof. Let $f_{\tau_0}(z,u) = \sum_{i+j=\tau_0} a_{ij} z^i u^j$. It can be factored into $\prod_{t\in T}(\alpha_t z + \beta_t u)^{m_t}$, where m_t denotes the multiplicities with $\sum_t m_t = \tau_0$.

Case 1. $f_{\tau_0}(z,u)$ is not a perfect square.

Depending on the parity of $\tau_0/2$, we first consider $\mathrm{wBl}_v : Y \to X$ with weights $v = \frac{1}{2}(\frac{\tau_0}{2}, \frac{\tau_0}{2}+1, 1, 1)$ or $\frac{1}{2}(\frac{\tau_0}{2}+1, \frac{\tau_0}{2}, 1, 1)$. It is a divisorial contraction with minimal discrepancy $\frac{1}{2}$ (cf. [4, Theorem 8.4]). Without loss of generality, we study the first case.

Now $E = (\mathbf{x}^2 + f_{\tau_0}(\mathbf{z},\mathbf{u}) = 0) \subset \mathbb{P}(\frac{\tau_0}{2}, \frac{\tau_0}{2}+1, 1, 1)$. Easy computation yields the following:

(1) $Y \cap U_1$ is non-singular and $Y \cap U_2$ is singular only at Q_2, which is a terminal quotient singularity of index $\frac{\tau_0}{2}+1$.

(2) $\mathrm{Sing}(Y)_{\mathrm{ind}=1} \cap U_3 \subset \{R_t\}_{m_t \geq 2, (\alpha_t,\beta_t)\neq(1,0)}$. Each R_t is defined by $x^2 + y^2 z + \mathit{unit} \cdot \bar{u}^{m_t} + \tilde{f}_{\geq \tau_0}(z,\bar{u}) = 0 \subset \mathbb{C}^4$, where $\bar{u} := u + \alpha_t/\beta_t$. This is a cD point with $\mu^\flat(R_t) \leq m_t$.

(3) Similarly, $\mathrm{Sing}(Y)_{\mathrm{ind}=1} \cap U_4 \subset \{R_t\}_{m_t \geq 2, (\alpha_t,\beta_t)\neq(0,1)}$. Each R_t is defined by $x^2 + y^2 u + \mathit{unit} \cdot \bar{z}^{m_t} + \tilde{f}_{\geq \tau_0}(\bar{z},u) = 0 \subset \mathbb{C}^4$, where $\bar{z} := z + \beta_t/\alpha_t$. This is a cD point with $\mu^\flat(R_t) \leq m_t$.

As a summary, one sees that $\mathrm{Sing}(Y) \subset \{Q_2, R_t\}_{m_t \geq 2}$. Notice $|T| = 1$ would implies that f_{τ_0} is a perfect square, which is a contradiction. Hence we may assume that $|T| > 1$ and therefore $m_t < \tau_0$ for all t. We can take a feasible resolution for each R_t and Q_2 to obtain the required feasible resolution for $P \in X$.

Case 2. $f_{\tau_0}(z,u) = (h_{\tau_0/2}(z,u))^2$ is a perfect square.

We need to make a coordinate change so that $P \in X$ is rewritten as

$$(x^2 + 2xh_{\tau_0/2}(z,u) + y^2 + f_{\tau_0+1}(z,u) + f_{>\tau_0+1}(z,u) = 0) \subset \mathbb{C}^4.$$

Depending on the parity of τ_0, we consider $\mathrm{wBl}_v : Y_1 \to X$ with weights $v = \frac{1}{2}(\frac{\tau_0}{2}+2, \frac{\tau_0}{2}+1, 1, 1)$ or $\frac{1}{2}(\frac{\tau_0}{2}+1, \frac{\tau_0}{2}+2, 1, 1)$. Without loss of generality, we study the first case. Now $E = (\mathbf{y}^2 + 2\mathbf{x}h_{\tau_0/2} + f_{\tau_0+1}(\mathbf{z},\mathbf{u}) = 0) \subset \mathbb{P}(\frac{\tau_0}{2}+2, \frac{\tau_0}{2}+1, 1, 1)$.

Easy computation yields the following:

(1) $Y \cap U_2$ is non-singular and $Y_1 \cap U_1$ is singular only at Q_1, which is a terminal quotient singularity of index $\frac{\tau_0}{2} + 2$.

(2) $Y \cap U_3$ is defined by

$$(\tilde{\varphi} : x^2 z + 2xh_{\tau_0/2}(1, u) + y^2 + \tilde{f}_{\geq \tau_0+1} = 0) \subset \mathbb{C}^4.$$

For any singularity $R \in \mathrm{Sing}(Y)_{\mathrm{ind}=1} \cap U_3$, we write $R = (0, 0, \alpha, \beta)$ and consider that coordinate change that $\bar{x} := y, \bar{y} := x, \bar{z} := z - \alpha, \bar{u} := u - \beta$. Then R is at worst a cD-like point and $\mu^\flat(R) \leq \tau_0 - 2$.

(3) The same holds for singularity in $\mathrm{Sing}(Y)_{\mathrm{ind}=1} \cap U_4$.

As a summary, one sees that $\mathrm{Sing}(Y)$ consist of a terminal quotient singularity Q_1 and possibly some cD-like points R_t with $\mu^\flat(R_t) < \tau_0$. We can take a feasible resolution for each R_t and Q_1 to obtain the required feasible resolution for $P \in X$. Q.E.D.

Proposition 19. *Given a cD-like point with $\mu^\flat(P \in X) = \mu_0$. Suppose that feasible resolutions exist for cAx/2-like point with $\tau \leq \mu_0$ and feasible resolutions exist for cD-like point with $\mu^\flat < \mu_0$. Then there is a feasible resolution for $P \in X$.*

Proof. We always fix a normal form once and for all. By Lemma 14, we may assume that $\mu_0 \geq 4$. We set $\mu' := \lfloor \frac{\mu_0}{2} \rfloor$. We consider divisorial contraction $Y \to X$ with weights $(\mu', \mu' - 1, 2, 1)$. Recall that $P \in X$ is given by

$$(\varphi : x^2 + y^2 z + \lambda y u^l + \sum a_{ij} z^i u^j = 0).$$

We may write $f_{2\mu'} := \sum_{2i+j=2\mu'} a_{ij} z^i u^j = \prod_t (\alpha_t z + \beta_t u^2)^{m_t}$.

Case 1. $\lambda = 0$.

It is straightforward to see that only singularity on $U_1 \cup U_2$ is Q_2, which is a terminal quotient singularity. On $U_3 \cup U_4$, for any singularity $R \in \mathrm{Sing}(Y)_{\mathrm{ind}=1} \subset \mathrm{Sing}(E)$, then R correspond to a factor $(\alpha_t z + \beta_t u^2)^{m_t}$ with $m_t \geq 2$. We distinguishes the following three subcases.

1-1. $R \neq Q_3, Q_4$.

By changing coordinates $\bar{z} := z + \frac{\beta_t}{\alpha_t}$, one sees that R is a cA point. Hence feasible resolution over R exists.

1-2. $R = Q_3$.

Since $Y \cap U_3$ is defined by

$$(\tilde{\varphi} : x^2 + y^2 + \sum a_{ij} z^{2i+j-2\mu'} u^j = 0) \subset \mathbb{C}^4 / \frac{1}{2}(1, 0, 1, 1).$$

One sees that Q_3 is a $cAx/2$-like point with

$$\tau(Q_4) = \min\{2i + j - 2\mu' + j | a_{ij} \neq 0\} \leq j_0 \leq \mu(P) = \mu^\flat(P) = \mu_0.$$

Feasible resolution over Q_3 exists by our hypothesis.

1-3. $R = Q_4$.

Since $Y \cap U_4$ is defined by

$$(\tilde{\varphi} : x^2 + y^2 z + \sum a_{ij} z^i u^{2i+j-2\mu'} = 0) \subset \mathbb{C}^4.$$

Then Q_4 is a cD-like point with $\lambda = 0$ as well. Since $\mu(P) = 2i_0 + j_0$ for some $z^{i_0} u^{j_0} \in \varphi$, one sees that

$$\mu(Q_4) = \min\{2i + j - 2\mu' + 2i | a_{ij} \neq 0\} \leq \mu(P) - 2\mu' + 2i_0 \leq \mu(P), \ (\dagger)$$

where the last inequality follows from $\mu' = \lfloor \frac{\mu_0}{2} \rfloor \geq \lfloor \frac{2i_0}{2} \rfloor$.

One can easily check that

$$\mu^\flat(Q_4) = \mu(Q_4) \leq \mu(P) = \mu^\flat(P);$$
$$\mu^\sharp(Q_4) \leq \mu^\sharp(P) + 2 - 2\mu' < \mu^\sharp(P).$$

By inductively on $\mu^\sharp$, there exist feasible resolution for $P \in X$.

Case 2. $\lambda \neq 0$.

Subcase 2-1. $2l - 2 = \mu(P)$.

We proceed as in Case 1 and see that $\mathrm{Sing}(Y) = \{Q_2, Q_3\}$, where Q_2 is a terminal quotient singularity and $Q_3 \in Y \cap U_3$ is given by

$$(\tilde{\varphi} : x^2 + y^2 - \frac{1}{4}\lambda^2 u^{2l} + \sum a_{ij} z^{2i+j-2\mu'} u^j = 0) \subset \mathbb{C}^4 / \frac{1}{2}(1, 0, 1, 1),$$

after completing the square. One sees that Q_3 is a $cAx/2$-like point with

$$\tau(Q_4) \leq \min\{2i + j - 2\mu' + j | a_{ij} \neq 0\} \leq \mu(P) = \mu^\flat(P) = \mu_0.$$

Feasible resolution over Q_3 exists by our hypothesis.

Subcase 2-2. $2l - 2 > \mu(P)$.

We can proceed as in Case 1 all the way to equation $\dagger$. Therefore, $l(Q_4) = l(P) - \mu' - 1$ and

$$\mu^\flat(Q_4) = \min\{2l - 2\mu' - 4, \mu(Q_4)\} \leq \min\{2l - 2, \mu(P)\} = \mu^\flat(P).$$

Inductively, we are reduced to either $\mu^\flat < \mu_0$ or $2l - 2 = \mu$. Hence feasible resolution exists.

Subcase 2-3. $2l - 2 < \mu(P)$.

For any $z^i u^j \in f$, one has $2i + j > 2l - 2$ and hence $i + j \geq l$. We consider $\mathrm{wBl}_v : Y \to X$ with $v = (l, l, 1, 1)$ instead.

By Lemma 7, $Y \cap U_4$ is nonsingular and hence $Y \to X$ is a divisorial contraction by Theorem 5. One sees that $\mathrm{Sing}(Y) = \{Q_2, Q_3\}$, where Q_2 is a terminal quotient singularity and $Q_3 \in Y \cap U_3$ is given by

$$(\tilde{\varphi} : x^2 + y^2 + \lambda y u^l + \sum a_{ij} z^{i+j-2l} u^j = 0) \subset \mathbb{C}^4,$$

which is a cD-like point. For a cD-like point, we introduce

$$\rho^\sharp(P \in X) := \min\{i + j | z^i u^j \in \varphi, j = 0 \text{ or } 1\},$$

which is finite and $\frac{1}{2}\mu \leq \rho^\sharp \leq \mu$. Compare Q_4 with P, we have $\rho^\sharp(Q_4) \leq \rho^\sharp(P) + 1 - 2l$. Repeat the process t-times, we have $Q_{t,4} \in Y_t \to \ldots \to Y_1 = Y \to X \ni P$ such that for t sufficiently large

$$\mu(Q_{t,4}) \leq 2(\mu(P) + t(1 - 2l)) < \mu_0(P).$$

Hence we are reduced to the situation $\mu^\flat(Q_{t,4}) \leq \mu(Q_{t,4}) < \mu_0$, or Subcase 2-1, or Subcase 2-2 in finite steps and feasible resolution over $P \in X$ exists by our hypothesis.

Q.E.D.

Combining all the above results in this section, we have the following:

Theorem 20. *There is a feasible resolution for any singularity of type cD or $cAx/2$.*

§5. resolution of $cAx/4, cD/2, cD/3$ points

In [5], Hayakawa shows that there is a partial resolution

$$X_n \to \ldots \to X_1 \to X \ni P,$$

for a point $P \in X$ of index $r > 1$ such that X_n has only terminal singularities of index 1 and each map is a divisorial contraction with minimal discrepancies. If $\mathrm{Sing}(X_n)_{\mathrm{ind}=1}$ is either of type cA or cD, then feasible resolution exists by the result of previous sections.

In fact, the partial resolution was constructed by picking any divisorial contraction with minimal discrepancy at each step. Therefore, for our purpose, it suffices to pick one divisorial contraction $Y \to X$ over a given higher index point $P \in X$ of type $cAx/4$, $cD/2$, $cD/3$, or $cE/2$ and verify that $\mathrm{Sing}(Y)_{\mathrm{ind}=1}$ is either of type cA or cD.

Lemma 21. *Given $P \in X$ of type $cAx/4$, there is a divisorial contraction $Y \to X$ with discrepancy $\frac{1}{4}$ such that $\mathrm{Sing}(Y)_{\mathrm{ind}=1}$ is of type cA or cD.*

Proof. We may write $P \in X$ as

$$(\varphi : x^2+y^2+f(z,u) = x^2+y^2+\sum_{i+j=2l+1\geq 3} a_{ij}z^{2i}u^j = 0) \subset \mathbb{C}^4/\frac{1}{4}(1,3,1,2).$$

Let $\sigma(P \in X) := \min\{i+j|a_{ij} \neq 0\}$, then we may write $f(z,u) = f_\sigma(z,u) + f_{>\sigma}(z,u)$.

Case 1. $f_\sigma(z,u)$ is not a perfect square.

Depending on parity of $\frac{\sigma-1}{2}$, we consider $\mathrm{wBl}_v : Y \to X$ with weights $v = \frac{1}{4}(\sigma+2, \sigma, 1, 2)$ or $\frac{1}{4}(\sigma, \sigma+2, 1, 2)$. By [4, Theorem 7.4], this is the only divisorial contraction.

Without loss of generality, we study the first weight. By Lemma 7, 8, we have $\mathrm{Sing}(Y)_{\mathrm{ind}=1} \cap U_i$ is empty for $i = 1, 2$. Moreover, Q_4 is a $cD/2$-like point of index 2. It suffices to consider U_3.

In U_3, $Y \cap U_3$ is defined by

$$(\tilde{\varphi} : x^2z + y^2 + \tilde{f}(z,u) = 0) \subset \mathbb{C}^4.$$

Therefore, it is immediate to see that $\mathrm{Sing}(Y) \cap U_3$ is either of type cA or cD.

Case 2. $f_\sigma(z,u) = -h(z,u)^2$ is a perfect square.

Depending on parity of $\frac{\sigma-1}{2}$, we need to make a coordinate change so that $P \in X$ is written as

$$(\varphi : x^2 + 2xh(z,u) + y^2 + f_{>\sigma}(z,u) = 0) \subset \mathbb{C}^4/\frac{1}{4}(1,3,1,2),$$

or

$$(\varphi : y^2 + 2yh(z,u) + x^2 + f_{>\sigma}(z,u) = 0) \subset \mathbb{C}^4/\frac{1}{4}(1,3,1,2).$$

We consider weighted blowup $Y \to X$ with weights $\frac{1}{4}(\sigma+4, \sigma+2, 1, 2)$ or $\frac{1}{4}(\sigma+2, \sigma+4, 1, 2)$ respectively. By [4, Theorem 7.4], this is a divisorial contraction.

Without loss of generality, we study the first weight. By Lemma 7, we have $\mathrm{Sing}(Y)_{\mathrm{ind}=1} \cap U_i$ is empty for $i = 1, 2$. Then $R \neq Q_4$ for Q_4 is a $cD/2$-like point of index 2. It suffices to consider U_3. Indeed, $Y \cap U_3$ is defined by

$$(\tilde{\varphi} : x^2z + 2xh_\sigma(1,u) + y^2 + \tilde{f}(z,u) = 0) \subset \mathbb{C}^4.$$

Therefore, it is immediate to see that $\mathrm{Sing}(Y) \cap U_3$ is either of type cA or cD. Q.E.D.

Lemma 22. *Given $P \in X$ of type $cD/2$, there is a divisorial contraction $Y \to X$ with discrepancy $\frac{1}{2}$ such that $\mathrm{Sing}(Y)_{\mathrm{ind}=1}$ is of type cA or cD.*

Proof. By Mori's classification [17, 22], one has that $P \in X$ is given by $(\varphi = 0) \subset \mathbb{C}^4/\frac{1}{2}(1,1,0,1)$ with φ being one the following

$$\begin{cases} x^2 + yzu + y^{2a} + u^{2b} + z^c, & a \geq b \geq 2, c \geq 3 \\ x^2 + y^2 z + \lambda y u^{2l+1} + f(z, u^2). & \end{cases}$$

Case 1. $\varphi = x^2 + yzu + y^{2a} + u^{2b} + z^c$.
We take weighted blowup $Y \to X$ with weights $v = \frac{1}{2}(3,1,2,1)$ (resp. $\frac{1}{2}(3,1,2,3)$) if $a = b = 2$ (resp. $a \geq 3$), which is a divisorial contraction by [5]. Note that $wt_v(yzu) = wt_v(\varphi)$. Hence uz (resp. yu, yz) appears in the equation of $Y \cap U_2$ (resp. U_3, U_4). By Corollary 9, we conclude that $\mathrm{Sing}(Y)_{\mathrm{ind}=1} \cap U_i$ is of type cA for $i = 2,3,4$. Together with Lemma 6, then we are done with this case.
Case 2. $\varphi = x^2 + y^2 z + \lambda y u^{2l+1} + f(z, u^2)$.
We may write $f(z, u^2) = \sum a_{ij} z^i u^{2j} \in (z^3, z^2 u^2, u^4)\mathbb{C}\{z, u^2\}$. We define

$$\begin{cases} \sigma := \min\{2i + 2j | z^i u^{2j} \in f\}; \\ \sigma^\flat := \min\{2l - 1, \sigma\}. \end{cases}$$

Note that we have $l \geq 1$ and $\sigma \geq 2$ for $P \in X$ is a $cD/2$ point.
Subcase 2-1. $l = 1$.
We consider weighted blowup $Y \to X$ with weight $v = \frac{1}{2}(2,1,2,1)$, which is a divisorial contraction by [5]. Now

$$E = (\mathbf{x}^2 + \mathbf{y}^2\mathbf{z} + \lambda \mathbf{y}\mathbf{u}^3 + f_{wt_v=2} = 0) \subset \mathbb{P}(2,1,2,1).$$

By Lemma 6, 7, one sees that $\mathrm{Sing}(Y)_{\mathrm{ind}=1} \cap U_i$ is empty for $i = 1,2,4$. Since Q_3 is at worst of type $cAx/2$. We are done.
Subcase 2-2. $l \geq 2$ and $\sigma = 2$.
One has $f_{\sigma=2} = u^4$, in particular, $u^4 \in f$. We consider weighted blowup $Y \to X$ with weight $v = \frac{1}{2}(2,1,2,1)$ again. Now

$$E = (\mathbf{x}^2 + \mathbf{y}^2\mathbf{z} + \mathbf{u}^4 = 0) \subset \mathbb{P}(2,1,2,1).$$

We thus have $\mathrm{Sing}(Y)_{\mathrm{ind}=1} \subset \mathrm{Sing}(E) \subset \{Q_3\}$. However, Q_3 is a point of index 2. We are done.
Subcase 2-3. $\sigma^\flat \geq 3$.
Let

$$\sigma' := 2\lfloor \frac{\sigma^\flat - 1}{2} \rfloor + 1 = \begin{cases} \sigma^\flat & \text{if } \sigma^\flat \text{ is odd;} \\ \sigma^\flat - 1 & \text{if } \sigma^\flat \text{ is even;} \end{cases}$$

We consider weighted blowup $Y \to X$ with weight $v = \frac{1}{2}(\sigma', \sigma' - 2, 4, 1)$, which is a divisorial contraction by [5]. Clearly, by Lemma 6, 7, one sees that $\mathrm{Sing}(Y)_{\mathrm{ind}=1} \cap U_i$ is empty for $i = 1, 2$. Since Q_3 is a point of index 4, it remains to consider U_4. Now $Y \cap U_4$ is given by

$$(\tilde{\varphi} : x^2 + y^2 z + \lambda y u^{(2l-1-\sigma')/2} + \tilde{f} = 0) \subset \mathbb{C}^4.$$

It follows that $Y \cap U_4$ is at worst of type cD. We are done. Q.E.D.

Lemma 23. *Given $P \in X$ of type $cD/3$, there is a divisorial contraction $Y \to X$ with discrepancy $\frac{1}{3}$ such that* $\mathrm{Sing}(Y)_{\mathrm{ind}=1}$ *is of type cA or cD.*

Proof. By Mori's classification [17, 22], one has that $P \in X$ is given as $(\varphi = 0) \subset \mathbb{C}^4/\frac{1}{3}(0, 2, 1, 1)$ with φ being one of the following:

$$\begin{cases} x^2 + y^3 + zu(z+u); \\ x^2 + y^3 + zu^2 + yg(z,u) + h(z,u); & g \in \mathfrak{m}^4, h \in \mathfrak{m}^6; \\ x^2 + y^3 + z^3 + yg(z,u) + h(z,u); & g \in \mathfrak{m}^4, h \in \mathfrak{m}^6. \end{cases}$$

Case 1. φ is one of the first two cases.
By [4, Theorem 9.9, 9.14, 9.20], the weighted blowup $Y \to X$ with weight $\frac{1}{3}(3, 2, 4, 1)$ is a divisorial contraction. Now

$$E = \left\{ \begin{array}{l} \mathbf{x}^2 + \mathbf{y}^3 + \mathbf{z}\mathbf{u}^2 = 0 \text{ or} \\ \mathbf{x}^2 + \mathbf{y}^3 + \mathbf{z}\mathbf{u}^2 + \lambda \mathbf{y}\mathbf{u}^4 + \lambda' \mathbf{u}^6 = 0 \end{array} \right\} \subset \mathbb{P}(3, 2, 4, 1),$$

for some λ, λ' respectively. It is easy to check that $\mathrm{Sing}(E) = Q_3$ and hence $\mathrm{Sing}(Y)_{\mathrm{ind}=1}$ is empty for the first two cases.
Case 2. $\varphi = x^2 + y^3 + z^3 + yg(z,u) + h(z,u)$.
Subcase 2-1. Either $u^4 \in g$ or $u^6 \in h$.
Then we consider the weighted blowup with weight $\frac{1}{3}(3, 2, 4, 1)$ again, which is a divisorial contraction (cf. [4, Theorem 9.20]. Now

$$E = (\mathbf{x}^2 + \mathbf{y}^3 + \lambda \mathbf{y}\mathbf{u}^4 + \lambda' \mathbf{u}^6 = 0) \subset \mathbb{P}(3, 2, 4, 1),$$

for some $(\lambda, \lambda') \neq (0, 0)$. One sees that $\mathrm{Sing}(E) = Q_3$ and hence $\mathrm{Sing}(Y)_{\mathrm{ind}=1}$ is empty.
Subcase 2-2. $u^4 \notin g$, $u^6 \notin h$ and either $zu^5 \in h$ or $u^9 \in h$.
Then we consider the weighted blowup with weight $\frac{1}{3}(3, 2, 4, 1)$ which is a divisorial contraction. Now

$$E = (\mathbf{x}^2 + \mathbf{y}^3 = 0) \subset \mathbb{P}(3, 2, 4, 1).$$

One sees that $\mathrm{Sing}(E) \subset U_3 \cup U_4$. However, the equation of $Y \cap U_4$ contains the term zu or u and hence contains at worst cA points by

Lemma 8. Together with the fact that Q_3 is a $cAx/4$ point, we are done with this case.

Subcase 2-3. $u^4 \notin g$, all $zu^5, u^6, u^9 \notin h$.
Then we consider the weighted blowup $Y \to X$ with weight $\frac{1}{3}(6,5,4,1)$, which is a divisorial contraction by [4, Theorem 9.25].

By Lemma 7, $Y \cap U_2$ is nonsingular away from Q_2, which is a quotient singularity of index 5. Together with $\mathrm{Sing}(Y) \cap U_1 = \emptyset$ and $Q_3 \notin Y$, it remains to check $Y \cap U_4$, which is defined by

$$(\tilde{\varphi} : x^2 + y^3 u + z^3 + \text{others} = 0) \subset \mathbb{C}^4.$$

which is at worst of type cE_6. In fact, this corresponds to Case 1 and 2 of the proof of Theorem 34. Notice that in the proof, we use weighted blowups wBl_v with $v = (2,2,1,1), (3,2,1,1)$ or $(3,2,2,1)$. After weighted blowup, there could have singularities of type $cA, cD, cA/2$, $cAx/2$, and terminal quotients. We thus concludes that feasible resolution exists for this case. Q.E.D.

We thus conclude the section by the following:

Theorem 24. *There is a feasible resolution for any singularity of type $cAx/4$, $cD/3$, or $cD/2$.*

§6. resolution of cE and $cE/2$ points

Recall that a cE point has the following description.

$$(\varphi : x^2 + y^3 + f(y,z,u) = x^2 + y^3 + yg(z,u) + h(z,u) = 0) \subset \mathbb{C}^4.$$

An isolated singularity with the above desription is called a cE-like singularity.

For a polynomial (resp. formal power series) $G(z,u) \in \mathbb{C}[z,u]$ (resp. $\mathbb{C}[[z,u]]$), we define

$$\tau(G) := \min\{j + k | z^j u^k \in G\}.$$

For cE singularity, one has $\tau(g) \geq 3$ and $\tau(h) \geq 4$. Moreover, either $\tau(g) = 3$ or $\tau(h) \leq 5$. More precisely,

(1) It is cE_6 if $\tau(h) = 4$ and $\tau(g) \geq 3$.
(2) It is cE_7 if $\tau(h) \geq 5$ and $\tau(g) = 3$.
(3) It is cE_8 if $\tau(h) = 5$ and $\tau(g) \geq 4$.

Remark 25. An isolated cE-like singularity is at worst of type cD (resp. cE_6, cE_7, cE_8) if $\tau(g) \leq 2$ or $\tau(h) \leq 3$ (resp. $\tau(h) \leq 4$, $\tau(g) \leq 3$, $\tau(h) \leq 5$).

26. Notations and Conventions

1. We fix the notation that $g_3(z,u) := g_{\tau=3}(z,u), h_4(z,u) := h_{\tau=4}(z,u)$ and $h_5(z,u) := h_{\tau=5}(z,u)$. In the case of cE_6, $\tau(h) = 4$. By replacing z, u and up to a constant, we may and do assume that

$$h_4 \in \{z^4, z^4 + z^3u, z^4 + 2z^3u + z^2u^2, z^4 + z^2u^2, z^4 + zu^3\}.$$

In particular, $z^4 \in h_4$.

In the case of cE_7, $\tau(g) = 3$. We may and do assume that

$$g_3 \in \{z^3, z^3 + z^2u, z^3 + zu^2\}.$$

In particular, $z^3 \in g_3$.

In the case of cE_8, $\tau(h) = 5$. We may and do assume that

$$\begin{aligned} h_5 \in \{&z^5, z^5 + z^4u, z^5 + 2z^4u + z^3u^2, z^5 + z^3u^2, \\ &z^5 + 2z^4u - z^3u^2 - 2z^2u^3, z^5 + z^2u^3, z^5 + zu^4\}. \end{aligned}$$

In particular, $z^5 \in h_5$.

2. We define

$$\tau^*(\varphi) := \min\{p | y^i z^j u^p \in \varphi \text{ with } i + j \leq 1\}.$$

Since $P \in X$ is isolated, there is a term yu^p, zu^p or u^p in φ otherwise P is singular along a line $(x = y = z = 0)$. Hence $\tau^*(\varphi)$ is a well-defined integer.

3. For a weight $v = (a, b, k, 1)$, we denote it v_l with $l = a+b+k-1$. In our discussion, we always consider weight v_l such that $v_l(\varphi) = l$.

4. Fix a weight $v = \frac{1}{r}(a, b, k, 1)$ with $r = 1, 2$, we write

$$\varphi = x^2 + y^3 + yg_v + yg_{v+1} + yg_> + h_v + v_{v+1} + h_>,$$

where g_v (resp. g_{v+1}) is the homogeneous part of $g(z, u)$ such that $wt_v(yg_v) = wt_v(\varphi)$ (resp. $wt_v(yg_v) = wt_v(\varphi) + 1$) and $g_>$ is the remaining part with greater weight, and h_v (resp. h_{v+1}) is the homogeneous part of $h(z, u)$ with v-weight equal to $wt_v(\varphi)$ (resp. $wt_v(\varphi) + 1$) and $h_>$ is the remaining part with greater weight.

5. For simplicity of notation, sometime we may denote by g_m or h_m for the v-homogeneous part with v-weight equal to m. This notation should not be confused with g_3 nor g_v.

6.1. Some preparation

The general strategy is as following. For a given cE or $cE/2$ singularity $P \in X$. We consider weighted blowup $Y \to X$ with weight

$v = \frac{1}{r}(a, b, k, 1)$ and $r = 1, 2$ such that $\frac{1}{2}(a+b+k+1) - wt_v(\varphi) = 1 + \frac{1}{r}$. This is a weighted blowup with discrepancy $\frac{1}{r}$ if E is irreducible. We check that $\mathrm{Sing}(Y) \cap U_4$ is isolated and each $R \in \mathrm{Sing}(Y) \cap U_4$ is terminal. Then the weighted blowup $Y \to X$ is a divisorial contraction with discrepancy $\frac{1}{r}$ by Theorem 5.

Moreover, we check that each singular point $R \in \mathrm{Sing}(Y)_{\mathrm{ind}=1}$ is "milder" than $P \in X$ in the sense that either it is of milder type, or it can only admit smaller weight. We can prove the existence of feasible resolution by induction on types and weights.

27. We work on $Y \cap U_4$.

Now $Y \cap U_4$ is defined by $\tilde{\varphi}$, which can be written as

$$\begin{aligned}\tilde{\varphi} &= x^2 u^{wt_v(x^2)-wt_v(\varphi)} + y^3 u^{wt_v(y^3)-wt_v(\varphi)} \\ &+ y g_v(z,1) + yu g_{v+1}(z,1) + \widetilde{yg_>} + h_v(z,1) + u h_{v+1}(z,1) + \widetilde{h_>},\end{aligned}$$

such that $u^2 | \widetilde{g_>}$ and $u^2 | \widetilde{h_{>v}}$.

Lemma 28. *Suppose that* $wt_v(x^2) = wt_v(\varphi)$ *or* $wt_v(\varphi) + 1$ *and* $wt_v(y^3) = wt_v(\varphi)$. *Then* $\mathrm{Sing}(Y) \cap U_4$ *is isolated UNLESS:*

$$\text{There is } s(z,u) \text{ such that} \begin{cases} g_v = -3s(z,u)^2, \\ h_v = 2s(z,u)^3, \\ h_{v+1} = -s(z,u) g_{v+1}. \end{cases} \quad \natural$$

Proof. If $2wt_v(x) = wt_v(\varphi)$, we have $\tilde{\varphi}_x = 2x$. If $2wt_v(x) = wt_v(\varphi)+1$, then $Y \cap U_1$ is non-singular away from Q_1 by Lemma 7. Hence we have $\mathrm{Sing}(Y) \cap U_4 \subset (x = 0)$ in both cases. Moreover, $\mathrm{Sing}(Y) \subset E$, hence we have $\mathrm{Sing}(Y) \cap U_4 \subset (u = 0)$.

Therefore, we have

$$\begin{aligned}\mathrm{Sing}(Y) \cap U_4 &\subset (x = u = 0) \cap (\tilde{\varphi} = \tilde{\varphi}_y = \tilde{\varphi}_u = 0) \\ &\subset (x = u = 0) \cap \Sigma,\end{aligned}$$

where Σ is defined as

$$\begin{cases} y^3 + y g_v + h_v = 0, \\ 3y^2 + g_v = 0, \\ y g_{v+1} + h_{v+1} = 0. \end{cases}$$

If g_v is not a perfect square, then $3y^2 + g_v$ is irreducible and hence Σ is finite. If g_v is a perfect square, then we write it as $g_v = -3s^2$. One sees that Σ is finite unless $y - s$ or $y + s$ divides the above three polynomials. The statement now follows. Q.E.D.

Lemma 29. *Suppose more generally that*

$$\varphi = x^2 + y^3 + s(z,u)y^2 + yg_v + yg_{v+1} + yg_> + h_v + h_{v+1} + h_>.$$

Suppose that $wt_v(x^2) = wt_v(\varphi)$ *and* $wt_v(y^3) = wt_v(\varphi) + 1$. *Then* $\mathrm{Sing}(Y) \cap U_4$ *is isolated UNLESS* $g_v = h_v = h_{v+1} = 0$.

Proof. Since $wt_v(y^3) = wt_v(\varphi) + 1$, then $Y \cap U_2$ is non-singular away from Q_2. Hence we have $\mathrm{Sing}(Y) \cap U_4 \subset (x = y = 0)$. Moreover, $\mathrm{Sing}(Y) \subset E$, hence we have $\mathrm{Sing}(Y) \cap U_4 \subset (u = 0)$.

Therefore, we have

$$\begin{aligned} Siny(Y) \cap U_4 &\subset (x = y = u = 0) \cap (\tilde{\varphi} = \tilde{\varphi}_y = \tilde{\varphi}_u = 0) \\ &\subset (x = y = u = 0) \cap (h_v(z,1) = g_v(z,1) = \tilde{\varphi}_u = 0) \\ &\subset (x = y = u = 0) \cap (h_v(z,1) = g_v(z,1) = h_{v+1}(z,1) = 0). \end{aligned}$$

The statement now follows. Q.E.D.

30. We study the most common case that

$$wt_v(x^2) = wt_v(y^3) = wt_v(\varphi).$$

Suppose furthermore that ♮ does not hold, then $\mathrm{Sing}(Y) \cap U_4$ is isolated. We now study the possible type of these singularities.

Notice that we have at least one of g_v, h_v, h_{v+1} is non-zero, otherwise ♮ holds.

Case 1. $h_v \neq 0$.

We write

$$h_v(z,u) = \lambda z^{m'} u^{n'} \prod (z - \alpha_t u^k)^{l'_t}.$$

Then

$$k \cdot \tau(h) \geq wt_v(h_\tau) \geq wt_v(\varphi) = n' + k(m' + \sum l'_t). \qquad \dagger_h$$

In particular,

$$\tau(h) \geq m' + \sum l'_t.$$

Then $E \cap U_4$ is defined by

$$\begin{aligned} &(x^2 + y^3 + yg_v(z,1) + h_v(z,1) = 0) \\ &= (x^2 + y^3 + yg_v(z,1) + \lambda' z^{m'} \prod (z - \alpha'_t)^{l'_t} = 0) \subset U_4 \cong \mathbb{C}^4, \end{aligned}$$

which is irreducible.

It is easy to see that

- if $m' + \sum l'_t \leq 2$ then $E \cap U_4$ is at worst Du Val of A-type and hence $\mathrm{Sing}(Y) \cap U_4$ is at worst of cA type;

- if $m' + \sum l'_t = 3$ then $E \cap U_4$ is at worst Du Val of D-type and hence $\mathrm{Sing}(Y) \cap U_4$ is at worst of cD type;
- if $m' + \sum l'_t = 4$ then $E \cap U_4$ is at worst Du Val of E_6-type and hence $\mathrm{Sing}(Y) \cap U_4$ is at worst of cE_6 type;
- if $m' + \sum l'_t = 5$ then $E \cap U_4$ is at worst Du Val of E_8-type and hence $\mathrm{Sing}(Y) \cap U_4$ is at worst of cE_8 type;

Notice also that if $P \in X$ is of type cE_6 (resp. cE_8), then $\mathrm{Sing}(Y) \cap U_4$ is at worst of cE_6 (resp. cE_8).

In any event, $Y \cap U_4$ is terminal. By Theorem 5, $Y \to X$ is a divisorial contraction with discrepancy 1.

Case 2. $g_v \neq 0$.

We write

$$g_v(z,u) = \lambda z^m u^n \prod (z - \alpha_t u^k)^{l_t}.$$

Then

$$k \cdot \tau(g) \geq wt_v(g_\tau) \geq wt_v(\varphi) - b = n + k(m + \sum l_t). \qquad \dagger_g$$

In particular,

$$\tau(g) \geq m + \sum l_t.$$

Then $E \cap U_4$ is defined by

$$\begin{aligned}&(x^2 + y^3 + y g_v(z,1) + h_v(z,1) = 0)\\ &= (x^2 + y^3 + \lambda y z^m \prod (z - \alpha_t)^{l_t} + h_v(z,1) = 0) \subset U_4 \cong \mathbb{C}^4,\end{aligned}$$

which is irreducible.

It is easy to see that

- if $m + \sum l_t \leq 1$ then $E \cap U_4$ is at worst Du Val of A-type and hence $\mathrm{Sing}(Y) \cap U_4$ is at worst of cA type;
- if $m + \sum l_t = 2$ then $E \cap U_4$ is at worst Du Val of D-type and hence $\mathrm{Sing}(Y) \cap U_4$ is at worst of cD type;
- if $m + \sum l_t = 3$ then $E \cap U_4$ is at worst Du Val of E_7-type and hence $\mathrm{Sing}(Y) \cap U_4$ is at worst of cE_7 type.

Notice also that if $P \in X$ is of type cE_7, then $\mathrm{Sing}(Y) \cap U_4$ is at worst of cE_7.

Case 3. $h_v = 0, h_{v+1} \neq 0$.

We write

$$h_{v+1}(z,u) = \lambda'' z^{m''} u^{n''} \prod (z - \alpha_t u^k)^{l''_t}.$$

Then

$$k \cdot \tau(h) \geq wt_v(h_\tau) \geq wt_v(\varphi) = n'' + k(m'' + \sum l''_t) - 1. \qquad \dagger'_h$$

In particular, if $k > 1$, then we still have

$$\tau(h) \geq m'' + \sum l''_t.$$

The same conclusion as in Case 1 still holds.

In any event, $Y \cap U_4$ is terminal. By Theorem 5, $Y \to X$ is a divisorial contraction.

As a summary, we conclude that

Theorem 31. *Given $P \in X$ a cE point defined by ($\varphi : x^2 + y^3 + yg(z,u) + h(z,u) = 0$). Let $Y \to X$ be a weight blowup with weight $v = (a,b,k,1)$ that $k > 1$. Suppose that $wt_v(x^2) = wt_v(y^3) = wt_v(\varphi)$ and $\natural$ does not hold. Then $Y \to X$ is a divisorial contraction. Also, any singularity on $Y \cap U_4$ is at worst of type cE_6 (resp. cE_7, cE_8) if $P \in X$ is of type cE_6 (resp. cE_7, cE_8).*

Remark 32. Consider the case that $P \in X$ is of type cE_6. Suppose the worst case that Y has a singularity R of type cE_6. This happens only when $h_v = z^4$ or $h_v = (z - \alpha_t u)^4$ for $m' + \sum l'_t \leq 4$. If $h_v = (z - \alpha'_t u^k)^4$. By considering the weight-invariant coordinate change that $\bar{z} = z - \alpha'_t u^k$, we may and do assume that $R = Q_4$ and Q_4 is the unique singularity in U_4.

We can make the same assumption if $P \in X$ is of type cE_7, cE_8.

Proposition 33. *Let $Y \to X$ be a weighted blowup of a cE point with weight $v = (a,b,k,1)$. Suppose that $wt_v(x^2) = wt_v(y^3) = wt_v(\varphi)$. If any one of $g_v = 0$, $h_v = 0$, $n > 0$, or $n' > 0$ holds, then* $\mathrm{Sing}(E) \cap U_2 - U_4 = \emptyset$.

In particular, if $\natural$ does not hold and $v = v_{30}, v_{24}, v_{18}, v_{12}$, then $Y \to X$ is a divisorial contraction and $\mathrm{Sing}(Y)_{\mathrm{ind}=1} \subset U_4$.

Proof. In affine coordinate U_2, E is defined by

$$\tilde{\phi} : x^2 + 1 + g_v(z,u) + h_v(z,u) = 0,$$

with $g_v(z,u)$ (resp. $h_v(z,u)$) being homogeneous with respect to the weight v of weight $wt_v(\varphi) - b$ (resp. $wt_v(\varphi)$). It follows that

$$\begin{cases} kz\frac{\partial g_v(z,u)}{\partial z} + u\frac{\partial g_v(z,u)}{\partial u} = (wt_v(\varphi) - b) \cdot g_v(u,v), \\ kz\frac{\partial h_v(z,u)}{\partial z} + u\frac{\partial h_v(z,u)}{\partial u} = wt_v(\varphi) \cdot h_v(u,v) \end{cases}$$

It follows that

$$\psi_1 := wt_v(\varphi)\tilde{\phi} - ax\tilde{\phi}_x - kz\tilde{\phi}_z - u\tilde{\phi}_u = wt_v(\varphi) + bg_v(u,v),$$

where $g_v(u,v) = \lambda z^m u^n \prod(z-\alpha_t u^k)^{l_t}$. Note that ψ_1 must be satisfied at any singular point $\mathrm{Sing}(E)\cap U_2$. If $\lambda = 0$, then one sees that $\mathrm{Sing}(E)\cap U_2 = \emptyset$.

If $n > 0$, then $\mathrm{Sing}(E)\cap U_2 \not\subset U_4$ otherwise $u = 0$ will leads to a contradiction.

If we consider

$$\psi_2 := (wt_v(\varphi) - b)\tilde{\phi} - (a - b/2)x\tilde{\phi}_x - kz\tilde{\phi}_z - u\tilde{\phi}_u$$
$$= (wt_v(\varphi) - b) - bh_v(z,u),$$

where $h_v(u,v) = \lambda' z'^m u'^n \prod(z - \alpha'_t u^k)^{l'_t}$. Then one sees similarly that $\mathrm{Sing}(E)\cap U_2 = \emptyset$ if $\lambda' = 0$ and $\mathrm{Sing}(E)\cap U_2 \not\subset U_4$ if $n' > 0$.

We now prove the second statement. Since $\natural$ does not hold, hence $Y \to X$ is a divisorial contraction. Therefore, $\mathrm{Sing}(Y)_{\mathrm{ind}=1} \subset \mathrm{Sing}(E)$. We have that $\mathrm{Sing}(E)\cap U_1 = \emptyset$. Notice that either $g_v = 0$ or $n > 0$ for v_{12}, v_{24}, v_{30}. Also one has either $h_v = 0$ or $n' > 0$ for v_{18}. Therefore, $\mathrm{Sing}(Y)_{\mathrm{ind}=1} \cap (U_1 \cup U_2 \cup U_4) \subset U_4$. Finally Q_3 is of index > 1. This completes the proof. Q.E.D.

6.2. Resolution of cE_6 points

In this subsection, we shall prove that

Theorem 34. *There is a feasible resolution for any cE_6 singularity.*

Proof. We will need to consider weighted blowup $Y \to X$ with the following weights $v_{12} = (6,4,3,1)$, $v_8 = (4,3,2,1)$, $v_6 = (3,2,2,1)$, $v_4 = (2,2,1,1)$ and $v_5 = (3,2,1,1)$. It is sufficient to show that $\mathrm{Sing}(Y)_{\mathrm{ind}>1}$ is not of type $cE/2$ and there exists feasible resolution on $\mathrm{Sing}(Y)_{\mathrm{ind}=1}$.

Case 1. $wt_{v_{12}}(f) < 12$, $wt_{v_8}(f) < 8$ and $wt_{v_6}(f) < 6$.

Subcase 1-1. h_4 is not a perfect square.

We consider the weighted blowup $Y \to X$ with weight $v_4 = (2,2,1,1)$. It is clear that E is irreducible if h_4 is not a perfect square.

Since $wt_{v_6}(f) < 6$, we must have a term $\theta \in f$ such that $wt_{v_6}(\theta) < 6$. One has that $\theta = yu^3$ or $\theta = u^5$.

Claim. $Y \to X$ is a divisorial contraction.

To see this, if $\theta = u^5$, then it follows that $Y \cap U_4$ is nonsingular and thus $Y \to X$ is a divisorial contraction by Theorem 5. If $\theta = yu^3$, then $\natural$ does not hold and hence $Y \cap U_4$ has at worst singularities of type cA. Therefore, $Y \to X$ is a divisorial contraction by Theorem 5.

Clearly, $Y \cap U_1$ is nonsingular. Moreover, $Y \cap U_2$ has singularity of type $cAx/2$ at Q_2 and at worst of type cA for points other than Q_2. Since $z^4 \in h_4$, we have $Q_3 \notin Y$. Therefore, feasible resolution exists for this case.

Subcase 1-2. h_4 is a perfect square, i.e. $h_4 = z^4$ or $z^4 + 2z^3u + z^2u^2$.

Since $wt_{v_6}(f) < 6$, we have either yu^3 or u^5 in f. Write $h_4 = -q(z,u)^2$. Consider the coordinate change $\bar{x} := x - q(z,u)$, we have

$$\bar{\varphi} := \bar{x}^2 + 2\bar{x}q(z,u) + y^3 + yg(z,u) + h_{\tau\geq 5}(z,u).$$

We consider weighted blowup with weight $v_5 = (3,2,1,1)$ instead. Note that we still have either yu^3 or $u^5 \in \bar{\varphi}$.

By Lemma 7, $Y \cap U_i$ is nonsingular away from Q_i for $i = 1,2,4$ and Q_1, Q_2 are terminal quotient singularity of index $3,2$ respectively. By Theorem 5, $Y \to X$ is a divisorial contraction. It remains to consider Q_3. Since $z^4 \in h_4$, we have $\bar{x}z^2 \in \bar{\varphi}$. Hence $Y \cap U_3$ is also nonsingular by Lemma 7. We thus conclude that $\mathrm{Sing}(Y) = \mathrm{Sing}(Y)_{\mathrm{ind}>1}$ consists of Q_1, Q_2, which are terminal quotient singularities of index $3,2$ respectively.

Case 2 $wt_{v_{12}}(f) < 12$, $wt_{v_8}(f) < 8$ and $wt_{v_6}(f) \geq 6$.

We consider the weighted blowup $Y \to X$ with weight $v_6 = (3,2,2,1)$. It is clear that E is irreducible. There is a term $\theta \in f$ with $wt_{v_8}(\theta) < 8$ and $wt_{v_6}(\theta) \geq 6$. One sees that

$$\theta \in \{yzu^2, yu^4, z^3u, z^2u^2, z^2u^3, zu^4, zu^5, u^6, u^7\}.$$

It follows in particular that at least one of g_v, h_v, h_{v+1} is non-zero.

Claim. $Y \to X$ is a divisorial contraction.

To see this, suppose first that $\natural$ holds, then $g_v = -3s(z,u)^2$ for some $s(z,u) \neq 0$. We may assume that $s(z,u) = u^2$ and hence $yu^4 \in \varphi$. Then $Y \cap U_4$ is nonsingular by Lemma 7 and hence $Y \to X$ is a divisorial contraction by Theorem 5.

Suppose that $\natural$ does not hold. Then $\mathrm{Sing}(Y) \cap U_4$ is isolated. In U_4, the corresponding term $\tilde{\varphi}$ of θ in $\tilde{\varphi}$ is

$$\tilde{\theta} \in \{yz, y, z^3, z^2, z^2u, z, zu, 1, u\}.$$

Hence $\mathrm{Sing}(Y) \cap U_4$ is at worst of type cD. By Theorem 5, $Y \to X$ is a divisorial contraction. This proved the Claim.

We consider $Y \cap U_3$. We have that $z^4 \in h_4$ and hence Q_3 is at worst of type $cA/2$. By Corollary 9, $\mathrm{Sing}(Y)_{\mathrm{ind}=1} \cap U_3$ is at worst of cA type. By Lemma 6, $\mathrm{Sing}(Y)_{\mathrm{ind}=1} \cap U_1 = \emptyset$. Together with $Q_2 \notin Y$, we concludes that $\mathrm{Sing}(Y)_{\mathrm{ind}=1}$ is at worst of type cD and $\mathrm{Sing}(Y)_{\mathrm{ind}>1} = \{Q_3\}$, of type $cA/2$. Feasible resolution exists for this case.

Case 3 $wt_{v_{12}}(f) < 12$ and $wt_{v_8}(f) \geq 8$.

We consider the weighted blowup $Y \to X$ with weight $v_8 = (4,3,2,1)$.

1. Note that $\tau(h) = 4$ and $wt_{v_8}(h) \geq 8$, we thus have $h_4 = z^4 \in h_v \neq 0$. By Lemma 29, $\mathrm{Sing}(Y) \cap U_4$ is isolated. Also, one has $Q_3 \notin Y$.

2. Since $wt_{v_{12}}(f) < 12$ and $wt_{v_8}(f) \geq 8$, there is a term $\theta = y^i z^j u^k \in f$ with $wt_{v_{12}}(\theta) < 12$ and $wt_{v_8}(\theta) \geq 8$. Hence the corresponding term $\tilde{\theta} = y^i z^j u^{k'} \in \tilde{\varphi}$ satisfying

$$i + j + k' = i + j + (3i + 2j + k - 8) \leq 3.$$

One can verify that $\mathrm{Sing}(Y) \cap U_4$ is at worst of type cE_6 with h_4 has at least two factors. Hence if there is a cE_6 points then it is in Case 1 or 2. By Theorem 5, $Y \to X$ is a divisorial contraction.

3. By Lemma 7, $\mathrm{Sing}(Y) \cap U_2 = \{Q_2\}$ and Q_2 is a terminal quotient singularity of index 3. Also $\mathrm{Sing}(Y)_{\mathrm{ind}=1} \cap U_1 = \emptyset$ by Lemma 6.

4. We summarize that $\mathrm{Sing}(Y)_{\mathrm{ind}>1}$ consists of Q_2, which is a quotient singularity of index 3 and two terminal quotient singularities of index 2 in the line $(\mathbf{y} = \mathbf{u} = 0) \subset E$ and $\mathrm{Sing}(Y)_{\mathrm{ind}=1} \subset U_4$ are at worst of type cE_6 in Case 1 or 2.

Case 4. $wt_{v_{12}}(f) \geq 12$.

We consider $Y \to X$ the weighted blowup with weight $v_{12} = (6, 4, 3, 1)$.

1. Since $wt_{v_{12}}(h) \geq 12$ and $\tau(h) = 4$, we have $z^4 \in h_v$. It follows that $Q_3 \notin Y$, $h_v \neq 2s^3$, and thus $\natural$ does not holds. By Theorem 31 and Proposition 33, the weighted blowup $Y \to X$ is a divisorial contraction with discrepancy 1. Moreover, $\mathrm{Sing}(Y)_{\mathrm{ind}=1} \subset U_4$ are at worst of type cD or there is only a unique point $R \in Y$ of type of type cE_6.

2. It follows that feasible resolution exists for $P \in X$ unless that $\mathrm{Sing}(Y)_{\mathrm{ind}=1} = R \in Y$ is of type cE_6. In fact, if it is of type cE_6, we may assume that $R = Q_4$ (cf. Remark 32).

Clearly,

$$\begin{cases} \tau^*(\tilde{\varphi}) < \tau^*(\varphi), \\ wt_{v_{12}}(\tilde{\varphi}) \leq wt_{v_{12}}(\varphi). \end{cases}$$

The existence of feasible resolution is thus reduced to cE_6 singularities with $wt_{v_{12}} < 12$ by induction on τ^*.

3. We remark that $\mathrm{Sing}(Y)_{\mathrm{ind}>1}$ consists of terminal quotient singularities on the line $(\mathbf{z} = \mathbf{u} = 0)$ and $(\mathbf{y} = \mathbf{u} = 0)$ of index $2, 3$ respectively.

This exhausts all cases of type cE_6. We thus conclude that for a given $P \in X$ of type cE_6, there is a feasible partial resolution $Y_s \to \ldots \to Y_1 = Y \to X$ such that $\mathrm{Sing}(Y_s)_{\mathrm{ind}=1}$ are at worst cD and $\mathrm{Sing}(Y_s)_{\mathrm{ind}>1}$ can only be of type $cA/2, cA/2$ or terminal quotient. Hence feasible resolution exists for Y_s and hence for $P \in X$. Q.E.D.

6.3. Resolution of $cE/2$ points

It is convenient to consider $cE/2$ points before we move into the cE_7 and cE_8 singularities. Given a $cE/2$ point $P \in X$, which is given by

$$(\varphi = x^2 + y^3 + \sum a_{ij} y z^j u^k + \sum b_{jk} z^j u^k = 0) \subset \mathbb{C}^4 / \frac{1}{2}(1, 0, 1, 1),$$

with $h_4 := \sum_{j+k=4} b_{jk} z^j u^k \neq 0$.

We will consider weighted blowup with weights $v_1 = \frac{1}{2}(3,2,3,1)$ or $v_2 = \frac{1}{2}(5,4,3,1)$. Note that $wt_{v_1}(z) = wt_{v_2}(z), wt_{v_1}(u) = wt_{v_2}(u)$. Hence may simply denote it as $wt_{3,1}(G)$ for $G \in \mathbb{C}[[z,u]]$.

Theorem 35. *There is a feasible resolution for any* $cE/2$ *singularity.*

Proof. We first consider weighted blowup $Y \to X$ with weight $v = \frac{1}{2}(3,2,3,1)$. As before, we can rewrite φ as

$$\varphi = x^2 + y^3 + yg_v + yg_{v+1} + yg_> + h_v + h_{v+1} + h_>.$$

Notice that Lemma 28 still holds in the current situation.

Case 1. $wt_{3,1}(h_4) = 3$, i.e. $u^4 \notin h_4, zu^3 \in h_4$.
It is straightforward to see that E is irreducible and $Y \cap U_4$ is nonsingular, hence $Y \to X$ is a divisorial contraction. Also $Y \cap U_3$ has singularity Q_3 of type $cD/3$, might have terminal quotient singularity of index 3 along the line $\mathbf{y} = \mathbf{u} = 0$ and might have singularity at worst of type cD. There is no other singularity.

Case 2. $wt_{3,1}(h_4) = 4$, i.e. $u^4, zu^3 \notin h_4, z^2u^2 \in h_4$.
Since $z^2u^2 \in h_{v+1}$, one sees that $\natural$ does not hold and hence $Y \cap U_4$ has only isolated singularities.

It is straightforward to see that $Y \cap U_4$ might have singularities at worst of type cD, hence $Y \to X$ is a divisorial contraction. On $Y \cap U_3$, there are a singularity Q_3 of type $cD/3$, possibly terminal quotient singularities of index 3 along the line $(\mathbf{y} = \mathbf{u} = 0)$ and possibly singularities at worst of type cD. There is no other singularity outside $U_3 \cup U_4$.

Case 3. $wt_{3,1}(h_4) \geq 5$ and $\natural$ does not hold.
The similar argument works. Indeed, there is a term $\theta \in \varphi$ among $\{yu^4, yzu^3, yu^6, zu^5, u^6, u^8\}$. The corresponding term in $\tilde{\varphi}$ the equation of $Y \cap U_4$ is among $\{y, yzu, yu, zu, 1, u\}$. It is easy to see that singularities are at worst of type cD or $cD/3$ as in Case 2.

Case 4. $wt_{3,1}(h_4) \geq 5$ and $\natural$ holds.
We then consider a coordinate change that $\bar{y} := y - \lambda u^2$ for some λ so that we may rewrite $P \in X$ as

$$\bar{\varphi} = x^2 + \bar{y}^3 + 3s\bar{y}^2 + \bar{y}g_{v+1} + \bar{y}g_{v+2} + \bar{y}g_> + \bar{h}_{v+2} + \bar{h}_{v+3} + \bar{h}_>$$

similarly.

Since $wt_{3,1}(h_4) \geq 5$, one has either z^3u or $z^4 \in \varphi$. It follows that either z^3u or $z^4 \in \bar{\varphi}$.

Subcase 4-1. Suppose that there is a term $\theta = y^i z^j u^k \in \bar{\varphi}$ such that $6i + 5j + k \leq 16$. We consider weighted blowup $Y \to X$ with weight

$\frac{1}{2}(5,4,3,1)$ instead. By Lemma 29, $Y \cap U_4$ is isolated. The corresponding term $\tilde{\theta} = y^i z^j u^{k'}$ in $Y \cap U_4$ satisfying

$$i + j + k' = j + (3j + k - 10)/2 \leq 3.$$

One sees that $Y \cap U_4$ has at worst cE_6 singularities. Hence $Y \to X$ is a divisorial contraction.

Moreover, $\mathrm{Sing}(Y) \cap U_i = \{Q_i\}$ for $i = 2, 3$, which is a terminal quotient singularity of index 4 and 3. Also $Y \cap U_1$ is non-singular. Therefore feasible resolutions exist.

Subcase 4-2. Suppose that there is no term $\theta = y^i z^j u^k \in \bar{\varphi}$ such that $6i + 5j + k \leq 16$. We consider weighted blowup $Y \to X$ with weight $v_3 = \frac{1}{2}(9,6,5,1)$ instead. Note that in this situation, $wt_{v_3}\bar{\varphi} = 9$ and $z^4 \in \bar{\varphi}$. It is easy to see that $\mathrm{Sing}(Y) \cap U_4$ is isolated by Lemma 28 or by direct computation. Indeed, $Y \cap U_4$ has at worst singularities of type cE_6. Hence $Y \to X$ is a divisorial contraction.

Moreover, $\mathrm{Sing}(Y)_{\mathrm{ind}>1} = \{Q_3\}$ which is of index 5. Another higher index point is a point $R \in (\mathbf{z} = \mathbf{u} = 0)$, which is terminal quotient of index 3. We thus conclude that a feasible resolution exists for any $cE/2$ point by Theorem 34 and results in previous sections. Q.E.D.

6.4. Resolution of cE_7 points

In this subsection, we consider cE_7 points.

Theorem 36. *There is a feasible resolution for any cE_7 singularity.*

Proof. We shall consider weights $v_{18} = (9,6,4,1)$, $v_{14} = (7,5,3,1)$, $v_{12} = (6,4,3,1)$, $v_9 = (5,3,2,1)$, $v_8 = (4,3,2,1)$, $v_6 = (3,2,2,1)$, $v_5 = (3,2,1,1)$ and discuss as in cE_6 case.

Case 1. $wt_{v_{18}}(f) < 18, \ldots, wt_{v_6}(f) < 6$.

We consider weighted blowup with weight $v_5 = (3,2,1,1)$.

Since $z^3 \in g_v$, we have that $yz^3 \in f$, E is irreducible, and $Y \cap U_3$ is non-singular. Hence, $Y \to X$ is a divisorial contraction by Theorem 5.

By Lemma 7, $Y \cap U_i$ is non-singular away from Q_i for $i = 1, 2, 3$. Notice that there is a term θ with $wt_{v_6}(\theta) < 6$ and $wt_{v_5}(\theta) \geq 5$. It follows that θ is yu^3 or u^5. Hence Q_4 is either non-singular or $Q_4 \notin Y$. Therefore, $\mathrm{Sing}(Y) = \mathrm{Sing}(Y)_{\mathrm{ind}>1} = \{Q_1, Q_2\}$, which are terminal quotient points of index 3 and 2 respectively.

Case 2.$wt_{v_{18}}(f) < 18, \ldots, wt_{v_8}(f) < 8$, $wt_{v_6}(f) \geq 6$.

We consider weighted blowup with weight $v_6 = (3,2,2,1)$ and proceed as in Case 2 of cE_6, then $\mathrm{wBl}_{v_6} : Y \to X$ is a divisorial contraction and $\mathrm{Sing}(Y) \cap U_4$ is at worst of type cD.

We consider $Y \cap U_3$. Since $yz^3 \in \varphi$, one sees that Q_3 is at worst of type $cD/2$. By Corollary 10, $\mathrm{Sing}(Y)_{\mathrm{ind}=1} \cap U_3$ is at worst of type cD.

By Lemma 6, $\mathrm{Sing}(Y)_{\mathrm{ind}=1} \cap U_1 = \emptyset$. Together with $Q_2 \notin Y$, we conclude that $\mathrm{Sing}(Y)_{\mathrm{ind}=1}$ is at worst of type cD and $\mathrm{Sing}(Y)_{\mathrm{ind}>1} = \{Q_3\}$, of type $cD/2$. Feasible resolution exists for this case.

Case 3. $wt_{v_{18}}(f) < 18, \ldots, wt_{v_9}(f) < 9$, $wt_{v_8}(f) \geq 8$.
We consider weighted blowup with weight $v_8 = (4,3,2,1)$.

There is a term $\theta = y^i z^j u^k$ satisfying $wt_{v_9}(\theta) < 9$ and $wt_{v_8}(\theta) \geq 8$. Hence either g_v or h_v contains θ and is non-zero. By Lemma 29, $\mathrm{Sing}(Y) \cap U_4$ is isolated.

By the same argument as in Case 3 of cE_6, one sees that $\mathrm{Sing}(Y) \cap U_4$ is at worst of type cE_6. This implies in particular that $Y \to X$ is a divisorial contraction.

Since both y^3, yz^3 are in φ, by Lemma 7, one has $Y \cap U_i$ is nonsingular away from Q_i for $i = 2, 3$. Together with $\mathrm{Sing}(Y)_{\mathrm{ind}=1} \cap U_1 = \emptyset$, we are done.

Case 4. $wt_{v_{18}}(f) < 18, \ldots, wt_{v_{12}}(f) < 12$, $wt_{v_9}(f) \geq 9$.
We consider weighted blowup with weight $v_9 = (5,3,2,1)$. One has $z^3 \in g_v \neq -3s^2$. Hence $\natural$ does not hold and $\mathrm{Sing}(Y) \cap U_4$ is isolated by Lemma 28.

We consider $Y \cap U_4$. Since $wt_{v_{12}}(\theta) < 12$ and $wt_{v_9}(\theta) \geq 9$ for some $\theta = y^i z^j u^k \in \varphi$, we have $\tilde{\theta} = y^i z^j u^{k'} \in \tilde{\varphi}$ with $i+j+k' = 4i+3j+k-9 \leq 2$. It follows easily that $Y \cap U_4$ has at worst singularity of type cA by Lemma 8. Therefore, $Y \to X$ is a divisorial contraction with discrepancy 1.

By Lemma 7, one sees that $Y \cap U_i$ is non-singular away from Q_i for $i = 1, 3$. Moreover, $Q_2 \notin Y$. Feasible resolution exists for this case.

Case 5.$wt_{v_{18}}(f) < 18$, $wt_{v_{14}}(f) < 14$, $wt_{v_{12}}(f) \geq 12$.
We consider the weighted blowup with weight $v_{12} = (6,4,3,1)$. One has $z^3 \in g_{v+1} \neq 0$.

Subcase 5-1. Suppose that $\natural$ does not hold.
Then $Y \to X$ is a divisorial contraction and $\mathrm{Sing}(Y)_{\mathrm{ind}=1} \subset U_4$ by Proposition 33.

Indeed, by the discussion in 30, we may assume that $\mathrm{Sing}(Y)_{\mathrm{ind}=1}$ singularities at worst of type cD or a singularity of type cE_7 at Q_4. Clearly, we have

$$\begin{cases} \tau^*(\tilde{\varphi}) < \tau^*(\varphi), \\ wt_{v_{18}}(\tilde{\varphi}) \leq wt_{v_{18}}(\varphi), \\ wt_{v_{14}}(\tilde{\varphi}) \leq wt_{v_{14}}(\varphi). \end{cases}$$

By induction on τ^*, we are reduced to the case that $wt_{v_{12}} < 12$.

Subcase 5-2. Suppose that $\natural$ hold.
Notice that there is $\theta \in \varphi$ with $\varphi_{v_{14}}(\theta) < 14, \varphi_{v_{12}}(\theta) \geq 12$, it is easy to see that $\theta \in yg_v$, h_v or in h_{v+1}. This implies in particular that

$s \neq 0$. We consider a coordinate change that $\bar{y} := y - s(z, u)$ for some $s = \alpha zu + \beta u^4$ so that we may write $P \in X$ as

$$\bar{\varphi} = x^2 + \bar{y}^3 + 3s\bar{y}^2 + \bar{y}g_{v+1} + \bar{y}\bar{g}_> + \bar{h}_>$$

similarly.

We consider weighted blowup with weight $v_{14} = (7, 5, 3, 1)$ instead in this situation. Since $\bar{y}z^3 \in \bar{\varphi}$, by Lemma 29, one sees that $\mathrm{Sing}(Y) \cap U_4$ is isolated. One can check that $\mathrm{Sing}(Y) \cap U_4$ has at worst singularity of type cD for $s \neq 0$. Therefore, $Y \to X$ is a divisorial contraction.

One can easily check that for $i = 2, 3$, $\mathrm{Sing}(Y) \cap U_i = \{Q_i\}$, which is terminal quotient of index 5 and 3 respectively. Moreover, $Q_1 \notin Y$ and hence there exists a feasible resolution.

Case 6. $wt_{v_{18}}(f) < 18$, $wt_{v_{14}}(f) \geq 14$.
We consider the weighted blowup with weight $v_{14} = (7, 5, 3, 1)$. Similarly, one has $g_3 = z^3$ and $z^3 \in g_v \neq 0$. By Lemma 29, $\mathrm{Sing}(Y) \cap U_4$ is isolated. Since $wt_{v_{18}}(\theta) < 18$ and $wt_{v_{14}}(\theta) \geq 14$ for some $\theta = y^i z^j u^k \in \varphi$, we have $\tilde{\theta} = y^i z^j u^{k'} \in \tilde{\varphi}$ with $i + j + k' = 6i + 4j + k - 14 \leq 3$. One can verify that any $R \in \mathrm{Sing}(Y) \cap U_4$ is at worst of type cE_6. Therefore $Y \to X$ is a divisorial contraction.

By Lemma 7, we have that $Y \cap U_i$ is nonsingular away from Q_i for $i = 2, 3$. Moreover $Q_1 \notin Y$, hence feasible resolution exists for this case.

Case 7. $wt_{v_{18}}(f) \geq 18$
We consider the weighted blowup with weight $v_{18} = (9, 6, 4, 1)$. Since $wt_{v_{18}}(g) \geq 18$ and $\tau(g) = 3$, we have $g_3 = z^3$ and $z^3 \in g_v \neq -3s^2$. It is clear that $\natural$ does not holds. By Proposition 33, $Y \to X$ is a divisorial contraction and $\mathrm{Sing}(Y)_{\mathrm{ind}=1} \subset U_4$.

$\mathrm{Sing}(Y) \cap U_4$ is at worst of type cE_7. If there is a singularity of type cE_7, then we proceed by induction in τ^*. Then it can be reduced to the cases with $wt_{v_{18}} < 18$.

This completes the proof that a feasible resolution exist for cE_7 singularity. Q.E.D.

6.5. Resolution of cE_8 points

In this subsection, we shall prove that

Theorem 37. *There is a feasible resolution for any cE_8 singularity.*

Proof. We will need to consider weights $v_{30} = (15, 10, 6, 1)$, $v_{24} = (12, 8, 5, 1)$,...etc.

Case 1.$wt_{v_{30}}(f) < 30, \ldots, wt_{v_8}(f) < 8$.
We consider weighted blowup with weight $v_6 = (3, 2, 2, 1)$. By 6.2, we have that $wt_{v_6}(f) \geq 6$ always holds and $z^5 \in h$. We consider weighted blowup with weight $v_6 = (3, 2, 2, 1)$ and proceed as in Case 2 of cE_6,

then $\mathrm{wBl}_{v_6} : Y \to X$ is a divisorial contraction and $\mathrm{Sing}(Y) \cap U_4$ is at worst of type cD.

We consider $Y \cap U_3$. We have that $z^5 \in \varphi$. Hence Q_3 is at worst of type $cE/2$ and $\mathrm{Sing}(Y)_{\mathrm{ind}=1} \cap U_3$ is at worst of type cE_6. By Lemma 6, $\mathrm{Sing}(Y)_{\mathrm{ind}=1} \cap U_1 = \emptyset$. Together with $Q_2 \notin Y$, we conclude that feasible resolution exists for this case.

Case 2.$wt_{v_{30}}(f) < 30, \ldots, wt_{v_9}(f) < 9$, $wt_{v_8}(f) \geq 8$

We consider weighted blowup with weight $v_8 = (4, 3, 2, 1)$. By the same argument as in Case 3 of cE_7, one has that $Y \to X$ is a divisorial contraction with $\mathrm{Sing}(Y) \cap U_4$ at worst of type cE_6.

Since y^3 is in φ, by Lemma 7, one has that $Y \cap U_2$ is nonsingular away from Q_2. Together with $\mathrm{Sing}(Y)_{\mathrm{ind}=1} \cap U_1 = \emptyset$ and Q_3 is of type $cAx/2$, we are done.

Case 3. $wt_{v_{30}}(f) < 30, \ldots, wt_{v_{12}}(f) < 12$, $wt_{v_9}(f) \geq 9$

We consider weighted blowup with weight $v_9 = (5, 3, 2, 1)$. Since $h_5 \neq 0$, we have either z^4u or $z^5 \in h$. Now the same argument as in Case 4 of cE_7 goes through.

Case 4.$wt_{v_{30}}(f) < 30, \ldots, wt_{v_{14}}(f) < 14$, $wt_{v_{12}}(f) \geq 12$.

We consider weighted blowup with weight $v_{12} = (6, 4, 3, 1)$. The proof is essentially parallel to Case 5 of cE_7. Note that we have $h_5 = z^5$ or $z^4u \in f$.

Subcase 4-1. Suppose $\natural$ does not hold.

Then $Y \to X$ is a divisorial contraction and $\mathrm{Sing}(Y)_{\mathrm{ind}=1} \subset U_4$ by Theorem 31 and Proposition 33.

Indeed by the discussion in 30, we know that either $\mathrm{Sing}(Y)_{\mathrm{ind}=1}$ consists of singularities at worst of type cE_6 or we may assume that Q_4, the only singularity in U_4, is of type cE_8.

Clearly, we have

$$\begin{cases} \tau^*(\tilde{\varphi}) < \tau^*(\varphi), \\ wt_{v_l}(\tilde{\varphi}) \leq wt_{v_l}(\varphi), \end{cases}$$

for all $l \geq 12$. By induction on τ^*, we are done.

Subcase 4-2. Suppose that $\natural$ hold.

As in Subcase 5-2 of cE_7, we consider a coordinate change and then the weighted blowup with weight $v_{14} = (7, 5, 3, 1)$ instead in this situation. Since $z^5 \in \bar{\varphi}$, by Lemma 29, one sees that $\mathrm{Sing}(Y) \cap U_4$ is isolated. One can check that $\mathrm{Sing}(Y) \cap U_4$ has at worst singularity of type cD for $s \neq 0$. Therefore, $Y \to X$ is a divisorial contraction.

One can easily check that for $i = 2, 3$, $\mathrm{Sing}(Y) \cap U_i = \{Q_i\}$, which is terminal quotient of index 5 and 3 respectively. Moreover, $Q_1 \notin Y$ and hence there exists a feasible resolution.

Case 5.$wt_{v_{30}}(f) < 30, \dots, wt_{v_{18}}(f) < 18,\ wt_{v_{14}}(f) \geq 14$
We can proceed as in Subcase 6 of cE_7. Since $z^5 \in h_{v+1} \neq 0$, we still have that $\mathrm{Sing}(Y) \cap U_4$ has isolated singularities by Lemma 29. Thus the same conclusion holds.
Case 6.$wt_{v_{30}}(f) < 30,\ wt_{v_{24}}(f) < 24, wt_{v_{20}}(f) < 20,\ wt_{v_{18}}(f) \geq 18$
We consider weighted blowup with weight $v_{18} = (9, 6, 4, 1)$.

Since there is a term $\theta \in f$ with $wt_{v_{20}}(\theta) < 20$ and $wt_{v_{18}}(\theta) \geq 18$. This implies $\tilde{\theta}$ is in g_v, h_v or h_{v+1}.
Subcase 6-1. Suppose $\natural$ does not hold.
Then $Y \to X$ is a divisorial contraction by Theorem 31.

One sees that $Y \cap U_3$ is given by $(\tilde{\varphi} : x^2 + y^3 + z^2 + \text{other terms} = 0) \subset \mathbb{C}^4/\frac{1}{4}(1, 2, 3, 1)$. Therefore, $\mathrm{Sing}(Y) \cap U_3$ is type $cAx/4$ or cA. We also have $\mathrm{Sing}(Y)_{\mathrm{ind}=1} \cap U_1 = \emptyset$, and $Q_2 \notin Y$.

It remains to consider $Y \cap U_4$. Notice that the corresponding term $\tilde{\theta} = y^i z^j u^{k'}$ has $i + j + k' \leq 4$ unless $\theta = z^4 u^3, \tilde{\theta} = z^4 u$. If $i + j + k' \leq 4$, then $Y \cap U_4$ has singularities at worst of type cE_7 and hence feasible resolution exists. Suppose that $\tilde{\theta} = z^4 u \in \tilde{\varphi}$. Hence

$$wt_{v_l}(\tilde{\varphi}) < l,$$

for $l = 30, 24, 20, 18$. Therefore, $Y \cap U_4$ has singularities at worst of type cE_8 in Subcase 1-4.
Subcase 6-2. Suppose that $\natural$ hold.
We first consider a coordinate change that $\bar{y} := y - s(z, u)$ with $s(z, u) \neq 0$ since there is θ in yg_v, h_v or h_{v+1}. Now $P \in X$ is defined as

$$\bar{\varphi} = x^2 + \bar{y}^3 + 3s\bar{y}^2 + \bar{y}g_{v+1} + \bar{y}\bar{g}_> + \bar{h}_>$$

and we consider weighted blowup with weight $v_{20} = (10, 7, 4, 1)$ instead in this situation.

Since $z^5 \in \varphi$ and hence $z^5 \in \bar{\varphi}$, one sees that $\mathrm{Sing}(Y) \cap U_4$ is isolated, by Lemma 29. Since $s = (\alpha z u^2 + \beta u^6) \neq 0$, we have either $\bar{y}^2 z u^2$ or $\bar{y}^2 u^6 \in \bar{\varphi}$. One can check that $Y \cap U_4$ has at worst singularities of type cD and thus $Y \to X$ is a divisorial contraction.

Together with the fact that $\mathrm{Sing}(Y)_{\mathrm{ind}=1} \cap U_1$ is empty, $Y \cap U_2$ is non-singular away from Q_2 and $Q_3 \in \mathrm{Sing}(Y)_{\mathrm{ind}>1}$, one sees that a feasible resolution exists.
Case 7. $v_{30}(f) < 30,\ wt_{v_{24}}(f) < 24,\ wt_{v_{20}}(f) \geq 20$
We consider the weighted blowup with weight $v_{20} = (10, 7, 4, 1)$. Since $z^5 \in h_v$, we have $Q_3 \notin Y$ and $\mathrm{Sing}(Y) \cap U_4$ is isolated by Lemma 29.

We work on U_4. There is a term $\theta = y^i z^j u^k \in f$ with $wt_{v_{24}}(\theta) < 24$ and $wt_{v_{20}}(\theta) \geq 20$. Hence $\tilde{\theta} = y^i z^j u^{k'}$ with $i + j + k' \leq 3$. It follows that $\mathrm{Sing}(Y) \cap U_4$ is at worst of type cE_6.

Therefore, $Y \to X$ is a divisorial contraction. Together with the fact that $\mathrm{Sing}(Y)_{\mathrm{ind}=1} \cap U_1$ is empty, $Y \cap U_2$ is non-singular away from Q_2 and $Q_3 \in \mathrm{Sing}(Y)_{\mathrm{ind}>1}$, one sees that a feasible resolution exists.

Case 8.$wt_{v_{30}}(f) < 30$, $wt_{v_{24}}(f) \geq 24$.
We consider the weighted blowup with weight $v_{24} = (12, 8, 5, 1)$. One notices that $\tau(h) = 5$ implies that $z^5 \in h_{v+1}$. Since $wt_{v_{24}}(g_{v+1}) = 17$ and hence $u^2 | g_{v+1}$. It follows that $z^5 \notin s(z,u)g_{v+1}$ and $\natural$ does not hold. Therefore $Y \to X$ is a divisorial contraction and $\mathrm{Sing}(Y)_{\mathrm{ind}=1} \subset U_4$ by Theorem 31 and Proposition 33.

Unless $\mathrm{Sing}(Y) \cap U_4 = \{Q_4\}$ is of type cE_8, we have feasible resolution of Y. If $\mathrm{Sing}(Y) \cap U_4 = \{Q_4\}$ is of type cE_8, then we have

$$\begin{cases} \tau^*(\tilde{\varphi}) < \tau^*(\varphi); \\ wt_{v_{30}}(\tilde{\varphi}) \leq wt_{v_{30}}(\varphi); \\ wt_{v_{24}}(\tilde{\varphi}) \leq wt_{v_{24}}(\varphi). \end{cases}$$

By induction on τ^*, the existence of feasible resolution is thus reduced to the existence of feasible resolution of milder singularity or to the existence of feasible resolution of cE_8 singularities with $wt_{v_{24}} < 24$.

Case 9. $wt_{v_{30}}(f) \geq 30$.
We consider the weighted blowup with weight $v_{30} = (15, 10, 6, 1)$. The similar argument as in Case 8 works.

This completes the proof. Q.E.D.

Proof of Main Theorem. This follows from Theorem 12, 13, 20, 24, 34, 35, 36, 37. Q.E.D.

References

[1] M. Artin, *Algebraic approximation of structures over complete local rings.* Inst. Hautes E'tudes Sci. Publ. Math. No. **36**, 1969 23–58.

[2] J. A. Chen and C. D. Hacon, *Factoring birational maps in dimension three*, J. Reine Angew. Math. **657** (2011), 173–197.

[3] S. D. Cutkosky, *Elementary contractions of Gorenstein threefolds*, Math. Ann. **280** (1988), no. 3, 521–525.

[4] T. Hayakawa, *Blowing ups of 3-dimensional terminal singularities*, Publ. Res. Inst. Math. Sci. **35** (1999), no. 3, 515–570.

[5] T. Hayakawa, *Blowing ups of 3-dimensional terminal singularities. II*, Publ. Res. Inst. Math. Sci. **36** (2000), no. 3, 423–456.

[6] T. Hayakawa, *Divisorial contractions to cD points*, preprints.

[7] T. Hayakawa, *Divisorial contractions to cE points*, preprints.

[8] M. Kawakita, *Divisorial contractions in dimension three which contract divisors to smooth points*, Invent. Math. **145** (2001), no. 1, 105–119.

[9] M. Kawakita, *Three-fold divisorial contractions to singularities of higher indices*, Duke Math. J. **130** (2005), no. 1, 57–126.
[10] M. Kawakita, *Supplement to classification of three-fold divisorial contractions*, Nagoya Math. J. **206** (2012), 67–73.
[11] Y. Kawamata, *Divisorial contractions to 33-dimensional terminal quotient singularities*, Higher-dimensional complex varieties (Trento, 1994), 241–246, de Gruyter, Berlin, 1996.
[12] J. Kollár, *Flops*. Nagoya Math. J. **113** (1989), 15–36.
[13] J. Kollár and S. Mori, *Classification of three-dimensional flips*, J. Amer. Math. Soc. **5** (1992), no. 3, 533–703.
[14] J. Kollár, *Real algebraic threefolds. II. Minimal model program*. J. Amer. Math. Soc. **12**, (1999), no. 1, 33–83.
[15] J. Kolla'r, and N.I. Shepherd-Barron, *Threefolds and deformations of surface singularities*, Invent. Math. **91** (1988), no. 2, 299–338.
[16] D. Markushevich, *Minimal discrepancy for a terminal cDV singularity is* 1, J. Math. Sci. Univ. Tokyo **3** (1996), no. 2, 445–456.
[17] S. Mori, *Threefolds whose canonical bundles are not numerically effective*, Ann. Math. **116** (1982), 133–176.
[18] S. Mori, *On* 3*-dimensional terminal singularities*, Nagoya Math. J. **98** (1985), 43–66.
[19] S. Mori, *Flip theorem and the existence of minimal models for* 3*-folds*, J. Amer. Math. Soc. **1** (1988), no. 1, 117–253.
[20] M. Reid, *Canonical 3-folds*, Journeés de Géometrie Algébrique d'Angers, Juillet 1979/Algebraic Geometry, Angers, 1979, pp. 273–310, Sijthoff & Noordhoff, Alphen aan den Rijn—Germantown, Md., 1980.
[21] M. Reid, *Minimal models of canonical 3-folds*, Adv. Stud. Pure Math. **1**(1983), 131-180.
[22] M. Reid, *Young person's guide to canonical singularities*, Proc. Symposia in pure Math. **46**(1987), 345-414.
[23] V. V. Shokurov, *A nonvanishing theorem*, Izv. Akad. Nauk SSSR Ser. Mat. **49** (1985), no. 3, 635–651.

Department of Mathematics, National Taiwan University, Taipei, 106, Taiwan
E-mail address: jkchen@math.ntu.edu.tw

Research Institute for Mathematical Sciences, Kyoto University, Kyoto 606-8502, Japan

National Center for Theoretical Sciences, Taipei, 106, Taiwan

Advanced Studies in Pure Mathematics 70, 2016
Minimal Models and Extremal Rays (Kyoto, 2011)
pp. 361–377

On finiteness of B-representations and semi-log canonical abundance

Christopher D. Hacon and Chenyang Xu

Dedicated to Professor Shigefumi Mori on his 60th Birthday

Abstract.

We give a new proof of the finiteness of **B**-representations. As a consequence of the finiteness of **B**-representations and Kollár's gluing theory on lc centers, we prove that the (relative) abundance conjecture for slc pairs is implied by the abundance conjecture for log canonical pairs.

§1. Introduction

Throughout this note, the ground field will be the field $\mathbb{C}$ of complex numbers. It is well known that even though the log minimal model program is focused on the study of log pairs (X, Δ) where X is a normal variety, for technical reasons it is often necessary to deal with log pairs (X, Δ) where X is a semi-normal variety. This naturally occurs in proofs by induction on the dimension where, for example, we restrict to the reduced part of the boundary of a dlt pair (X, Δ) (cf. e.g. [Kolláretal92, KMM94, Birkar12, HX13, FG11]) or when we study moduli of pairs, as normal varieties can degenerate to

Received August 23, 2011.
Revised January 12, 2013; September 18, 2013.
2010 *Mathematics Subject Classification.* 14E07, 14E30.
Key words and phrases. **B**-representations, semi-log canonical abundance.
We are indebted to Y. Gongyo, J. McKernan, J. Kollár, V. Shokurov and R. Zong for helpful conversations. Especially, the main ideas in Section 4 follow [Kollár11]. Part of this work was done while the second author was visiting RIMS, which he would like to thank for the hospitality. We would like to thank the referee for many useful suggestions. The first author was partially supported by NSF research grants no: 0757897, 1300750 and a grant from the Simons foundation, the second author was partially supported by the grant 'The Recruitment Program of Global Experts'.

non-normal ones (cf. [Kollár13, HX13]). In [Fujino00], O. Fujino first used the **B**-representation to study the semi-log canonical abundance conjecture and proved the conjecture in the 3 dimensional case. Recently, J. Kollár has developed a useful technique for gluing log canonical centers (cf. [Kollár13, Kollár11]) that reduces many questions on semi-normal pairs to questions on their normalizations. An important result used in Kollár's theory and in Fujino's work (cf. [Fujino00]), is the finiteness of **B**-representations, which was first proved by Deligne-Nakamura-Ueno in the klt case, and then generalized to the log canonical case by Fujino-Gongyo (cf. [FG11]). Recall the following.

Definition 1.1. Let (X, Δ) be a projective dlt pair. We define the birational automorphism group $\mathrm{Bir}(X, \Delta)$ of (X, Δ) to be the group of all birational maps g of X such that if we take a common resolution

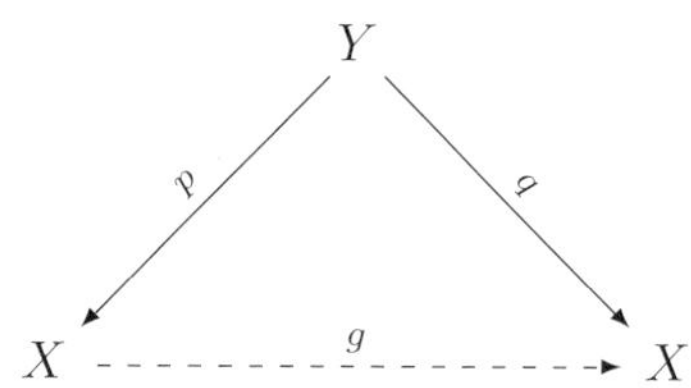

then $p^*(K_X + \Delta) = q^*(K_X + \Delta)$. We call the induced homomorphism

$$\rho_m : \mathrm{Bir}(X, \Delta) \to \mathrm{Aut}(H^0(X, \mathcal{O}_X(m(K_X + \Delta))))$$

the **B-representation** of the pair (X, Δ). As far as we know, **B**-birational maps and **B**-representations for general log pairs were first explicitly introduced in [Fujino00].

In this note we first aim to give a new proof of the following result.

Theorem 1. *Given a projective dlt pair (X, Δ) such that $K_X + \Delta$ is a semi-ample $\mathbb{Q}$-divisor. There exists $m \in \mathbb{N}$, such that the image of the **B**-representation*

$$\rho_M : \mathrm{Bir}(X, \Delta) \to \mathrm{Aut}(H^0(X, \mathcal{O}_X(M(K_X + \Delta))))$$

is finite for any positive integer M divisible by m.

Remark 1.2. Theorem 1 was proven by different methods in [FG11, 1.1]. The argument in the current note was originally contained in [HMX14]. During the preparation of [HMX14] we were informed of [FG11] and we decided to include Theorem 1 in a separated paper. Our proof uses the case when the Kodaira dimension is 0, which was proved by Gongyo (cf. [Gongyo13]) using ideas from [Fujino00].

The second part of our paper is focused on the study of semi-log canonical abundance. One of the main applications of the finiteness of **B**-representations is to prove that abundance for semi-log canonical pairs, follows from abundance for log canonical pairs (cf. [Fujino00, FG11]). Using Kollár's gluing theory, as a consequence of Theorem 1, we prove the following.

Theorem 2. *Let (X, Δ) be a semi-log canonical pair, $f : X \to S$ a projective morphism, $n : \bar{X} \to X$ the normalization and write $n^*(K_X + \Delta) = K_{\bar{X}} + \bar{\Delta} + \bar{D}$, where $\bar{D}$ is the double locus. If $K_{\bar{X}} + \bar{\Delta} + \bar{D}$ is semi-ample over S, then $K_X + \Delta$ is semi-ample over S.*

As a corollary we recover the following result conjectured by C. Birkar (cf. [Birkar12, 1.2]), which is known to be a natural step of the log canonical minimal model program in the relative case (cf. [KMM94, Section 7]).

Corollary 1.3. *Let (X, Δ) be a $\mathbb{Q}$-factorial dlt pair which is projective over a variety S, and $T := \lfloor \Delta \rfloor$ where Δ is a $\mathbb{Q}$-divisor. Suppose that*

(1) *$K_X + \Delta$ is nef over S,*
(2) *$(K_X + \Delta)|_{T_i}$ is semi-ample over S for each component T_i of T,*
(3) *$K_X + \Delta - \epsilon P$ is semi-ample over S for some $\mathbb{Q}$-divisor $P \geq 0$ with $\mathrm{Supp}(P) = T$ and for any sufficiently small rational number $\epsilon > 0$.*

Then, $K_X + \Delta$ is semi-ample over S.

Another corollary is the following result, which answers a question raised by J. Kollár in the problem session in MSRI in March 2009.

Corollary 1.4. *Let (X, Δ) be a semi-log canonical pair and $f : X \to S$ a projective morphism, such that $K_X + \Delta \equiv_{\mathbb{Q},S} 0$. Then $K_X + \Delta \sim_{\mathbb{Q},S} 0$.*

Remark 1.5. A result due to O. Fujino and Y. Gongyo (cf. [FG11, 4.13]), implies Theorem 2 when the base S is projective. Recent work of J. Kollár on gluing lc centers provides us with a technique to prove the general case.

The absolute case of Corollary 1.4 or the case when S is projective is also known (cf. [Gongyo13]). However, the general relative case does not seem to be available anywhere in the literature (see the remark in [FG11, 4.16]).

§2. Hodge theoretic construction

Construction 2.1 (cf. [Kollár07a, 8.4.6]). Let (X, Δ) be a log canonical pair, and $f : X \to Y$ a proper surjective morphism of normal varieties with connected fibers such that $n = \dim X - \dim Y$ and $K_X + \Delta \sim_{\mathbb{Q},Y} 0$.

Let $p : W \to X$ be a log resolution of (X, Δ). Write

$$p^*(K_X + \Delta) = K_W + E + F - G,$$

where E and G are integral effective divisors with no common components and $F = \{F\}$. Let a be an integer such that aF is an integral divisor. Denote by $\phi = f \circ p : W \to Y$.

Let $Y^0 \subset Y$ be a smooth open subset such that ϕ is smooth over Y^0. We denote by $\bullet^0$ the base change over Y^0. Replacing Y^0 by an open subset, we may assume that $(K_X + \Delta)|_{X^0} \sim_{\mathbb{Q}} 0$. We define the line bundle $V^0 = \omega_{W^0}^{-1}(G^0 - E^0)$ so that $V^{0\otimes a} \cong \mathcal{O}_{W^0}(aF^0)$. This data defines a local system $\mathbb{V}^0$ on $W^0 \setminus \operatorname{Supp}(E^0 \cup F^0)$ (cf. [Kollár07a, 8.4.6], [EV92, 3.2] and its proof).

Consider the normalization of the corresponding μ_a-cover $\pi : W' \to W^0$, and denote by E' the reduced divisor supported on $\pi^* E^0$. The push-forward $\pi_*(\mathbb{C}|_{W' \setminus E'})$ has a μ_a-action. If we decompose

$$\pi_*(\mathbb{C}|_{W' \setminus E'}) = \bigoplus_i \pi_*(\mathbb{C}|_{W' \setminus E'})^{(i)}$$

into the corresponding eigenspaces, then $\mathbb{V}^0$ is isomorphic to the restriction of $\pi_*(\mathbb{C}|_{W' \setminus E'})^{(1)}$ to $W^0 \setminus (E^0 \cup F^0)$. We denote $\pi_*(\mathbb{C}|_{W' \setminus E'})^{(1)}$ by $\mathbb{V}$. So $\mathbb{V}$ is determined up to the choice of a unit in $\mathcal{O}_{Y^0}$. However, $(R^n \phi_* \mathbb{V})^{\otimes a}$ (we will sloppily denote $\phi|_{W^0 \setminus E^0}$ by ϕ) is a well defined local system on Y^0 (cf. [Kollár07a, 8.4.7]).

Denote by $\phi' : W' \to Y^0$ the composite morphism $\phi \circ \pi$ (and its restriction to open subsets). Then $R^n \phi'_* \mathbb{V}$ is a direct summand of $R^n \phi'_* \mathbb{C}|_{W' \setminus E'}$ which carries a variation of mixed Hodge structure. We remark that even though W' has quotient singularities, locally (W', E') is a finite quotient of a log smooth pair and hence Hodge theoretically it behaves as a smooth variety with a simple normal crossing divisor, at least for $\mathbb{Q}$-coefficients (cf. [Steenbrink77, Section 1]).

The local system $R^n \phi_* \mathbb{V}$ gives a variation of mixed Hodge structure on Y^0 (see [FF12, FFS14] and the references therein). It follows from [Steenbrink77, 1.18] that the bottom piece of the Hodge filtration gives an isomorphism

$$F^n R^n \phi'_*(\mathbb{C}|_{W' \setminus E'}) \cong \phi'_* \omega_{W'/Y}(E').$$

By taking the corresponding eigenspaces, we conclude that $F^n R^n \phi_*(\mathbb{V}|_{W^0 \setminus E^0}) \cong \phi_* \mathcal{O}_{W^0}(G^0)$ is a line bundle over Y^0 (cf. [Kollár07a, 8.4.5(7')]).

Denote this line bundle by L. Since L is of rank 1, there exists an integer i such that, $L \subset W_{n+i}(R^n\phi_*\mathbb{V})$ but $L \not\subset W_{n+i-1}(R^n\phi_*\mathbb{V})$. Let $\mathbb{H}$ be the smallest pure sub-$\mathbb{Q}$-VHS of $\mathrm{Gr}^W_{n+i}(R^n\phi_*\mathbb{V})$ which contains L.

Lemma 2.2. *$\mathbb{H}$ does not depend on the choice of the resolution of W.*

Remark 2.3. Let $\mathbb{H}_1$ and $\mathbb{H}_2$ be two local systems defined over Zariski dense open subsets $U_1, U_2 \subset Y$ such that $\mathbb{H}_1|_{U_1 \cap U_2} \cong \mathbb{H}_2|_{U_2 \cap U_2}$. Since Y is normal, then there is a unique local system $\mathbb{H}$ (up to a unique isomorphism) defined over $U_1 \cup U_2$, such that its restriction to U_i is isomorphic to $\mathbb{H}_i$.

Proof of Lemma 2.2. From the above remark, it suffices to verify the lemma for any nonempty open set $Y^0 \subset Y$. For two resolutions W_1, W_2, we can assume that there exists a morphism $\psi : W_2 \to W_1$. By shrinking Y^0, we can also assume that W_1 and W_2 are both smooth over Y^0 (cf. Remark 2.3). If $p_1^*(K_X + \Delta) = K_{W_1} + E_1 + F_1 - G_1$, and aF_1 is integral, then

$$p_2^*(K_X + \Delta) = \psi^*(K_{W_1} + E_1 + F_1 - G_1) = K_{W_2} + E_2 + F_2 - G_2,$$

and aF_2 is easily seen to be integral. Applying the construction 2.1 for both W_i (here we choose the same unit in $\mathcal{O}_{Y^0}$ for W_i), we have that W_2' is the normalization of $W_2^0 \times_{W_1^0} W_1'$ and we denote by $\psi' : W_2' \to W_1'$. Let $E^* = \mathrm{Supp}(\psi'^{-1}E_1')$, then we know that $E_2' \subset E^*$.

$$\begin{array}{ccc} E_2' \subset E^* \subset W_2' & \longrightarrow & E_2 \subset W_2 \\ \Big\downarrow \psi' & & \Big\downarrow \psi \\ E_1' \subset W_1' & \longrightarrow & E_1 \subset W_1 \end{array}$$

Therefore, there exist morphisms between mixed Hodge structures on Y^0

$$i : R^n\phi'_{2*}(\mathbb{C}|_{W_2' \setminus E_2'}) \to R^n\phi'_{2*}(\mathbb{C}|_{W_2' \setminus E^*}) \qquad \text{and}$$

$$j : R^n\phi'_{1*}(\mathbb{C}|_{W_1' \setminus E_1'}) \to R^n\phi'_{2*}(\mathbb{C}|_{W_2' \setminus E^*}).$$

Applying the Hodge filtration $F^n(\cdot)$, it follows from [Steenbrink77, 1.18] that

$$F^n R^n\phi'_{2*}(\mathbb{C}|_{W_2' \setminus E_2'}) \cong \phi'_{2*}\omega_{W_2'/Y}(E_2'),$$

and similarly for the other pairs. Thus we have morphisms

$$F^n i : \phi'_{2*}\omega_{W'_2/Y}(E'_2) \to \phi'_{2*}\omega_{W'_2/Y}(E^*) \qquad \text{and}$$

$$F^n j : \phi'_{1*}\omega_{W'_1/Y}(E'_1) \to \phi'_{2*}\omega_{W'_2/Y}(E^*).$$

which are isomorphisms by Lemma 2.4.

Replacing Y^0 by a smaller open set, by the discussion in 2.1, there are morphisms between mixed Hodge structures,

$$R^n\phi_{1*}\mathbb{V}_1 \to R^n\phi_{2*}(\mathbb{C}|_{W'_2\setminus E^*}) \leftarrow R^n\phi_{2*}\mathbb{V}_2.$$

We conclude that if we take $F^n(\cdot)$ of each term above, then we obtain an induced isomorphism $F^n(R^n\phi_{1*}\mathbb{V}_1) \to F^n(R^n\phi_{2*}\mathbb{V}_2)$. Since polarized-VHSs form a semi-simple category, considering $\mathrm{Gr}^W_{n+i}R^n\phi_{j*}\mathbb{V}_j$, we see that the images of $\mathbb{H}_1$ and $\mathbb{H}_2$ are mapped to the same pure Hodge substructure of $\mathrm{Gr}^W_{n+i}R^n\phi_{2*}(\mathbb{C}|_{W'_2\setminus E^*})$. Q.E.D.

Lemma 2.4. *Let $f : Y_2 \to Y_1$ be a birational morphism between normal projective varieties. Let Δ_1 be reduced divisor on Y_1, such that (Y_1, Δ_1) is log canonical. Let $f_*^{-1}(\Delta_1) \leq \Delta_2 \leq f_*^{-1}(\Delta_1) + \mathrm{Ex}(f)$ be an effective Weil divisor on Y_2 whose support contains all divisors of discrepancy -1 with respect to the pair (Y_1, Δ_1). Then there is a natural morphism $f_*\mathcal{O}_{Y_2}(K_{Y_2}+\Delta_2) \to \mathcal{O}_{Y_1}(K_{Y_1}+\Delta_1)$ inducing an isomorphism*

$$H^0(Y_2, \mathcal{O}_{Y_2}(K_{Y_2} + \Delta_2)) \cong H^0(Y_1, \mathcal{O}_{Y_1}(K_{Y_1} + \Delta_1)).$$

Proof. By assumption, we can write

$$F + f^*(K_{Y_1} + \Delta_1) = K_{Y_2} + \Delta_2 + E,$$

where $F, E \geq 0$, F is exceptional and $\lfloor E \rfloor = 0$. Therefore,

$$\begin{aligned} H^0(Y_1, \mathcal{O}_{Y_1}(K_{Y_1} + \Delta_1)) &\cong H^0(Y_2, \mathcal{O}_{Y_2}(\lfloor F + f^*(K_{Y_1} + \Delta_1)\rfloor)) \\ &\cong H^0(Y_2, \mathcal{O}_{Y_2}(K_{Y_2} + \Delta_2 + \lfloor E \rfloor)) \cong H^0(Y_2, \mathcal{O}_{Y_2}(K_{Y_2} + \Delta_2)). \end{aligned}$$

Q.E.D.

§3. Finiteness of B-representations

In this section we prove Theorem 1. By assumption, there exists a positive integer m, such that $|m(K_X + \Delta)|$ is base point free and it induces an algebraic fibration structure $f : X \to Y$, so that

$$Y = \mathrm{Proj} \bigoplus_{d \geq 0} H^0(X, \mathcal{O}_X(dm(K_X + \Delta))).$$

It follows from Kawamata's canonical bundle formula (cf. [Kawamata98, Ambro99]) that we can write

$$K_X + \Delta \sim_{\mathbb{Q}} f^*(K_Y + B + J),$$

where B is the boundary part and J is the moduli part. Let $P \in Y$ be a codimension 1 point, then the coefficient of P in B is defined by

$$1 - \mathrm{lct}(X, \Delta; f^{-1}(P))$$

where the log canonical threshold is computed over a neighborhood of the generic point of P. In particular, B is an effective $\mathbb{Q}$-divisor. The moduli part $J = J(X, \Delta)$ is defined as an equivalence class of $\mathbb{Q}$-divisors, coming from Hodge theory.

Remark 3.1. Note that [Kawamata98] defines J only in the case when the restriction $K_F + \Delta_F = (K_X + \Delta)|_F$ to a general fiber F is klt. If $K_F + \Delta_F$ is only dlt, then consider S a minimal stratum of $\lfloor \Delta \rfloor$ that dominates Y. Let $(\pi \circ f_S) : S \to Y_S \to Y$ be the Stein factorization. Restricting to this stratum we obtain a pair $K_S + \Delta_S = (K_X + \Delta)|_S$ which is klt when restricted to a general fiber F_S of $S \to Y_S$. We may then define the boundary and moduli parts $B_S = B(S, \Delta_S)$ and $J_S = J(S, \Delta_S)$ on Z_S by applying [Kawamata98] to (S, Δ_S). We claim that B_S is the pullback of the divisor $B := B(X, \Delta)$ on Y. To see this, note that replacing Y by a big open subset and X by an appropriate birational model, we may assume that for any codimension 1 point $P \in Y$, $(X, \mathrm{Supp}(\Delta + f^{-1}(P)))$ is log smooth. Let b_P be the coefficient of $B(X, \Delta)$ at P, then $(X, \Delta + b_P f^*(P))$ is dlt and there is a divisor E on X dominating P of coefficient 1 in $\Delta + b_P f^*(P)$. By [Kollár13, 4.42], every irreducible component of $\pi^{-1}(P)$ is dominated by a divisor F on S of coefficient 1 in $\Delta_S + f^*(P)|_S = \Delta_S + f_S^*\pi^*(P)$. The claim now follows easily.

Alternatively, one can define $J(X, \Delta)$ directly by using VMHS as in [Kawamata11], [FF12] and [FFS14]. By what we have observed above, these two definitions coincide, that is $J(X, \Delta) = \frac{1}{\deg \pi}\pi_* J(S, \Delta_S)$.

For any $g \in \mathrm{Bir}(X, \Delta)$ and any $d \geq 0$ we have homomorphisms $\rho_{dm} : \mathrm{Bir}(X, \Delta) \to \mathrm{Aut}_{\mathbb{C}} H^0(X, \mathcal{O}_X(dm(K_X + \Delta)))$ where

$$\rho_{dm}(g) = g^* : H^0(X, \mathcal{O}_X(dm(K_X + \Delta))) \to H^0(X, \mathcal{O}_X(dm(K_X + \Delta))).$$

Consider the induced homomorphism

$$\chi : \mathrm{Bir}(X, \Delta) \to \mathrm{Aut}(Y).$$

For any element $g \in \mathrm{Bir}(X, \Delta)$, there is a commutative diagram

Let F be the geometric generic fiber of f and $n = \dim F = \dim X - \dim Y$ be the relative dimension. Then for any d, we have a short exact sequence

$$1 \to G \to \rho_{dm}(\mathrm{Bir}(X, \Delta)) \to \chi(\mathrm{Bir}(X, \Delta)) \to 1,$$

where $G \subset H^0(Y, \mathcal{O}_Y^*) = \mathbb{C}^*$. We know that G is finite (cf. [Gongyo13, 4.9]), so it suffices to show the following result.

Theorem 3. *The image* $\chi(\mathrm{Bir}(X, \Delta)) \subset \mathrm{Aut}(Y)$ *is finite.*

First it is easy to see the following result.

Lemma 3.2. *The image of* $\chi(\mathrm{Bir}(X, \Delta))$ *is contained in* $\mathrm{Aut}(Y, B)$.

Proof. Let $P \in Y$ be a codimension 1 point, as we noted, the coefficient of P in B is defined by $1 - \mathrm{lct}(X, \Delta; f^{-1}(P))$. This number is unchanged if we replace (X, Δ) by any log resolution. Let W be a log resolution such that the following diagram is commutative

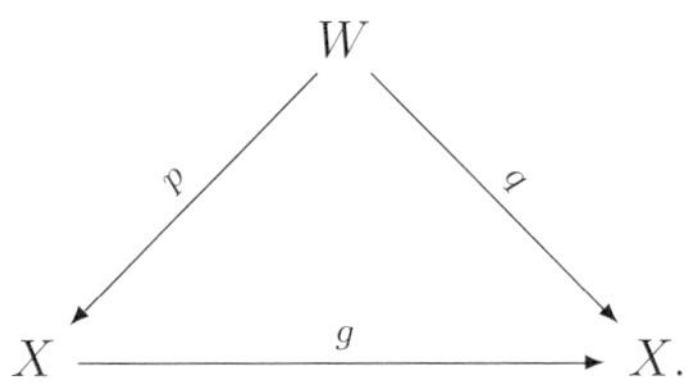

Since $g \in \mathrm{Bir}(X, \Delta)$, we know that we can write $K_W + \Delta_W = p^*(K_X + \Delta) = q^*(K_X + \Delta)$. The coefficient of P in $\chi(g)^* B$ is the same as the coefficient of $\chi(g)_* P$ in B, which is

$$\begin{aligned} 1 - \mathrm{lct}(X, \Delta; f^{-1}(\chi(g)_* P)) &= 1 - \mathrm{lct}(W, \Delta_W; (f \circ q)^{-1}(\chi(g)_* P)) \\ &= 1 - \mathrm{lct}(W, \Delta_W; (\chi(g) \circ f \circ p)^{-1}(\chi(g)_* P)). \end{aligned}$$

Since $\chi(g) \in \mathrm{Aut}(Y)$, the right hand side is the same as $1 - \mathrm{lct}(W, \Delta_W; (f \circ p)^{-1}(P))$, which is the coefficient of P in B. Thus $\chi(g) \in \mathrm{Aut}(Y, B)$. Q.E.D.

By the following result, we see that the birational maps also preserve $\mathbb{H}$, where $\mathbb{H}$ is the local system defined in 2.1,

Proposition 3.3. *(1) Let* $g \in \mathrm{Bir}(X, \Delta)$. *If we assume* $\mathbb{H}$ *is defined over an open set* $Y^0 \subset Y$, *then over* $Y^0 \cap \chi(g)^{-1}(Y^0)$, *there exists an isomorphism* $i_g : \chi(g)^* \mathbb{H} \cong \mathbb{H}$.
(2) Let $g_1, g_2 \in \mathrm{Bir}(X, \Delta)$, *then over* $Y^0 \cap \chi(g_1)^{-1}(Y^0) \cap \chi(g_2)^{-1}(Y^0)$ *we have* $i_{g_1} \circ i_{g_2} = i_{g_1 \circ g_2}$.

Proof. (1) Let W be a common log resolution,

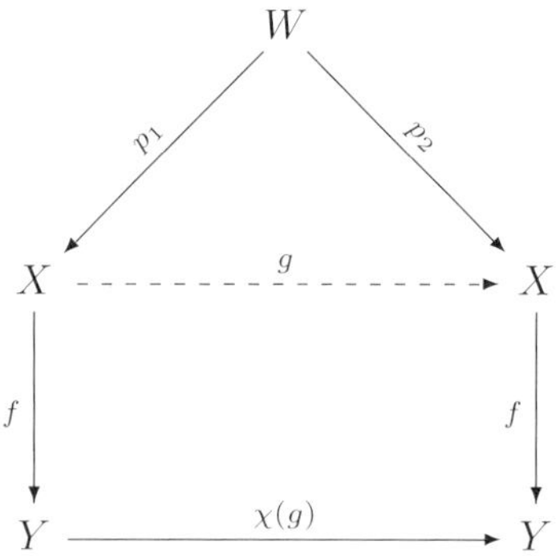

Since $p_1^*(K_X+\Delta) = p_2^*(K_X+\Delta)$, we obtain the same local system on some open subset $W^0 \subset W$, whose image is an open subset $Y^0 \subset Y$. We can shrink Y^0, so that $R^n\phi_{1*}(\mathbb{V})$ (resp. $R^n\phi_{2*}(\mathbb{V})$) is defined over $\chi(g)^{-1}(Y^0)$ (resp. Y^0), where we denote $f \circ p_i$ by ϕ_i.

Since $p_1^*(K_X+\Delta) = p_2^*(K_X+\Delta)$, we conclude $\chi(g)^*(L) \cong L$. Therefore, because of the semi-simplicity of VHS, the isomorphism

$$\chi(g)^* : R^n\phi_{1*}(\mathbb{V}) \to R^n\phi_{2*}(\mathbb{V})$$

which sends L to L, will send $\mathbb{H}$ to $\mathbb{H}$.

(2) This also follows by a similar argument since $(\chi(g_1 \circ g_2))^*L = L$. Q.E.D.

Since $\chi(\mathrm{Bir}(X,\Delta))$ preserves the polarization on Y, it is a subgroup of $\mathrm{PGL}(N)$ for some N. Let G be the Zariski closure of $\chi(\mathrm{Bir}(X,\Delta))$ in $\mathrm{PGL}(N)$, in particular G is an algebraic group. Thus, to show that $\chi(\mathrm{Bir}(X,\Delta))$ is finite, we only need to verify that G does not contain $\mathbb{G}_a$ or $\mathbb{G}_m$. We will show that G does not contain $\mathbb{G}_m$. (The argument for $\mathbb{G}_a$ is similar and we leave it to the reader.)

Choose $Y_0 \subset Y \setminus B$ to be an open set which the Hodge structure $\mathbb{H}$ does not degenerate, and define an open set

$$\widetilde{Y} = \bigcup_{g \in \chi(\mathrm{Bir}(X,\Delta))} g(Y_0).$$

Therefore, $\widetilde{Y}$ is invariant under the action of G as it is invariant under the Zariski dense subgroup $\chi(\mathrm{Bir}(X,\Delta))$ of G. By Proposition 3.3 and Remark 2.3, $\mathbb{H}$ is non-degenerate over $\widetilde{Y}$.

Then for a general point $z \in \widetilde{Y}$, we consider the closure of the orbit $o : \mathbb{P}^1 \to Y$ of $\mathbb{G}_m \cdot z$. Since $\widetilde{Y}$ is G-invariant, we know that the pull back $o^*\mathbb{H}$ is well defined over $\mathbb{G}_m \subset \mathbb{P}^1$. Let $\pi : Y' \to Y$ be a G-equivariant resolution of $(Y, Y \setminus \widetilde{Y})$, thus $\mathbb{H}' = \pi^*\mathbb{H}$ is defined outside a simple normal crossing divisor (see [Kollár07b]). We can write $\pi^*(K_Y+B+J) = K_{Y'}+B'+J'$ where $(Y', B'+J')$ is the decomposition which the Kawamata subadjunction formula yields on Y'. In particular, (Y', B') is sub log canonical.

Let $\phi : Z \to Y'$ be a branched covering such that $\phi^*\mathbb{H}'$ has unipotent monodromies and hence admits a canonical extension over Z (see [Kollár07a, 8.10.10]). Let $\phi_C : C \to \mathbb{P}^1$ be the normalization of $\mathbb{P}^1 \times_Y Z$. As $o(\mathbb{P}^1)$ is the compactification of a general orbit, we know that ϕ_C is of degree d. Then we know that $\phi_C^*(o^*\mathbb{H})$ has a canonical extension from the preimage of $\phi_C^{-1}(\mathbb{G}_m)$ to C which coincides with the restriction of the canonical extension of $\phi^*\mathbb{H}'$ over C (see [Deligne70], [Kollár07a, 8.10.8]). Therefore, if $J_Z = \bar{E}^{n+i,0}(\phi^*\mathbb{H}')$, then

$$J_C = \bar{E}^{n+i,0}(\phi^*\mathbb{H}'|_C) = \bar{E}^{n+i,0}(\phi^*\mathbb{H}')|_C = J_Z|_C.$$

As $\mathbb{G}_m \cong \mathbb{A}^1 \setminus \{0\} \subset \mathbb{P}^1$. There exists a ramified cover $d : \mathbb{P}^1 \to \mathbb{P}^1$, such that if we denote by $o_d = o \circ d$, then $o_d^*(\mathbb{H})$ is defined over $\mathbb{A}^1 \setminus \{0\}$, and it has unipotent monodromies near 0 and ∞. As $o_d^*(\mathbb{H})$ is non-degenerate on $\mathbb{P}^1 \setminus \{0, \infty\}$, $o_d^*\mathbb{H}$ is trivial by Rigidity Theorem (see [Deligne71, 4.1.2], [Schmid73, 7.24]). Since $\bar{E}^{n+i,0}(o_d^*(\mathbb{H})) = \mathcal{O}_{\mathbb{P}^1}(J_d)$, we have

$$\frac{1}{d} d_* J_d \sim_{\mathbb{Q}} \frac{1}{\deg \phi_C} \phi_{C*} J_C \sim_{\mathbb{Q}} J'|_{\mathbb{P}^1} = 0,$$

where the first equality follows from the fact that the moduli part is well defined for any choice of base change such that the monodromy is unipotent, and since ϕ is finite the second equality follows from restricting both sides of $\frac{1}{\deg \phi}(\phi_* J_Z) \sim_{\mathbb{Q}} J'$ under the lifting $\mathbb{P}^1 \to Y'$ of o.

On the other hand, by Lemma 3.4, we have

$$\deg(o^*(K_Y + B + J)) = \deg((K_{Y'} + B')|_{\mathbb{P}^1}) + \deg(J'|_{\mathbb{P}^1}) \leq 0.$$

Since $K_Y + B + J$ is ample, this is a contradiction and so this completes the proof of Theorem 3.

Lemma 3.4. *Let (Y, B) be a sub log canonical pair and $\psi : \mathbb{G}_m \times (Y, B) \to (Y, B)$ a faithful action. For general $t \in Y$, if we denote by $\psi_t : \mathbb{P}^1 \times \{t\} \to Y$ the closure of the orbit, then* $\deg \psi_t^*(K_Y + B) \leq 0$.

The following lemma is due to Iitaka (cf. [Iitaka81]). Since we could not find an explicit reference, we include a short argument for the reader's convenience.

Proof. Let $\pi : Y' \to Y$ be a $\mathbb{G}_m$-equivariant log resolution of (Y, B) and $B' = \mathrm{Supp}(\pi_*^{-1}(B) + \mathrm{Ex}(\pi))$. Thus, for a general point $t \in Y$, we have that

$$\mathbb{G}_m \cdot t \cap (\mathrm{Supp}(B) \cup \pi(\mathrm{Ex}(\pi))) = \emptyset.$$

We then have the lifting $\phi_s : \mathbb{P}^1 \to Y'$ of ψ_t. Since (Y, B) is log canonical, $\pi^*(K_Y + B) \leq K_{Y'} + B'$. We may assume that there is a quasi-projective smooth variety T with a point $s \in T$ and a generically finite dominating morphism, $\phi_T : \mathbb{P}^1 \times T \to Y'$ such that $\phi_T^{-1}(B') \subset \{0, \infty\} \times T$, and $\phi_T|_{\mathbb{P}^1 \times \{s\}} = \phi_s$. Then by the log pull back formula,we conclude that

$$\phi_T^*(K_{Y'} + B') \leq K_{\mathbb{P}^1 \times T} + \{0\} \times T + \{\infty\} \times T.$$

Therefore, we conclude that

$$\deg \psi_t^*(K_Y + B) \leq \deg \phi_s^*(K_{Y'} + B') \leq \mathbb{P}^1 \cdot (K_{\mathbb{P}^1 \times T} + \{0\} \times T + \{\infty\} \times T),$$

which is computed by $\deg(K_{\mathbb{P}^1} + \{0\} + \{\infty\}) = 0$. Q.E.D.

§4. Abundance for slc pairs

In this section, we will study semi-log canonical abundance. By [Fujino00], it is known that if S is a point, then Theorem 1 implies Theorem 2. By [FG11], Theorem 2 is also known when S is projective. The gluing theory for log canonical centers, developed by J. Kollár (cf. [Kollár12, Kollár13]), provides a very powerful tool to study semi-log canonical varieties and allows us to prove Theorem 2 in full generality.

More specifically, Kollár's recent theory of giving any lc center a source and a spring provides a useful tool for checking that the pro-finite equivalence relation is finite (cf. [Kollár11]). We note that this is not true for general pro-finite equivalence relations. The fact that we are gluing lc centers for relatively projective pairs plays an essential role here.

First, the main theorem for abstract gluing theory is the following.

Theorem 4 ([Kollár13, 9.21]). *Let (X, S_*) be an excellent scheme or algebraic space over a field of characteristic 0 with a stratification. Assume that (X, S_*) satisfies the conditions (HN) and (HSN). Let $R \rightrightarrows X$ be a finite, set theoretic, stratified equivalence relation. Then*

(1) *the geometric quotient X/R exists,*
(2) *$\pi : X \to X/R$ is stratifiable and*
(3) *$(X/R, \pi_* S_*)$ also satisfies the conditions (HN) and (HSN).*

We now recall the following notation in [Kollár13, 4.28], which was built on the earlier work of F. Ambro (cf. [Ambro03]) and O. Fujino (cf. [Fujino99]).

Definition 4.1. *We call $f : (X, \Delta) \to Y$ a crepant log structure if*

(1) *(X, Δ) is log canonical,*
(2) *f is projective, surjective, with connected fibers,*
(3) *$K_X + \Delta \sim_{\mathbb{Q},f} 0$.*

We note that, by taking a dlt modification of (X, Δ) (cf. [KK10, 3.1]), we can assume that (X, Δ) is dlt.

The proof of Theorem 2 needs the stratification we develop in [HX13, Section 3]. More precisely, we let $\bar{g} : \bar{X} \to \bar{Y}$ be the morphism over S induced by the relatively semiample $\mathbb{Q}$-divisor $K_{\bar{X}} + \bar{\Delta} + \bar{D}$ and we consider the pro-finite equivalence relation

$$(\sigma_1, \sigma_2) : \bar{T} \rightrightarrows \bar{Y}$$

given by the image of the pro-finite relation $D^n \rightrightarrows \bar{X}$. Recall that D^n denotes the normalization of $\bar{D}$ on $\bar{X}$. Note that $\bar{T}$ and $\bar{Y}$ are normal. We may assume that $\bar{g}$ is induced by $|m(K_{\bar{X}} + \bar{\Delta} + \bar{D})|$ for some sufficiently divisible positive integer m. If we write $K_{D^n} + \Theta = (K_{\bar{X}} + \bar{\Delta} + \bar{D})|_{D^n}$, then (D^n, Θ) is also log canonical.

As in [HX13, Section 3], we consider the minimal qlc stratifications $S_*\bar{T}$ and $S_*\bar{Y}$ given by the crepant log structures $(D^n, \Theta) \to \bar{T}$ and $(\bar{X}, \bar{\Delta} + \bar{D}) \to \bar{Y}$. We note that if (X^d, Δ^d) is a dlt modification of $(\bar{X}, \bar{\Delta} + \bar{D})$ (cf. [KK10, 3.1]), then

$$(X^d, \Delta^d) \to (\bar{X}, \bar{\Delta} + \bar{D}) \to \bar{Y}$$

gives the same minimal qlc stratification on $\bar{Y}$.

Proposition 4.1. *We have the following facts.*

(1) *The stratifications $S_*\bar{T}$ and $S_*\bar{Y}$ satisfy conditions (HN) and (HSN).*
(2) *The pro-finite relation $(\sigma_1, \sigma_2) : \bar{T} \rightrightarrows \bar{Y}$ is stratified.*

Proof. The first statement follows from [KK10, 5.7] and the second one follows from [HX13, 3.11] which is a consequence of [KK10, 1.7]. Q.E.D.

By Theorem 4, to prove that the quotient $\bar{Y}/\bar{T}$ exists, we only need to verify that $\bar{T} \rightrightarrows \bar{Y}$ generates a finite relation $R \to \bar{Y}$. By [Kollár13, 9.55], it suffices to check this over the generic point of each stratum V^0

of $S_i\bar{Y}$. We work over the generic point of $\bar{f}(V^0)$, where $\bar{f}:\bar{Y}\to S$ is the induced morphism. Let $\eta\in S$ be one such point and $\bar{\eta}$ its algebraic closure. We must verify that the equivalence relation $R\times_S\bar{\eta}\subset\bar{Y}\times\bar{Y}\times_S\bar{\eta}$ is finite over $V^0\times\bar{\eta}$.

Let Z be one of the minimal lc centers of (X^d,Δ^d) which dominates the closure V of V^0. $(Z,\mathrm{Diff}^*_Z\Delta^d)$ is dlt and its restriction over V^0 is klt. We take the Stein factorization $Z\to\widetilde{V}\to V$. We will need the following results from [Kollár11].

Proposition 4.2 ([Kollár11, 1]). *For different choices of the minimal non-klt centers Z and Z', we have*

(1) $(Z,\mathrm{Diff}^*_Z\Delta^d)$ *is* **B***-birational to* $(Z',\mathrm{Diff}^*_{Z'}\Delta^d)$ *over* V, *and*
(2) $\widetilde{V}$ *is isomorphic to* $\widetilde{V}'$ *over* V.

Definition 4.2 ([Kollár11, 18]). *We denote by* $\mathrm{Src}(V,X^d,\Delta^d):=(Z,\mathrm{Diff}^*_Z\Delta^d)$ *a source of the lc center* V *and* $\mathrm{Spr}(V,X^d,\Delta^d):=\widetilde{V}$ *its spring. As in Proposition 4.2,* $\mathrm{Src}(V,X^d,\Delta^d)$ *is determined up to* **B***-equivalence.*

Let V^0_{ij} be an i-dimensional stratum, and $\widetilde{V}^0_{ij}$ the pre-image of V^0_{ij} under the morphism $\widetilde{V}_{ij}\to V_{ij}$. We define $\mathrm{Spr}_i(\bar{Y},X^d,\Delta^d)=\amalg_j\widetilde{V}^0_{ij}$, where the disjoint union runs over all i-dimensional strata V^0_{ij}. Then the main structural result is the following.

Theorem 4.3 ([Kollár11, 28]). *Let $R\subset\bar{Y}\times\bar{Y}$ be the relation generated by $\bar{T}\rightrightarrows\bar{Y}$ as above. Let $p_i:\mathrm{Spr}_i(\bar{Y},X^d,\Delta^d)\to S_i\bar{Y}$ be the induced finite morphisms. Let $\bar{\eta}_{ij}$ be the algebraic closure of the generic point of $\bar{f}(V_{ij})$. Then*

$$((p_i\times p_i)^{-1}(R\cap(S_i\bar{Y}\times S_i\bar{Y}))\times_S\bar{\eta}_{ij})\cap(\widetilde{V}^0_{ij}\times\widetilde{V}^0_{ij}\times_S\bar{\eta}_{ij})$$

is a subset of the graph $\cup_g\Gamma(\chi(g))$ for all $g\in\mathrm{Bir}(Z_{\bar{\eta}_{ij}},\mathrm{Diff}^*_{Z_{\bar{\eta}_{ij}}}\Delta^d)$.

Proof of Theorem 2. By Theorem 3, we have that $\cup_g\Gamma(\chi(g))$ is finite over $\widetilde{V}^0_{ij}\times\bar{\eta}_{ij}$, and so the hypotheses of Theorem 4 are satisfied. Thus there exists a quotient Y of $\bar{T}\rightrightarrows\bar{Y}$. Following the proof of [HX13, 3.1], we see that there exists a line bundle L on Y whose pull back to X is isomorphic to $\mathcal{O}_X(m(K_X+\Delta))$ for some integer $m>0$. Then it is easy to see that L is relatively ample over S, which means that $K_X+\Delta$ is relatively semi-ample over S. Q.E.D.

Proof of Corollary 1.3. The argument is the same as in [KMM94, 7.4]. We include it here for the reader's convenience. It follows from Theorem 2 that $m(K_T+\Delta_T):=m(K_X+\Delta)|_T$ is base point free over

S for any sufficiently divisible integer $m > 0$. By assumption (3), the relative base locus of $m(K_X + \Delta)$ is contained in the support of T. We write

$$m(K_X+\Delta)-T=(m-1)(K_X+\Delta-\epsilon P)+K_X+\Delta-(T-(m-1)\epsilon P).$$

Since $K_X+\Delta-\epsilon P$ is semi-ample over S, and $(X, \Delta-(T-(m-1)\epsilon P))$ is klt for $0<\epsilon \ll 1$, by Kollár's injectivity theorem (cf. [Kollár86]), we have that

$$R^1 f_*\mathcal{O}_X(m(K_X+\Delta)-T) \xrightarrow{T} R^1 f_*\mathcal{O}_X(m(K_X+\Delta))$$

is an injection and hence $f_*\mathcal{O}_X(m(K_X+\Delta)) \to f_*\mathcal{O}_T(m(K_T+\Delta_T))$ is surjective. We have a commutative diagram.

$$\begin{array}{ccc} f^*f_*\mathcal{O}_X(m(K_X+\Delta)) & \longrightarrow & f^*f_*\mathcal{O}_T(m(K_T+\Delta_T)) \\ \downarrow & & \downarrow \\ \mathcal{O}_X(m(K_X+\Delta)) & \longrightarrow & \mathcal{O}_T(m(K_T+\Delta_T)) \end{array}$$

Since the upper arrow and the right arrow are surjective, $m(K_X + \Delta)$ is relatively globally generated along T over S and so $m(K_X + \Delta)$ is relatively globally generated over S. Q.E.D.

Proof of Corollary 1.4. By Theorem 2, we may assume that X is irreducible, i.e., we only need to treat the case that (X, Δ) is log canonical. We may assume that S is normal and dominated by X. By induction, we may assume that $Corollary$1.4 is known for lower dimensions. Replacing (X, Δ) by a dlt modification, we may assume that (X, Δ) is $\mathbb{Q}$-factorial and dlt. Let $T = \lfloor \Delta \rfloor$. If we write $(K_X+\Delta)|_T = K_T + \Delta_T$, then (T, Δ_T) is a dslt pair.

We run a $(K_X + \Delta - \epsilon T)$-MMP with scaling by an ample divisor H for a small positive integer ϵ. If T has a component which dominates S, then the MMP ends with a Fano contraction, and we can apply the same argument as in [Gongyo13]. Otherwise, all components of T are vertical over S. Since over the generic point of S, we have that $K_X + \Delta$ is klt, then we know that it is $\mathbb{Q}$-linearly equivalent to 0 (cf. [Nakayama04, V, 4.9]).

Therefore, $(X, \Delta - \epsilon T)$ is a klt pair, whose generic fiber has a good model. By, [HX13, 1.1 and 2.9], we conclude that the MMP with scaling terminates with a good model X' of $(X, \Delta - \epsilon T)$ over S. Since X' is indeed a minimal model for all $(X, \Delta - \epsilon' T)$ with $0 < \epsilon' < \epsilon$, which is good again by [HX13, 1.1].

The MMP sequence is $(K_X + \Delta)$-trivial, so (X', Δ') is a lc pair. We let $\mu : (X^d, \Delta^d) \to (X', \Delta')$ be a dlt model. Let T' be the strict transform of T on X', then $(X', \Delta' - \epsilon T')$ is klt and the non-klt locus of (X', Δ') is contained in $\mathrm{Supp}(T')$. Since each μ-exceptional divisor has coefficient 1 in Δ^d, it follows that $\lfloor \Delta^d \rfloor = \mathrm{Supp}(\mu^* T')$. Thus if $T^d = \mu^* T'$, then $K_{X^d} + \Delta^d - \epsilon T^d = \mu^*(K_{X'} + \Delta' - \epsilon T')$ is semiample over S. By induction on the dimension, $K_\Sigma + \Delta_\Sigma$ is semiample over S for all components Σ of $\lfloor \Delta^d \rfloor = \mathrm{Supp}(T^d)$. By Corollary 1.3, $K_{X^d} + \Delta^d$ is semi-ample over S and hence so is $K_{X'} + \Delta'$. Since each step of the MMP is $(K_X + \Delta)$-trivial, this implies that $K_X + \Delta$ is semi-ample over S, i.e., $K_X + \Delta \sim_{\mathbb{Q},S} 0$. Q.E.D.

References

[Ambro99] F. Ambro; *The adjunction conjecture and its applications*, math.AG/9903060.

[Ambro03] F. Ambro; *Quasi-log varieties.* Tr. Mat. Inst. Steklova **240** (2003), Biratsion. Geom. Linein. Sist. Konechno Porozhdennye Algebry, 220-239; translation in Proc. Steklov Inst. Math. 2003, no. 1 **240**, 214-233.

[Birkar12] C. Birkar; *Existence of log canonical flips and a special LMMP.* Publ. Math. Inst. Hautes Études Sci. **115** (2012), Issue 1, 325-368.

[Deligne70] P. Deligne: *Equations Différentielles à Points Singuliers Réguliers*, Lecture Notes in Math. **163**, Springer-Verlag 1970.

[Deligne71] P. Deligne; *Théorie de Hodge. II.* Inst. Hautes Études Sci. Publ. Math. No. **40** (1971), 5–57.

[EV92] H. Esnault, E. Viehweg; *Lectures on Vanishing Theorems.* DMV Seminar, Band 20, Birkhäuser Verlag Basel, 1992

[FF12] O. Fujino, T. Fujisawa; *Variations of mixed Hodge structure and semipositivity theorems*, arXiv:1203.6697. To appear in Publ. Res. Inst. Math. Sci..

[FFS14] O. Fujino, T. Fujisawa, M. Saito, *Some Remarks on the Semipositivity Theorems*, Publ. Res. Inst. Math. Sci. **50** (2014), no. 1, 85–112.

[FG11] O. Fujino, Y. Gongyo; *Log pluricanonical representations and abundance conjecture.* arXiv:1104.0361. To appear in Compos. Math..

[Fujino99] O. Fujino; *Applications of Kawamata's positivity theorem.* Proc. Japan Acad. Ser. A Math. Sci. 75 (1999), no. 6, 75–79.

[Fujino00] O. Fujino; *Abundance theorem for semi log canonical threefolds.* Duke Math. J. **102** (2000), no. 3, 513–532.

[Gongyo13] Y. Gongyo; *Abundance theorem for numerical trivial log canonical divisors of semi-log canonical pairs.* J. Algebraic Geom. **22** (2013), 549-564.

[HMX14] C. Hacon, J. McKernan, C. Xu; *Boundedness of the moduli functor.* Preprint in preparation.

[HX13] C. Hacon, C. Xu; *Existence of log canonical closures.* Invent. Math. **192** (2013), Issue 1, 161-195.

[Iitaka81] S. iitaka, *Birational geometry for open varieties*, Volume 76 of Séminaire de mathématiques supérieures, Université de Montreal Departement de Mathématiques Séminaire scientifique OTAN. Presses de l'Université de Montréal, 1981

[Kawamata98] Y. Kawamata; *Subadjunction of log canonical divisors. II.* Amer. J. Math. **120** (1998), no. 5, 893–899.

[Kawamata11] Y. Kawamata; *Semipositivity theorem for reducible algebraic fiber spaces*, Pure Appl. Math. Q. 7 (2011), no. 4, Special Issue: In memory of Eckart Viehweg, 1427-1447.

[KK10] J. Kollár, S. Kovács; *Log canonical singularities are Du Bois.* J. Amer. Math. Soc. **23** (2010), no. 3, 791-813.

[KMM94] S. Keel; K. Matsuki; J. McKernan; *Log abundance theorem for threefolds.* Duke Math. J. **75** (1994), no. 1, 99–119.

[Kollár86] J. Kollár; *Higher direct images of dualizing sheaves. I.* Ann. of Math. (2) **123** (1986), no. 1, 11–42.

[Kollár07a] J. Kollár; *Kodaira's canonical bundle formula and adjunction.* Flips for 3-folds and 4-folds, 134–162, Oxford Lecture Ser. Math. Appl., 35, Oxford Univ. Press, Oxford, 2007.

[Kollár07b] J. Kollár; *Lectures on resolution of singularities.* Annals of Mathematics Studies, **166**. Princeton University Press, Princeton, NJ, 2007.

[Kollár11] J. Kollár; *Sources of log canonical centers.* arXiv:1107.2863.

[Kollár12] J. Kollár; *Quotients by finite equivalence relations.* With an appendix by Claudiu Raicu. Math. Sci. Res. Inst. Publ., **59**, *Current developments in algebraic geometry*, 227-256, Cambridge Univ. Press, Cambridge, 2012.

[Kollár13] J. Kollár; *Singularities of the minimal model program.* With a collaboration of Sándor Kovács. *Cambridge Tracts in Mathematics*, **200**. Cambridge University Press, Cambridge, 2013.

[Kolláretal92] J. Kollár et al.; *Flips and abundance for algebraic threefolds.* Papers from the Second Summer Seminar on Algebraic Geometry held at the University of Utah, Salt Lake City, Utah, August 1991. Astérisque No. **211** (1992). *Société Mathématique de France, Paris*, 1992.

[Nakayama04] N. Nakayama; *Zariski-decomposition and abundance.* MSJ Memoirs, **14**. Mathematical Society of Japan, Tokyo, 2004.

[Schmid73] W. Schmid, *Variation of Hodge structure: the singularities of the period mapping.* Invent. Math. **22** (1973), 211-319.

[Steenbrink77] J. Steenbrink; *Mixed Hodge structure on the vanishing cohomology.* Real and complex singularities (Proc. Ninth Nordic Summer School/NAVF Sympos. Math., Oslo, 1976), 525–563. Sijthoff and Noordhoff, Alphen aan den Rijn, 1977.

Department of Mathematics
University of Utah
155 South 1400 East
JWB 233
Salt Lake City, UT 84112, USA
E-mail address: hacon@math.utah.edu

Beijing International Center of Mathematics Research,
5 Yiheyuan Road,
Haidian District,
Beijing 100871,
China
E-mail address: cyxu@math.pku.edu.cn

Advanced Studies in Pure Mathematics 70, 2016
Minimal Models and Extremal Rays (Kyoto, 2011)
pp. 379–396

K3 surfaces of genus sixteen

Shigeru Mukai

Abstract.

The generic polarized $K3$ surface (S, h) of genus 16, that is, $(h^2) = 30$, is described in a certain compactified moduli space $\mathcal{T}$ of twisted cubics in $\mathbb{P}^3$, as a complete intersection with respect to an almost homogeneous vector bundle of rank 10. As corollary we prove the unirationality of the moduli space $\mathcal{F}_{16}$ of such K3 surfaces.

§1. Introduction

Let $\mathcal{F}_g$ be the moduli space of quasi-polarized $K3$ surface (S, h) of genus g, *i.e.*, $(h^2) = 2g - 2$. $\mathcal{F}_g$ is an arithmetic quotient of the 19-dimensional bounded symmetric domain of type IV, and a quasi-projective variety. It is shown in [3] that $\mathcal{F}_g$ is of general type for $g \geq 63$ but the birational classification of $\mathcal{F}_g$ is still far from being complete. For $g \leq 10$ and $g = 12, 13, 18, 20$, the generic (S, h) is a complete intersection in a suitable homogeneous space with respect to a suitable homogeneous vector bundle. As corollary the unirationality of $\mathcal{F}_g$ is proved for those values of g in [5, 7, 8]. In this article we shall describe the generic member of $\mathcal{F}_{16}$ using the EPS moduli space $\mathcal{T} := G(2, 3; \mathbb{C}^4)$ of twisted cubics in $\mathbb{P}^3$.

The EPS moduli space $\mathcal{T}$ is constructed by Ellingsrud-Piene-Strømme [2] as the GIT quotient of the tensor product $\mathbb{C}^2 \otimes \mathbb{C}^3 \otimes V$, V being a 4-dimensional vector space, by the obvious action of $GL(2) \times GL(3)$. $\mathcal{T}$ is a smooth equivariant compactification of the 12-dimensional homogeneous space $PGL(V)/PGL(2)$. A point $t \in \mathcal{T}$ represents an equivalence class of 2×3 matrices whose entries belong to V. Its three minors define a subscheme R_t of the projective space $\mathbb{P}(V)$. R_t is a cubic curve mostly and a plane with an embedded point in exceptional cases. By

Received February 29, 2012.
Revised June 28, 2014.
Supported in part by the JSPS Grant-in-Aid for Scientific Research (B) 22340007, (S) 19104001, (A)22244003, (S)22224001 and (S)25220701.

construction there exists two natural vector bundles $\mathcal{E}, \mathcal{F}$ of rank 3, 2, respectively, with $\det \mathcal{E} \simeq \det \mathcal{F}$, and the tautological homomorphism

$$\mathcal{E} \otimes V^{\vee} \longrightarrow \mathcal{F}$$

on $\mathcal{T}$. (Be cautioned that the same letters $\mathcal{E}$ and $\mathcal{F}$ are used for the dual vector bundles in [2].) The tautological homomorphism induces linear maps

$$(S^2V)^{\vee} \longrightarrow H^0(\mathcal{E}) \quad \text{and} \quad (S^{2,1}V)^{\vee} \longrightarrow H^0(\mathcal{F}). \tag{1}$$

(See §2.) Here S^2V is the second symmetric tensor product, and

$$S^{2,1}V = \ker[V \otimes S^2V \to S^3V] \tag{2}$$

is the space of linear syzygies among second symmetric tensors. $S^{2,1}V$ is of dimension 20.

For two subspaces $M \subset (S^2V)^{\vee}$ and $N \subset (S^{2,1}V)^{\vee}$, we consider the common zero locus

$$\bigcap_{s \in \bar{M}} (s)_0 \cap \bigcap_{t \in \bar{N}} (t)_0 \subset \mathcal{T}. \tag{3}$$

of global sections $s \in \bar{M}$ and $t \in \bar{N}$, where $\bar{M}$ and $\bar{N}$ are the images of M and N in $H^0(\mathcal{E})$ and $H^0(\mathcal{F})$, respectively. The case $\dim M = \dim N = 2$ is most interesting. We denote the common zero locus (3) by $S_{M,N}$ in this case.

Theorem 1.1. *If M and N are general, then $S_{M,N}$ is a (smooth) $K3$ surface, and the restriction of $H := c_1(\mathcal{E})$ is a polarization of genus* 16.

For general M and N, $S_{M,N}$ is a complete intersection in $\mathcal{T}$ with respect to the vector bundle $\mathcal{E}^{\oplus 2} \oplus \mathcal{F}^{\oplus 2}$ of rank 10. Furthermore the following converse also holds:

Theorem 1.2. *Generic $K3$ surface of genus* 16 *is isomorphic to the complete intersection $S_{M,N}$.*

A twisted cubic

$$R : \text{rank} \begin{pmatrix} f_{11} & f_{12} & f_{13} \\ f_{21} & f_{22} & f_{23} \end{pmatrix} \leq 1, \quad f_{ij} \in V \tag{4}$$

in $\mathbb{P}(V) = \mathbb{P}^3$ is *apolar to M* if all minors of the matrix are perpendicular to M. Similarly R is *apolar to N* if all linear syzygies among the three minors are perpendicular to N. The K3 surface $S_{M,N}$ in Theorem 1.1 parametrizes all R which are apolar to both M and N.

If a 2-dimensional subspace $M \subset (S^2V)^\vee$ is general, then the kernel Syz_M of the multiplication map $V \otimes M^\perp \to S^3V$ is of dimension 12, where $M^\perp \subset S^2V$ is the space of quadratic forms apolar to M. Hence the totality of $S_{M,N}$ are parametrized by an open subset of a generic $G(2,12)$-bundle $\mathcal{P}$ over the 16-dimensional Grassmannian $G(2,(S^2V)^\vee)$ which parametrizes M. By Theorem 1.1, we have the rational map

$$\Psi_{16} : \mathcal{P} \cdots \to \mathcal{F}_{16}, \quad (M, \bar{N}) \mapsto (S_{M,N}, H|_S), \tag{5}$$

whose dominance is Theorem 1.2, where $\bar{N} \subset (Syz_M)^\vee$ is the image of N by the linear map $S^{2,1}V^\vee \to (Syz_M)^\vee$. Therefore, as bi-product, we have

Corollary *The moduli space $\mathcal{F}_{16}$ of polarized K3 surface of genus 16 is unirational.*

In order to prove the theorems, we study a certain special case in detail. More explicitly, we consider the space $M_0 \subset (S^2V)^\vee \simeq S^2(V^\vee)$ spanned by two *reducible* quadratic forms $q_1 = XY, q_2 = ZT$, and study the common zero locus $\mathcal{T}_{M_0} := (q_1)_0 \cap (q_2)_0$ of M_0 in $\mathcal{T}$. $\mathcal{T}_{M_0}$ parametrizes all twisted cubics whose defining quadratic forms do not contain the term xy or zt, where $(x:y:z:t)$ is a homogeneous coordinate $\mathbb{P}^3$ and $(X:Y:Z:T)$ is the dual coordinate of $\mathbb{P}^{3,*}$.

If N is general, then $S_{M_0,N}$ is a quartic surface in $\mathbb{P}^3$ which contains two quintic elliptic curves E_1 and E_2 with $(E_1.E_2) = 3$. In particular we have Theorem 1.1. Moreover, the restriction of $\mathcal{E}$ to $S_{M_0,N}$ is an extension of three line bundles $\mathcal{O}_S(E_1), \mathcal{O}_S(E_2)$ and $\mathcal{O}_S(H - E_1 - E_2)$ ((21) in §6), and the restriction of $\mathcal{F}$ contains $\mathcal{O}_S(E_1)^{\oplus 2}$ as a subsheaf (Proposition 3.1). These give us the following vanishing of higher cohomology groups which is the key of the proof of Theorem 1.2.

Proposition 1.3. *If both M and N are general, then the restrictions of $\mathcal{E}, \mathcal{F}$ to $S := S_{M,N}$ are simple and satisfy*

$$\mathrm{Ext}^i(\mathcal{E}|_S, \mathcal{F}|_S) = H^i(S, \mathcal{E}|_S) = H^i(S, \mathcal{F}|_S) = 0, \text{ for all } \ i > 0.$$

After preparing some basic facts on the EPS moduli space $\mathcal{T} = G(2,3;\mathbb{C}^4)$ in §2, we first study the locus $\mathcal{T}_Q$ of twisted cubics apolar to one reducible quadric in §3. We next study the locus $\mathcal{T}_{B_1,B_2}$ of twisted cubics which have two skew lines as their bisecants in §4 and the above $\mathcal{T}_{M_0}$ in §5. We prove Theorem 1.1 at the end of §6 and Theorem 1.2 in §7 using *doubly octagonal* K3 surfaces $S_{M_0,N}$. The final §8 is logically unnecessary but explains how Theorem 1.2 originates from the description of Fano 3-fold of genus 12 in [6].

Notations and convention All varieties are considered over the complex number field $\mathbb{C}$. The projective space $\mathbb{P}(V)$ associated to a vector space V is that in Grothendieck's sense. The Grassmann variety of s-dimensional subspaces of V is denoted by $G(s, V)$. The isomorphism class of $G(s, V)$ is denoted by $G(s, n)$ when $\dim V = n$. The dual vector space (and more generally the dual vector bundle) is denoted by $V^\vee$. *Twisted cubic* is used in the generalized sense of [2]. But the locus where twisted cubics are not curves is of sufficiently large codimension, and hence is never crucial in our argument.

§2. Pair of vector bundles whose ranks differ by one

Let (E, F) be a pair of vector bundles on a scheme S such that

$$\det E \simeq \det F, \quad \operatorname{rank} E = \operatorname{rank} F + 1. \tag{6}$$

Let r be the rank of F. r homomorphisms $f_1, \ldots, f_r \in \operatorname{Hom}(E, F)$ give rise the homomorphism

$$f_1 \wedge \cdots \wedge f_r : \wedge^r E \to \wedge^r F \simeq \det F$$

which can be regarded as a global section of E by our assumption (6). Since $f_1 \wedge \cdots \wedge f_r$ is symmetric with respect to $f_1, \ldots, f_r$, we have a linear map

$$S^r \operatorname{Hom}(E, F) \to \operatorname{Hom}(\wedge^r E, \wedge^r F) \simeq H^0(E). \tag{7}$$

If $g : E \to F$ is a homomorphism, then $g(f_1 \wedge \cdots \wedge f_r)$ is a global section of F. Hence we have another linear map

$$S^r \operatorname{Hom}(E, F) \otimes \operatorname{Hom}(E, F) \to H^0(F), \quad ((f_1, \ldots, f_r), g) \mapsto g(f_1 \wedge \ldots \wedge f_r).$$

Since $S^{r+1} \operatorname{Hom}(E, F)$ lies in the kernel of this linear map, we have

$$S^{r,1} \operatorname{Hom}(E, F) \to H^0(F). \tag{8}$$

Let V be a vector space and let $G(r, r+1; V)$ be the GIT quotient of the tensor product $\mathbb{C}^r \otimes \mathbb{C}^{r+1} \otimes V$ by $GL(r) \times GL(r+1)$. There are two natural vector bundles $\mathcal{E}, \mathcal{F}$ of rank $r+1$, r, respectively, with $\det \mathcal{E} \simeq \det \mathcal{F}$, and the tautological homomorphism

$$\mathcal{E} \otimes V^\vee \to \mathcal{F} \tag{9}$$

on $G(r, r+1; V)$. This has the following universal property.

(*) If a homomorphism $E \otimes V^\vee \to F$ satisfies (6) and if the induced homomorphism $S^r V^\vee \otimes \mathcal{O}_S \to E$ is surjective, then there exists a unique morphism $\Phi : S \to G(r, r+1; V)$ such that $E \otimes V^\vee \to F$ coincides with the pull-back of (9). This Φ will be denoted by $\Phi_{E,F,V^\vee} : S \to G(r, r+1; V)$, or $\Phi_{E,F}$ if $V^\vee = \mathrm{Hom}(E, F)$.

Remark 2.1. If E, F are vector bundles of rank $r+1, r$, respectively. Then, putting $L = (\det E)^{-1} \otimes \det F$, we have

$$\mathrm{Hom}(E, F) \simeq \mathrm{Hom}(E \otimes L, F \otimes L) \quad \text{and} \quad \det(E \otimes L) \simeq \det(F \otimes L).$$

Hence, the assumption (6) is not restrictive.

In the sequel we apply the case $r = 2, \dim V = 4$ to $K3$ surfaces of genus 16. $G(2, 3; V)$ is regarded as a subvariety of the Grassmannian $G(3, S^2 V)$ by $R \mapsto H^0(\mathbb{P}^3, \mathcal{O}_\mathbb{P}(2 - R))$, where $H^0(\mathbb{P}^3, \mathcal{O}_\mathbb{P}(2 - R))$ is the 3-dimensional space of quadratic forms vanishing on R. $G(2, 3; V)$ is also a subvariety of another Grassmannian $G(2, S^{2,1} V)$ by $R \mapsto Syz_R$, where Syz_R is the 2-dimensional space of linear syzygies among $H^0(\mathbb{P}^3, \mathcal{O}_\mathbb{P}(2 - R))$.

Let $S_{M,N} \subset \mathcal{T} = G(2, 3; V)$ be as in the introduction for general 2-dimensional subspaces M and N.

Proposition 2.2. 1) *$S_{M,N}$ is the disjoint union of $K3$ surfaces and abelian surfaces.*

2) *The degree of $S_{M,N}$ with respect to $H := c_1(\mathcal{E})$ is equal to* 30.

3) *The second Chern number of the restrictions of $\mathcal{E}$ and $\mathcal{F}$ to $S_{M,N}$ are equal to* 13 *and* 9, *respectively.*

Proof. 1) The vector bundles $\mathcal{E}$ and $\mathcal{F}$ are generated by the global sections coming from $S^2 V^\vee$ and $S^{2,1} V^\vee$, respectively. Hence by the Bertini type theorem (see Remark 2.4 below), the general complete intersection $S_{M,N}$ is smooth of expected dimension, which is equal to $\dim \mathcal{T} - 2 \cdot \mathrm{rank}\, \mathcal{E} - 2 \cdot \mathrm{rank}\, \mathcal{F} = 2$. The canonical bundle of $S_{M,N}$ is trivial by the adjunction formula [7, (1.5)] since $c_1(\mathcal{T}) = 4H$.

2) The degree of $S_{M,N}$ is equal to

$$(H^2.c_{\mathrm{top}}(\mathcal{E}^{\oplus 2} \oplus \mathcal{F}^{\oplus 2})) = (H^2.c_3(\mathcal{E})^2.c_2(\mathcal{F})^2),$$

which is equal to $(c_1^2 c_3^2 d_2^2) = 30$ by [1, Table 1].

3) The Chern numbers are equal to

$$c_2(\mathcal{E}|_S) = (c_2 c_3^2 d_2^2) = 13 \quad \text{and} \quad c_2(\mathcal{F}|_S) = (c_3^2 d_2^3) = 9,$$

respectively, again by [1, Table 1]. Q.E.D.

Remark 2.3. A computation using the description of the tangent bundle of $\mathcal{T}$ in [1, (4.4)] shows that the Euler number of $S_{M,N}$ is equal to 24. This shows that a K3 surface appears in $S_{M,N}$ and it is unique.

Remark 2.4. The Bertini type theorem proved in [7, Theorem 1.10] holds in the following more general form: if a subspace $W \subset H^0(\mathcal{G})$ generates a vector bundle $\mathcal{G}$ of rank r and if a global section $s \in W$ is general, then the scheme of zeroes of s is smooth of codimension r.

§3. Twisted cubics apolar to a reducible quadric

We fix a line l in $\mathbb{P}(V) = \mathbb{P}^3$ and consider the subvariety

$$\mathcal{T}_B := \{R \,|\, \mathrm{length}(R \cap l) \geq 2\} \subset \mathcal{T}$$

consisting of twisted cubics which have l as a bisecant line. Here "B" stands for bisecant. $\mathcal{T}_B$ is a 10-dimensional variety. Assigning the intersection $l \cap R$ to l, we obtain the rational map

$$f_B : \mathcal{T}_B \cdots \to \mathbb{P}^2 = \mathrm{Sym}^2 l. \tag{10}$$

Let $\mathcal{D}$ be the subvariety of $\mathcal{T}$ consisting of reducible twisted cubics. $\mathcal{D}$ is a divisor. Let $\mathcal{D}_B$ be the intersection $\mathcal{D} \cap \mathcal{T}_B$. $\mathcal{D}_B$ decomposes into the union of two irreducible components $\mathcal{D}_{B,1}$ and $\mathcal{D}_{B,2}$ according as the intersection of the conical part of R and l. Every general member R of $\mathcal{D}_{B,2}$ is the union of a line and a conic which meets l at two points. Oppositely every general member R of $\mathcal{D}_{B,1}$ is the union of a line and a conic both of which meet l.

The restriction of the syzygy bundle $\mathcal{F}$ to $\mathcal{T}_B$ is described using this former divisor $\mathcal{D}_{B,2}$.

Proposition 3.1. *The restriction $\mathcal{F}|_{\mathcal{T}_B}$ contains the rank 2 vector bundles $f_B^* \mathcal{O}_{\mathbb{P}}(1)^{\oplus 2}$ as a subsheaf, and the quotient $(\mathcal{F}|_{\mathcal{T}_B})/(f_B^* \mathcal{O}_{\mathbb{P}}(1)^{\oplus 2})$ is a line bundle on the divisor $\mathcal{D}_{B,2}$.*

Proof. We take a homogeneous coordinate $(x : y : z : t)$ of $\mathbb{P}^3$ and assume that the line l is defined, say, by $x = y = 0$. We describe the syzygy space Syz_R of a twisted cubic R in $\mathcal{T}_B$ using the two quadrics containing the union $R \cup l$.

First we construct a homomorphism $f_B^* \mathcal{O}_{\mathbb{P}}(1)^{\oplus 2} \to \mathcal{F}|_{\mathcal{T}_B}$. Since l is a bisecant of R, the union $R \cup l$ is contained in two different quadrics, say, $cx - ay = 0$ and $dx - by = 0$ with $a, b, c, d \in V = \langle x, y, z, t\rangle_{\mathbb{C}}$. The third quadric containing R is defined by $ad - bc = 0$. Hence R is defined by the three minors of the matrix $\begin{pmatrix} x & a & b \\ y & c & d \end{pmatrix}$. Therefore, the syzygy space Syz_R of R is spanned by

$$x \otimes (ad - bc) - a \otimes (dx - by) + b \otimes (cx - ay) \tag{11}$$

and

$$y \otimes (ad - bc) - c \otimes (dx - by) + d \otimes (cx - ay). \tag{12}$$

When R runs over $\mathcal{T}_B$, these syzygies generate a subspace $Syz_1 \subset S^{2,1}V$ of codimension 2. More precisely, $S^{2,1}V$ has 20 tensors of the form (monomial) $\otimes$ (monomial) $-$ (monomial) $\otimes$ (monomial) as its basis, and Syz_1 is generated by all except $z \otimes zt - t \otimes z^2, t \otimes zt - z \otimes t^2$. The syzygies

$$a \otimes by - b \otimes ay, \quad c \otimes dx - d \otimes cx, \quad a, b, c, d \in V$$

are contained in the vector space Syz_1, and generate a subspace Syz_2 isomorphic to $(\bigwedge^2 V)^{\oplus 2}$. The quotient Syz_1/Syz_2 is canonically isomorphic to $\langle x, y\rangle_{\mathbb{C}} \otimes S^2(V/\langle x, y\rangle_{\mathbb{C}})$. Since the quadric $ad - bc = 0$ cut the two points $f_B(R)$ from l, we have a homomorphism $f_B^*\mathcal{O}_{\mathbb{P}}(1)^{\oplus 2} \to \mathcal{F}|_{\mathcal{T}_B}$ on $\mathcal{T}_B$.

If R does not belong to the divisor $\mathcal{D}_{B,2}$, then the union $R \cup l$ is the intersection of two quadrics $cx - ay = 0$ and $dx - by = 0$. Moreover, the residual classes of (11) and (12) are $x \otimes \overline{ad - bc}$ and $y \otimes \overline{ad - bc}$, respectively. Hence $f_B^*\mathcal{O}_{\mathbb{P}}(1)^{\oplus 2} \to \mathcal{F}|_{\mathcal{T}_B}$ is an isomorphism outside $\mathcal{D}_{B,2}$.

On the contrary assume that $[R] \in \mathcal{D}_{B,2}$. Then the intersection of two quadrics containing $R \cup l$ is the union of a plane containing l, say $x = 0$, and a line. R is defined by the three minors of the matrix of the form $\begin{pmatrix} x & a & b \\ 0 & c & d \end{pmatrix}$, and Syz_R is spanned by

$$x \otimes (ad - bc) - a \otimes dx + b \otimes cx, \tag{13}$$

which is a specialization of (11), and $-c \otimes dx + d \otimes cx \in Syz_2$, a specialization of (12). Since (13) is not contained in the subspace Syz_2, the cokernel of $f_B^*\mathcal{O}_{\mathbb{P}}(1)^{\oplus 2} \hookrightarrow \mathcal{F}|_{\mathcal{T}_B}$ is a line bundle on $\mathcal{D}_{B,2}$. Q.E.D.

Now we study the locus $\mathcal{T}_Q$ of twisted cubics which are apolar to a quadric $Q : q = 0 \subset \mathbb{P}^{3,*}$ when $q \in (S^2V)^\vee \simeq S^2(V^\vee)$ is of rank 2. The quadric Q is the union of two distinct planes P_1 and P_2. Let l be the line joining the two points $[P_1]$ and $[P_2] \in \mathbb{P}^3 = (\mathbb{P}^{3,*})^*$. q is the pull-back of a quadratic form $\bar{q}$ on $l^* \simeq \mathbb{P}^1$ by the projection $\mathbb{P}^{3,*} \cdots \to l^*$. A twisted cubic R is apolar to q if and only if the restriction of $H^0(\mathbb{P}^3, \mathcal{O}_{\mathbb{P}}(2 - R))$ to l is apolar to $\bar{q}$.

Proposition 3.2. *The following are equivalent to each other.*
1) *A twisted cubic $R \subset \mathbb{P}^3$ is apolar to q.*
2) *l is a bisecant line of R and the intersection $R \cap l$ is apolar to $\bar{q}$.*

Proof. 2) $\Longrightarrow$ 1) If l is a bisecant of R, then the union $R \cup l$ is contained in two distinct quadrics. Hence the restriction map

$$H^0(\mathbb{P}^3, \mathcal{O}_{\mathbb{P}}(2-R)) \to H^0(l, \mathcal{O}_l(2)) \tag{14}$$

is of rank ≤ 1. Hence, if furthermore $R \cap l$ is apolar to $q|_l$, then R is apolar to q.

1) $\Longrightarrow$ 2) Let $W \subset H^0(l, \mathcal{O}_l(2))$ be the space of quadratic forms apolar to $\bar{q}$. If R is apolar to q, then the image of the restriction map (14) is contained in W. Since $\dim W = 2$, the linear map (14) is not injective, that is, the union $R \cup l$ is contained in a quadric. Hence $R \cap l$ is non-empty. Since the quadratic forms in W has no common zero, the rank of (14) is at most one, which shows (2). Q.E.D.

By the proposition, $\mathcal{T}_Q$ is contained in $\mathcal{T}_B$. More precisely, it coincides with the pull-back of a line by the rational map (10). In particular, we have the rational map

$$f_Q : \mathcal{T}_Q \cdots \to \mathbb{P}^1 \subset \mathbb{P}^2 = \mathrm{Sym}^2 l, \quad R \mapsto R \cap l. \tag{15}$$

§4. Twisted cubics with two fixed bisecant lines

We fix a pair of skew lines l_1 and l_2 in $\mathbb{P}(V) = \mathbb{P}^3$ and consider the (8-dimensional) subvariety

$$\mathcal{T}_{B_1,B_2} := \{R \,|\, \mathrm{length}(R \cap l_1) \ge 2,\ \mathrm{length}(R \cap l_2) \ge 2\} \subset \mathcal{T}$$

consisting of twisted cubics which have both l_1 and l_2 as bisecant lines. Restricting (10) we have two rational maps

$$f_{B_i} : \mathcal{T}_{B_1,B_2} \cdots \to \mathbb{P}^2 = \mathrm{Sym}^2 l_i, \quad i = 1, 2. \tag{16}$$

Now we consider the correspondence

$$Y = \{(R, Q) \,|\, R \subset Q\} \subset \mathcal{T}_{B_1,B_2} \times \Lambda \tag{17}$$

between $\mathcal{T}_{B_1,B_2}$ and the linear web $\Lambda := |\mathcal{O}_{\mathbb{P}}(2 - l_1 - l_2)|$ of quadrics Q containing l_1 and l_2. Assume that a twisted cubic R belongs to $\mathcal{T}_{B_1,B_2}$. As we saw in the proof of Poposition 3.2, the restriction maps

$$H^0(\mathbb{P}^3, \mathcal{O}_{\mathbb{P}}(2-R)) \to H^0(l_i, \mathcal{O}_l(2)), \quad i = 1, 2$$

are of rank at most one. Hence there exists a quadric which contains $R \cup l_1 \cup l_2$. Therefore, the first projection $\pi : Y \to \mathcal{T}_{B_1,B_2}$ is surjective. π is not an isomorphism at $[R]$ if and only if $\dim |\mathcal{O}_{\mathbb{P}}(2 - l_1 - l_2 - R)| > 0$.

Proposition 4.1. *The following are equivalent for a twisted cubic $[R]$ in $\mathcal{T}_{B_1,B_2}$.*

1) $\dim|\mathcal{O}_{\mathbb{P}}(2-l_1-l_2-R)|>0$.

2) $R\supset l_1$ *or* $R\supset l_2$.

Proof. 1) $\Rightarrow$ 2) There exist two distinct quadrics Q_1 and Q_2 which contains $C=l_1\cup l_2\cup R$. If $\deg C\leq 4$ then 2) follows. Otherwise, we have $\deg C>\deg Q_1\cdot\deg Q_2$, and Q_1 and Q_2 have a common component. Therefore, the intersection $Q_1\cap Q_2$ is the union of plane and a line. Hence 2) holds.

2) $\Rightarrow$ 1) If R contains both l_1 and l_2, then 1) is obvious. If $R\supset l_1$ and $R\not\supset l_2$, then $R\cup l_2$ is contained in two distinct quadrics. Hence 1) holds true. Similarly 1) holds in the case where $R\supset l_2$ and $R\not\supset l_1$. Q.E.D.

More explicitly we have the following whose proof is straightforward.

Proposition 4.2. *If a twisted cubic $[R]\in\mathcal{T}_{B_1,B_2}$ satisfies the equivalent conditions of the preceding proposition, then it satisfies one of the following:*

(a) R is the union of l_1 and a conic which have l_2 as a bisecant line, or vice versa, or

(b) R is the union $m_1\cup m_2\cup l_i$ of three lines, with $i=1$ or 2, such that both m_1 and m_2 intersect l_1 and l_2, or

(c) R is the union $l_1\cup l_2\cup m$ of three lines such that m intersects both l_1 and l_2.

The twisted cubics satisfying (a) are parametrized by open subsets of two $\mathbb{P}^4$-bundles A_1 and A_2 over $\mathbb{P}^1$. More precisely, A_1 is a $\mathbb{P}^4$-bundle over $|\mathcal{O}_{\mathbb{P}}(1-l_1)|\simeq\mathbb{P}^1$, the pencil of planes P containing l_1, and its fiber over $[P]$ parametrizes the conics in P passing through the intersection point $P\cap l_2$. In particular, both A_1 and A_2 are of dimension 5. The twisted cubics satisfying (c) are parametrized by the intersection $A_1\cap A_2$, which is isomorphic to $l_1\times l_2$. The twisted cubics satisfying (b) are parametrized by two copies of $\mathrm{Sym}^2(\mathbb{P}^1\times\mathbb{P}^1)$. In particular they are 4-dimensional families. Therefore, the first projection $\pi:Y\to\mathcal{T}_{B_1,B_2}$ of (17) is birational and we have the rational map

$$\mathcal{T}_{B_1,B_2}\cdots\to\Lambda\simeq\mathbb{P}^3,\quad R\mapsto Q$$

assigning the unique quadric $Q\in|\mathcal{O}_{\mathbb{P}}(2-R-l_1-l_2)|$ to R. The correspondence Y in (17) is nothing but the graph of this rational map.

Proposition 4.3. *Y is an 8-dimensional irreducible variety, and a generic $\mathbb{P}^5$-bundle over $\Lambda=|\mathcal{O}_{\mathbb{P}}(2-l_1-l_2)|$.*

Proof. We denote the second projection $Y \to |\mathcal{O}_{\mathbb{P}}(2 - l_1 - l_2)|$ by g, and the locus of singular members of $|\mathcal{O}_{\mathbb{P}}(2 - l_1 - l_2)|$ by Λ_0. Every member of Λ_0 is the union of two distinct planes. If $Q \notin \Lambda_0$, the fiber of g over Q is $|\mathcal{O}_{\mathbb{P}\times\mathbb{P}}(1, 2)| \simeq \mathbb{P}^5$. The fiber over $Q \in \Lambda_0$ is reducible. But it is easily checked that it is also of dimension 5. Q.E.D.

Assume that a smooth member $Q \in |\mathcal{O}_{\mathbb{P}}(2 - l_1 - l_2)|$ is defined by $xt - yz = 0$ for a homogeneous coordinate $(x; y; z; t)$ of $\mathbb{P}^3$. Then Q contains two 5-dimensional families of twisted cubics. They correspond to the matrices of the form

$$\begin{pmatrix} x & z & f \\ -y & -t & g \end{pmatrix} \quad \text{and} \quad \begin{pmatrix} x & y & f \\ -z & -t & g \end{pmatrix},$$

where f and g are linear forms. The former family is characterized by the property that the $x = y = 0$ is a bisecant line, and the latter family has $x = z = 0$ as a bisecant line.

§5. Twisted cubics apolar to two reducible quadrics

In this section we study the locus $\mathcal{T}_{M_0}$ of twisted cubics apolar to $M_0 \subset (S^2V)^\vee$ when M_0 is spanned by two quadratic forms q_1 and q_2 of rank 2. q_i is the pull-back of a quadratic form $\bar{q}_1$ on a line l_i for $i = 1, 2$. We assume that two lines l_1 and l_2 are skew. By Proposition 3.2, $\mathcal{T}_{M_0}$ is the pull-back of $\mathbb{P}^1 \times \mathbb{P}^1$ by the rational map $\mathcal{T}_{M_0} \cdots\to \mathbb{P}^2 \times \mathbb{P}^2$ defined by (16). We denote the restriction of (16) by

$$f_i : \mathcal{T}_{M_0} \cdots\to \mathbb{P}^1 \subset \mathrm{Sym}^2 l_i, \quad i = 1, 2. \tag{18}$$

Similar to the previous section, we consider the correspondence

$$X = \{(R, Q) \mid R \subset Q\} \subset \mathcal{T}_{M_0} \times \Lambda \tag{19}$$

between T_{M_0} and Λ. We denote the second projection $X \to \Lambda$ by g. When a quadric Q in Λ is smooth, the fiber of g over $[Q]$ is a 3-dimensional projective subspace of $|\mathcal{O}_{\mathbb{P}\times\mathbb{P}}(1, 2)| \simeq \mathbb{P}^5$. Similar to Proposition 4.3, X is irreducible of dimension 6, and a generic $\mathbb{P}^3$-bundle over Λ.

Proposition 4.1 holds for $\mathcal{T}_{M_0}$ too, and we have the following by Proposition 4.2.

Proposition 5.1. *If the first projection $\pi : X \to \mathcal{T}_{M_0}$ is not an isomorphism at $[R]$, then one of the following holds:*

(a) R is the union of l_1 and a conic which have l_2 as a bisecant line, or vice versa, or

(b) *R is the union $m_1 \cup m_2 \cup l_i$ of three lines, with $i = 1$ or 2, such that both m_1 and m_2 intersect l_1 and l_2, or*

(c) *R is the union $l_1 \cup l_2 \cup m$ of three lines such that m intersects both l_1 and l_2.*

The twisted cubics satisfying (a) are parametrized by open subsets of A'_1 and A'_2 which are $\mathbb{P}^3$-bundles over $\mathbb{P}^1$. In particular, both A'_1 and A'_2 are of dimension 4. The twisted cubics satisfying (c) are parametrized by the intersection $A'_1 \cap A'_2 \simeq l_1 \times l_2$. Since the twisted cubics satisfying (b) forms a 3-dimensional family, the first projection π is birational, and we obtain the rational map

$$\mathcal{T}_{M_0} \cdots \to \Lambda \simeq \mathbb{P}^3$$

which assigns the unique quadric $Q \in |\mathcal{O}_{\mathbb{P}}(2 - R - l_1 - l_2)|$ to R. The correspondence X in (19) is nothing but the graph of this rational map. $\pi^{-1}(A'_1)$ is of dimension 5 and its image by g is $\Lambda_0 \simeq l_2 \times l_1$.

We need also the following information on the restriction of the syzygy bundle $\mathcal{F}$ to a general fiber of the second projection $g : X \to \Lambda$.

Lemma 5.2. *If Q in Λ is smooth, then the restriction of $\mathcal{F}$ to $g^{-1}[Q] \simeq \mathbb{P}^3$ is isomorphic to $\mathcal{O}_{\mathbb{P}}(1)^{\oplus 2}$.*

Proof. We take a homogeneous coordinate $(x : y : z : t)$ of $\mathbb{P}^3$ such that

$$Q_1 : XY = 0, \quad Q_2 : ZT = 0, \quad Q : xt - yz = 0,$$

where $(X : Y : Z : T)$ is the dual coordinate of $\mathbb{P}^{3,*}$. A twisted cubic in the fiber $g^{-1}[Q]$ is defined by the three minors of the matrix $\begin{pmatrix} x & z & by + b't \\ -y & -t & ax + a'z \end{pmatrix}$, where a, a', b, b' are constants. (See the argument at the end of §4.) The syzygy space Syz_R of R is generated by

$$x \otimes \{(ax + a'z)z + (by + b't)t\} - z \otimes \{(ax + a'z)x + (by + b't)y\} + (by + b't) \otimes q$$

and

$$-y \otimes \{(ax + a'z)z + (by + b't)t\} + t \otimes \{(ax + a'z)x + (by + b't)y\} + (ax + a'z) \otimes q,$$

where we put $q = xt - yz$. Hence when R runs over the fiber $g^{-1}[Q]$, Syz_R generates the vector space of dimension 8 with the following basis:

$$x \otimes xz - z \otimes x^2, x \otimes z^2 - z \otimes xz, x \otimes yt - z \otimes y^2 - y \otimes q, x \otimes t^2 - z \otimes yt - t \otimes q,$$

$$-y \otimes xz + t \otimes x^2 - x \otimes q, -y \otimes z^2 - t \otimes xz - z \otimes q, y \otimes yt - t \otimes y^2, -y \otimes t^2 - t \otimes yt.$$

Syz_R has a 1-dimensional intersection with the vector space spanned by the first four syzygies, and so does with that spanned by the last four. Hence the fiber $g^{-1}[Q]$ is the projective space with $(a : a' : b : b')$ as its homogeneous coordinate, and $\mathcal{F}|_{g^{-1}[Q]}$ is isomorphic to $\mathcal{O}_{\mathbb{P}}(1)^{\oplus 2}$.

Q.E.D.

§6. Doubly octagonal K3 surface and proof of Theorem 1.1

Let $S_{M_0,N} \subset T_{M_0}$ be the zero locus of the global section of $\mathcal{F}^{\oplus 2}$ corresponding to a 2-dimensional subspace $N \subset (S^{2,1}V)^{\vee}$.

Lemma 6.1. *If N is general, then $S_{M_0,N}$ is disjoint from A'_1 and A'_2, that is, a twisted cubic in $S_{M_0,N}$ does not contain the line l_1 or l_2 as a component.*

Proof. We may assume that $q_1 = XY$ and $q_2 = ZT$ for a homogeneous coordinate $(x : y : z : t)$ of $\mathbb{P}^3$, where $(X : Y : Z : T)$ is the dual coordinate of $\mathbb{P}^{3,*}$.

Since $\mathcal{F}^{\oplus 2}$ is of rank 4 and generated by its global sections, it suffices to show that a twisted cubic satisfying (a) does not belong to $S_{M_0,N}$. Assume that such a cubic R satisfies the first half of the statement (a) of Proposition5.1. Then R is defined by three minors of a matrix of the form $\begin{pmatrix} f & * & * \\ 0 & x & y \end{pmatrix}$ and has $x \otimes yf - y \otimes xf$ as its syzygy, where f is a linear commination of x and y. When R runs over A'_1 these syzygies span the 2-dimensional vector space $\langle x \otimes yz - y \otimes xz, x \otimes yt - y \otimes xt \rangle_{\mathbb{C}}$ in $S^{2,1}V$. Since $N \subset (S^{2,1}V)^{\vee}$ is a general 2-dimensional space, its intersection with $N^{\perp}$ is zero. Hence A'_1 is disjoint from $S_{M_0,N}$. The same holds for A'_2.

Q.E.D.

By the lemma, the morphism $\pi : X \to T_{M_0}$ is an isomorphism over $S_{M_0,N}$. Hence we denote its pull-back in X by the same symbol $S_{M_0,N} \subset X$. The restriction of the rational map f_i $(i = 1, 2)$ to $S_{M_0,N}$ is a morphism, which we also denote by the same symbol $f_i : S_{M_0,N} \to \mathbb{P}^1 \subset \mathrm{Sym}^2 l_i$.

Now we study the intersection of divisor $\mathcal{D}_{B,2}$ (§3) with $S_{M_0,N}$. Let $\mathcal{D}_1$ be the locus of reducible twisted cubics R whose conical component has l_1 as a bisecant line.

Lemma 6.2. *If N is general, then the intersection $Z := \mathcal{D}_1 \cap S$ is isomorphic to $\mathbb{P}^1$.*

Proof. More precisely, we show that the restriction of $f_2|_Z : Z \to \mathbb{P}^1 \subset \mathrm{Sym}^2 l_2$ is the double cover induced from $\mathbb{P}^1 \times \mathbb{P}^1 \to \mathrm{Sym}^2 l_2$.

Let (p_1, p_2) be an ordered pair of points of l_2 which is apolar to (or orthogonal with respect to) $\bar{q}_2$. It suffice to show that there exist a unique reducible twisted cubic $R = C \cup l$ in $S_{M_0,N} \cap \mathcal{D}_1$ whose linear part l passes through p_1 and conical part C through p_2. Such a twisted cubic is the common zero locus of the matrix of the form $\begin{pmatrix} f & f_1 & f_2 \\ 0 & g_1 & g_2 \end{pmatrix}$, where f is the equation of the plane spanned by l and p_2, and g_1, g_2 are linear forms vanishing at p_1. One syzygy of R is $s(R) := g_1 \otimes fg_2 - g_2 \otimes fg_1$ which belongs to the space of syzygies

$$\langle x \otimes fy - y \otimes fx, y \otimes fz - z \otimes fy, z \otimes fx - x \otimes z \rangle_{\mathbb{C}}, \tag{20}$$

where $\{x, y, z\}$ is a basis of linear forms vanishing at p_2. Since N is of dimension 2, $s(R)$ belongs to $N^{\perp}$ for suitable choice of g_1 and g_2. Similarly another syzygy of R independent from $s(R)$ belongs to $N^{\perp}$ for suitable choice of f_1 and f_2. This shows the existence of the required $R = C \cup l$.

When an unordered pair $\{p_1, p_2\}$ runs over $\mathbb{P}^1 \subset \mathrm{Sym}^2 l_2$, the image of f_2, (20) is a 1-dimensional family of 3-dimensional subspaces. Hence the usual dimension count argument shows that the linear part l is unique for a given (p_1, p_2) if we choose N general enough. Similarly the conical part C is unique also if N is general. Q.E.D.

We now compute the intersection numbers of several divisor classes on S. We denote the restriction of $H = c_1(\mathcal{E})$ to S by h, and the divisor class of a general fiber of $f_i : S \to \mathbb{P}^1$ by a_i for $i = 1, 2$.

For every R in S, $H^0(\mathcal{O}_{\mathbb{P}}(2-R))$ has 1-dimensional intersection with $H^0(\mathcal{O}_{\mathbb{P}}(2 - l_1 - l_2))$ and 2-dimensional intersection with $H^0(\mathcal{O}_{\mathbb{P}}(2 - l_i))$, $i = 1, 2$, by Proposition 4.1 and Lemma 6.1. Hence we have an exact sequence

$$0 \to \mathcal{O}_S(a_1) \oplus \mathcal{O}_S(a_2) \to \mathcal{E}|_S \to \mathcal{O}_S(b) \to 0 \tag{21}$$

on S, where we put $b = h - a_1 - a_2$.

Lemma 6.3. 1) $(h.a_1) = (h.a_2) = 8$.
2) $(a_1.a_2) = 3$.

Proof. 1) A general fiber of the morphism (15) consists of all twisted cubics passing through two points $p_1, p_2 \in l$. Hence its fundamental cohomology class is $(c_2 - d_2)^2$ by [1, Section 7]. Hence $(h.a_1)$ and $(h.a_2)$ are equal to the intersection number $(c_1(c_2 - d_2)^2 d_2^2 c_3)$, which is equal to $82 - 2 \cdot 57 + 40 = 8$ by [1, Table 1].

2) By Proposition 2.2 and the exact sequence (21), we have $c_2(\mathcal{E}|_S) = (a_1.a_2) + (b.a_1 + a_2) = 13$. Hence 2) follows from 1). Q.E.D.

By the lemma, the a_1, a_2 and b spans an integral sublattice of rank 3 in the Picard lattice of S with inner product $\begin{pmatrix} 0 & 3 & 5 \\ 3 & 0 & 5 \\ 5 & 5 & 4 \end{pmatrix}$. Since the discriminant is equal to 14 and square free, $\langle a_1, a_2, b\rangle_{\mathbb{Z}}$ is a primitive sublattice. Theorem 1.1 follows from Proposition 2.2 and the following

Lemma 6.4. *$S = S_{M_0,N}$, for general N, is mapped to a quartic surface by the morphism $g : \mathcal{T}_{M_0} \to \mathbb{P}^3$.*

Proof. The pull-back of the tautological line bundle of $\mathbb{P}^3$ by g is $\mathcal{O}_S(b)$. By Lemma 6.3, we have $(b^2) = (h - a_1 - a_2)^2 = 4$. Hence the restricted morphism $g|_S : S \to \mathbb{P}^3$ is of degree 4. By Lemma 5.2, every general fiber of $g|_S$ is a linear subspace of $\mathbb{P}^3$. Hence $g|_S$ cannot be a double cover of a quadric or a quartic cover of a plane. Hence $g|_S$ is birational onto a quartic surface. Q.E.D.

Since $(a_1.a_2)$ and $(a_1.b)$ are coprime, the divisor class a_1 is primitive. Hence the fiber of f_1 is connected. Therefore, f_1 is an elliptic fibration of degree 8 of the polarized K3 surface $(S_{M_0,N}, h)$. The same holds for f_2. We call $S_{M_0,N}$ *doubly octagonal* for this reason. The Mukai vectors of $\mathcal{E}|_S$ and $\mathcal{F}|_S$ are $(3, h, 5)$ and $(2, h, 8)$, respectively, by Proposition 2.2. Hence, we have $\chi(\mathcal{E}|_S, \mathcal{F}|_S) = 24 - 30 + 10 = 4$, $(v(\mathcal{E}|_S)^2) = 30 - 30 = 0$ and $(v(\mathcal{F}|_S)^2) = 30 - 32 = -2$ ([4, §2]).

§7. Proof of Proposition 1.3 and Theorem 1.2

We prove Proposition 1.3 step by step. Let S be $S_{M_0,N}$ for general N as in the previous section.

claim 1. *$H^i(S, \mathcal{E}|_S) = 0$ for all $i > 0$.*

Proof. Since $\mathcal{O}_S(b)$ is the pull-back of $\mathcal{O}_{\mathbb{P}}(1)$ by g, $H^i(S, \mathcal{O}_S(b)) = 0$ for all $i > 0$. Since $|a_j|$ contains a smooth elliptic curve, $H^i(S, \mathcal{O}_S(a_j)) = 0$, for all $i > 0$ and $j = 1, 2$. Hence the claim follows from the exact sequence (21). Q.E.D.

We need to investigate the restriction of the syzygy bundle $\mathcal{F}$ to S. By Proposition 3.1, we have an exact sequence

$$0 \to \mathcal{O}_S(a_1) \oplus \mathcal{O}_S(a_1) \to \mathcal{F}|_S \to j_*\gamma \to 0, \tag{22}$$

where $j : Z = \mathcal{D}_1 \cap S \hookrightarrow S$ is a natural inclusion and γ is a line bundle on Z. We have $\deg \gamma = 5$ by Lemma 6.2 and Proposition 2.2.

claim 2. *$H^i(S, \mathcal{F}|_S) = 0$ for all $i > 0$.*

Proof. Obvious from (22) and the vanishing $H^1(Z,\gamma) = 0$ and $H^i(S,\mathcal{O}_S(a_1)) = 0$ for $i > 0$. Q.E.D.

claim 3. $\mathrm{Ext}^i(\mathcal{E}|_S,\mathcal{F}|_S) = 0$ *for all* $i > 0$.

Proof. We denote $\mathcal{E}|_S,\mathcal{F}|_S$ by E and F, respectively. Since $\chi(E,F) = 4$, it suffice to show $\dim \mathrm{Hom}(E,F) = 4$ and $\mathrm{Hom}(F,E) = 0$. Since E is extension of three line bundles $\mathcal{O}_S(a_1),\mathcal{O}_S(a_2),\mathcal{O}_S(b)$, it suffice to show

$$h^0(F(-a_1)) + h^0(F(-a_2)) + h^0(F(-b)) \leq 4.$$

Taking dual and twisting by $\mathcal{O}_S(a_1)$, the exact sequence (22) induces an exact sequence

$$0 \to F(-a_2 - b) \to \mathcal{O}_S \oplus \mathcal{O}_S \to j_*\alpha \to 0, \tag{23}$$

where α is a line bundle of degree 1 on Z. The induced linear map $H^0(\mathcal{O}_S \oplus \mathcal{O}_S) \to H^0(\alpha)$ is an isomorphism. Tensoring with $\mathcal{O}_S(a_2)$, we have the exact sequence

$$0 \to F(-b) \to \mathcal{O}_S(a_2) \oplus \mathcal{O}_S(a_2) \to (j_*\alpha) \otimes \mathcal{O}_S(a_2) \to 0$$

The restriction of the linear system $|a_2|$ to Z is of degree 2 and free. Hence

$$H^0(\mathcal{O}_S(a_2) \oplus \mathcal{O}_S(a_2)) \to H^0(j_*\alpha \otimes \mathcal{O}_S(a_2))$$

is injective. Therefore, we have $H^0(F(-b)) = 0$.

The exact sequence (22) twisted by $\mathcal{O}_S(-a_1)$ is

$$0 \to \mathcal{O}_S \oplus \mathcal{O}_S \to F(-a_1) \to j_*\beta \to 0.$$

for a line bundle β of degree -3. Hence we have $h^0(F(-a_1)) = 2$, and similarly $h^0(F(-a_2)) = 2$. This shows $\dim \mathrm{Hom}(E,F) = 4$.

$\mathrm{Hom}(F,E) = 0$ follows from $H^0(F(-a_1 - b)) = H^0(F(-a_2 - b)) = H^0(F(-a_1 - a_2)) = 0$. Q.E.D.

claim 4. *The natural linear map* $V = \mathbb{C}^4 \to \mathrm{Hom}(\mathcal{E}|_S,\mathcal{F}|_S)$ *(via* $\mathrm{Hom}(\mathcal{E},\mathcal{F})$*) is an isomorphism.*

Proof. It suffice to show the linear map is injective. Assume the contrary. Then there exists a point $p \in \mathbb{P}^3$ such that every R belonging to S is the union of three lines passing through p. This is obviously impossible. Q.E.D.

claim 5. $\mathcal{F}|_S$ *is simple.*

Proof. By the exact sequence (23), $\mathcal{F}|_S$ is the reflection of $j_*\alpha$ by the rigid bundle $\mathcal{O}_S$ Since $j_*\alpha$ is simple so is $\mathcal{F}|_S$ by [4, Proposition 2.25]. (See also the remark below.) Q.E.D.

Remark 7.1. In the terminology of [9], $\mathcal{F}|_S$ is the spherical twist $T_{\mathcal{O}_S}(j_*\alpha)$ of $j_*\alpha$ by the spherical object $\mathcal{O}_S$. Since $T_{\mathcal{O}_S}$ is an autoequivalence of the derived category of coherent sheaves on S by [9, Theorem 1.2], $\mathcal{F}|_S$ is simple.

Proof of Proposition 1.3. We already proved it mostly in the above claims 1–5 taking $S_{M_0,N}$ as S, except for the simpleness of $\mathcal{E}|_S$. We need an extra argument, since the restriction of $\mathcal{E}$ on $S_{M_0,N}$ is not simple. In fact, the 6-fold $\mathcal{T}_{M_0}$ has an action of the 3-dimensional torus, and the restriction of $\mathcal{E}$ to there is not simple.

By the exact sequence (21), $\mathcal{E}|_S$ is an extension of the direct sum of two line bundles by the line bundle $\mathcal{O}_S(b)$. Now we replace the direct sum by nontrivial extension G of $\mathcal{O}_S(a_1)$ by $\mathcal{O}_S(a_2)$. This is possible since $(a_1 - a_2)^2 = -6$. Furthermore, we take a nontrivial extension E' of G by $\mathcal{O}_S(b)$. This is possible since $(a_1 - b)^2 = (a_2 - b)^2 = -6$. Since $|b - a_i| = |a_i - b| = \emptyset$ for $i = 1, 2$ and since $|a_1 - a_2| = |a_2 - a_1| = \emptyset$, E' is simple. (The emptyness of linear systems follows easily since a_1, a_2 and b are nef.) There is a family of vector bundles $\{E_t\}$ on S parametrized by the affine line $\mathbb{A}^1$ such that $E_0 \simeq \mathcal{E}|_S$ and $E_t \simeq E'$ for every $t \neq 0$.

By the upper semi-continuity of cohomology, E' satisfies the same vanishing as $\mathcal{E}|_S$. In particular, we have $\dim \mathrm{Hom}(E', \mathcal{F}|_S) = \chi(E', \mathcal{F}|_S) = 4$. The universal property (*) in §2 applies to the pair $(E', \mathcal{F}|_S)$ and $V' := \mathrm{Hom}(E', \mathcal{F}|_S)$, and we have a morphism from S to the EPS compactification $\mathcal{T}'(\simeq \mathcal{T})$ of twisted cubics. The morphism is an embedding since it is so for the pair $(\mathcal{E}|_S, \mathcal{F}|_S)$. The image is a complete intersection with respect to $\mathcal{E}^{\oplus 2} \oplus \mathcal{F}^{\oplus 2}$ by virtue of the cohomology vanishing since it is so for the pair $(\mathcal{E}|_S, \mathcal{F}|_S)$. S is isomorphic to $S_{M',N'}$ for a pair (M', N'), which is a deformation of the pair (M_0, N), by the claims 1–4. This K3 surface $S_{M',N'}$ satisfies all the requirement of the proposition. Q.E.D.

Proof of Theorem 1.2. We denote the non-empty open subset of $\mathcal{P}$ (see Introduction) consisting of (M, N) such that the restriction of $\mathcal{E}$ and $\mathcal{F}$ to $S_{M,N}$ satisfies the requirement of Proposition 1.3 by $\mathcal{P}_0$. Let (S, h) be a small deformation of $(S_{M,N}, H|_{S_{M,N}})$ as polarized $K3$ surface. Then by Proposition 1.3 and the proposition below $\mathcal{E}|_{S_{M,N}}$ and $\mathcal{F}|_{S_{M,N}}$ deforms to vector bundles E and F, with $\det E \simeq \det F \simeq \mathcal{O}_S(h)$, on S. Since (E, F) is a small deformation of $(\mathcal{E}|_{S_{M,N}}, \mathcal{F}|_{S_{M,N}})$, it embeds S into $\mathcal{T}$ and the image of S is a complete intersection with respect to $\mathcal{E}^{\oplus 2} \oplus \mathcal{F}^{\oplus 2}$, again by Proposition 1.3 and a similar argument in its proof. Therefore, the image of the classification morphism

$$\mathcal{P}_0 \to \mathcal{F}_{16}, \quad (M, \bar{N}) \mapsto (S_{M,N}, \mathcal{O}_S(1)),$$

is open. Q.E.D.

Proposition 7.2. ([7, Proposition 4.1]) *Let E be a simple vector bundle on a $K3$ surface S and (S', L') be a small deformation of $(S, \det E)$. Then there is a deformation (S', E') of the pair (S, E) such that* $\det E' \simeq L'$.

Remark 7.3. The rational map (5) factors through the birational quotient $\mathcal{P}/PGL(4)$, which is of dimension 21(=20+16-15). Every general fiber of $\mathcal{P}/PGL(4) \cdots \to \mathcal{F}_{16}$ is birational to a (K3) surface. In fact, it is the moduli space $M_S(3, h, 5)$ of semi-rigid vector bundles E on S with Mukai vector $(3, h, 5)$. See [8, Corollary to Theorem 2] for a similar result in the case of genus 13.

§8. Comparison with the case of genus 12

It is worth recalling here the description of a general K3 surface (S, h) of genus 12, that is, $(h^2) = 22$, which is the origin of our description in this article.

There exist two rigid vector bundles E and F on S with Mukai vector $(3, h, 4)$ and $(2, h, 6)$, respectively. The vector space $V^\vee = \mathrm{Hom}(E, F)$ is of dimension 4 and the universal property (*) applies. Hence we have a morphism from S to the EPS compactification $\mathcal{T}$. The induced linear map $S^2V^\vee \to H^0(E)$ is surjective and its kernel N is of dimension 3. Hence the image of S is contained in the moduli $\mathcal{T}_N$ of twisted cubics apolar to the net of quadrics $|N|$. $\mathcal{T}_N$ is a Fano 3-fold of genus 12 ([6, §3]). The image of S is an anticanonical member of $\mathcal{T}_N$.

Theorem 1.2 concerning on K3 surfaces of genus 16 was found replacing two vector bundles E and F above by a semi-rigid vector bundle ([4, §3]) with Mukai vector $(3, h, 5)$ and a rigid vector bundle with Mukai vector $(2, h, 8)$, respectively.

References

[1] G. Ellingsrud, and S.A. Strømme, The number of twisted cubic curves on the general quintic threefold, Math. Scand. **76**(1995), 5–34.

[2] G. Ellingsrud, R. Piene, and S.A. Strømme, On the variety of nets of quadrics defining twisted cubic curves. In F. Ghione, C. Peskine and E. Sernesi, editors, *Space Curves*, Lecture Notes in Math. **1266** (1987), Springer-Verlag, pp. 84–96.

[3] V.A. Gritsenko, K. Hulek and G.A. Sankaran, The Kodaira dimension of the moduli spaces of K3 surfaces, Invent. math. **169** (2007), 519–567.

[4] S. Mukai, On the moduli space of bundles on K3 surfaces I, in *Vector Bundles on Algebraic Varieties*, Oxford University Press, 1987, pp.341-413.

[5] S. Mukai, Curves, K3 surfaces and Fano 3-folds of genus ≤ 10, in '*Algebraic Geometry and Commutative Algebra in Honor of Masayoshi Nagata*', pp. 357–377, Kinokuniya, Tokyo, 1988.

[6] S. Mukai, Biregular classification of Fano threefolds and Fano manifolds of coindex 3, Proc. Nat. Acad. Sci. **86** (1989), 3000-3002.

[7] S. Mukai, Polarized K3 surfaces of genus 18 and 20, in *Complex Projective Geometry*, ed. G. Ellingsrud et. al., London Math. Soc. 1992, pp. 264–276.

[8] S. Mukai: Polarized K3 surfaces of genus thirteen, In S. Mukai et. al, editors, *Moduli spaces and Arithmetic Geometry (Kyoto, 2004)*, Adv. Stud. Pure Math. **45** (2006), Math. Soc. Japan and Amer. Math. Soc., pp. 315–326.

[9] P. Seidel and R.P. Thomas, Braid group actions on derived categories of coherent sheaves, Duke Math. J. **108**(2001), 37–108.

Research Institute for Mathematical Sciences, Kyoto University, Kyoto 606-8502, Japan

E-mail address: mukai@kurims.kyoto-u.ac.jp

Advanced Studies in Pure Mathematics 70, 2016
Minimal Models and Extremal Rays (Kyoto, 2011)
pp. 397–420

On $\mathbb{Q}$-Fano 3-folds of Fano index 2

Yuri Prokhorov and Miles Reid

To Shigefumi Mori in friendship and admiration

§1. Introduction

1.1. $\mathbb{Q}$-Fano 3-folds

A $\mathbb{Q}$-*Fano* 3-*fold* is a projective 3-fold X with at worst terminal singularities and ample anticanonical divisor $-K_X$. Here, bearing in mind Mori's fundamental notion of extremal ray, we assume also that X is $\mathbb{Q}$-factorial and has rank 1, that is, $\operatorname{Pic} X \simeq \mathbb{Z}$ or equivalently, $\operatorname{Cl} X \otimes \mathbb{Q} \simeq \mathbb{Q}$. We define the *Fano* and $\mathbb{Q}$-*Fano index* of X by:

$$\mathrm{q_F}(X) := \max\{q \in \mathbb{Z} \mid -K_X \sim qA \text{ with } A \text{ a Weil divisor}\},$$

$$\mathrm{q_{\mathbb{Q}}}(X) := \max\{q \in \mathbb{Z} \mid -K_X \sim_{\mathbb{Q}} qA \text{ with } A \text{ a Weil divisor}\},$$

where $\sim$ is linear equivalence and $\sim_{\mathbb{Q}}$ is $\mathbb{Q}$-linear equivalence. Clearly, $\mathrm{q_F}(X)$ divides $\mathrm{q_{\mathbb{Q}}}(X)$, and the two coincide unless $K_X + qA \in \operatorname{Cl} X$ is a nontrivial torsion element. An important invariant of a $\mathbb{Q}$-Fano 3-fold is its *genus* $\mathrm{g}(X) := \dim |-K_X| - 1$.

1.2. Background facts

Kaori Suzuki [Suz04] restricts the $\mathbb{Q}$-Fano index of X to one of

(1.2.1) $$\mathrm{q_{\mathbb{Q}}}(X) \in \{1, \dots, 11, 13, 17, 19\}.$$

See also [Pro10b, Lemma 3.3]. Moreover, the following results are due to the first author.

Received March 5, 2012.
Revised August 6, 2014.
Y.P. acknowledges partial support from RFBR grants No. 11-01-92613-KO_a, 11-01-00336-a, the grant of Leading Scientific Schools No. 2998.2014.1, and Russian Academic Excellence Project '5–100'.
M.R. is partly supported as Scholar of KIAS.

1.2.2. Theorem ([Pro10b]). *Let X be a $\mathbb{Q}$-Fano 3-fold of $\mathbb{Q}$-Fano index $q := \mathrm{q}_{\mathbb{Q}}(X) \geq 9$. Then $\mathrm{Cl}\, X \simeq \mathbb{Z}$.*

(i) *If $q = 19$ then $X \simeq \mathbb{P}(3,4,5,7)$.*
(ii) *If $q = 17$ then $X \simeq \mathbb{P}(2,3,5,7)$.*
(iii) *If $q = 13$ and $\mathrm{g}(X) > 4$ then $X \simeq \mathbb{P}(1,3,4,5)$.*
(iv) *If $q = 11$ and $\mathrm{g}(X) > 10$ then $X \simeq \mathbb{P}(1,2,3,5)$.*
(v) *$q \neq 10$.*

1.2.3. Theorem ([Pro10c]). *Let X be a $\mathbb{Q}$-Fano 3-fold of $\mathbb{Q}$-Fano index q.*

(vi) *If $q = 9$ and $\mathrm{g}(X) > 4$ then $X \simeq X_6 \subset \mathbb{P}(1,2,3,4,5)$.*
(vii) *If $q = 8$ and $\mathrm{g}(X) > 10$ then $X \simeq X_6 \subset \mathbb{P}(1,2,3^2,5)$ or $X_{10} \subset \mathbb{P}(1,2,3,5,7)$.*
(viii) *If $q = 7$ and $\mathrm{g}(X) > 17$ then $X \simeq \mathbb{P}(1^2,2,3)$.*
(ix) *If $q = 6$ and $\mathrm{g}(X) > 15$ then $X \simeq X_6 \subset \mathbb{P}(1^2,2,3,5)$.*
(x) *If $q = 5$ and $\mathrm{g}(X) > 18$ then $X \simeq \mathbb{P}(1^3,2)$ or $X_4 \subset \mathbb{P}(1^2,2^2,3)$.*
(xi) *If $q = 4$ and $\mathrm{g}(X) > 21$ then $X \simeq \mathbb{P}^3$ or $X_4 \subset \mathbb{P}(1^3,2,3)$.*
(xii) *If $q = 3$ and $\mathrm{g}(X) > 20$ then $X \simeq X_2 \subset \mathbb{P}^4$ or $X_3 \subset \mathbb{P}(1^4,2)$.*

Here we study the case $\mathrm{q}_{\mathbb{Q}}(X) = 2$.

1.2.4. Theorem ([BS07b]). *The Hilbert series of $\mathbb{Q}$-Fano 3-folds with $q = \mathrm{q}_{\mathbb{Q}}(X) = \mathrm{q}_{\mathrm{F}}(X) = 2$ belong to at most 1492 cases.*

The online database [GRDB] lists the numerical type of candidates (the data going into the Hilbert series of their graded rings).

1.3. Main results

1.3.1. Main Theorem. *Let X be a $\mathbb{Q}$-Fano 3-fold of rank 1 with $\mathrm{q}_{\mathbb{Q}}(X) = \mathrm{q}_{\mathrm{F}}(X) = 2$ and K_X not Cartier. Let A be a Weil divisor on X such that $-K_X = 2A$.*

Then $\dim |A| \leq 4$. Moreover, if $\dim |A| = 4$, then X belongs to the single irreducible family constructed in 6.3.7 (see also 1.5).

1.3.2. Corollary. *A $\mathbb{Q}$-Fano 3-fold with $\mathrm{q}_{\mathbb{Q}}(X) = \mathrm{q}_{\mathrm{F}}(X) = 2$ and K_X not Cartier has $\mathrm{g}(X) \leq 16$.*

Remark 1.3.3. If K_X is Cartier and $\mathrm{q}_{\mathrm{F}}(X) = 2$, then X is a *del Pezzo variety* [Fuj90]. Two cases with $\mathrm{Cl}\, X \simeq \mathbb{Z}$ have $\dim |A| > 4$:

(a) the complete intersection of two quadrics $X = X_{2\cdot 2} \subset \mathbb{P}^5$, with $\dim |A| = 5$ and $\mathrm{g}(X) = 19$; and
(b) $X = X_5 \subset \mathbb{P}^6$ a section of the Grassmannian $\mathrm{Gr}(2,5) \subset \mathbb{P}^9$ by a subspace of codimension 3, with $\dim |A| = 6$ and $\mathrm{g}(X) = 23$. In this case X must be smooth by [Pro10a, Cor. 5.3].

1.4. Strategy of proof

Sections 4–5 contain the proof of Main Theorem 1.3.1. The Kawamata blowup of a $\frac{1}{r}(2, a, r-a)$ point initiates a Sarkisov link ending in a fibre space or a $\mathbb{Q}$-Fano 3-fold with $q \geq 3$; the assumption $\dim |A| \geq 4$ leads to a manageable case division. The auxiliary Section 3 treats the cases with $q \geq 3$, most of which lead to a contradiction, with just one surviving in Section 5 to characterize our Main Example.

1.5. The Main Example

Section 6 gives several constructions of the exceptional family of Main Theorem 1.3.1, $\mathbb{Q}$-Fanos X with $\operatorname{Cl} X = \mathbb{Z} \cdot A$, $K_X = -2A$ and $\dim |A| = 4$. They arise from the simplest type of Sarkisov link:

$$\begin{array}{ccc} & E \subset X_1 \supset F & \\ & \swarrow \qquad \searrow & \\ P \in X & & Q \supset \Gamma_5 \ni P_0 \end{array} \tag{1.5.1}$$

starting from the nonsingular quadric hypersurface $Q \subset \mathbb{P}^4$, a point $P_0 \in Q$ and an irreducible curve $\Gamma_5 \subset E_0$ of degree 5 contained in the tangent hyperplane section $E_0 = Q \cap T_{P_0}Q$, with $\operatorname{mult}_{P_0} \Gamma_5 = 3$. We make the symbolic blowup $X_1 \to Q$ of Γ_5, then contract the birational transform $E_1 \simeq E_0 \simeq \mathbb{P}(2,1,1) \subset X_1$ of E_0 to a $\frac{1}{3}(1,2,2)$ orbifold point.

The symbolic blowup of $\Gamma \subset E_0 \subset Q$ is the relative Proj of the symbolic algebra $\mathcal{A} = \bigoplus \mathcal{I}_\Gamma^{[n]}$. For a singular curve Γ contained in a nodal surface, this is a local graded ring construction with a universal description, studied in much more detail and generality in Tom Ducat's thesis [Du15]. Compare [Du14].

1.6. Discussion

The study of $\mathbb{Q}$-Fanos divides into birational and biregular considerations. Biregular methods study projective embedding by multiples of A, or more precisely, generators and relations for the Gorenstein graded ring $R(X, A)$. This is effective when $R(X, A)$ has small codimension, especially if it is a hypersurface or codimension 2 complete intersection, etc. In contrast, birational methods are powerful when the linear system $|A|$ is large, implying a low canonical threshold, and allowing us to impose noncanonical singularities on $|A|$ and study X via the resulting Sarkisov link, aiming for a birational construction or a nonexistence result. The interest of this paper is as a meeting point of the two methods.

1.7. The fabulous half-elephant; more cases with $q = 2$

A surface section $F \in |A|$ of a $\mathbb{Q}$-Fano 3-fold X of index 2 is a del Pezzo surface (sometimes very singular). In a few cases where F has the simplest orbifold points such as $\frac{1}{3}(2,2)$ or $\frac{1}{5}(2,4)$, Reid and Suzuki [RS03] study such surfaces in terms of cascades of projections from nonsingular points. This foreshadows one construction of our Main Example in Section 6, and hints at other cases that might make interesting challenges, especially the X with $\dim|A| = 3$ or 2. Del Pezzo surfaces with only $\frac{1}{3}(2,2)$ orbifold points are classified in current work of Alessio Corti and Liana Heuberger [CH15]. Kuzma Khrabrov [Kh14] has partial results on $\mathbb{Q}$-Fano 3-folds X of index 2 with $\dim|A| \geq 2$.

§2. The method

2.1. Construction of a Sarkisov link [Ale94]

Let $\mathcal{M}$ be a linear system on X with no fixed part, and canonical threshold $c := \operatorname{ct}(X, \mathcal{M})$. Assume $-(K_X + c\mathcal{M})$ is ample. Then $(X, c\mathcal{M})$ is canonical but not terminal, so we can pull out an irreducible divisor $E \subset \widetilde{X}$ by an extremal divisorial extraction $f\colon \widetilde{X} \to X$, such that $\widetilde{X}$ has only terminal $\mathbb{Q}$-factorial singularities, $\rho(\widetilde{X}/X) = 1$, and f is $(K + c\mathcal{M})$-crepant:

$$K_{\widetilde{X}} + c\widetilde{\mathcal{M}} = f^*(K_X + c\mathcal{M}). \tag{2.1.1}$$

As in [Ale94], running a $(K + c\mathcal{M})$-MMP on $\widetilde{X}$ gives a Sarkisov link of type I or II:

$$\begin{array}{ccccc} & \widetilde{X} & \dashrightarrow & \overline{X} & \\ {}^{f}\swarrow & & & & \searrow^{\overline{f}} \\ X & & & & \widehat{X} \end{array} \tag{2.1.2}$$

where $\widetilde{X}$ and $\overline{X}$ have only $\mathbb{Q}$-factorial terminal singularities, $\rho(\widetilde{X}) = \rho(\overline{X}) = 2$, $\widetilde{X} \dashrightarrow \overline{X}$ is a chain of log flips, and $\overline{f}$ is a Mori extremal contraction, either a divisorial contraction to a $\mathbb{Q}$-Fano 3-fold $\widehat{X}$, or a Mori fibre space over a curve or surface $\widehat{X}$. In either case, $\rho(\widehat{X}) = 1$. We write $\widetilde{D}$ and $\overline{D}$ for the birational transform on $\widetilde{X}$ and $\overline{X}$ of a divisor or linear system D on X.

Assume that $K_X + \lambda\mathcal{M} + \Xi \sim_{\mathbb{Q}} 0$ for some $\lambda > c$ and an effective $\mathbb{Q}$-divisor Ξ. We can write

$$K_{\widetilde{X}} + \lambda\widetilde{\mathcal{M}} + \widetilde{\Xi} + aE \sim_{\mathbb{Q}} f^*(K_X + \lambda\mathcal{M} + \Xi) \sim_{\mathbb{Q}} 0, \tag{2.1.3}$$

where $a > 0$ is the log discrepancy of f. Note that if $K_X + \lambda\mathcal{M} + \Xi \sim 0$ then it is a Cartier divisor; we can assume that the mobile system $\widetilde{\mathcal{M}}$ has no common divisor with $\widetilde{\Xi}$. Then $K_{\widetilde{X}}$, $\lambda\widetilde{\mathcal{M}}$ and $\widetilde{\Xi}$ are all integral Weil divisors, and therefore a is an integer.

Remark 2.1.4. We use the extremal extraction $\widetilde{X} \to X$ with ray R and exceptional surface E to initiate a Sarkisov link. By (2.1.1), $K_{\widetilde{X}} + c\widetilde{\mathcal{M}}$ is nef and big, with

$$K_{\widetilde{X}} \cdot R < 0, \quad (K_{\widetilde{X}} + c\widetilde{\mathcal{M}}) \cdot R = 0, \quad \text{so that } \widetilde{\mathcal{M}} \cdot R > 0. \tag{2.1.5}$$

The MMP that constructs the Sarkisov link proceeds by increasing λ in $K + \lambda\mathcal{M}$. Each step makes $K + \lambda\mathcal{M}$ bigger on the ray R, so on the exceptional surface E and its birational transforms. Thus the MMP can never contract the birational transform of E.

2.2. Case $\overline{f}$ not birational

Assume that $\overline{f}$ is not birational. Then $\widehat{X}$ is either a smooth rational curve or a del Pezzo surface with at worst Du Val singularities and $\rho(\widehat{X}) = 1$ [MP08]. We also have $\overline{f}(\overline{E}) = \widehat{X}$ by Remark 2.1.4, or because no multiple $n\overline{E}$ of the exceptional divisor $\overline{E}$ of f moves on $\overline{X}$. In this case we write $\overline{F}$ for a general fiber of $\overline{f}$. Let Θ be an ample Weil divisor on $\widehat{X}$ whose class generates $\operatorname{Cl}\widehat{X}$ modulo torsion. If $\widehat{X}$ is a surface with $K_{\widehat{X}}^2 = 1$, we take $\Theta = -K_{\widehat{X}}$.

2.2.1. For $\widehat{X}$ a surface, one of the following holds:

(i) $-K_{\widehat{X}} \cdot \Theta = 3$, $-K_{\widehat{X}} \sim 3\Theta$, $\widehat{X} \simeq \mathbb{P}^2$ and $\dim|\Theta| = 2$;

(ii) $-K_{\widehat{X}} \cdot \Theta = 2$, $-K_{\widehat{X}} \sim 4\Theta$, $\widehat{X} \simeq \mathbb{P}(1,1,2)$ and $\dim|\Theta| = 1$;

(iii) $-K_{\widehat{X}} \cdot \Theta = 1$, $-K_{\widehat{X}} \sim d\Theta$, where $d := K_{\widehat{X}}^2 \le 6$, and the minimal resolution of $\widehat{X}$ is a blowup of $\mathbb{P}^2$ at $9-d$ points in *almost general position*. In this case, $\dim|\Theta| = 0$ or 1. Moreover, by Kawamata–Viehweg vanishing and orbifold Riemann–Roch [YPG], for an ample Weil divisor $B \sim_{\mathbb{Q}} t\Theta$ we have

$$\dim|B| \le \frac{t(t+d)}{2d}. \tag{2.2.2}$$

2.3. Case $\overline{f}$ birational

Assume that the contraction $\overline{f}$ is birational. In this case, $\widehat{X}$ is a $\mathbb{Q}$-Fano 3-fold and $\overline{f}$ contracts a unique exceptional divisor $\overline{F}$. Remark 2.1.4 implies that $\overline{E} \neq \overline{F}$ (or argue that $\overline{E} = \overline{F}$ would imply that $X \dashrightarrow \widehat{X}$ is an isomorphism in codimension, leading to a contradiction). Write $\widetilde{F} \subset \widetilde{X}$ and $F := f(\widetilde{F})$ for its birational transform. Set $\widehat{q} := \mathrm{q}_{\mathbb{Q}}(\widehat{X})$. For a divisor $\overline{D}$ on $\overline{X}$, we put $\widehat{D} := \overline{f}_*\overline{D}$.

2.4. Computer search for $\mathbb{Q}$-Fano 3-folds

All $\mathbb{Q}$-Fano 3-folds belong to a finite number of algebraic families [Kaw92]. In fact, Kawamata's proof implies that the possible "candidate" $\mathbb{Q}$-Fano 3-folds can be listed, although the volume of computation makes computer searches inevitable. This method was used in [Suz04], [BS07a], [BS07b], [Pro07], [Pro10b], [Pro10c]. See [GRDB] for explicit lists.

We outline the algorithm, starting with a useful remark.

Remark 2.4.1. The local analytic Weil divisor class group of a 3-fold $\mathbb{Q}$-factorial terminal point $P \in X$ is cyclic $\mathrm{Cl}(X,P) \simeq \mathbb{Z}/r$, and is generated by the canonical divisor K_X [Kaw88, Lemma 5.1]. In particular, if X is a $\mathbb{Q}$-Fano 3-fold, its local Gorenstein index r at every terminal point is coprime to the $\mathbb{Q}$-Fano index $q = \mathrm{q_F}(X)$.

2.4.2. Let X be a $\mathbb{Q}$-Fano 3-fold. For simplicity we assume that $q := \mathrm{q}_{\mathbb{Q}}(X) = \mathrm{q_F}(X) \geq 3$ (the only case we need in this section). Let A be a Weil divisor such that $-K_X \sim qA$ and $\mathcal{B}(X) = \{(r_P, b_P)\}$ the basket of orbifold points of X [YPG].

Step 1. We have the equality

$$-K_X \cdot c_2(X) + \sum_{P\in\mathcal{B}} \frac{r_P - 1}{r_P} = 24, \tag{2.4.3}$$

where $-K_X \cdot c_2(X) > 0$ [Kaw92]. Hence there is only a finite (but huge) number of possibilities for the basket $\mathcal{B}(X)$ and $-K_X \cdot c_2(X)$. Let $r := \mathrm{lcm}(\{r_P\})$ be the Gorenstein index of X.

Step 2. (1.2.1) says that $q \in \{3, \ldots, 11, 13, 17, 19\}$. Remark 2.4.1 implies that $\gcd(q, r) = 1$, which eliminates some possibilities.

Step 3. In each case we compute A^3 by the formula

$$A^3 = \frac{12}{(q-1)(q-2)}\Big(1 - \frac{A \cdot c_2}{12} + \sum_{P\in B} c_P(-A)\Big).$$

(see [Suz04]), where c_P is the correction term in the orbifold Riemann–Roch formula [YPG]. The number rA^3 must be an integer [Suz04, Lemma 1.2].

Step 4. Next, the Bogomolov–Miyaoka inequality (see [Kaw92]) implies that

$$(4q^2 - 3q)\, A^3 \leq -4K_X \cdot c_2(X) \tag{2.4.4}$$

[Suz04, Prop. 2.2].

Step 5. Finally, the Kawamata–Viehweg vanishing theorem gives $\chi(tA) = h^0(tA) = 0$ for $-q < t < 0$. We check this condition using orbifold Riemann–Roch [YPG], [BRZ].

§3. On $\mathbb{Q}$-Fano 3-folds of Fano index ≥ 3

3.1. A result of Fujita

A *polarized variety* is a pair (X, S) consisting of a projective variety X and an ample Cartier divisor S on X. Its Δ-*genus* is defined as follows [Fuj90]:

$$\Delta(X, S) = \dim X + S^{\dim X} - h^0(X, \mathcal{O}_X(S)). \tag{3.1.1}$$

It is known that $\Delta(X, S) \geq 0$ and Fujita [Fuj90] classifies polarized varieties of small Δ-genera. We use the following easy consequence of Fujita's classification.

3.1.2. Lemma. *Let X be a $\mathbb{Q}$-Fano 3-fold and S an ample Weil divisor on X such that $\dim |S| > 0$, $|S|$ has no fixed components, and $-K_X \sim_{\mathbb{Q}} \lambda S$ with $\lambda \geq 2$. Assume that the pair $(X, |S|)$ is terminal. Then one of the following holds:*

(i) $X \simeq \mathbb{P}^3$, $\lambda = 4$, $\dim |S| = 3$;
(ii) $X \simeq \mathbb{P}^3$, $\lambda = 2$, $\dim |S| = 9$;
(iii) $X \simeq X_2 \subset \mathbb{P}^4$ *is a smooth quadric*, $\lambda = 3$, $\dim |S| = 4$;
(iv) X *is a del Pezzo 3-fold of degree* $1 \leq d \leq 5$, $\lambda = 2$, $\dim |S| = d + 1$;
(v) $X \simeq \mathbb{P}(1^3, 2)$, $\lambda = 5/2$, $\dim |S| = 6$.

Proof. Replace S with a general member of $|S|$. Since $(X, |S|)$ is terminal, the surface S is smooth and contained in the smooth locus of X [Ale94, 1.22]. By the adjunction formula we have $-K_S \sim (\lambda - 1)S_{|S}$. Hence S is a (smooth) del Pezzo surface and $(\lambda - 1)^2 S^3 = K_S^2$. Since $H^i(X, \mathcal{O}_X) = 0$ and $H^i(S, \mathcal{O}_S(S)) = 0$ for $i > 0$, by Riemann–Roch we have

$$h^0(X, \mathcal{O}_X(S)) = h^0(S, \mathcal{O}_S(S)) + 1 = \frac{\lambda}{2} S^3 + 2. \tag{3.1.3}$$

Therefore,

$$\Delta(X, S) = 3 + S^3 - \frac{\lambda}{2} S^3 - 2 = 1 + \frac{(2 - \lambda) S^3}{2} = 1 + \frac{(2 - \lambda) K_S^2}{2(\lambda - 1)^2}. \tag{3.1.4}$$

If $S \simeq \mathbb{P}^2$, then $\mathcal{O}_S(S) = \mathcal{O}_{\mathbb{P}^2}(l)$, where $3 = (\lambda - 1) l \geq l$. Then $\Delta(X, S) = 0$ and [Fuj90, Th. 5.10 and 5.15] gives cases (i) and (v). If $S \simeq \mathbb{P}^1 \times \mathbb{P}^1$, then $\mathcal{O}_S(S) = \mathcal{O}_{\mathbb{P}^1\times\mathbb{P}^1}(k,k)$, where $k(\lambda-1)=2$. So, $\lambda = 2$ or 3, $\Delta(X,S)=0$, and [Fuj90, Th. 5.10 and 5.15] gives cases (ii) or (iii). Finally, if $S \not\simeq \mathbb{P}^2$, $\mathbb{P}^1\times\mathbb{P}^1$, then K_S is a primitive element of $\operatorname{Pic} S$. Hence $\lambda = 2$ and $\Delta(X,S)=1$. Then we have case (iv) [Fuj90, Ch. 1, §9]. Q.E.D.

Lemma 3.2 ([Pro10b, Th. 1.4 (vii)]). *Let X be a $\mathbb{Q}$-Fano 3-fold with terminal singularities and with $q := \mathrm{q}_{\mathbb{Q}}(X) \ge 5$. Let A be a Weil divisor such that $-K_X \sim_{\mathbb{Q}} qA$. If $\dim|A| \ge 2$, then $X \simeq \mathbb{P}(1^3,2)$.*

Proof. We first consider the case $\operatorname{rk}\operatorname{Cl} X = 1$ and $\mathrm{q}_{\mathbb{Q}}(X) = \mathrm{q}_{\mathrm{F}}(X)$ (in particular, X is a $\mathbb{Q}$-Fano 3-fold and $-K_X \sim qA$). Running a computer search as in 2.4 gives $-K_X^3 \ge 125/2$. Then by [Pro07] we have $X \simeq \mathbb{P}(1^3,2)$.

Next consider the case $\operatorname{rk}\operatorname{Cl} X > 1$ and $\mathrm{q}_{\mathbb{Q}}(X) = \mathrm{q}_{\mathrm{F}}(X)$. We get a contradiction in this case. Run the MMP. The property $-K_X \sim qA$ is preserved. At the end we get a $\mathbb{Q}$-Fano 3-fold $\overline{X}$ with $K_{\overline{X}} \sim q\overline{A}$ and $\dim|\overline{A}| \ge 2$. By (1.2.1) we have $\mathrm{q}_{\mathbb{Q}}(\overline{X}) = \mathrm{q}_{\mathrm{F}}(\overline{X}) = q$. By the above $\overline{X} \simeq \mathbb{P}(1^3,2)$ and $\dim|A| = \dim|\overline{A}| = 2$. Let $\overline{P} \in \overline{X}$ be the point of type $\frac{1}{2}(1,1,1)$. Consider the final step $g\colon \widetilde{X} \to \overline{X}$ of the MMP, a divisorial contraction, and let $\widetilde{E} \subset \widetilde{X}$ be its exceptional divisor. There are the following possibilities:

(a) $g(\widetilde{E}) = \overline{P}$. Then $K_{\widetilde{X}} \sim_{\mathbb{Q}} g^*K_{\overline{X}} + \frac{1}{2}\widetilde{E}$, $\widetilde{E} \simeq \mathbb{P}^2$, and $\mathcal{O}_{\widetilde{E}}(\widetilde{E}) \simeq \mathcal{O}_{\mathbb{P}^2}(-2)$ [Kaw96]. Hence, $\mathcal{O}_{\tilde{E}}(-K_{\widetilde{X}}) \simeq \mathcal{O}_{\mathbb{P}^2}(1)$. We get a contradiction because $-K_{\widetilde{X}}$ is divisible by $q \ge 5$.

(b) $g(\widetilde{E})$ is either a smooth point or a curve. In this case $g(\widetilde{E}) \not\subset \operatorname{Bs}|\overline{A}| = \{\overline{P}\}$. On the other hand, g is a $K_{\widetilde{X}}$-negative contraction, a contradiction.

Finally assume that the torsion part of $\operatorname{Cl} X$ is nontrivial. Every torsion element $\xi_1 \in \operatorname{Cl} X$ of order $n_1 > 1$ defines a $\boldsymbol{\mu}_{n_1}$-cover $\pi_1\colon X_1 \to X$ that is étale in codimension 2. Repeating the procedure we get a sequence

$$X_m \xrightarrow{\pi_m} X_{m-1} \xrightarrow{\pi_{m-1}} \cdots \xrightarrow{\pi_2} X_1 \xrightarrow{\pi_1} X. \tag{3.2.1}$$

with each π_k a $\boldsymbol{\mu}_{n_k}$-cover that is étale in codimension 2 and $\operatorname{Cl} X_m$ torsion free. By the above, $X_m \simeq \mathbb{P}(1^3,2)$. Since

$$h^0(X_m, \pi^*A) = h^0(X,A) = 3, \tag{3.2.2}$$

$\boldsymbol{\mu}_{n_m}$ acts trivially on $H^0(X_m,\pi^*A) = H^0(\mathcal{O}_{\mathbb{P}(1^3,2)}(1))$. On the other hand, we can take independent sections $x_1, x_2, x_3 \in H^0(\mathcal{O}_{\mathbb{P}(1^3,2)}(1))$ as orbinates at the $\frac{1}{2}(1,1,1)$-point $P_m \in X_m$. This contradicts that the point $(X_m, P_m)/\boldsymbol{\mu}_{n_m}$ is terminal. Q.E.D.

In a similar way to Lemma 3.2, one can prove the following.

Lemma 3.3 ([Pro10b, Th. 1.4 (vi)]). *Let X be a $\mathbb{Q}$-Fano 3-fold with terminal singularities and with $q := \mathrm{q}_{\mathbb{Q}}(X) \geq 7$. Let A be a Weil divisor such that $-K_X \sim_{\mathbb{Q}} qA$. If $\dim |A| \geq 1$, then $X \simeq \mathbb{P}(1^2, 2, 3)$.*

Proposition 3.4. *Let X be a $\mathbb{Q}$-Fano 3-fold and let $q := \mathrm{q}_{\mathbb{Q}}(X)$. Let $\mathcal{M}$ be a linear system on X such that $\dim \mathcal{M} \geq 4$ and $-K_X \sim 2\mathcal{M} + \Xi$, where Ξ is a nonzero effective Weil divisor. Then $\operatorname{Cl} X \simeq \mathbb{Z}\cdot\Xi$, $q = 2n+1$ is odd, and $\mathcal{M} \sim n\Xi$. Moreover, one of the following holds:*

(i) $q = 13$, $X \simeq \mathbb{P}(1,3,4,5)$;
(ii) $q = 11$, $X \simeq \mathbb{P}(1,2,3,5)$;
(iii) $q = 9$, $X \simeq X_6 \subset \mathbb{P}(1,2,3,4,5)$;
(iv) $q = 7$, $X \simeq \mathbb{P}(1^2,2,3)$;
(v) $q = 5$, $X \simeq X_4 \subset \mathbb{P}(1^2,2^2,3)$;
(vi) $q = 5$, $X \simeq \mathbb{P}(1^3,2)$;
(vii) $q = 3$, $X \simeq X_2 \subset \mathbb{P}^4$.

Proof. By assumption $q \geq 3$. If $q \geq 9$, then the assertion follows by [Pro10b, Prop. 3.6] and Theorem 1.2.3 (vi). So assume that $3 \leq q \leq 8$.

Let A be a Weil divisor with $-K_X \sim_{\mathbb{Q}} qA$ and n the integer such that $\mathcal{M} \sim_{\mathbb{Q}} nA$. If $\operatorname{Cl} X$ is torsion free, we can run the computer search 2.4. We get $q \neq 4$ and $\mathrm{g}(\widehat{X}) \geq 21$. Then by Theorem 1.2.3 we get one of cases (iv)–(vii). Thus from now on we assume that $\operatorname{Cl} X$ contains a nontrivial torsion element.

We may assume that $\mathcal{M}$ has no fixed part. If the pair $(X, \mathcal{M})$ is terminal, then X is in (vi) or (vii) by Lemma 3.1.2. Assume that $(X, \mathcal{M})$ is not terminal. Apply Construction 2.1 to $(X, \mathcal{M})$. We can write

$$K_{\widetilde{X}} + 2\widetilde{\mathcal{M}} + \widetilde{\Xi} + a\widetilde{E} \sim f^*(K_X + 2\mathcal{M} + \Xi) \sim 0,$$

where $a \in \mathbb{Z}_{>0}$. Hence,

$$K_{\overline{X}} + 2\overline{\mathcal{M}} + \overline{\Xi} + a\overline{E} \sim 0. \tag{3.4.1}$$

First consider the case of 2.2 where $\overline{f}$ is not birational. In particular, $\widehat{X}$ is either $\mathbb{P}^1$ or a del Pezzo surface as in 2.2.1.

Assume that $\overline{\mathcal{M}}$ is $\overline{f}$-horizontal. Restricting the relation (3.4.1) to a general fiber $\overline{F}$ of $\overline{f}$ we get

$$-K_{\overline{F}} \sim_{\mathbb{Q}} 2\overline{\mathcal{M}}_{|\overline{F}} + \overline{\Xi}_{|\overline{F}} + a\overline{E}_{|\overline{F}}, \tag{3.4.2}$$

where the divisors $\overline{\mathcal{M}}_{|\overline{F}}$ and $\overline{E}_{|\overline{F}}$ are ample. This is possible only if $\overline{F} \simeq \mathbb{P}^2$, $\widehat{X} \simeq \mathbb{P}^1$, $\mathcal{O}_{\overline{F}}(\overline{\mathcal{M}}) \simeq \mathcal{O}_{\overline{F}}(\overline{E}) \simeq \mathcal{O}_{\mathbb{P}^2}(1)$, and $a = 1$. From the exact sequence

$$(3.4.3) \qquad 0 \to \mathcal{O}_{\overline{X}}(\overline{\mathcal{M}} - \overline{F}) \to \mathcal{O}_{\overline{X}}(\overline{\mathcal{M}}) \to \mathcal{O}_{\overline{F}}(\overline{\mathcal{M}}) \to 0$$

we get

$$(3.4.4) \qquad h^0(\mathcal{O}_{\overline{X}}(\overline{\mathcal{M}} - \overline{F})) \geq h^0(\mathcal{O}_{\overline{X}}(\overline{\mathcal{M}})) - h^0(\mathcal{O}_{\overline{F}}(\overline{\mathcal{M}})) \geq 2.$$

Thus $\overline{\mathcal{M}} \ni \overline{F} + \overline{L}$, where $\overline{L} \in |\mathcal{M} - \overline{F}|$ is a mobile divisor. Hence there is a decomposition $-K_X \sim 2F + 2L + \Xi$. In particular, $q \geq 5$ and $F \sim_{\mathbb{Q}} L \sim_{\mathbb{Q}} A$. This implies that $\overline{f}$ has no multiple fibers. So, the group $\operatorname{Cl} \overline{X}$ is torsion free. Since $\mathcal{O}_{\overline{F}}(\overline{E}) \simeq \mathcal{O}_{\mathbb{P}^2}(1)$, the class of $\overline{E}$ is not divisible in $\operatorname{Cl} \overline{X}$. Hence $\operatorname{Cl} X$ is also torsion free, a contradiction.

Therefore, $\overline{\mathcal{M}}$ is $\overline{f}$-vertical. Then $\overline{\mathcal{M}} = \overline{f}^* \mathcal{B}$, where $\mathcal{B}$ is a linear system of Weil divisors on $\widehat{X}$ with $\dim \mathcal{B} \geq 4$. We use the notation of 2.2. Let $\overline{G} = \overline{f}^* \Theta$. We can write $\mathcal{B} \sim_{\mathbb{Q}} t\Theta$ for some $t \in \mathbb{Z}_{>0}$. Then

$$(3.4.5) \qquad -K_{\overline{X}} \sim_{\mathbb{Q}} 2t\overline{G} + \overline{\Xi} + a\overline{E},$$

so $8 \geq q \geq 2t + 1$ and $t \leq 3$. If $\widehat{X} \simeq \mathbb{P}^1$, we obviously have $\dim \mathcal{B} \leq 2$. Therefore, $\widehat{X}$ is a surface. Now we use 2.2.1.

If $t = 1$, then $\dim \mathcal{B} \leq 2$, a contradiction. Consider the case $t = 2$. Then $\dim \mathcal{B} \geq 4$ only in the case $\widehat{X} \simeq \mathbb{P}^2$. Then $q \geq 5$, $G \sim_{\mathbb{Q}} A$, and $m = 2$. Since $\dim |G| \geq 2$, by Lemma 3.2 we have $X \simeq \mathbb{P}(1^3, 2)$. Consider the case $t = 3$. Then $q \geq 7$ and $G \sim_{\mathbb{Q}} A$. Since $\dim \mathcal{B} \geq 4$, we have either $\widehat{X} \simeq \mathbb{P}^2$, $\widehat{X} \simeq \mathbb{P}(1, 1, 2)$, or $K^2_{\widehat{X}} = 1$. In either case $\dim |G| \geq 1$ (recall that if $K^2_{\widehat{X}} = 1$, we take $\Theta = -K_{\widehat{X}}$). By Lemma 3.3 we get $X \simeq \mathbb{P}(1^2, 2, 3)$.

Now assume that $\overline{f}$ is birational. We have

$$(3.4.6) \qquad -K_{\widehat{X}} \sim 2\widehat{\mathcal{M}} + \widehat{\Xi} + a\widehat{E},$$

where, as usual, we write $\widehat{\Lambda} = \overline{f}_* \overline{\Lambda}$ for the birational transform of $\widehat{X}$ of a divisor (or a linear system) $\overline{\Lambda}$ on $\overline{X}$.

Clearly, $\dim \widehat{\mathcal{M}} \geq \dim \mathcal{M}$. If $(\widehat{X}, \widehat{\mathcal{M}})$ is not terminal, we can repeat the procedure 2.1 and continue. Thus we may assume that $(\widehat{X}, \widehat{\mathcal{M}})$ is as in (i)–(vii). In particular, $\operatorname{Cl} \widehat{X}$ is torsion free and $\widehat{\Xi} + a\widehat{E} \sim \widehat{\Theta}$, where $\widehat{\Theta}$ is the ample generator of $\operatorname{Cl} \widehat{X}$. So, $\widehat{\Xi} = 0$, $a = 1$, and $\widehat{E} \sim \widehat{\Theta}$. In particular, the class of $\widetilde{E}$ is a primitive element of $\operatorname{Cl} \widetilde{X} \simeq \mathbb{Z} \oplus \mathbb{Z}$. In this case, $\operatorname{Cl} X$ is also torsion free. Q.E.D.

§4. Proof of Main Theorem 1.3.1

Let X be a $\mathbb{Q}$-Fano 3-fold such that $-K_X \sim 2A$ for a primitive element $A \in \operatorname{Cl} X$. Assume that $\dim |A| \geq 4$ and K_X is not Cartier. We apply Construction 2.1 with $\mathcal{M} := |A|$, $\lambda = 2$ and $\Xi = 0$. By Lemma 3.1.2 the pair $(X, \mathcal{M})$ is not terminal. Hence in the notation of (2.1.3), the discrepancy $a > 0$. On the other hand, a is an integer. Therefore, $a \geq 1$.

Lemma 4.1. *The map $\overline{f}$ in* (2.1.2) *is birational.*

Proof. Suppose that $\overline{f}$ is not birational. Let $\overline{F}$ be a general fiber of $\overline{f}$. If $\overline{\mathcal{M}}$ is $\overline{f}$-vertical, then $\overline{\mathcal{M}} = \overline{f}^* \widehat{\mathcal{B}}$, where $\widehat{\mathcal{B}}$ is a linear system on $\widehat{X}$ whose class generates $\operatorname{Cl} \widehat{X} / \operatorname{Tors}$. But then $\dim \overline{\mathcal{M}} = \dim \widehat{\mathcal{B}} \leq 2$ by 2.2.1, contradicting our assumption.

Thus $\overline{\mathcal{M}}$ is $\overline{f}$-horizontal. Then $-K_{\overline{F}} = 2\overline{\mathcal{M}}_{|\overline{F}} + a\overline{E}_{|\overline{F}}$. This implies that $\overline{F} \simeq \mathbb{P}^2$, that is, $\overline{f}$ is generically a $\mathbb{P}^2$-bundle and $\mathcal{O}_{\overline{F}}(\overline{\mathcal{M}}) \simeq \mathcal{O}_{\mathbb{P}^2}(1)$. From the exact sequence

$$0 \to \mathcal{O}_{\overline{X}}(\overline{\mathcal{M}} - \overline{F}) \to \mathcal{O}_{\overline{X}}(\overline{\mathcal{M}}) \to \mathcal{O}_{\overline{F}}(\overline{\mathcal{M}}) \to 0 \tag{4.1.1}$$

we get

$$h^0(\mathcal{O}_{\overline{X}}(\overline{\mathcal{M}} - \overline{F})) \geq 2. \tag{4.1.2}$$

Therefore, $\overline{\mathcal{M}} \ni \overline{F} + \overline{L}$, where $\overline{F}$ and $\overline{L}$ are mobile divisors. This contradicts $\mathrm{q}_{\mathbb{Q}}(X) = 2$. Q.E.D.

4.2. Notation

When $\overline{f}$ is birational, $\widehat{X}$ is a $\mathbb{Q}$-Fano. Recall that we write $\widehat{\Lambda} = \overline{f}_* \overline{\Lambda}$ for the birational transform on $\widehat{X}$ of a divisor (or a linear system) $\overline{\Lambda}$ on $\overline{X}$. We have

$$-K_{\widehat{X}} \sim 2\widehat{\mathcal{M}} + a\widehat{E} \quad \text{with} \quad a > 0, \quad \dim \widehat{\mathcal{M}} \geq 4. \tag{4.2.1}$$

By Proposition 3.4 the class of $\widehat{E}$ is the ample generator of $\operatorname{Cl} \widehat{X} \simeq \mathbb{Z}$, $\widehat{q} = 2n + 1$, and $\widehat{\mathcal{M}} \subset |n\widehat{E}|$. Moreover, $\widehat{X}$ belongs to one of the possibilities listed in Proposition 3.4.

Assume first that $\widehat{q} > 3$. We consider the case $\widehat{q} = 3$ in the next section. We make frequent use of the following easy observation.

Remark 4.2.2. In the notation of 4.2, assume that there is a member $\widehat{M} \in \widehat{\mathcal{M}}$ such that $\widehat{M} = \widehat{L}_1 + \widehat{L}_2$, where $\widehat{L}_1$ and $\widehat{L}_2$ are effective ample Weil divisors. Then either $\operatorname{Supp} \widehat{L}_1 = \widehat{E}$ or $\operatorname{Supp} \widehat{L}_2 = \widehat{E}$.

Indeed, we can write

$$\overline{\mathcal{M}} \sim_{\mathbb{Q}} \overline{L}_1 + \overline{L}_2 + \gamma\overline{F}, \tag{4.2.3}$$

where $\overline{L}_i$ is the birational transform of $\widehat{L}_i$ and $\gamma \geq 0$. Therefore,

$$\mathcal{M} \sim f_*\chi_*^{-1}\overline{\mathcal{M}} \sim_{\mathbb{Q}} f_*\chi_*^{-1}\overline{L}_1 + f_*\chi_*^{-1}\overline{L}_2 + \gamma F. \tag{4.2.4}$$

Since the class of A is a primitive element of $\operatorname{Cl} X$, we have either $f_*\chi_*^{-1}\overline{L}_1 = 0$ or $f_*\chi_*^{-1}\overline{L}_2 = 0$ (and $\gamma = 0$).

4.2.5. Corollary. *Assume that we have* $\dim|n\widehat{E}| = 4$ *in the notation of 4.2. Then for any partition* $n = n_1 + n_2$, $n_i \in \mathbb{Z}$ *either* $\dim|n_1\widehat{E}| \leq 0$ *or* $\dim|n_2\widehat{E}| \leq 0$.

Proof. In this case $\widehat{\mathcal{M}} = |n\widehat{E}|$ is a complete linear system. Hence, one can take $\widehat{L}_i \in |n_i\widehat{E}|$. Q.E.D.

We consider the cases of Proposition 3.4 separately.

4.2.6. Cases (i), (iii) and (v). Then $\dim|n\widehat{E}| = 4$ and n is even. Apply Corollary 4.2.5 with $n_1 = n_2 = n/2$. We get a contradiction because $\dim|n_i\widehat{E}| > 0$.

4.2.7. Case (ii), $\widehat{X} \simeq \mathbb{P}(1,2,3,5)$. Then $n = 5$ and $\dim|n\widehat{E}| = 5$. Thus $\widehat{\mathcal{M}} \subset |5\widehat{E}|$ is a subsystem of codimension ≤ 1. Since $\dim 2\widehat{E}| = 1$, we can take $\widehat{L}_1 \in |2\widehat{E}|$ so that $\widehat{L}_1 \neq 2\widehat{E}$. Since $\dim|3\widehat{E}| = 2$, there exists a one dimensional family of divisors $\widehat{L}_2 \in |3\widehat{E}|$ such that $\widehat{L}_1+\widehat{L}_2 \in \widehat{\mathcal{M}}$. So we may assume that $\widehat{L}_2 \neq 3\widehat{E}$. By Remark 4.2.2 we get a contradiction.

4.2.8. Case (iv), that is, $\widehat{X} \simeq \mathbb{P}(1^2,2,3)$. Then $n = 3$ and $\dim|n\widehat{E}| = 6$. Thus $\widehat{\mathcal{M}} \subset |3\widehat{E}|$ is a subsystem of codimension ≤ 2. Since $\dim|\widehat{E}| = 1$, —we can take $\widehat{L}_1 \in |\widehat{E}|$ so that $\widehat{L}_1 \neq \widehat{E}$. Since $\dim|2\widehat{E}| = 3$, there exists a one dimensional family of divisors $\widehat{L}_2 \in |2\widehat{E}|$ such that $\widehat{L}_1 + \widehat{L}_2 \in \widehat{\mathcal{M}}$. So we may assume that $\widehat{L}_2 \neq 2\widehat{E}$. By Remark 4.2.2 we get a contradiction.

4.2.9. Case (vi), $\widehat{X} \simeq \mathbb{P}(1^3,2)$. Then $n = 2$ and $\dim|n\widehat{E}| = 6$. Thus $\widehat{\mathcal{M}} \subset |2\widehat{E}|$ is a subsystem of codimension ≤ 2.

Assume that $\overline{f}(\overline{F})$ is a curve. Then

$$K_{\overline{X}} = \overline{f}^*K_{\widehat{X}} + \overline{F} \quad \text{and} \quad \overline{E} = \overline{f}^*\widehat{E} - \gamma\overline{F}.$$

Since any member of $|\widehat{E}|$ is smooth in codimension one, $\gamma \leq 1$. Moreover, since $n\overline{E}$ is not mobile for any n, we have $\gamma > 0$. Hence, $\gamma = 1$. So,

$$K_{\overline{X}} + 5\overline{E} + 4\overline{F} = \overline{f}^*(K_{\widehat{X}} + 5\widehat{E}) \sim 0.$$

This implies that $-K_X$ is divisible by 4, a contradiction.

Hence $\overline{f}(\overline{F}) \in \widehat{X}$ is a point, say $\widehat{P}$. If $\widehat{P} \in \widehat{X}$ is the point of index 2, then $\overline{f}$ is the blowup of the maximal ideal [Kaw96]. In this case $\overline{X}$ has exactly two extremal contractions: $\overline{f}$ and the $\mathbb{P}^1$-bundle induced by the projection $\mathbb{P}(1^3, 2) \dashrightarrow \mathbb{P}^2$. On the other hand, the second contraction must be birational, a contradiction. Hence $\overline{P} \in \widehat{X}$ is a smooth point.

Let $\widehat{\mathcal{L}} := |\widehat{E}|$. Take a general member $\widehat{L}_1 \in \widehat{\mathcal{L}}$. Dimension counting shows that there exists $\widehat{L}_2 \in \widehat{\mathcal{L}}$ such that $\widehat{L}_1 + \widehat{L}_2 \in \widehat{\mathcal{M}}$. If $\widehat{L}_2 \neq \widehat{E}$, we get a contradiction by Remark 4.2.2. Thus $\widehat{L}_2 = \widehat{E}$ for any choice of $\widehat{L}_1 \in \widehat{\mathcal{L}}$. Therefore, $\widehat{E} + \widehat{\mathcal{L}} \subset \widehat{\mathcal{M}}$ and we can write $\overline{\mathcal{M}} \sim_{\mathbb{Q}} \overline{\mathcal{L}} + \overline{E} + \gamma\overline{F}$, where $\gamma \geq 0$. Then

$$0 \sim K_{\overline{X}} + 2\overline{\mathcal{M}} + \overline{E} \sim_{\mathbb{Q}} K_{\overline{X}} + 2\overline{\mathcal{L}} + 3\overline{E} + 2\gamma\overline{F}. \tag{4.2.10}$$

Note that the only base point of $\widehat{\mathcal{L}}$ is the point of index 2. Hence, $\overline{\mathcal{L}} \sim_{\mathbb{Q}} \overline{f}^*\widehat{\mathcal{L}}$. Let $\widehat{\mathcal{L}}' \subset \widehat{\mathcal{L}}$ be the subsystem consisting of elements passing through $\widehat{P}$. Then we can write

$$\overline{\mathcal{L}}' \sim_{\mathbb{Q}} \overline{f}^*\widehat{\mathcal{L}}' - \delta\overline{F} \sim_{\mathbb{Q}} \overline{\mathcal{L}} - \delta\overline{F}, \quad \text{with } \delta > 0. \tag{4.2.11}$$

Therefore,

$$0 \sim_{\mathbb{Q}} K_{\overline{X}} + 2\overline{\mathcal{L}} + 3\overline{E} + 2\gamma\overline{F} \sim_{\mathbb{Q}} K_{\overline{X}} + 2\overline{\mathcal{L}}' + 3\overline{E} + 2(\delta + \gamma)\overline{F}. \tag{4.2.12}$$

This gives us $-K_X \sim_{\mathbb{Q}} 2\mathcal{L}' + 2(\delta + \gamma)F$ which contradicts $\mathrm{q}_{\mathbb{Q}}(X) = 2$.

§5. Conclusion of the proof of Main Theorem 1.3.1

This section considers Case (vii), when $\widehat{X} = Q \subset \mathbb{P}^4$ is a smooth quadric. Then $\widehat{\mathcal{M}} = |\mathcal{O}_Q(1)|$ is a complete linear system, and in particular is base point free. Thus $\overline{\mathcal{M}} \sim_{\mathbb{Q}} \overline{f}^*\widehat{\mathcal{M}}$. We also have $\widehat{E} \in |\mathcal{O}_Q(1)|$ and $\overline{f}(\overline{F}) \subset \widehat{E}$.

Lemma 5.1. *$\Gamma := \overline{f}(\overline{F})$ is a curve.*

Proof. Assume that $\overline{f}(\overline{F})$ is a point. Let $\widehat{\mathcal{M}}' \subset \widehat{\mathcal{M}}$ be the subsystem consisting of elements passing through $\overline{f}(\overline{F})$. Then we can write

$$\overline{\mathcal{M}}' \sim_{\mathbb{Q}} \overline{f}^*\widehat{\mathcal{M}}' - \delta\overline{F} \sim_{\mathbb{Q}} \overline{\mathcal{M}} - \delta\overline{F}, \quad \text{with } \delta > 0. \tag{5.1.1}$$

Therefore,

$$\begin{aligned} 0 &\sim_{\mathbb{Q}} \overline{f}^*(K_{\widehat{X}} + 2\widehat{\mathcal{M}}' + \widehat{E}) \sim_{\mathbb{Q}} \overline{f}^*(K_{\widehat{X}} + 2\widehat{\mathcal{M}} + \widehat{E}) \\ &\sim_{\mathbb{Q}} K_{\overline{X}} + 2\overline{\mathcal{M}} + \overline{E} \sim_{\mathbb{Q}} K_{\overline{X}} + 2\overline{\mathcal{M}}' + \overline{E} + 2\delta\overline{F}. \end{aligned} \tag{5.1.2}$$

This gives us $-K_X \sim_{\mathbb{Q}} 2\mathcal{M}' + 2\delta F$ which contradicts $q_{\mathbb{Q}}(X) = 2$. Q.E.D.

Lemma 5.2. *$\overline{E} \simeq \mathbb{P}(1,1,2)$, $\mathbb{F}_2$, or $\mathbb{P}^1 \times \mathbb{P}^1$.*

Proof. Clearly, $\widehat{E} \simeq \mathbb{P}(1,1,2)$ or $\mathbb{P}^1 \times \mathbb{P}^1$. In particular, the pair $(\widehat{X}, \widehat{E})$ is plt. Since $K_{\overline{X}} \sim_{\mathbb{Q}} \overline{f}^* K_{\widehat{X}} + \overline{F}$ and $\widehat{E}$ is smooth at the generic point of Γ, we have

$$K_{\overline{X}} + \overline{E} \sim \overline{f}^*(K_{\widehat{X}} + \widehat{E}). \tag{5.2.1}$$

Hence the pair $(\overline{X}, \overline{E})$ is plt and the divisor $K_{\overline{X}} + \overline{E}$ is Cartier. By adjunction, the surface $\overline{E}$ has at worst Du Val singularities. Moreover, $K_{\overline{E}} = \overline{f}^*_{|\overline{E}} K_{\widehat{E}}$, that is, the restriction $\overline{f}_{\overline{E}}$ is either an isomorphism or the minimal resolution of $\widehat{E}$. Q.E.D.

Lemma 5.3. *$-K_{\overline{X}}$ is nef.*

Proof. Recall that by our construction $\overline{X}$ has exactly two extremal rays. Denote them by R_1 and R_2. One of them, say R_1, is generated by nontrivial fibers of $\overline{f}$. Let C be an extremal curve on $\overline{X}$ that generates R_2. Assume that $-K_{\overline{X}}$ is not nef. Then $K_{\overline{X}} \cdot C > 0$ and C must be a flipped curve (because a divisorial contraction must be K-negative in our situation). Since $-K_{\overline{X}} \sim_{\mathbb{Q}} \overline{E} + 2\overline{f}^* \widehat{E}$, we have $\overline{E} \cdot C < 0$. In particular, $C \subset \overline{E}$. Since C is a flipped curve, it cannot be mobile on $\overline{E}$, that is, $\dim |C| = 0$. By Lemma 5.2, the only possibility is that $\overline{E} \simeq \mathbb{F}_2$ and C is the negative section of $\mathbb{F}_2$. But in this case C is contracted by $\overline{f}$ to a point, that is, the class of C lies in R_1, a contradiction. Q.E.D.

Lemma 5.4. *$K_{\overline{X}}$ is not Cartier at some point of $\overline{E}$.*

Proof. By (5.2.1) the divisor $K_{\overline{X}}$ is Cartier outside $\overline{E}$. Assume that $K_{\overline{X}}$ is Cartier near $\overline{E}$. Since $-K_{\overline{X}}$ is nef, the map $\overline{X} \dashrightarrow \widetilde{X}$ is either an isomorphism or a flop. In either case $\widetilde{X}$ has the same type of singularities as $\overline{X}$, that is, $K_{\widetilde{X}}$ is Cartier. By the classification of extremal contractions of Gorenstein terminal 3-folds [Cut88] the divisor $2K_X$ is Cartier. This contradicts the following remark. Q.E.D.

5.4.1. Corollary. *The curve Γ has a singular point that is not a local complete intersection.*

Proof. Indeed otherwise by [KM92, Prop. 4.10.1] the map $\overline{f}$ is the blowup of Γ and $K_{\overline{X}}$ is Cartier. Q.E.D.

5.4.2. Corollary. *$\widehat{E} \simeq \mathbb{P}(1,1,2)$, the curve Γ is not a Cartier divisor on $\widehat{E}$, and Γ is singular at the vertex of $\mathbb{P}(1,1,2)$.*

Lemma 5.5. $\deg \Gamma = 5$.

Proof. Let $\widehat{C} \subset \widehat{E}$ be a general hyperplane section. Since $-K_{\overline{X}}$ is nef,

$$0 \le -K_{\overline{X}} \cdot \overline{C} = -K_{\widehat{X}} \cdot \widehat{C} - (\Gamma \cdot \widehat{C})_{\widehat{E}} = 6 - \deg \Gamma. \tag{5.5.1}$$

Since Γ is not a Cartier divisor on $\widehat{E}$, its degree should be odd. If $\deg \Gamma \neq 5$, then Γ is either a line or a twisted cubic. In particular, it is smooth, a contradiction. Q.E.D.

5.6. Thus $\deg \Gamma = 5$ and Γ is singular. Then Γ can be given, in coordinates u_1, u_2, v for $E \simeq \mathbb{P}(1,1,2)$, by an equation $\gamma = v\alpha_3 + \beta_5$, where $\alpha_3(u_1,u_2)$ and $\beta_5(u_1,u_2)$ are homogeneous polynomial of the indicated degree. Thus P is a triple point of Γ and is its only singularity. Thus Γ is as in Main Example 1.5 and the conclusion of Theorem 1.3.1. By [KM92, Th. 4.9] the extraction $\overline{f}\colon \overline{X} \to Q = \widehat{X}$ of Γ is unique up to isomorphism over Q. Since $\rho(\overline{X}/Q) = 1$, the Sarkisov link (1.5.1) is uniquely determined. This completes the proof of Theorem 1.3.1.

§6. Examples

6.1. Symbolic blowup

This section is closely related to parts of Tom Ducat's thesis [Du15], and we acknowledge his help with our treatment.

Let $\Gamma \subset M$ be a reduced singular curve in a nonsingular 3-fold. The *symbolic blowup* of Γ in M is the relative Proj of the *symbolic algebra*, the graded algebra $\bigoplus_{n \ge 0} \mathcal{I}_\Gamma^{[n]}$, where $\mathcal{I}_\Gamma^{[n]}$ is the nth symbolic power, that is, the ideal in $\mathcal{O}_Q$ of functions vanishing n times at the generic point of Γ. In other words, in the primary decomposition of $\mathcal{I}_\Gamma^n$, ignore the embedded component at singular points P of Γ. (Primary decomposition is built into the computer algebra packages.)

In our case, Γ is a curve contained in a $\frac{1}{2}(1,1)$ orbifold point $P \in E_0 \subset M$ as a Weil divisor, whose class generates the local class group $\operatorname{Cl}_P E_0 \simeq \mathbb{Z}/2$. For simplicity, we treat $\Gamma \subset E_0 \subset M$ as germs around a singular point $P \in \Gamma$ in local analytic coordinates (but see 6.4). Write $\mathbb{C}^2_{\langle u_1,u_2\rangle}$ for the orbifold double cover of $P \in E_0$, and

$$(x_1, x_2, x_3) = (u_1^2, u_1u_2, u_2^2) \tag{6.1.1}$$

for the invariant monomials. Then M has local coordinates x_1, x_2, x_3, with $g = x_1x_3 - x_2^2$ the local equation of E_0, and $\Gamma \subset E_0$ corresponds to an invariant curve $\Gamma : (\gamma = 0) \subset \mathbb{C}^2$, with equation $\gamma = \gamma(u_1, u_2)$ an odd function of the orbinates.

To see $\Gamma \subset M$ in equations, first render into x_i the invariant multiples $u_1\gamma$ and $u_2\gamma$ of γ, say as

$$u_1\gamma = bx_1 - ax_2 = -f_2 \quad \text{and} \quad u_2\gamma = bx_2 - ax_3 = f_1, \tag{6.1.2}$$

where a, b are functions of x_1, x_2, x_3. Taking into account that a and b are in the maximal ideal (x_1, x_2, x_3) (because the curve Γ is singular at P, and not locally planar), and with a little massaging, we can put the generators of $\mathcal{I}_\Gamma$ and the syzygies between them in the determinantal form $\bigwedge^2 M = (f_1, f_2, g)$, where

$$M = \begin{pmatrix} x_1 & x_2 & a_2x_2 + a_3x_3 \\ x_2 & x_3 & b_1x_1 + b_2x_2 \end{pmatrix} \quad \text{and} \quad M \begin{pmatrix} f_1 \\ f_2 \\ g \end{pmatrix} \equiv 0. \tag{6.1.3}$$

In this case, the symbolic algebra needs just one further generator in degree 2, whose restriction to E_0 is the local equation

$$b^2x_1 - 2abx_2 + a^2x_3 \tag{6.1.4}$$

of the Cartier divisor $2\Gamma \subset E_0$. Rather than primary decomposition, we derive this final generator and its relations by unprojection.

Replacing f_1, f_2, g by ξ_1, ξ_2, η in (6.1.3) gives

$$\begin{pmatrix} x_1 & x_2 & a_2x_2 + a_3x_3 \\ x_2 & x_3 & b_1x_1 + b_2x_2 \end{pmatrix} \begin{pmatrix} \xi_1 \\ \xi_2 \\ \eta \end{pmatrix} = 0. \tag{6.1.5}$$

Equations (6.1.5) define the blowup of the ideal $\mathcal{I}_\Gamma = (f_1, f_2, g)$ as a codimension 2 complete intersection in $M \times \mathbb{P}^2_{\langle \xi_1, \xi_2, \eta \rangle}$, containing the "irrelevant" codimension 3 complete intersection $V(\xi_1, \xi_2, \eta)$. However, M has entries in (x_1, x_2, x_3), so it also contains the codimension 2 complete intersection $V(x_1, x_2, x_3)$ – the blowup of Γ must contain $\mathbb{P}^2$ over the origin (because Γ is not a local complete intersection). We rearrange (6.1.5) as

$$\begin{pmatrix} \xi_1 & \xi_2 + a_2\eta & a_3\eta \\ b_1\eta & \xi_1 + b_2\eta & \xi_2 \end{pmatrix} \begin{pmatrix} x_1 \\ x_2 \\ x_3 \end{pmatrix} = 0. \tag{6.1.6}$$

The unprojection of $V(x_1, x_2, x_3)$ is given by the 4×4 Pfaffians of

$$\begin{pmatrix} \zeta & \xi_1 & \xi_2 + a_2\eta & a_3\eta \\ & b_1\eta & \xi_1 + b_2\eta & \xi_2 \\ & & x_3 & -x_2 \\ & & & x_1 \end{pmatrix}. \tag{6.1.7}$$

Geometrically, this is the blowup of $\mathcal{I}_\Gamma$ followed by the unprojection contracting $\mathbb{P}^2$ over the origin to a point. We can also view it simply as a practical means of writing down the generator h in degree 2 satisfying

$$(6.1.8) \qquad (x_1, x_2, x_3)h = \bigwedge^2 \begin{pmatrix} f_1 & f_2 + a_2 g & a_3 g \\ b_1 g & f_1 + b_2 g & f_2 \end{pmatrix},$$

so that clearly $h \in \mathcal{I}_\Gamma^{[2]}$, without computer algebra. In computational terms, this means that we can modify $f_1^2, f_1 f_2, f_2^2$ modulo $g \cdot \mathcal{I}_\Gamma$ to make them identically divisible by x_3, x_2, x_1 respectively, with (say)

$$(6.1.9) \qquad f_1 f_2 - b_1 a_3 g^2 = -x_2 h,$$

where $h_{|E_0}$ is the equation (6.1.4) defining $2\Gamma \subset E_0$.

Proposition 6.2. *The symbolic algebra of $\mathcal{I}_\Gamma$ is generated by ξ_1, ξ_2, η in degree* 1 *(corresponding to f_1, f_2, g), and ζ in degree* 2 *(corresponding to h). The ideal of relations is generated by the maximal Pfaffians of* (6.1.7).

Thus the symbolic blowup $M_1 \to M$ of Γ is the codimension 3 *Gorenstein subvariety*

$$(6.2.1) \qquad M_1 \subset M \times \mathbb{P}(1,1,1,2)_{\langle \xi_1, \xi_2, \eta, \zeta \rangle}$$

defined by the Pfaffians of (6.1.7). *It has the following properties. If Γ is nonsingular at P it is not applicable. If Γ is singular it defines a morphism $M_1 \to M$ which is the ordinary blowup of $\mathcal{I}_\Gamma$ outside the origin. The birational transform $E_1 \subset M_1$ is isomorphic to E_0.*

The fibre of M_1 over the origin is $\mathbb{P}(2,1)_{\langle \zeta, \eta \rangle}$ passing through P_ζ, which is a $\frac{1}{2}(1,1,1)$ orbifold point, and at most one more singular point. If $\operatorname{mult}_P \Gamma \geq 5$ *then $P_\eta \in M_1$ has embedding dimension* 3, *so is not terminal. If* $\operatorname{mult}_P \Gamma = 3$ *then M_1 is terminal, and in fact:*

1. *M_1 is quasismooth if Γ has* 3 *distinct tangent branches.*
2. *M_1 has a cA_1 point if Γ has a double tangent branch.*
3. *M_1 has a cA_2 point if Γ has a triple tangent branch.*

In the local description, the cA_1 and cA_2 points of M_1 are arbitrary.

Proof. Although the precise statement is somewhat involved, the proof is easy. The generators and relations follow from the hyperplane section principle: indeed, the symbolic algebra restricted to E_0 is just an algebra of $\mathbb{Z}/2$ invariants, generated by $u_1\gamma, u_2\gamma$ and γ^2, and the restriction map is surjective by our choice of generators ξ_1, ξ_2 and ζ.

The birational transform $E_1 \to E_0$ is an isomorphism because the symbolic algebra of the singular (nonplanar) curve $\Gamma \subset M$ maps onto that of the $\mathbb{Q}$-Cartier divisor on $\Gamma \subset E_0$. The analysis of the singularities is straightforward. The cases correspond to the different possibilities for the cubic leading terms in γ coming from

$$b_1 u_1^3 + b_2 u_1^2 u_2 - a_2 u_1 u_2^2 - a_3 u_2^3. \tag{6.2.2}$$

Q.E.D.

Remark 6.2.3. In the case that Γ has distinct tangent branches, its symbolic blowup can be done as an explicit construction in the nonsingular category that is folklore in the subject: blowing up P gives an exceptional $\Pi = \mathbb{P}^2$ with normal bundle $\mathcal{O}(-1)$. The 3 branches of Γ meet Π at noncollinear points, and blowing up Γ produces a dP_6 with a hexagon formed of the 3 blown up lines on Π together with the 3 lines joining them in Π, that are $(-1,-1)$ curves. Flopping these takes Π into $\Pi' = \mathbb{P}^2$ with normal bundle $\mathcal{O}(-2)$ by a standard quadratic transformation $\Pi \dashrightarrow \Pi'$, and Π' contracts to a $\frac{1}{2}(1,1,1)$ point.

6.3. Two examples

We apply this to contruct two families of $\mathbb{Q}$-Fano 3-folds X and Y of index 2 with $\mathrm{Cl} = \mathbb{Z} \cdot A$, each with a single $\frac{1}{3}(1,2,2)$ orbifold point and invariants

$$\begin{aligned} -K_X &= 2A_X \quad \text{with} \quad A_X^3 = \tfrac{10}{3}, \quad \dim |A_X| = 4, \\ -K_Y &= 2A_Y \quad \text{with} \quad A_Y^3 = \tfrac{7}{3}, \quad \dim |A_Y| = 3. \end{aligned} \tag{6.3.1}$$

Their Hilbert series come from this by the Ice Cream formula of [BRZ]:

$$\begin{aligned} P_{X,A_X}(t) &= \tfrac{1+t+t^2}{(1-t)^4} + \tfrac{t^2}{(1-t)^3(1-t^3)} = \tfrac{1+2t+4t^2+2t^3+t^4}{(1-t)^3(1-t^3)} \\ P_{Y,A_X}(t) &= \tfrac{1+t^2}{(1-t)^4} + \tfrac{t^2}{(1-t)^3(1-t^3)} = \tfrac{1+t+3t^2+t^3+t^4}{(1-t)^3(1-t^3)}. \end{aligned} \tag{6.3.2}$$

Example 6.3.3. Let $E_0 \subset \mathbb{P}^3$ be the ordinary quadratic cone and $\Gamma_7 \subset E_0 \subset \mathbb{P}^3$ a curve of degree 7 that is singular at the node $P \in E_0$, with $\mathrm{mult}_P \Gamma_7 = 3$. The symbolic blowup of Γ_7 defines an extremal extraction $Y_1 \to \mathbb{P}^3$, with the birational transform $E_1 \subset Y_1$ isomorphic to E_0 and to $\mathbb{P}(2,1,1)$.

Write B for the polarizing $\mathcal{O}(1)$ of $\mathbb{P}^3$ and its pullback to Y_1, so that $E_0 \sim 2B$, and let $F \subset Y_1$ be the scroll over Γ_7. Note that $2\Gamma_7 \sim 7B$ in $\mathrm{Pic}\, E_0$, so $2F \sim 7B$ in $\mathrm{Pic}\, E_1$. In $\mathrm{Cl}\, Y_1$ we have

$$2B = E_1 + F \quad \text{and} \quad K_{Y_1} = -4B + F = -2B - E_1. \tag{6.3.4}$$

We give Y_1 the polarizing divisor $A_1 = B + \frac{2}{3}E_1$. Then

$$3A_1 = 3B + 2E_1 = 7B - 2F \tag{6.3.5}$$

so that $3A_1$ is a Cartier divisor restricting to a linearly trivial divisor on E_1, that is, $\mathcal{O}_{E_1}(3A_1) \simeq \mathcal{O}_{E_1}$. Standard use of vanishing gives that $H^0(Y_1, \mathcal{O}_{Y_1}(3A_1)) \twoheadrightarrow H^0(\mathcal{O}_{E_1})$ is surjective, so that $|3A_1|$ is a free linear system, ample outside E_1, so contracts E_1 to a $\frac{1}{3}(2,1,1)$ orbifold point on a 3-fold Y.

Also, (6.3.4) gives

$$K_{Y_1} - \tfrac{1}{3}E_1 = -2B - \tfrac{4}{3}E_1 = -2A_1. \tag{6.3.6}$$

Hence $-K_Y = 2A$, with A an ample Weil divisor on Y. The contraction $Y_1 \to Y$ is the Kawamata blowup of $\frac{1}{3}(2,1,1)$, with discrepancy $\frac{1}{3}E_1$.

Example 6.3.7. The Main Example of Theorem 1.3.1 is almost the same. We start from the nonsingular quadric $Q \subset \mathbb{P}^4$ and the ordinary quadratic cone obtained as the intersection $E_0 = T_{P,Q} \cap Q$ with its tangent hyperplane at a point $P \in Q$. Let $\Gamma_5 \subset E_0$ be a irreducible quintic curve, assumed singular at P (it follows that $\operatorname{mult}_P \Gamma = 3$).

The symbolic blowup $X_1 \to Q$ of Γ_5 has exceptional scroll F, and birational transform $E_1 \simeq E_0$. As before, write $B = \mathcal{O}(1)$ for the polarizing divisor of Q, so that $E_0 \sim B$, and also for its pullback to X_1. Thus in $\operatorname{Cl} X_1$ we have

$$B = E_1 + F \quad \text{and} \quad K_{X_1} = -3B + F = -2B - E_1. \tag{6.3.8}$$

We give X_1 the polarising divisor $A_1 = B + \frac{2}{3}E_1$. Then $3A_1 = 3B + 2E_1 = 5B - 2F$ is a Cartier divisor with $\mathcal{O}_{E_1}(3A_1) \simeq \mathcal{O}_{E_1}$ with surjective restriction $H^0(\mathcal{O}_{X_1}(3A_1)) \twoheadrightarrow H^0(\mathcal{O}_{E_1})$. Thus $|3A_1|$ is a free linear system contracting E_1 to a $\frac{1}{3}(1,2,2)$ point. Now $K_{X_1} = -2B - G1 = -2A_1 + \frac{1}{3}E_1$ so that $-K_X = 2A$ with A an ample Weil divisor, and $X_1 \to X$ is the Kawamata blowup, with discrepancy $\frac{1}{3}E_1$.

6.4. Alternative graded ring constructions

We can treat the examples of 6.3 in graded ring terms. This is how we originally discovered them. Moreover, the algebra is interesting in its own right, and displays features that are possibly typical for index 2 Fano constructions.

The construction of Y is immediate. Its Hilbert series (6.3.2) is

$$\frac{1 - 2t^2 - 3t^3 + 3t^4 + 2t^5 - t^7}{(1-t)^4(1-t^2)^2(1-t^3)}, \tag{6.4.1}$$

indicating the codimension 3 subvariety $Y \subset \mathbb{P}(1^4, 2^2, 3)_{\langle x_{0\dots 3}, y_1, y_2, z\rangle}$ defined by the maximal Pfaffians of a 5×5 matrix of degrees

$$(6.4.2) \qquad \begin{pmatrix} 3 & 2 & 2 & 2 \\ & 2 & 2 & 2 \\ & & 1 & 1 \\ & & & 1 \end{pmatrix}, \quad \text{typically} \quad \begin{pmatrix} z & y_1 & y_2 + a_2 & a_3 \\ & b_1 & y_1 + b_2 & y_2 \\ & & x_3 & -x_2 \\ & & & x_1 \end{pmatrix},$$

with a_2, a_3, b_1, b_2 general quadratic forms in x_0, x_1, x_2, x_3. Every Y in this family is given in this way.

One can follow the argument back to see that in this case $\Gamma_7 \subset \mathbb{P}^3$ is defined by $\bigwedge^2 M = 0$ where

$$(6.4.3) \qquad M = \begin{pmatrix} x_1 & x_2 & a_2x_2 + a_3x_3 \\ x_2 & x_3 & b_1x_1 + b_2x_2 \end{pmatrix},$$

and the y_i in degree 2 are the rational forms solving

$$(6.4.4) \qquad M \begin{pmatrix} y_1 \\ y_2 \\ 1 \end{pmatrix} = 0.$$

Plausible though it may seem at first sight, it is a mistake to confuse the codimension 2 variety $\overline{Y}_{4,4} \subset \mathbb{P}(1^4, 2^2)$ defined by these equations with the symbolic blowup Y_1 of Γ_7. The latter is a relative construction over $\mathbb{P}^3$, and is contained in $\mathbb{P}^3 \times \mathbb{P}^2$, so it has the ratios $f_1 : f_2 : g$ as regular functions, where $g = x_1x_3 - x_2^2$ is the equation of E_0. It is not simply polarized or projectively Gorenstein. As we have seen, Y_1 has just one orbifold point of type $\frac{1}{2}(1, 1, 1)$.

In contrast, $\overline{Y}_{4,4}$ contains $\mathbb{P}(1, 2, 2)_{\langle x_0, y_1, y_2\rangle}$ with ideal (x_1, x_2, x_3), and has $\mathbb{P}^1_{\langle y_1, y_2\rangle}$ as $\frac{1}{2}(1, 1)$ orbifold locus, so is not terminal. It is clearly obtained from $Y \subset \mathbb{P}(1^4, 2^2, 3)$ by eliminating z. Putting back z is a Type I or Kustin–Miller unprojection, with the Pfaffians of (6.4.2) giving the linear relations for z, so is perfectly valid as a construction of Y. However, the birational relation between Y and $\overline{Y}_{4,4}$ involves first the weighted blowup of the $\frac{1}{3}(1, 2, 2)$ point with the given weights $(1, 2, 2)$, not the Kawamata blowup with weights $(2, 1, 1)$, and this takes us outside the Mori category. A similar thing happens in many other constructions or attempted constructions of index 2 $\mathbb{Q}$-Fanos.

There is a similar narrative for the Main Example X, starting from $\Gamma_5 \subset E_0 \subset Q$. The Hilbert series P_X has the form $\frac{N(t)}{(1-t)^5(1-t^2)^2(1-t^3)}$

with numerator

$$(6.4.5)\qquad \begin{aligned} N(t) = 1 - t^2 - 4t^3 - 4t^4 & \\ + 4t^4 + 8t^5 + 4t^6 & \\ - 4t^6 - 4t^7 - t^8 + t^{10}. & \end{aligned}$$

We keep the masked terms $-4t^4 + 4t^4$ to indicates that $R(X, A)$ needs 4 relations in degree 4. In fact, in order to have $\frac{1}{3}(1, 2, 2)_{x_4, y_1, y_2}$ at P_z, there must be 4 relations $zx_i = c_i$ to eliminate $x_0, \dots, x_3$ there.

Now eliminating z projects X to $\overline{X} \subset \mathbb{P}(1^5, 2^2)$ in codimension 3, with the Hilbert series

$$(6.4.6)\qquad \frac{1 - t^2 - 4t^3 + 4t^4 + t^5 - t^7}{(1 - t)^5(1 - t^2)^2},$$

which corresponds to the Pfaffians of a skew 5×5 matrix of degrees

$$(6.4.7)\qquad \begin{pmatrix} 1 & 1 & 1 & 2 \\ & 1 & 1 & 2 \\ & & 1 & 2 \\ & & & 2 \end{pmatrix}, \quad \text{typically} \quad \begin{pmatrix} x_0 & x_1 & x_2 & a \\ & x_2 & x_3 & b \\ & & x_4 & -y_1 \\ & & & y_2 \end{pmatrix}.$$

Here we choose coordinates on $\mathbb{P}^4$ with $Q : x_0x_4 - x_1x_3 + x_2^2 = 0$ and $P = (1, 0, \dots, 0)$, making $T_{P,Q} : x_4 = 0$ and $E_0 : x_1x_3 = x_2^2$. Let $\Gamma_5 \subset E_0 \subset \mathbb{P}^3$ be an irreducible curve of degree 5, assumed to be singular at P.

In (6.4.7), a and b are quadratic forms in $x_{0\dots4}$. The conditions that $\overline{X}$ defined by the Pfaffians of (6.4.7) contains $\mathbb{P}(1, 2, 2)_{\langle x_4, y_1, y_2\rangle}$ defined by the ideal $(x_0, \dots, x_3)$ is that a, b do not contain x_4^2. Then the 7 entries in the first two rows of (6.4.7) are in the ideal (x_0, x_1, x_2, x_3), so it is a Jerry_{12}. At the same time, the equations of $\Gamma_5 \subset E_0$ take the determinantal format $\bigwedge^2 M$ with M as in (6.1.3).

As before, unprojecting $\mathbb{P}(1, 2, 2)_{\langle x_4, y_1, y_2\rangle} \subset \overline{X}$ is a contruction of X as a Type I unprojection from $\overline{X} \supset \mathbb{P}(1, 2, 2)$, but $\overline{X}$ itself again has a line of $\frac{1}{2}$ points, so is not Mori category.

The "double Jerry" calculations of [T&J, 9.2] gives the unprojection variable z and most of its unprojection equations $zx_i = c_i$. Eliminating the pivot $m_{12} = x_0$ from the Pfaffians of (6.4.7), gives two equations

$$(6.4.8)\qquad \begin{pmatrix} x_1 & x_2 & a \\ x_2 & x_3 & b \end{pmatrix} \begin{pmatrix} y_2 \\ y_1 \\ x_4 \end{pmatrix} = 0 \quad \text{with} \quad \begin{aligned} a &= a_2x_2 + a_3x_3, \\ b &= b_1x_1 + b_2x_2, \end{aligned}$$

that we rearrange

$$(6.4.9) \qquad \begin{pmatrix} b_1x_4 & y_2+b_2x_4 & y_1 \\ y_2 & y_1+a_2x_4 & a_3x_4 \end{pmatrix} \begin{pmatrix} x_1 \\ x_2 \\ x_3 \end{pmatrix} = 0.$$

From this we assemble a second Jerry$_{12}$ matrix

$$(6.4.10) \qquad \begin{pmatrix} -z & b_1x_4 & y_2+b_2x_4 & y_1 \\ & y_2 & y_1+a_2x_4 & a_3x_4 \\ & & x_3 & -x_2 \\ & & & x_1 \end{pmatrix},$$

whose maximal Pfaffians provide the equations for x_1z, x_2z, x_3z. The final unprojection equation for x_0z

$$(6.4.11) \quad -x_0z = (b_1x_1+b_2x_2-a_2x_3)y_1 + a_3(x_3y_2+b_1x_2x_4+b_2x_3x_4).$$

exists by the theory of Kustin–Miller unprojection, but we don't know any smart way of deducing it. It has to be calculated by a laborious primary decomposition or colon ideal calculation, or by writing out the Kustin–Miller complexes.

The Jerry$_{12}$ matrix (6.4.10) defines a codimension 3 subvariety in the family of our second example $Y \subset \mathbb{P}(1^4, 2^2, 3)$ (compare (6.4.2)), but specialized to contain $\mathbb{P}^2_{\langle x_1,x_2,x_3\rangle}$ defined by the codimension 4 ideal (x_4, y_1, y_2, z). This is a third construction of our Main Example X.

To do this from scratch: in Example 6.3.3, suppose that the curve $\Gamma_7 \subset E_0 \subset \mathbb{P}^3$ break up as the plane conic section $(x_0 = 0)$ plus a quintic Γ_5. Blowing up Γ_7 transforms the plane $(x_0 = 0)$ into a copy of $\mathbb{P}^2$ with normal bundle $\mathcal{O}(-1)$, that contracts to a point of the quadric $Q \subset \mathbb{P}^4$, taking E_0 and $\Gamma_5 \subset E_0$ isomorphically into the data for 6.3.7.

6.5. Summary: Three constructions of X

Our Main Example X can be obtained in three different ways

(1) The symbolic blowup of $\Gamma_5 \subset E_0 \subset Q$ followed by the contraction of E_1. Viewed from X, this is the Sarkisov link from its $\frac{1}{3}(1,2,2)$ point P of Section 1.5; it is initiated by the Kawamata blowup, that is the $(2,1,1)$ weighted blowup of P.

(2) Construct the codimension 3 variety $\overline{X} \subset \mathbb{P}(1^5, 2^2)$ given in the Pfaffian form (6.4.7), containing $\mathbb{P}(1,2,2)$, then unproject this plane. Viewed from X, this starts from the $(1,2,2)$ weighted blowup of P, which introduces a line of $\frac{1}{2}(1,1)$ orbifold points, so takes us out of the Mori category.

(3) Construct the codimension 3 variety $Y' \subset \mathbb{P}(1^4, 2^2, 3)$ as in Example 6.3.3, but specialized to contain $\mathbb{P}^2_{\langle x_1, x_2, x_3 \rangle}$. Its equations are the maximal Pfaffians of the Jerry_{12} matrix (6.4.10). One checks that Y' has 4 ordinary nodes on $\mathbb{P}^2$ as its only singularities for general choices of (a_2, a_3, b_1, b_2), so that it unprojects to a quasismooth X. Viewed from X, this starts from the ordinary blowup of a general point.

References

[Ale94] Valery Alexeev, General elephants of **Q**-Fano 3-folds, *Compositio Math.* **91** (1994), 91–116

[GRDB] Gavin Brown et al., Graded ring database, `www.grdb.co.uk`

[T&J] G. Brown, M. Kerber and M. Reid, Fano 3-folds in codimension 4, Tom and Jerry, Part I, *Compositio Math.* **148** (2012), 1171–1194

[BS07a] Gavin Brown and Kaori Suzuki, Computing certain Fano 3-folds, *Japan J. Indust. Appl. Math.* **24** (2007), 241–250

[BS07b] Gavin Brown and Kaori Suzuki, Fano 3-folds with divisible anticanonical class, *Manuscripta Math.* **123** (2007), 37–51

[BRZ] Anita Buckley, Miles Reid and Shengtian Zhou, Ice cream and orbifold Riemann–Roch, *Izv. Math.* **77** (2013), 461–486

[CH15] Alessio Corti and Liana Heuberger, Del Pezzo surfaces with 1/3(1,1) points, arXiv:1505.02092, 42 pp.

[CPR] Alessio Corti, Aleksandr Pukhlikov and Miles Reid, Fano 3-fold hypersurfaces, In *Explicit birational geometry of 3-folds*, *LMS Lecture Notes* **281**, CUP, 2000, pp. 175–258

[CR02] Alessio Corti and Miles Reid, Weighted Grassmannians, In *Algebraic geometry*, de Gruyter, Berlin, 2002, pp. 141–163

[Cut88] Steven Cutkosky, Elementary contractions of Gorenstein 3-folds, *Math. Ann.* **280** (1988), 521–525

[Du14] Tom Ducat, Divisorial extractions from singular curves in smooth 3-folds, I, arXiv:1493.7614, 21 pp.

[Du15] Tom Ducat, Mori extractions from singular curves in a smooth 3-fold, Univ. of Warwick PhD thesis, 2015, 111 + xiv pp. Available from `wrap.warwick.ac.uk/73056`

[Fuj90] Takao Fujita, *Classification theories of polarized varieties*, *LMS Lecture Notes* **155**, CUP, 1990

[Kaw88] Yujiro Kawamata, Crepant blowing-up of 3-dimensional canonical singularities and its application to degenerations of surfaces, *Ann. of Math. (2)* **127** (1988), 93–163

[Kaw92] Yujiro Kawamata, Boundedness of **Q**-Fano 3-folds, In *Proceedings of the International Conference on Algebra, Part 3 (Novosibirsk, 1989)*, *Contemp. Math.* **131** Providence, RI, 1992, AMS, pp. 439–445

[Kaw96] Yujiro Kawamata, Divisorial contractions to 3-dimensional terminal quotient singularities, In *Higher-dimensional complex varieties (Trento, 1994)*, de Gruyter, Berlin, 1996, pp. 241–246

[Kh14] Kuzma Khrabrov, A small cage in the zoo of Fano 3-folds, first draft, available from `www.warwick.ac.uk/staff/Miles.Reid/3folds`, 7 pp.

[KM92] János Kollár and Shigefumi Mori, Classification of three-dimensional flips, *J. Amer. Math. Soc.* **5** (1992), 533–703

[Mor82] S. Mori, Threefolds whose canonical bundles are not numerically effective, *Ann. Math.* **115** (1982), 133–176

[MP08] Shigefumi Mori and Yuri Prokhorov, On **Q**-conic bundles, *Publ. Res. Inst. Math. Sci.*, 44 (2008), 315–369

[Pro07] Yuri Prokhorov, The degree of **Q**-Fano 3-folds, *Russian Acad. Sci. Sb. Math.* **198** (2007), 1683–1702

[Pro10a] Yuri Prokhorov, **Q**-Fano 3-folds, I, *Adv. Geom.* **13** (2013), 389–418

[Pro10b] Yuri Prokhorov, **Q**-Fano 3-folds of large Fano index, I, *Doc. Math., J. DMV* **15** (2010), 843–872

[Pro10c] Yuri Prokhorov, Fano threefolds of large Fano index and large degree, *Russian Acad. Sci. Sb. Math.* **204** (2013), 347-382

[Rei83] Miles Reid, Minimal models of canonical 3-folds, In *Algebraic varieties and analytic varieties (Tokyo, 1981)*, *Adv. Stud. Pure Math.* **1**, North-Holland, Amsterdam, 1983, pp. 131–180

[YPG] Miles Reid, Young person's guide to canonical singularities, In *Algebraic geometry, Bowdoin, 1985 (Brunswick, Maine, 1985)*, *Proc. Sympos. Pure Math.* **46**, AMS, Providence, RI, 1987, pp. 345–414

[RS03] Miles Reid and Kaori Suzuki, Cascades of projections from log del Pezzo surfaces, In *Number theory and algebraic geometry*, *LMS Lecture Notes* **303**, CUP, 2003, pp. 227–249

[Suz04] Kaori Suzuki, On Fano indices of **Q**-Fano 3-folds, *Manuscripta Math.* **114** (2004), 229–246

Yuri Prokhorov, Steklov Mathematical Institute,
8 Gubkina street, Moscow 119991, Russia
Department of Algebra, Faculty of Mathematics,
Moscow State University, Moscow 117234, Russia
Laboratory of Algebraic Geometry,
National Research University Higher School of Economics, Russia
E-mail address: `prokhoro@mi.ras.ru`

Miles Reid, Mathematics Institute, University of Warwick,
Coventry CV4 7AL, UK
E-mail address: `Miles.Reid@warwick.ac.uk`

Advanced Studies in Pure Mathematics

A SERIES OF UP-TO-DATE GUIDES OF LASTING INTEREST TO ADVANCED MATHEMATICS

Volume 1 Algebraic Varieties and Analytic Varieties.
Edited by S. Iitaka. February, 1983

Volume 2 Galois Groups and their Representations.
Edited by Y. Ihara. December, 1983.

Volume 3 Geometry of Geodesics and Related Topics.
Edited by K. Shiohama. June, 1984

Volume 4 Group Representations and Systems of Differential Equations.
Edited by K. Okamoto. March, 1985

Volume 5 Foliations.
Edited by I. Tamura. February, 1986

Volume 6 Algebraic Groups and Related Topics.
Edited by R. Hotta. March, 1985

Volume 7 Automorphic Forms and Number Theory.
Edited by I. Satake. February, 1986

Volume 8 Complex Analytic Singularities.
Edited by T. Suwa and P. Wagreich. February, 1987

Volume 9 Homotopy Theory and Related Topics.
Edited by H. Toda. February, 1987

Volume 10 Algebraic Geometry, Sendai, 1985.
Edited by T. Oda. July, 1987

Volume 11 Commutative Algebra and Combinatorics.
Edited by M. Nagata and H. Matsumura.
October, 1987

Volume 12 Galois Representations and Arithmetic Algebraic Geometry.
Edited by Y. Ihara. November, 1987

Volume 13 Investigations in Number Theory.
Edited by T. Kubota. March, 1988

Volume 14 Representations of Lie Groups, Kyoto, Hiroshima, 1986.
Edited by K. Okamoto and T. Oshima. February, 1989

Volume 15 Automorphic Forms and Geometry of Arithmetic Varieties.
Edited by K. Hashimoto and Y. Namikawa. July, 1989

Volume 16 Conformal Field Theory and Solvable Lattice Models.
Edited by M. Jimbo, T. Miwa and A. Tsuchiya.
June, 1988

Volume 17 Algebraic Number Theory—in honor of K. Iwasawa.
Edited by J. Coates, R. Greenberg, B. Mazur and I. Satake. August, 1989

Volume 18 I—Recent Topics in Differential and Analytic Geometry.
Edited by T. Ochiai. November, 1990
II—Kähler Metric and Moduli Spaces.
Edited by Edited by T. Ochiai. November, 1990

Volume 19 Integrable Systems in Quantum Field Theory and Statistical Mechanics.
Edited by M. Jimbo, T. Miwa and A. Tsuchiya.
November, 1989

Volume 20 Aspects of Low Dimensional Manifolds.
Edited by Y. Matsumoto and S. Morita.
December, 1992

Volume 21 Zeta Functions in Geometry.
Edited by N. Kurokawa and T. Sunada.
December, 1992

Volume 22 Progress in Differential Geometry.
Edited by K. Shiohama. September, 1993

Volume 23 Spectral and Scattering Theory and Applications.
Edited by K. Yajima. February, 1994

Volume 24 Progress in Algebraic Combinatorics.
Edited by E. Bannai and A. Munemasa. May, 1996

Volume 25 CR-Geometry and Overdetermined Systems.
Edited by T. Akahori, G. Komatsu, K. Miyajima, M. Namba and K. Yamaguchi. July, 1997

Volume 26 Analysis on Homogeneous Spaces and Representation Theory of Lie Groups, Okayama-Kyoto.
Edited by M. Kashiwara, T. Kobayashi, T. Matsuki, K. Nishiyama and T. Oshima. April, 2000

Volume 27 Arrangements – Tokyo 1998.
Edited by M. Falk and H. Terao. August, 2000

Volume 28 Combinatorial Methods in Representation Theory.
Edited by M. Kashiwara, K. Koike, S. Okada, I. Terada, and H.-F. Yamada. December, 2000

Volume 29 Singularities – Sapporo 1998.
Edited by J.-P. Brasselet and T. Suwa.
December, 2000

Volume 30 Class Field Theory – Its Centenary and Prospect.
Edited by K. Miyake. April, 2001

Volume 31 Taniguchi Conference on Mathematics Nara '98.
Edited by M. Maruyama and T. Sunada. May, 2001

Volume 32 Groups and Combinatorics — in memory of Michio Suzuki.
Edited by E. Bannai, H. Suzuki, H. Yamaki and T. Yoshida. October, 2001

Volume 33 Computational Commutative Algebra and Combinatorics.
Edited by T. Hibi. February, 2002

Volume 34 Minimal Surfaces, Geometric Analysis and Symplectic Geometry.
Edited by K. Fukaya, S. Nishikawa and J. Spruck.
June, 2002

Volume 35 Higher Dimensional Birational Geometry.
Edited by S. Mori and Y. Miyaoka. July, 2002

Volume 36 Algebraic Geometry 2000, Azumino.
Edited by S. Usui, M. Green, L. Illusie, K. Kato, E. Looijenga, S. Mukai and S. Saito. November, 2002

Volume 37 Lie Groups, Geometric Structures and Differential Equations — One Hundred Years after Sophus Lie —.
Edited by T. Morimoto, H. Sato and K. Yamaguchi.
December, 2002

Volume 38 Operator Algebras and Applications.
Edited by H. Kosaki. January, 2004

Volume 39 Stochastic Analysis on Large Scale Interacting Systems.
Edited by T. Funaki and H. Osada. March, 2004

Volume 40 Representation Theory of Algebraic Groups and Quantum Groups.
Edited by T. Shoji, M. Kashiwara, N. Kawanaka, G. Lusztig and K. Shinoda. March, 2004

Volume 41 Stochastic Analysis and Related Topics in Kyoto
In honour of Kiyosi Itô.
Edited by H. Kunita, S. Watanabe and Y. Takahashi.
March, 2004

Volume 42 Complex Analysis in Several Variables
– Memorial Conference of Kiyoshi Oka's Centennial Birthday, Kyoto/Nara 2001.
Edited by K. Miyajima, M. Furushima, H. Kazama, A. Kodama, J. Noguchi, T. Ohsawa, H. Tsuji and T. Ueda. June, 2004

Volume 43 Singularity Theory and Its Applications.
Edited by S. Izumiya, G. Ishikawa, H. Tokunaga, I. Shimada and T. Sano. January, 2007

Volume 44 Potential Theory in Matsue.
Edited by H. Aikawa, T. Kumagai, Y. Mizuta and N. Suzuki. August, 2006

Volume 45 Moduli Spaces and Arithmetic Geometry (Kyoto, 2004).
Edited by S. Mukai, Y. Miyaoka, S. Mori, A. Moriwaki and I. Nakamura. January, 2007

Volume 46 Singularities in Geometry and Topology 2004.
Edited by J.-P. Brassele and T. Suwa. February, 2007

Volume 47 Asymptotic Analysis and Singularities.
1—Hyperbolic and dispersive PDEs and fluid mechanics.
2—Elliptic and parabolic PDEs and related problems.
Edited by H. Kozono, T. Ogawa, K. Tanaka, Y. Tsutsumi and E. Yanagida. October, 2007

Volume 48 Finsler Geometry, Sapporo 2005
— In Memory of Makoto Matsumoto.
Edited by S. V. Sabau and H. Shimada.
November, 2007

Volume 49 Probability and Number Theory — Kanazawa 2005.
Edited by S. Akiyama, K. Matsumoto, L. Murata and H. Sugita. December, 2007

Volume 50 Algebraic Geometry in East Asia — Hanoi 2005.
Edited by K. Konno and V. Nguyen-Khac.
February, 2008

Volume 51 Surveys on Geometry and Integrable Systems.
Edited by M. Guest, R. Miyaoka and Y. Ohnita.
November, 2008

Volume 52 Groups of Diffeomorphisms
in honor of Shigeyuki Morita
on the occasion of his 60th birthday.
Edited by R. Penner, D. Kotschick, T. Tsuboi,
N. Kawazumi, T. Kitano and Y. Mitsumatsu.
November, 2008

Volume 53 Advances in Discrete Dynamical Systems.
Edited by S. Elaydi, K. Nishimura, M. Shishikura and
N. Tose. June, 2009

Volume 54 Algebraic Analysis and Around
in honor of Professor Masaki Kashiwara's 60th birthday.
Edited by T. Miwa, A. Matsuo, T. Nakashima and
Y. Saito. February, 2009

Volume 55 Noncommutativity and Singularities
Proceedings of French–Japanese symposia held at IHÉS in 2006.
Edited by J.-P. Bourguignon, M. Kotani, Y. Maeda and
N. Tose. July, 2009

Volume 56 Singularities — Niigata–Toyama 2007.
Edited by J.-P. Brasselet, S. Ishii, T. Suwa and
M. Vaquie. November, 2009

Volume 57 Probabilistic Approach to Geometry.
Edited by M. Kotani, M. Hino and T. Kumagai.
March, 2010

Volume 58 Algebraic and Arithmetic Structures of Moduli Spaces
(Sapporo 2007).
Edited by I. Nakamura and L. Weng. June, 2010

Volume 59 New Developments in Algebraic Geometry, Integrable
Systems and Mirror Symmetry (RIMS, Kyoto, 2008).
Edited by M.-H. Saito, S. Hosono and K. Yoshioka.
August, 2010

Volume 60 Algebraic Geometry in East Asia — Seoul 2008.
Edited by J. Keum, S. Kondō, K. Konno and K. Oguiso.
September, 2010

Volume 61 Exploring New Structures and Natural Constructions
in Mathematical Physics.
Edited by K. Hasegawa, T. Hayashi, S. Hosono and
Y. Yamada. October, 2011

Volume 62 Arrangements of Hyperplanes—Sapporo 2009.
Edited by H. Terao and S. Yuzvinsky. March, 2012

Volume 63 Galois–Teichmüller Theory and Arithmetic Geometry.
Edited by H. Nakamura, F. Pop, L. Schneps and A. Tamagawa. October, 2012

Volume 64 Nonlinear Dynamics in Partial Differential Equations.
Edited by S.-I. Ei, S. Kawashima, M. Kimura and T. Mizumachi. April, 2015

Volume 65 Algebraic Geometry in East Asia — Taipei 2011.
Edited by J. A. Chen, M. Chen, Y. Kawamata and J. Keum. April, 2015

Volume 66 Singularities in Geometry and Topology 2011.
Edited by V. Blanlœil and O. Saeki. May, 2015

Volume 67 Variational Methods for Evolving Objects.
Edited by L. Ambrosio, Y. Giga, P. Rybka and Y. Tonegawa. June, 2015

Volume 68 School on Real and Complex Singularities in São Carlos, 2012.
Edited by O. Saeki, R. N. Araújo dos Santos, T. Nishimura and V. H. Jorge Pérez. July, 2016

Volume 69 Development of Moduli Theory—Kyoto 2013.
Edited by O. Fujino, S. Kondō, A. Moriwaki, M.-H. Saito and K. Yoshioka. July, 2016

Volume 70 Minimal Models and Extremal Rays (Kyoto, 2011).
Edited by J. Kollár, O. Fujino, S. Mukai and N. Nakayama. September, 2016

TO BE CONTINUED